Synaptic Stress and Pathogenesis of Neuropsychiatric Disorders

Maurizio Popoli • David Diamond
Gerard Sanacora
Editors

Synaptic Stress and Pathogenesis of Neuropsychiatric Disorders

Springer

Editors
Maurizio Popoli
Department of Pharmacological
and Biomolecular Sciences
University of Milano
Milano
Italy

Gerard Sanacora
Psychiatry Depression Research Program
Yale School of Medicine
New Haven
Connecticut
USA

David Diamond
Department of Psychology
University of South Florida
Tampa
Florida
USA

ISBN 978-1-4939-1055-7 ISBN 978-1-4939-1056-4 (eBook)
DOI 10.1007/978-1-4939-1056-4
Springer New York Heidelberg Dordrecht London

Library of Congress Control Number: 2014939454

Printed on acid-free paper

Springer is part of Springer Science+Business Media (www.springer.com)

Preface

The term "stress" is universally recognized as difficult to define, and yet, people typically report experiencing stress as an almost everyday experience. A commonly used definition of stress is the perception of an event that threatens homeostasis, the normal equilibrium of bodily function, and an insufficient ability to cope with environmental challenges. The types of stressors humans have experienced have changed over the history of our species, such as the primarily physical challenges faced by our distant ancestors, as compared to purely psychological stressors, which are far more common in our modern societies. The body elaborates a complex response to stressors, the so-called stress response, which is orchestrated by the brain and involves multiple physiological systems, including interactions of the autonomic nervous system with central neuroendocrine systems. It is well-known to psychiatrists and neurologists (indeed, also to the layman) that stress is a major risk factor for neuropsychiatric, neurodegenerative and metabolic diseases, but exactly how stress may facilitate or trigger these diseases is still very much debated, and insufficiently understood. A further complication is represented by the observation that stress is bi-faced, and may have positive or negative influences on the bodily functions, depending on the type and duration of the stressor, as well as on the individual's response to the stressor.

The science of the study of stress can trace its origins to the pioneering work of Claude Bernard, who in the 19th century developed the idea of the constancy of the internal environment (le milieu intérieur) as the necessary condition for a free and stable life (la vie constante ou libre). According to Bernard "all the vital mechanisms … have only one object, that of preserving constant the conditions of life in the internal environment". Bernard's theorizing regarding the milieu intérieur was extended in the 1920's by Walter Cannon, who coined the term "homeostasis", which described processes by which physiological systems preserve the stability of the internal environment. Cannon's work addressed how perturbations from a setpoint, or the optimal physiological state, were corrected by negative feedback mechanisms. Cannon also coined the phrase "acute stress response" (ASR), which described his view that animals react to life threatening experiences with the now classic "fight or flight" response, produced by activation of the sympathetic nervous system. Cannon's description of the ASR and fight or flight responses were to become the activational stage of an individual's response to a threat, as described by Hans Selye, the first true stress researcher. Selye conceptualized stress in terms of a set of non-specific responses he referred to as the "general adaptation syndrome", which described the three stage process of activation, adaptation, and ultimately, exhaustion of resources, all of which contributed to stress-induced pathology.

Bernard and Cannon's seminal ideas on homeostasis and Selye's general adaptation syndrome provide a structure for categorizing the impressive body of research in the chapters of this book which were written by prominent neuroscientists. Selye's "activational" phase of stress is manifested as increased activity in the sympathetic nervous system and hypothalamic-pituitary-adrenal axis, and ultimately, as activation of brain emotion, memory and attention centers. The first section of this book addresses research on neural mechanisms underlying the activational phase of the stress response with techniques that were unimaginable in the times of Bernard, Cannon and Selye. The six chapters in section one are a compendium of state-of-the-art approaches which have characterized cellular, molecular and physiological responses to stress. Joels, Popoli, Yan, Campolongo, Hill, Bains and their co-authors have described how stress neuromodulators, with an emphasis on corticosterone and endocannabinoids, as well as stress effects on glutamate and GABA neurotransmitter systems, exert dramatic effects on synaptic physiology in diverse brain areas, including the hippocampus, amygdala, prefrontal cortex and hypothalamus.

The second of Selye's phases in the general adaptation syndrome can be considered the brain's attempt to adapt to the challenge of the stress experience. One feature of the neural adaptation to stress is the rapid development of synaptic and behavioral plasticity to adopt efficient behavioral responses to current, and future, stress challenges. The second section of this book focuses on this issue, with scholarly reviews that emphasize the capacity of the brain to generate synaptic plasticity underlying emotional memory processing. The five chapters by Segal, Kim, Diamond, Howland, Sandi and their co-authors describe the modulation of synaptic plasticity by behavioral stress and neuromodulators, with an emphasis on influence of corticosterone on the dorsal and ventral hippocampus, subiculum, prefrontal cortex and amygdala.

The condition in which homeostasis seems to fail is analogous to the "exhaustion" phase of Selye's general adaptation syndrome. This area of research, which has generated a vast amount of work on stress-induced psychopathology, is addressed in the third section of the book. Here, prominent clinicians and preclinical researchers have integrated basic stress research with findings from clinical studies to enhance our understanding of how acute and chronic stress are linked to pathological states, including common diseases of Western society, such as immune, cardiovascular and psychiatric disorders. The erudite chapters written by Sibille, Rajkowska, McCullumsmith, Reagan, Sanacora and their co-authors addressed diverse approaches to the study of how stress modulators, with an emphasis on glutamate and GABA, are linked to neural and glial involvement in major depressive disorder, schizophrenia and psychosis, as well as metabolic disorders, such as obesity, diabetes and metabolic syndrome. The chapter by Sanacora and co-authors analyzes how stress-related effects on the glutamate system can drive the development of novel therapeutic strategies.

Finally, a recent watershed event in the development of our appreciation of the complexity of the science of stress is the extension of the homeostasis concept to “allostasis”, which means “stability through change”. Whereas homeostasis was conceptualized as a relatively static process involving stability around a fixed setpoint, allostasis is a more dynamic, adaptive process in which a setpoint can change, for example, as a result of repeated acute stress experiences. Thus, in allostasis, the concept of negative feedback mechanisms and stability around a setpoint is maintained, but it is the setpoint, itself, that can change as a function of life’s experiences.

The editors are pleased to point out that Bruce McEwen, one of the most prolific and influential of all stress researchers, has provided his perspective on allostasis in the introduction to the book. For over four decades, Bruce has advanced the boundaries of our understanding of the neurobiology and neuroendocrinology of stress with his elegant and comprehensive research on behavioral and brain processes involved in the “good and bad” sides of the neuroendocrinology of stressful experiences. In the introduction, Bruce has discussed his conception of allostasis, and in particular his contribution to our understanding of allostatic load, which is the toll that chronic stress takes on the body. Finally, he has provided a balanced overview of the involvement of glucocorticoids in the behavioral and physiological responses of neuroendocrine and autonomic systems as a major component of lifestyle effects on behavior and brain health.

The editors are well-aware that the works reported in this volume are only a small part of the great scientific effort undertaken at present to understand the brain under stress, and wish to apologize for all findings and lines of evidence that could not be included or mentioned here. Although the title of this volume was restricted to the relationship between stress and neuropsychiatric disorders, undoubtedly the reviews and primary results provided here will be of interest to bench scientists, as well as clinicians, to learn of the latest research on fundamental neuroendocrine stress mechanisms and stress-related diseases.

Maurizio Popoli, David Diamond, Gerard Sanacora
September 2013

Contents

Contributors

Piray Atsak Department of Cognitive Neuroscience, Radboud University Medical Centre, Nijmegen, The Netherlands

Donders Institute for Brain, Cognition and Behaviour, Radboud University Nijmegen, Nijmegen, The Netherlands

Jaideep S. Bains Department of Physiology and Pharmacology, Hotchkiss Brain Institute, University of Calgary, Calgary, AB, Canada

Laura A. Bullard Medical Research Service, VA Hospital, Tampa, FL, USA

Department of Psychology, University of South Florida, Tampa, FL, USA

Center for Preclinical and Clinical Research on PTSD, University of South Florida, Tampa, FL, USA

Patrizia Campolongo Department of Physiology and Pharmacology, Sapienza University of Rome, Rome, Italy

Don A. Davies Department of Physiology, University of Saskatchewan, Saskatoon, SK, Canada

David M. Diamond Medical Research Service, VA Hospital, Tampa, FL, USA

Department of Psychology, University of South Florida, Tampa, FL, USA

Department of Molecular Pharmacology and Physiology, University of South Florida, Tampa, FL, USA

Center for Preclinical and Clinical Research on PTSD, University of South Florida, Tampa, FL, USA

George E. Farmer Medical Research Service, VA Hospital, Tampa, FL, USA

Department of Psychology, University of South Florida, Tampa, FL, USA

Center for Preclinical and Clinical Research on PTSD, University of South Florida, Tampa, FL, USA

Beverly French Department of Psychiatry, University of Pittsburgh, Pittsburgh, PA, USA

J.Megan Gray Department of Cell Biology and Anatomy, Hotchkiss Brain Institute, University of Calgary, Calgary, Canada

Claudia A. Grillo Department of Pharmacology, Physiology and Neuroscience, University of South Carolina School of Medicine, Columbia, SC, USA

J.C. Hammond Department of Psychiatry and Behavioral Neurobiology, University of Alabama at Birmingham, Birmingham, USA

Matthew N. Hill Department of Cell Biology and Anatomy, Hotchkiss Brain Institute, University of Calgary, Calgary, Canada

John G. Howland Department of Physiology, University of Saskatchewan, Saskatoon, SK, Canada

Marian Joëls Department of Translational Neuroscience, Brain Center Rudolf Magnus, University Medical Center, Utrecht, The Netherlands

Henk Karst Department of Translational Neuroscience, Brain Center Rudolf Magnus, University Medical Center, Utrecht, The Netherlands

Eun Joo Kim Department of Psychology, Program in Neurobiology and Behavior, University of Washington, Seattle, WA, USA

Jeansok J. Kim Department of Psychology, Program in Neurobiology and Behavior, University of Washington, Seattle, WA, USA

Carly Kiselycznyk Department of Psychiatry, Yale University, Suite 901, New Haven CT, USA

Harm Krugers Swammerdam Institute for Life Sciences—Center for Neuroscience, University of Amsterdam, Amsterdam, The Netherlands

Gediminas Luksys Division of Cognitive Neuroscience, University of Basel, Basel, Switzerland

Division of Molecular Psychology, University of Basel, Basel, Switzerland

Nicola Maggio Department of Neurology, Talpiot Medical Leadership Program, J. Sagol Neuroscience Center, The Chaim Sheba Medical Center, Ramat Gan, Israel

R.E. McCullumsmith Department of Psychiatry and Behavioral Neuroscience, University of Cincinnati, Cincinnati, OH, USA

Bruce S. McEwen Harold and Margaret Milliken Hatch, Laboratory of Neuroendocrinology, The Rockefeller University, New York, NY, USA

J.H. Meador-Woodruff Department of Psychiatry and Behavioral Neurobiology, University of Alabama at Birmingham, Birmingham, USA

Laura Musazzi Laboratory of Neuropsychopharmacology and Functional Neurogenomics—Dipartimento di Scienze Farmacologiche e Biomolecolari and CEND, Università di Milano, Milano, Italy

Nicoletta Nava Stereology and Electron Microscopy Laboratory, Centre for Stochastic Geometry and Advanced Bioimaging, Aarhus University Hospital, Aarhus C, Denmark

Jens R Nyengaard Stereology and Electron Microscopy Laboratory, Centre for Stochastic Geometry and Advanced Bioimaging, Aarhus University Hospital, Aarhus C, Denmark

Collin R. Park Medical Research Service, VA Hospital, Tampa, FL, USA

Department of Psychology, University of South Florida, Tampa, FL, USA

Center for Preclinical and Clinical Research on PTSD, University of South Florida, Tampa, FL, USA

Blake A. Pellman Department of Psychology, Program in Neurobiology and Behavior, University of Washington, Seattle, WA, USA

Carla Perego Laboratory of Cell Physiology—Dipartimento di Scienze Farmacologiche e Biomolecolari, Università di Milano, Milano, Italy

Maurizio Popoli Laboratory of Neuropsychopharmacology and Functional Neurogenomics—Dipartimento di Scienze Farmacologiche e Biomolecolari and CEND, Università di Milano, Milano, Italy

Grazyna Rajkowska Department of Psychiatry and Human Behavior, University of Mississippi Medical Center, Jackson, MS, USA

Lawrence P. Reagan Department of Pharmacology, Physiology and Neuroscience, University of South Carolina School of Medicine, Columbia, SC, USA

Benno Roozendaal Department of Cognitive Neuroscience, Radboud University Medical Centre, Nijmegen, The Netherlands

Donders Institute for Brain, Cognition and Behaviour, Radboud University Nijmegen, Nijmegen, The Netherlands

Gerard Sanacora Department of Psychiatry, Yale University, New Haven, CT, USA

Carmen Sandi Laboratory of Behavioral Genetics, Brain Mind Institute, Ecole Polytechnique Federale de Lausanne (EPFL), Lausanne, Switzerland

Menahem Segal Department of Neurobiology, Weizmann Institute of Science, Rehovot, Israel

Marianne L. Seney Department of Psychiatry, University of Pittsburgh, Pittsburgh, PA, USA

D. Shan Department of Psychiatry and Behavioral Neurobiology, University of Alabama at Birmingham, Birmingham, USA

Etienne Sibille Department of Psychiatry, University of Pittsburgh, Pittsburgh, PA, USA

Center for Neuroscience, University of Pittsburgh, Pittsburgh, PA, USA

Giulia Treccani Laboratory of Neuropsychopharmacology and Functional Neurogenomics—Dipartimento di Scienze Farmacologiche e Biomolecolari and CEND, Università di Milano, Milano, Italy

Haley A. Vecchiarelli Department of Cell Biology and Anatomy, Hotchkiss Brain Institute, University of Calgary, Calgary, Canada

Zhen Yan Department of Physiology and Biophysics, School of Medicine and Biomedical Sciences, State University of New York at Buffalo, Buffalo, NY, USA

Chapter 1
The Brain on Stress: The Good and the Bad

Bruce S. McEwen

Abstract Stress is a universal human experience and the word "stress" has many connotations and meanings. This review is intended to give a balanced overview of the good and bad sides of the response to stressful experiences. The brain is the central organ of stress and adaptations and has the capacity for considerable structural and functional plasticity which, though diminishing over the lifecourse, is nevertheless present in the adult brain. The brain not only perceives what is stressful but it determines the behavioral and physiological responses of neuroendocrine and autonomic systems that directly and indirectly regulate the metabolic and immune systems. The brain is also the target of circulating hormones and mediators of immune and metabolic systems. Glucocorticoids play a key role in most, if not all, of these actions and their positive, as well as negative effects will be discussed. As a way of avoiding ambiguity of the word "stress," the concepts of allostasis and allostatic load and overload will be introduced to provide biological basis for understanding the interactions of brain and body and influences of stressful experiences and resulting "lifestyle" on both brain and body. Early life experiences have lasting effects on brain and body and emerging evidence suggest that the reactivation of plasticity mechanisms in the brain may be useful in modifying and even reversing effects of experiences in early life, as well as in adult life.

1.1 Introduction

"Stress" is a commonly used word in daily life that refers to experiences that cause feelings of anxiety and frustration because they push us to the limits of our ability to successfully cope. Besides time pressures and daily hassles at work and home, there are stressors related to economic insecurity, poor health, and interpersonal conflict. There are also situations that are life-threatening—accidents, natural disasters, violence—and these evoke the classical "fight or flight" response. In contrast to daily

B. S. McEwen (✉)
Harold and Margaret Milliken Hatch, Laboratory of Neuroendocrinology, The Rockefeller University, 1230 York Avenue, New York, NY, 10065 USA
Tel.: 212 327 8624
e-mail: Bruce.McEwen@rockefeller.edu

M. Popoli et al. (eds.), *Synaptic Stress and Pathogenesis of Neuropsychiatric Disorders*,
DOI 10.1007/978-1-4939-1056-4_1, © Springer Science+Business Media New York 2014

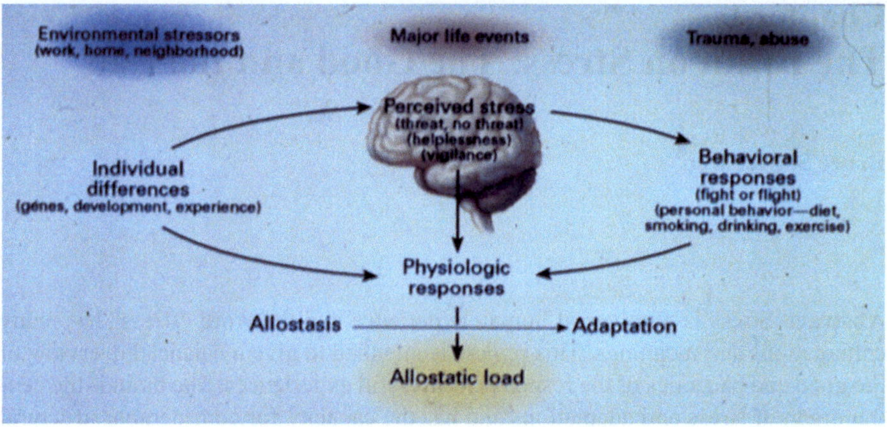

Fig. 1.1 Central role of the brain in allostasis and the behavioral and physiological response to stressors. (From McEwen 1998 by permission)

hassles, these stressors are acute, and yet they also usually lead to chronic stress in the aftermath of the tragic event.

The most common stressors are ones that operate chronically, often at a low level, and that cause us to alter the way we live. For example, being "stressed out" may cause us to be anxious and/or depressed, to lose sleep at night, to eat comfort foods and take in more calories than our bodies need, and to smoke or drink alcohol excessively. Being "stressed out" may also cause us to neglect seeing friends, or to take time off or engage in regular physical activity as we, for example, sit at a computer and try to get out from under the burden of "too much to do in so little time." Often we are tempted to take medications—anxiolytics, sleep promoting agents—to help us cope, and, with time, our bodies may increase in weight….

The brain is the organ that decides what is stressful and determines the behavioral and physiological responses, whether health promoting or health damaging (Fig. 1.1). And the brain is a biological organ that changes under acute and chronic stress and directs many systems of the body—metabolic, cardiovascular, immune—that are involved in the short- and long-term consequences of being stressed out. What does chronic stress do to the body and the brain? This chapter summarizes some of the current information placing emphasis on how the stress hormones can play both protective and damaging roles in brain and body, depending on how tightly their release is regulated, and it discusses some of the approaches for dealing with stress in our complex world.

1.2 Types of Stress

"Stress" can be classified into three types: good stress, tolerable stress, and toxic stress (http://developingchild.harvard.edu/library/reports_and_working_papers/policy_framework/). Good stress is a term used in popular language to refer to the experience of rising to a challenge, taking a risk and feeling rewarded by an often positive outcome. A related term is "eustress." Good self-esteem and good impulse control and decision-making capability, all functions of a healthy architecture of the brain, are important here! Even adverse outcomes can be "growth experiences" for individuals with such positive, adaptive characteristics.

"Tolerable stress" refers to those situations where bad things happen, but the individual with healthy brain architecture is able to cope, often with the aid of family, friends, and other individuals who provide support. Here, "distress" refers to the uncomfortable feeling related to the nature of the stressor and the degree to which the individual feels a lack of ability to influence or control the stressor (Lazarus and Folkman 1984).

Finally, "toxic stress" refers to the situation in which bad things happen to an individual who has limited material and social support; this person may also have brain architecture that reflects effects of adverse early life events, such as growing up in a chaotic home, as well as abuse and neglect, that have impaired the development of good impulse control and judgment and adequate self-esteem. Here, the degree and/or duration of "distress" may be greater and the ability to cope and show resilience is impaired. With toxic stress, the inability to cope is likely to have adverse effects on behavior and physiology, and this will result in a higher degree of allostatic overload, as will be explained below.

1.2.1 Allostasis and Allostatic Load: Protection versus Damage in the Response to Stressors

In spite of the refinement, the word "stress" is still an ambiguous term and has connotations in common usage that make it less useful in understanding how the body handles the events that are stressful, and insight into these processes can lead to a better understanding of how best to intervene, a topic that will be discussed at the end of this chapter. There are two sides to this story: on the one hand, the body responds to almost any event or challenge by releasing chemical mediators—e.g., catecholamines that increase heart rate and blood pressure—and help us cope with the situation; on the other hand, chronic elevation of these same mediators—e.g., chronically increased heart rate and blood pressure—produce a chronic wear and tear on the cardiovascular system that can result, over time, in disorders such as strokes and heart attacks. For this reason, the term "allostasis" was introduced by Sterling and Eyer (Sterling and Eyer 1988) to refer to the active process by which the body responds to daily events and maintains homeostasis (allostasis literally means "achieving stability through change"). Because chronically increased allostasis can lead to disease, we introduced the term "allostatic load or overload" to refer to the wear and tear that results from either too much stress or from inefficient management of allostasis, e.g., not turning off the response when it is no longer needed. Other forms of allostatic load are summarized in Fig. 1.2 and involve not turning on an adequate response in the first place or not habituating to the recur-

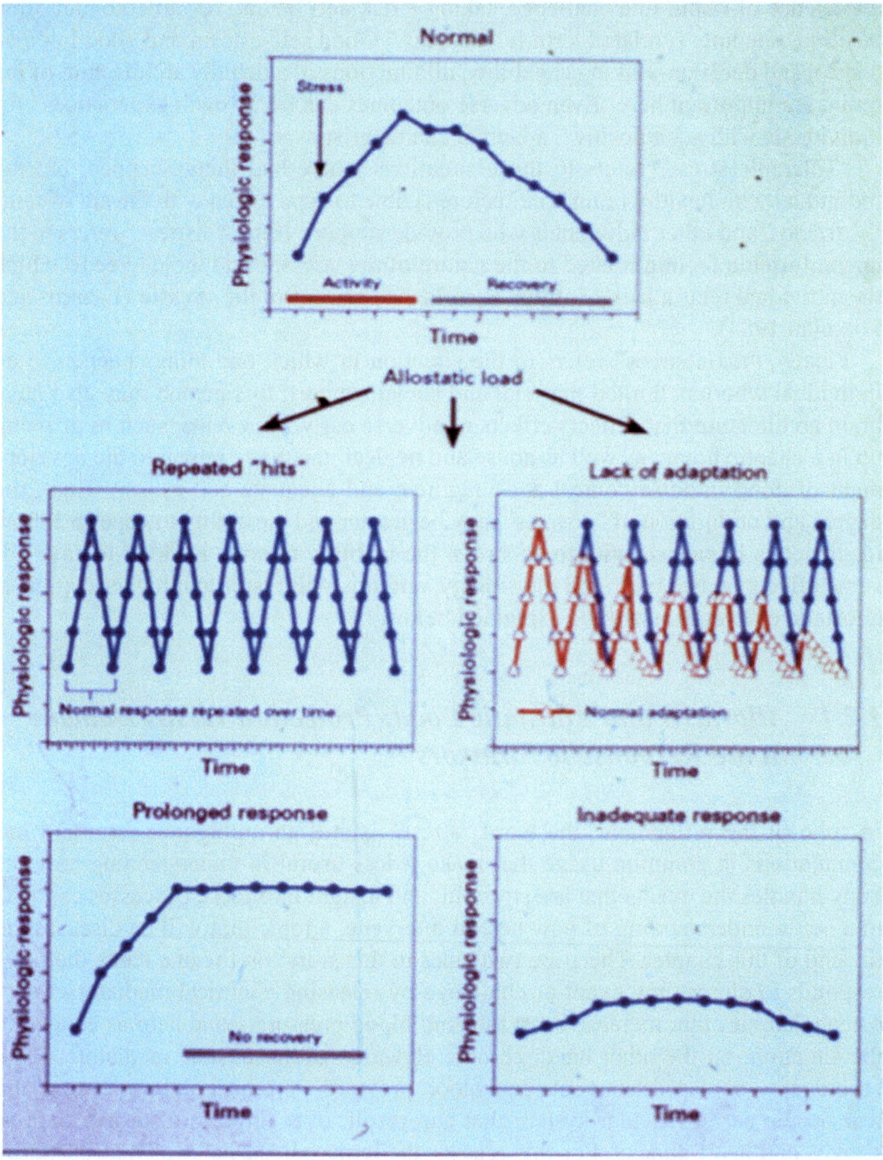

Fig. 1.2 Four types of allostatic load. The *top panel* illustrates the normal allostatic response, in which a response is initiated by a stressor, sustained for an appropriate interval, and then turned off. The remaining panels illustrate four conditions that lead to allostatic load: *top, left* repeated "hits" from multiple stressors; *top, right* lack of adaptation; *bottom, left* prolonged response due to delayed shut down; and *bottom, right* inadequate response that leads to compensatory hyperactivity of other mediators (e.g., inadequate secretion of glucocorticoid, resulting in increased levels of cytokines that are normally counter-regulated by glucocorticoids). (From McEwen 1998 by permission)

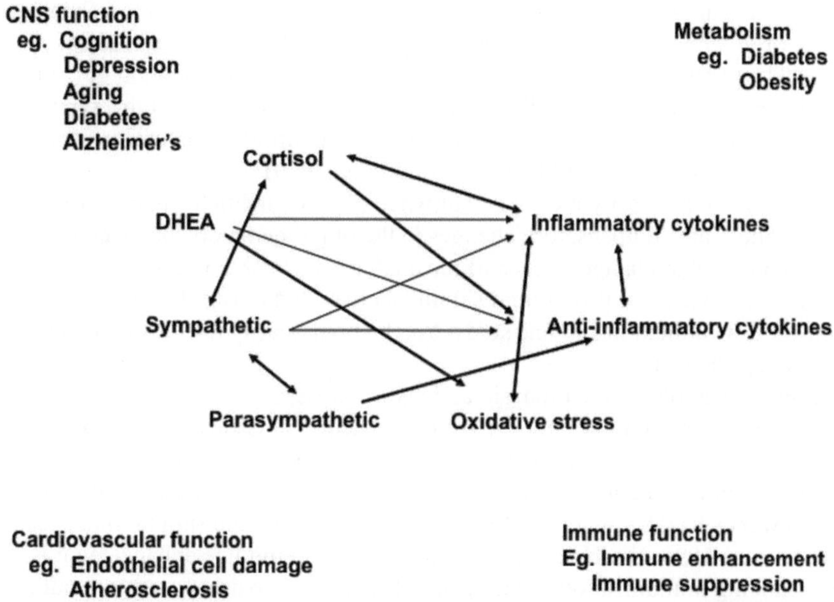

Fig. 1.3 Nonlinear network of mediators of allostasis involved in the stress response. *Arrows* indicate that each system regulates the others in a reciprocal manner, creating a nonlinear network. Moreover, there are multiple pathways for regulation—e.g., inflammatory cytokine production is negatively regulated via anti-inflammatory cytokines as well as via parasympathetic and glucocorticoid pathways, whereas sympathetic activity increases inflammatory cytokine production. Parasympathetic activity, in turn, contains sympathetic activity. *CNS* central nervous system, *DHEA* dehydroepiandrosterone. (Modified from McEwen 2006)

rence of the same stressor and thus leading to a persistent response rather than dampening the allostatic response. This is well illustrated by the lack of habituation of the salivary cortisol response to a repeated public speaking challenge in individuals with low self-esteem (Kirschbaum et al. 1995). Such individuals are reported to have a smaller hippocampus and have low self-esteem and locus of control (Pruessner et al. 2005).

Thus, protection and damage are the two contrasting sides of the physiology involved in defending the body against the challenges of daily life, whether or not we call them "stressors." Besides adrenalin and noradrenalin, there are many mediators that participate in allostasis, and they are linked together in a network of regulation that is nonlinear (Fig. 1.3), meaning that each mediator has the ability to regulate the activity of the other mediators, sometimes in a biphasic manner.

Glucocorticoids produced by the adrenal cortex in response to adrenocorticotropic hormone (ACTH) from the pituitary gland are the other major "stress hormones." Pro- and anti-inflammatory cytokines are produced by many cells in the body, and they regulate each other and are, in turn, regulated by glucocorticoids and catecholamines (Bierhaus et al. 2003). Whereas catecholamines can increase proinflammatory cytokine production, glucocorticoids are known to inhibit this production (Sapolsky et al. 2000). And yet, there are exceptions—proinflammatory effects of glucocorticoids that depend on dose and cell or tissue type (Dinkel et al. 2003). The parasympathetic nervous system also plays an important regulatory role in this nonlinear network of allostasis, since it generally opposes the sympathetic nervous system and, for example, slows the heart and also has anti-inflammatory effects (Borovikova et al. 2000; Thayer and Lane 2000).

What this nonlinearity means is that when any one mediator is increased or decreased, there are compensatory changes in the other mediators that depend on time course and level of change of each of the mediators. Unfortunately, we cannot measure all components of this system simultaneously and must rely on measurements of only a few of them in any one study. Yet the nonlinearity must be kept in mind in interpreting the results.

A good example of the biphasic actions of stress, i.e., "protection versus damage" is in the immune system, in which an acute stressor activates an acquired immune response via mediation by catecholamines and glucocorticoids and locally produced immune mediators and, yet, a chronic exposure to the same stressor over several weeks has the opposite effect and results in immune suppression (Dhabhar and McEwen 1999; Dhabhar et al. 2012a). The acute immune enhancement is good for enhancing immunization, fighting an infection, or repairing a wound, but is deleterious to health for an autoimmune condition such as psoriasis or Krohn's disease; on the other hand, the immune suppression is good in the case of an autoimmune disorder and deleterious for fighting an infection or repairing a wound. In an immune sensitive skin cancer, acute stress is effective in inhibiting tumor progression while chronic stress exacerbates progression.

1.3 Brain Response to Stressors

The discovery of receptors for glucocorticoids in the hippocampus (McEwen et al. 1968) has led to many investigations in animal models and translation to the human brain using modern imaging methods. The most striking findings from animal models have identified structural plasticity in the hippocampus, consisting of ongoing neurogenesis in the dentate gyrus (Cameron and Gould 1996) and remodeling of dendrites and synapses in the major neurons of Ammon's horn (McEwen 1999). The mediators of this plasticity include excitatory amino acids and glucocorticoids, along with a growing list of other mediators, such as oxytocin, corticotrophin releasing factor, brain-derived neurotrophic factor (BDNF), lipocalin-2 and tissue plasminogen activator (tPA) (McEwen 2007; Mucha et al. 2011). Moreover, glucocorticoid actions involve both genomic and nongenomic mechanisms that implicate mineralocorticoid, as well as glucocorticoid receptors and their translocation to mitochondria, as well as cell nuclei, and, an as-yet unidentified G-protein coupled membrane-associated glucocorticoid receptor related to endocannabinoid production (Du et al. 2009a, Hill and McEwen 2010).

Studies of the human hippocampus have demonstrated shrinkage of the hippocampus not only in mild cognitive impairment and Alzheimer's (de Leon et al. 1997), but also in type 2 diabetes (Gold et al. 2007), prolonged major depression (Sheline 2003), Cushing's disease (Starkman et al. 1999), and posttraumatic stress disorder (PTSD) (Gurvits et al. 1996). Moreover, in nondisease conditions, such as chronic stress (Gianaros et al. 2007), chronic inflammation (Marsland et al. 2008), lack of physical activity (Erickson et al. 2009), and jet lag (Cho 2001), smaller hippocampal or temporal lobe volumes have been reported. As noted above, smaller hippocampal volumes are also reported in individuals with low self-esteem and locus of control (Pruessner et al. 2005).

So far there is no indication as to whether these changes are due to volume reduction in dentate gyrus due to inhibited neuronal replacement or to dendritic shrinkage or glial cell loss, or a combination of all three. Autopsy studies on depression-suicide have indicated loss of glial cells and smaller neuron soma size (Stockmeier et al. 2004), which is indicative of a smaller dendritic tree. With regard to type 2 diabetes, it should be emphasized that the hippocampus has receptors for, and the ability to take up and respond to insulin, ghrelin, insulin-like growth factor-1 (IGF1) and leptin, and that IGF1 mediates exercise-induced neurogenesis (McEwen 2007). Thus, besides its response to glucocorticoids, the hippocampus is an important target of metabolic hormones that have a variety of adaptive actions in the healthy brain which is perturbed in metabolic disorders, such as diabetes (McEwen 2007).

1.4 Structural Plasticity in Other Brain Regions

The discovery and implications of stress and glucocorticoid effects in the hippocampus have led to exploration of other brain regions involved in cognition, mood, and behavioral self-regulation. The amygdala shows quite different responses to acute and chronic stress than the hippocampus. The amygdala responds to glucocorticoids in the formation of emotionally charged memories (Roozendaal et al. 2004), and acute stress causes a delayed formation of dendritic spines in basolateral amygdala neurons and an increase of anxiety after 10 days (Mitra et al. 2005). Chronic stress of the same type that impairs dentate gyrus neurogenesis and causes dendritic shrinkage and spine loss in Ammon's horn neurons, also causes expansion of dendrites in the basolateral amygdala (Vyas et al. 2002), while inducing spine downregulation in the medial amygdala (Bennur et al. 2007). The latter is dependent on tPA while the former is not (Bennur et al. 2007).

Translating to the human brain, amygdala hyperactivity is reported in major depression, as well as in anxiety disorders, such as PTSD (Drevets 2000) and enlargement of the amygdala has been reported in acute depression (Frodl et al. 2003). With respect to PTSD, a novel approach after acute trauma is the administration of glucocorticoids, based on the counter-intuitive findings that low normal glucocorticoid levels at the time of open heart surgery, as well as accident trauma, predispose towards development of PTSD symptoms (Schelling et al. 2004; Zohar et al. 2011).

It is, therefore, of interest that glucocorticoid administration before, during, or right after trauma protects against PTSD-like symptoms in animal models and PTSD symptoms in people (Rao et al. 2012; Schelling et al. 2004; Zohar et al. 2011).

Increased amygdala reactivity to angry and sad faces is reported in individuals with early signs of cardiovascular disease (Gianaros et al. 2009), suggesting that the increased sympathetic activity and blood pressure reactivity may be a cause of allostatic load resulting from increased reactivity to daily experiences over time. Increased amygdala reactivity to faces has also been reported in individuals traumatized by 9/11 (Ganzel et al. 2008), as well as after sleep deprivation (Yoo et al. 2007).

The prefrontal cortex is another, now well-studied, target of chronic stress. In the same chronic stress models that lead to amygdala neuronal hypertrophy and shrinkage of dendrites in hippocampus, there is shrinkage of dendrites and loss of spines throughout the medial prefrontal cortex while dendrites expand in the orbitofrontal cortex (OFC) (Liston et al. 2006). Because the OFC is involved in determining the saliency of reward or punishment (Schoenbaum and Roesch 2005), this may reinforce the changes in the basolateral amygdala. For the medial prefrontal cortex, stress-induced impairment has been linked to poor cognitive flexibility in both animal and human studies (Dias-Ferreira et al. 2009; Liston et al. 2009; Liston et al. 2006). Moreover, circadian disruption impairs cognitive flexibility and causes shrinkage of medial prefrontal cortical dendrites (Karatsoreos et al. 2011). These studies complement those on the hippocampus/temporal lobe noted above in flight crews suffering from chronic jet lag (Cho 2001) and raise important questions about how the brain handles shift work, jet lag, and chronic sleep deprivation. Furthermore, aging in rats is associated with loss of recovery of stress-induced shrinkage of dendrites of medial prefrontal cortical dendrites (Bloss et al. 2010), and this harkens back to the glucocorticoid cascade hypothesis (Sapolsky et al. 1986), since the mechanism for medial prefrontal cortical dendritic remodeling is likely to involve the same mechanisms as those in the hippocampus, namely, excitatory amino acids and glucocorticoids (Cerqueira et al. 2005; Martin and Wellman 2011).

1.5 Deleterious Effects of Early Life Adversity

Lifetime experiences have a profound impact on the brain, both as a target of stress and allostatic load and as a determinant of physiological and behavioral response to stressors. Animal models have taught us that prenatal stress of the mother can impair features of normal brain development (Maccari and Morley-Fletcher 2007) and that prolonged separation of infant from mother also impairs other aspects of brain development and function (Eiland and McEwen 2012; Francis et al. 2002; Plotsky et al. 2005). On the positive side, good maternal care and consistency of that care is a powerful determinant of life-long patterns of reduced anxiety and efficient stress reactivity, as well as social, physical, and cognitive development (Akers et al. 2008; Caldji et al. 2000; Tang et al. 2011, 2012). Moreover, there are transgenerational effects that appear to be behaviorally transmitted by the mother to the female offspring (Francis et al. 1999). In contrast, inconsistent maternal care and maternal anxiety, for example, from food insecurity, produce anxiety in offspring and appear to contribute to metabolic syndrome and predisposition to diabetes, which itself has adverse effects on the brain (Kaufman et al. 2007, 2005). Thus, the behavioral and physiological consequences of early life abuse and neglect are profound, and the epigenetic concept of behavioral transmission of abuse and its effects on human brain function are being explored at the level of epigenetic regulation of gene expression (McGowan et al. 2009).

Genotype is an important factor in determining the response to experiences (Caspi et al. 2002, 2003). An important addition to the new emphasis on gene x environment interactions is the notion of reactive alleles, as opposed to "bad genes," since alleles that can lead to pathophysiology under adverse conditions can also lead to superior outcomes when the individual with that reactive allele experiences a nurturing environment (Boyce and Ellis 2005; Suomi 2006).

During the last 10–15 years, a number of studies have documented that stress becomes bodily inscribed also in human fetuses and children, with major implications for health throughout the lifespan (e.g., Entringer et al. 2011). Allostatic overload and epigenetic mechanisms shape the developing brain and body's biological vulnerability to disease, as well as its responsiveness to potential interventions (McEwen 1998). Of particular relevance for children are experiences of abuse and neglect (Anda et al. 2010). On the physiological level, adverse childhood experiences are associated with dysregulated cardiovascular, metabolic, and immunological function, which in turn feed into numerous disease conditions both in the somatic and psychiatric domains (Anda et al. 2010). Chaos in the home and inconsistent parenting impairs brain development. This can lead to disturbed cognitive function, instable mood, low self-esteem, and numerous unhealthy activities, including overeating, substance abuse, sexual acting-out, and other forms of legal or illegal risk-taking (Evans et al. 2004).

As to the mechanisms of effects of stressful and other experiences, it is clear from the discussion above and from Fig. 1.3, that there are many interacting mediators. However, glucocorticoids stand out as having particularly important roles in the middle of all of these interactions, both positive and negative.

1.6 Diverse Role of Glucocorticoids

Glucocorticoid actions may be classified as direct genomic, indirect genomic, and nongenomic (Popoli et al. 2012; Yamamoto 1985), and all of these mechanisms may be involved in these two studies (see Fig. 1.4). Glucocorticoid and mineralocorticoid receptors are found in membrane-associated sites and are associated with release of glutamate (Karst et al. 2005; Popoli et al. 2012; Prager and Johnson 2009), translocation to mitochondria where calcium sequestration and free radical balance is regulated (Du et al. 2009b), and stimulation of the release of endocannabinoids

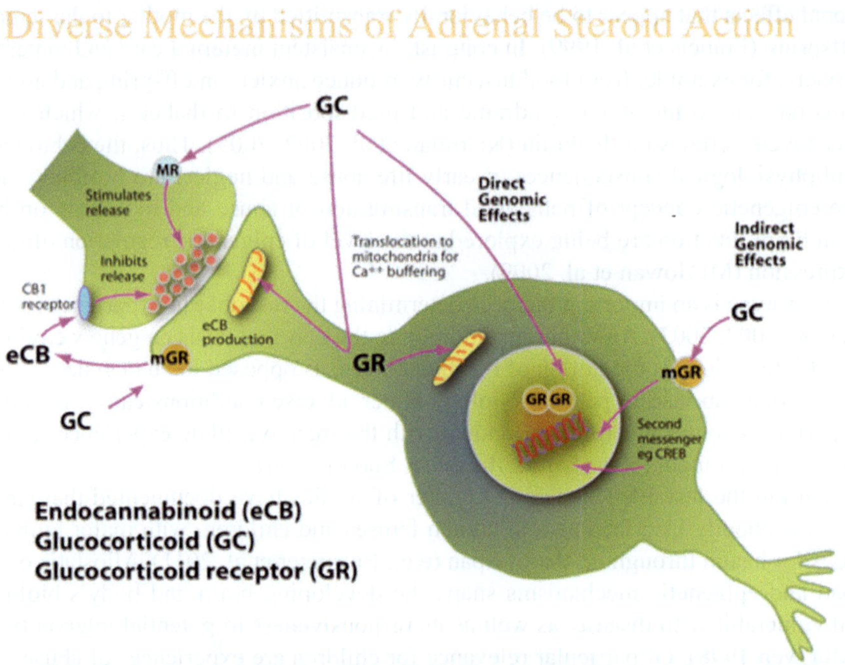

Fig. 1.4 Adrenal steroids produce multiple effects, both rapid and delayed, via multiple mechanisms. Besides direct genomic effects via classical glucocorticoid receptors (*GR*), there are also indirect genomic effects with other transcription factors. Glucocorticoids also translocate GR to mitochondria, and there are membrane-associated forms of both GR and mineralocorticoid receptors (*MR*) that effect glutamate release and stimulate endocannabinoid synthesis

(eCB) (Hill and McEwen 2010; Tasker et al. 2006). There are trophic actions by low physiological levels of glucocorticoids to maintain turnover of spine synapses (Liston and Gan 2011) and dendritic growth (Gould et al. 1990), suggesting a previously unappreciated role in maintaining a dynamic brain architecture. And glucocorticoids have been shown to promote plasticity induced by binocular visual stimulation in reversing amblyopia in adult life produced by monocular deprivation during development (Spolidoro et al. 2011). Moreover, glucocorticoid actions on processes such as neurogenesis in the dentate gyrus and contextual learning involve concurrent activity of other mediator systems, such as oxytocin for neurogenesis (Leuner et al. 2012) and adrenergic mechanisms for contextual learning (Okuda et al. 2004).

Thus a key aspect of this view of glucocorticoid action is their dependence on other mediators and ongoing cellular processes. For example, glucocorticoid stimulation of direct release of glutamate, on the one hand, is counterbalanced by glucocorticoid induction of eCB formation which can feedback from postsynaptic sites to inhibit presynaptic glutamate release in a homeostatic manner, although *gamma-aminobutyric acid* (GABA) release is also a target of eCB inhibition and can lead to a disinhibition when cannabinoid (CB1) receptors are expressed on inhibitory terminals (Hill and McEwen 2010; Popoli et al. 2012). Glucocorticoid (GC) action at the primary genomic levels also can involve synergy with other transcription regulation machinery, e.g., as in the case of GC-mediated activation of the mitogen-activated protein kinase (MAPK) pathway leading to phosphorylation of extracellular-signal-regulated kinases (ERKs) that then involves induction of protein mediators, such as Ras and Raf-1 along with indirect interactions with Stat5, Fos, Jun, Creb, and NF-kB (Revest et al. 2005). Clearly, our understanding of the complex and widespread actions of adrenal steroid hormones throughout the developing and adult nervous system is just beginning, and plasticity of neurons is emerging as a major topic of investigation, with considerable therapeutic potential!

1.7 Reactivation of Plasticity

What can be done to remediate the effects of chronic stress, as well as the biological embedding associated with early life adversity? Interventions may involve pharmaceutical, as well as behavioral, or "top-down," interventions (i.e., interventions that involve integrated central nervous system (CNS) activity, as opposed to pharmacological agents) that include cognitive-behavioral therapy, physical activity, and programs that promote social support and integration and meaning and purpose in life (McEwen and Gianaros 2011). More targeted interventions for emotional and cognitive dysfunction may arise from fundamental studies of such developmental processes as the reversal of amblyopia and other conditions by "releasing the brakes" that retard structural and functional plasticity (Bavelier et al. 2010). It should be noted that many of these interventions that are intended to promote plasticity and slow decline with age, such as physical activity and positive social interactions that give meaning and purpose, are also useful for promoting "positive health" and "eudamonia" (Ryff and Singer 1998; Singer et al. 2005) independently of any notable disorder and within the range of normal behavior and physiology.

Moreover, interventions towards changing physiology and brain function may be useful when adaptation to a particular environment, as in the Active Calibration Model (Del Giudice et al. 2011), has resulted in an individual who then chooses, or is forced, to adapt to a different, e.g., more or less threatening or nurturing, environment. Concerning biological embedding in neural architecture and the balance of neurochemical systems, in the case of adversity or shifting environments, one can hope at least to compensate, even if one cannot reverse, those effects of early life adversity (Caldji et al. 1998). However, it is perhaps premature to draw that conclusion, since the ultimate limits of adult brain plasticity are still unknown, as will be discussed below.

A powerful "top down" therapy (i.e., an activity, usually voluntary, involving activation of integrated nervous system activity, as opposed to pharmacological therapy which has a more limited target) is regular physical activity, which has actions that improve prefrontal and parietal cortex blood flow and enhance executive function (Colcombe et al. 2004). Moreover, regular physical activity, consisting of walking an hour a day, 5 out of 7 days a week, increases hippocampal volume in previously sedentary adults (Erickson et al. 2011). This finding complements work showing that fit individuals have larger hippocampal volumes than sedentary adults of the same age-range (Erickson et al. 2009). It is also well known that regular physical activity is an effective antidepressant and protects against cardiovascular disease, diabetes, and dementia (Babyak et al. 2000; Snyder et al. 2010). Moreover, intensive learning has also been shown to increase volume of the human hippocampus (Draganski et al. 2006).

Social integration and support, and finding meaning and purpose in life, are known to be protective against allostatic load (Seeman et al. 2002) and dementia (Boyle et al. 2010), and programs such as the Experience Corps that promote these along with increased physical activity, have been shown to slow the decline of physical and mental health and to improve prefrontal cortical blood flow in a similar manner to regular physical activity (Carlson et al. 2009; Fried et al. 2004).

Depression and anxiety disorders are examples of a loss of resilience, in the sense that changes in brain circuitry and function, caused by the stressors that precipitate the disorder, become "locked" in a particular state and thus need external intervention. Indeed, prolonged depression is associated with shrinkage of the hippocampus (Sheline 1996, 2003) and prefrontal cortex (Drevets et al. 1997). While there appears to be no neuronal loss, there is evidence for glial cell loss and smaller neuronal cell nuclei (Rajkowska 2000; Stockmeier et al. 2004), which is consistent with a shrinking of the dendritic tree described above after chronic stress. Indeed, a few studies indicate that pharmacological treatment may reverse the decreased hippocampal volume in unipolar (Vythilingam et al. 2004) and bipolar (Moore et al. 2000) depression, but the possible influence of concurrent cognitive-behavioral therapy in these studies is unclear.

Depression is more prevalent in individuals who have had adverse early life experiences (Anda et al. 2010). BDNF may be a key feature of the depressive state and elevation of BDNF by diverse treatments ranging from antidepressant drugs to regular physical activity may be a key feature of treatment (Duman and Monteggia 2006). Yet, there are other potential applications, such as the recently reported ability of fluoxetine to enhance recovery from stroke (Chollet et al. 2011). However, a key aspect of this new view (Castren and Rantamaki 2010) is that the drug is opening a "window of opportunity" that may be capitalized by a positive behavioral intervention, e.g., behavioral therapy in the case of depression or the intensive physiotherapy to promote neuroplasticity to counteract the effects of a stroke.

This is consistent with animal model work that shows that ocular dominance imbalance from early monocular deprivation can be reversed by patterned light exposure in adulthood that can be facilitated by fluoxetine, on the one hand (Vetencourt et al. 2008) and food restriction or intermittent glucocorticoid treatment, on the other hand (Spolidoro et al. 2011). Investigations of underlying mechanisms for the reestablishment of a new window of plasticity are focusing on the balance between excitatory and inhibitory transmission and removing molecules that put the "brakes" on such plasticity (Bavelier et al. 2010; Espinosa and Stryker 2012).

In this connection it is important to reiterate that successful behavioral therapy, which is tailored to individual needs, can produce volumetric changes in both prefrontal cortex in the case of chronic fatigue (de Lange et al. 2008), and in amygdala, in the case of chronic anxiety (Holzel et al. 2010). This reinforces two important messages: (1) that plasticity-facilitating treatments should be given within the framework of a positive behavioral or physical therapy intervention; and (2) that negative experiences during the window may even make matters worse (Castren and Rantamaki 2010). In that connection, it should be noted that BDNF also has the ability to promote pathophysiology, as in seizures (Heinrich et al. 2011; Kokaia et al. 1995; Scharfman 1997).

1.8 Conclusions

The ability of the brain and the body to adapt successfully to acute and chronic stress is an increasingly important topic in the modern world. This overview has emphasized the interplay between the good and the bad, namely, the cumulative wear and tear (allostatic load/overload) facilitated by the same mediators that are essential for adaptation and survival. The role of glucocorticoids deserves emphasis because of the multiple mechanisms and effects that they have throughout the brain and the body, both good and bad. The brain has a central role in the perception and the response to stressors, as well as being the target of allostatic load/overload along with the rest of the body (Fig. 1.1). Biological embedding of early experiences interacts with influences of the chemical and physical environment and sets the course for the body as it attempts to cope with challenges during the life course. All experiences in adult, as well as early life, leave an imprint via epigenetic influences and altered patterns of gene expression, as well as brain architecture and function that are modifiable. This review has noted that "top down" therapies, sometimes aided by pharmaceutical agents, have potential to treat disorders due to stressful and traumatic experiences because of an increased recognition that the mature brain is more malleable than previously believed. In this regard, there is growing awareness of the need to understand what constitutes optimal health, and, thus, a future research goal should be to provide a neurobiological framework for understanding underlying mechanisms for developing and maintaining positive affect and self-efficacy and self-esteem and how these are biologically embedded in a nurturing environment via epigenetic influences, including effects upon reactive alleles in the genome.

References

Akers KG, Yang Z, DelVecchio DP, Reeb BC, Romeo RD, et al. Social competitiveness and plasticity of neuroendocrine function in old age: influence of neonatal novelty exposure and maternal care reliability. PLoS ONE. 2008;3(7):e2840.

Anda RF, Butchart A, Felitti VJ, Brown DW. Building a framework for global surveillance of the public health implications of adverse childhood experiences. Am J Prev Med. 2010;39:93–8.

Babyak M, Blumenthal JA, Herman S, Khatri P, Doraiswamy M, et al. Exercise treatment for major depression: maintenance of therapeutic benefit at 10 months. Psychosom Med. 2000;62:633–8.
Bavelier D, Levi DM, Li RW, Dan Y, Hensch TK. Removing brakes on adult brain plasticity: from molecular to behavioral interventions. J Neurosci. 2010;30:14964–71.
Bennur S, Shankaranarayana Rao BS, Pawlak R, Strickland S, McEwen BS, Chattarji S. Stress-induced spine loss in the medial amygdala is mediated by tissue-plasminogen activator. Neuroscience. 2007;144:8–16.
Bierhaus A, Wolf J, Andrassy M, Rohleder N, Humpert PM, et al. A mechanism converting psychosocial stress into mononuclear cell activation. Proc Natl Acad Sci U S A. 2003;100:1920–5.
Bloss EB, Janssen WG, McEwen BS, Morrison JH. Interactive effects of stress and aging on structural plasticity in the prefrontal cortex. J Neurosci. 2010;30:6726–31.
Borovikova LV, Ivanova S, Zhang M, Yang H, Botchkina GI, et al. Vagus nerve stimulation attenuates the systemic inflammatory response to endotoxin. Nature. 2000;405:458–62.
Boyce WT, Ellis BJ. Biological sensitivity to context: I. An evolutionary-developmental theory of the origins and functions of stress reactivity. Dev Psychopathol. 2005;17:271–301.
Boyle PA, Buchman AS, Barnes LL, Bennett DA. Effect of a purpose in life on risk of incident Alzheimer disease and mild cognitive impairment in community-dwelling older persons. Arch Gen Psychiatry. 2010;67:304–10.
Caldji C, Tannenbaum B, Sharma S, Francis D, Plotsky PM, Meaney MJ. Maternal care during infancy regulates the development of neural systems mediating the expression of fearfulness in the rat. Proc Natl Acad Sci U S A. 1998;95:5335–40.
Caldji C, Diorio J, Meaney MJ. Variations in maternal care in infancy regulate the development of stress reactivity. Biol Psychiatry. 2000;48:1164–74.
Cameron HA, Gould E. The control of neuronal birth and survival. In: Shaw C, editor. Receptor dynamics in neural development. Boca Raton: CRC; 1996. pp. 141–57.
Carlson MC, Erickson KI, Kramer AF, Voss MW, Bolea N, et al. Evidence for neurocognitive plasticity in at-risk older adults: the experience corps program. J Gerontol A Biol Sci Med Sci. 2009;64:1275–82.
Caspi A, McClay J, Moffitt TE, Mill J, Martin J, et al. Role of genotype in the cycle of violence in maltreated children. Science. 2002;297:851–4.
Caspi A, Sugden K, Moffitt TE, Taylor A, Craig IW, et al. Influence of life stress on depression: moderation by a polymorphism in the *5-HTT* gene. Science. 2003;301:386–9.
Castren E, Rantamaki T. The role of BDNF and its receptors in depression and antidepressant drug action: reactivation of developmental plasticity. Dev Neurobiol. 2010;70:289–97.
Cerqueira JJ, Pego JM, Taipa R, Bessa JM, Almeida OFX, Sousa N. Morphological correlates of corticosteroid-induced changes in prefrontal cortex-dependent behaviors. J Neurosci. 2005;25:7792–800.
Cho K. Chronic 'jet lag' produces temporal lobe atrophy and spatial cognitive deficits. Nature Neurosci. 2001;4:567–8.
Chollet F, Tardy J, Albucher JF, Thalamas C, Berard E, et al. Fluoxetine for motor recovery after acute ischaemic stroke (FLAME): a randomised placebo-controlled trial. Lancet Neurol. 2011;10:123–30.
Colcombe SJ, Kramer AF, Erickson KI, Scalf P, McAuley E, et al. Cardiovascular fitness, cortical plasticity, and aging. Proc Natl Acad Sci U S A. 2004;101:3316–21.
de Lange FPK, Hagoort P, et al. Increase in prefrontal cortical volume following cognitive behavioural therapy in patients with chronic fatigue syndrome. Brain. 2008;131:2172–80.
de Leon MJG, Convit A, et al. Frequency of hippocampus atrophy in normal elderly and Alzheimer's disease patients. Neurobiol Aging. 1997;18:1–11.
Del Giudice M, Ellis BJ, Shirtcliff EA. The adaptive calibration model of stress responsivity. Neurosci Biobehav Rev. 2011;35:1562–92.
Dhabhar F, McEwen B. Enhancing versus suppressive effects of stress hormones on skin immune function. Proc Natl Acad Sci U S A. 1999;96:1059–64.

Dhabhar FS, Malarkey WB, Neri E, McEwen BS. Stress-induced redistribution of immune cells—from barracks to boulevards to battlefields: a tale of three hormones—Curt Richter Award winner. Psychoneuroendocrinology. 2012a;37:1345–68.

Dias-Ferreira E, Sousa JC, Melo I, Morgado P, Mesquita AR, et al. Chronic stress causes frontostriatal reorganization and affects decision-making. Science. 2009;325:621–5.

Dinkel K, MacPherson A, Sapolsky RM. Novel glucocorticoid effects on acute inflammation in the CNS. J Neurochem. 2003;84:705–16.

Draganski B, Gaser C, Kempermann G, Kuhn HG, Winkler J, et al. Temporal and spatial dynamics of brain structure changes during extensive learning. J Neurosci. 2006;26:6314–7.

Drevets WC. Neuroimaging studies of mood disorders. Biol Psychiatry. 2000;48:813–29.

Drevets WC, Price JL, Simpson JR Jr, Todd RD, Reich T, et al. Subgenual prefrontal cortex abnormalities in mood disorders. Nature. 1997;386:824–7.

Du J, McEwen BS, Manji HK. Glucocorticoid receptors modulate mitochondrial function. Commun Integr Biol. 2009a;2:1–3.

Du J, Wang Y, Hunter R, Wei Y, Blumenthal R, et al. Dynamic regulation of mitochondrial function by glucocorticoids. Proc Natl Acad Sci U S A. 2009b;106:3543–8.

Duman RS, Monteggia LM. A neurotrophic model for stress-related mood disorders. Biol Psychiatry. 2006;59:1116–27.

Eiland L, McEwen BS. Early life stress followed by subsequent adult chronic stress potentiates anxiety and blunts hippocampal structural remodeling. Hippocampus. 2012;22:82–91.

Entringer S, Epel ES, Kumsta R, Lin J, Hellhammer DH, et al. Stress exposure in intrauterine life is associated with shorter telomere length in young adulthood. Proc Natl Acad Sci U S A. 2011;108:E513–8.

Erickson KI, Prakash RS, Voss MW, Chaddock L, Hu L, et al. Aerobic fitness is associated with hippocampal volume in elderly humans. Hippocampus. 2009;19:1030–9.

Erickson KI, Voss MW, Prakash RS, Basak C, Szabo A, et al. Exercise training increases size of hippocampus and improves memory. Proc Natl Acad Sci U S A. 2011;108:3017–22.

Espinosa JS, Stryker MP. Development and plasticity of the primary visual cortex. Neuron. 2012;75:230–49.

Evans GW, Gonnella C, Marcynyszyn LA, Gentile L, Salpekar N. The role of chaos in poverty and children's socioemotional adjustment. Psychol Science. 2004;16:560–5.

Francis D, Diorio J, Liu D, Meaney MJ. Nongenomic transmission across generations of maternal behavior and stress responses in the rat. Science. 1999;286:1155–8.

Francis DD, Diorio J, Plotsky PM, Meaney MJ. Environmental enrichment reverses the effects of maternal separation on stress reactivity. J Neurosci. 2002;22:7840–3.

Fried LP, Carlson MC, Freedman M, Frick KD, Glass TA, et al. A social model for health promotion for an aging population: initial evidence on the experience corps model. J Urban Health. 2004;81:64–78.

Frodl T, Meisenzahl EM, Zetzsche T, Born C, Jager M, et al. Larger amygdala volumes in first depressive episode as compared to recurrent major depression and healthy control subjects. Biol Psychiatry. 2003;53:338–44.

Ganzel BL, Kim P, Glover GH, Temple E. Resilience after 9/11: multimodal neuroimaging evidence for stress-related change in the healthy adult brain. Neuroimage. 2008;40:788–95.

Gianaros PJ, Jennings JR, Sheu LK, Greer PJ, Kuller LH, Matthews KA. Prospective reports of chronic life stress predict decreased grey matter volume in the hippocampus. Neuroimage. 2007;35:795–803.

Gianaros PJ, Hariri AR, Sheu LK, Muldoon MF, Sutton-Tyrrell K, Manuck SB. Preclinical atherosclerosis covaries with individual differences in reactivity and functional connectivity of the amygdala. Biol Psychiatry. 2009;65:943–50.

Gold SM, Dziobek I, Sweat V, Tirsi A, Rogers K, et al. Hippocampal damage and memory impairments as possible early brain complications of type 2 diabetes. Diabetologia. 2007;50:711–9.

Gould E, Woolley C, McEwen BS. Short-term glucocorticoid manipulations affect neuronal morphology and survival in the adult dentate gyrus. Neuroscience. 1990;37:367–75.

Gurvits TV, Shenton ME, Hokama H, Ohta H, Lasko NB, et al. Magnetic resonance imaging study of hippocampal volume in chronic, combat-related posttraumatic stress disorder. Biol Psychiatry. 1996;40:1091–9.

Heinrich C, Lahteinen S, Suzuki F, Anne-Marie L, Huber S, et al. Increase in BDNF-mediated TrkB signaling promotes epileptogenesis in a mouse model of mesial temporal lobe epilepsy. Neurobiol Dis. 2011;42:35–47.

Hill MN, McEwen BS. Involvement of the endocannabinoid system in the neurobehavioural effects of stress and glucocorticoids. Prog Neuropsychopharmacol Biol Psychiatry. 2010;34:791–7.

Holzel BK, Carmody J, Evans KC, Hoge EA, Dusek JA, et al. Stress reduction correlates with structural changes in the amygdala. Soc Cogn Affect Neurosci. 2010;5:11–7.

Karatsoreos IN, Bhagat S, Bloss EB, Morrison JH, McEwen BS. Disruption of circadian clocks has ramifications for metabolism, brain, and behavior. Proc Natl Acad Sci U S A. 2011;108:1657–62.

Karst H, Berger S, Turiault M, Tronche F, Schutz G, Joels M. Mineralocorticoid receptors are indispensable for nongenomic modulation of hippocampal glutamate transmission by corticosterone. Proc Natl Acad Sci U S A. 2005;102:19204–7.

Kaufman D, Smith ELP, Gohil BC, Banerji MA, Coplan JD, et al. Early appearance of the metabolic syndrome in socially reared bonnet macaques. J Clin Endocrinol Metab. 2005;90:404–8.

Kaufman D, Banerji MA, Shorman I, Smith ELP, Coplan JD, et al. Early-life stress and the development of obesity and insulin resistance in juvenile bonnet macaques. Diabetes. 2007;56:1–5.

Kirschbaum C, Prussner JC, Stone AA, Federenko I, Gaab J, et al. Persistent high cortisol responses to repeated psychological stress in a subpopulation of healthy men. Psychosom Med. 1995;57:468–74.

Kokaia M, Ernfors P, Kokaia Z, Elmer E, Jaenisch R, Lindvall O. Suppressed epileptogenesis in BDNF mutant mice. Exp Neurol. 1995;133:215–24.

Lazarus RS, Folkman S, editors. Stress, appraisal and coping. New York:Springer; 1984.

Leuner B, Caponiti JM, Gould E. Oxytocin stimulates adult neurogenesis even under conditions of stress and elevated glucocorticoids. Hippocampus. 2012;22:861–8.

Liston C, Gan WB. Glucocorticoids are critical regulators of dendritic spine development and plasticity in vivo. Proc Natl Acad Sci U S A. 2011;108:16074–9.

Liston C, Miller MM, Goldwater DS, Radley JJ, Rocher AB, et al. Stress-induced alterations in prefrontal cortical dendritic morphology predict selective impairments in perceptual attentional set-shifting. J Neurosci. 2006;26:7870–4.

Liston C, McEwen BS, Casey BJ. Psychosocial stress reversibly disrupts prefrontal processing and attentional control. Proc Natl Acad Sci U S A. 2009;106:912–7.

Maccari S, Morley-Fletcher S. Effects of prenatal restraint stress on the hypothalamus-pituitary-adrenal axis and related behavioural and neurobiological alterations. Psychoneuroendo. 2007;32:S10–5.

Marsland AL, Gianaros PJ, Abramowitch SM, Manuck SB, Hariri AR. Interleukin-6 covaries inversely with hippocampal grey matter volume in middle-aged adults. Biol Psychiatry. 2008;64:484–90.

Martin KP, Wellman CL. NMDA receptor blockade alters stress-induced dendritic remodeling in medial prefrontal cortex. Cereb Cortex. 2011;21:2366–71.

McEwen BS. Protective and damaging effects of stress mediators. New Eng J Med. 1998;338:171–9.

McEwen BS. Stress and hippocampal plasticity. Annu Rev Neurosci. 1999;22:105–22.

McEwen BS. Protective and damaging effects of stress mediators: central role of the brain. Dialogues Clin Neurosci. 2006;8:367–81.

McEwen BS. Physiology and neurobiology of stress and adaptation: central role of the brain. Physiol Rev. 2007;87:873–904.

McEwen BS, Gianaros P. Stress- and allostasis-induced brain plasticity. Annu Rev Med 2011;62:431–45.

McEwen BS, Weiss J, Schwartz L. Selective retention of corticosterone by limbic structures in rat brain. Nature. 1968;220:911–2.

McGowan PO, Sasaki A, D'Alessio AC, Dymov S, Labonte B, et al. Epigenetic regulation of the glucocorticoid receptor in human brain associates with childhood abuse. Nature Neurosci. 2009;12:241–3.

Mitra R, Jadhav S, McEwen BS, Vyas A, Chattarji S. Stress duration modulates the spatiotemporal patterns of spine formation in the basolateral amygdala. Proc Nat Acad Sci U S A. 2005;102:9371–6.

Moore GJ, Bebehuk JM, Wilds IB, Chen G, Manji HK. Lithium-induced increase in human brain grey matter. Lancet. 2000;356:1241–2.

Mucha M, Skrzypiec AE, Schiavon E, Attwood BK, Kucerova E, Pawlak R. Lipocalin-2 controls neuronal excitability and anxiety by regulating dendritic spine formation and maturation. Proc Natl Acad Sci U S A. 2011;108:18436–41.

Okuda S, Roozendaal B, McGaugh JL. Glucocorticoid effects on object recognition memory require training-associated emotional arousal. Proc Natl Acad Sci U S A. 2004;101:853–8.

Plotsky PM, Thrivikraman KV, Nemeroff CB, Caldji C, Sharma S, Meaney MJ. Long-term consequences of neonatal rearing on central corticotropin-releasing factor systems in adult male rat offspring. Neuropsychopharmacology. 2005;30:2192–204.

Popoli M, Yan Z, McEwen BS, Sanacora G. The stressed synapse: the impact of stress and glucocorticoids on glutamate transmission. Nat Rev Neurosci. 2012;13:22–37.

Prager EM, Johnson LR. Stress at the synapse: signal transduction mechanisms of adrenal steroids at neuronal membranes. Sci Signal 2009;2:re5.

Pruessner JC, Baldwin MW, Dedovic K, Renwick RM NK, Lord C, et al. Self-esteem, locus of control, hippocampal volume, and cortisol regulation in young and old adulthood. Neuroimage. 2005;28:815–26.

Rajkowska G. Postmortem studies in mood disorders indicate altered numbers of neurons and glial cells. Biol Psychiatry. 2000;48:766–77.

Rao RP, Anilkumar S, McEwen BS, Chattarji S. Glucocorticoids protect against the delayed behavioral and cellular effects of acute stress on the amygdala. Biol Psychiatry. 2012;72:466–75.

Revest JM, Di Blasi F, Kitchener P, Rouge-Pont F, Desmedt A, et al. The MAPK pathway and Egr-1 mediate stress-related behavioral effects of glucocorticoids. Nat Neurosci. 2005;8:664–72.

Roozendaal B, Hahn EL, Nathan SV, de Quervain DJ-F, McGaugh JL. Glucocorticoid effects on memory retrieval require concurrent noradrenergic activity in the hippocampus and basolateral amygdala. J Neurosci. 2004;24:8161–9.

Ryff CD, Singer B. The contours of positive human health. Psychol Inquiry. 1998;9:1–28.

Sapolsky RM, Krey LC, McEwen BS. The neuroendocrinology of stress and aging: the glucocorticoid cascade hypothesis. Endocr Rev 1986;7:284–301.

Sapolsky RM, Romero LM, Munck AU. How do glucocorticoids influence stress responses? Integrating permissive, suppressive, stimulatory, and preparative actions. Endocrine Rev. 2000;21:55–89.

Scharfman HE. Hyperexcitability in combined entorhinal/hippocampal slices of adult rat after exposure to brain-derived neurotrophic factor. J Neurophysiol. 1997;78:1082–95.

Schelling G, Kilger E, Roozendaal B, de Quervain DJ-F, Briegel J, et al. Stress doses of hydrocortisone, traumatic memories, and symptoms of posttraumatic stress disorder in patients after cardiac surgery: a randomized study. Biol Psychiatry. 2004;55:627–33.

Schoenbaum G, Roesch M. Orbitofrontal cortex, associative learning, and expectancies. Neuron. 2005;47:633–6.

Seeman TE, Singer BH, Ryff CD, Dienberg G, Levy-Storms L. Social relationships, gender, and allostatic load across two age cohorts. Psychosom Med. 2002;64:395–406.

Sheline YI. Hippocampal atrophy in major depression: a result of depression-induced neurotoxicity? Mol Psychiatry. 1996;1:298–9.

Sheline YI. Neuroimaging studies of mood disorder effects on the brain. Biol Psychiatry. 2003;54:338–52.

Singer B, Friedman E, Seeman T, Fava GA, Ryff CD. Protective environments and health status: cross-talk between human and animal studies. Neurobiol Aging. 2005;26S:S113–8.

Snyder MA, Smejkalova T, Forlano PM, Woolley CS. Multiple ERbeta antisera label in ERbeta knockout and null mouse tissues. J Neurosci Methods 2010;188:226–34.
Spolidoro M, Baroncelli L, Putignano E, Maya-Vetencourt JF, Viegi A, Maffei L. Food restriction enhances visual cortex plasticity in adulthood. Nat Commun 2011;2:320.
Starkman MN, Giordani B, Gebrski SS, Berent S, Schork MA, Schteingart DE. Decrease in cortisol reverses human hippocampal atrophy following treatment of Cushing's disease. Biol Psychiatry. 1999;46:1595–602.
Sterling P, Eyer J. Allostasis: a new paradigm to explain arousal pathology. In: Fisher S, Reason J, editors. Handbook of life stress, cognition and health. New York: Wiley; 1988. pp. 629–49.
Stockmeier CA, Mahajan GJ, Konick LC, Overholser JC, Jurjus GJ, et al. Cellular changes in the postmortem hippocampus in major depression. Biol Psychiatry. 2004;56:640–50.
Suomi SJ. Risk, resilience, and gene x environment interactions in rhesus monkeys. Ann N Y Acad Sci 2006;1094:52–62.
Tang AC, Reeb-Sutherland BC, Yang Z, Romeo RD, McEwen BS. Neonatal novelty-induced persistent enhancement in offspring spatial memory and the modulatory role of maternal self-stress regulation. J Neurosci. 2011;31:5348–52.
Tang AC, Yang Z, Reeb-Sutherland BC, Romeo RD, McEwen BS. Maternal modulation of novelty effects on physical development. Proc Natl Acad Sci U S A. 2012;109:2120–5.
Tasker JG, Di S, Malcher-Lopes R. Minireview: rapid glucocorticoid signaling via membrane-associated receptors. Endocrinology. 2006;147:5549–56.
Thayer JF, Lane RD. A model of neurovisceral integration in emotion regulation and dysregulation. J Affect Disord. 2000;61:201–16.
Vetencourt JFM, Sale A, Viegi A, Baroncelli L, De Pasquale R, et al. The antidepressant fluoxetine restores plasticity in the adult visual cortex. Science. 2008;320:385–8.
Vyas A, Mitra R, Rao BSS, Chattarji S. Chronic stress induces contrasting patterns of dendritic remodeling in hippocampal and amygdaloid neurons. J Neurosci. 2002;22:6810–8.
Vythilingam M, Vermetten E, Anderson GM, Luckenbaugh D, Anderson ER, et al. Hippocampal volume, memory, and cortisol status in major depressive disorder: effects of treatment. Biol Psychiatry. 2004;56:101–12.
Yamamoto K. Steroid receptor regulated transcription of specific genes and gene networks. Ann Rev Genet. 1985;19:209–52.
Yoo S-S, Gujar N, Hu P, Jolesz FA, Walker MP. The human emotional brain without sleep—a prefrontal amygdala disconnect. Curr Biol. 2007;17:R877–8.
Zohar J, Yahalom H, Kozlovsky N, Cwikel-Hamzany S, Matar MA, et al. High dose hydrocortisone immediately after trauma may alter the trajectory of PTSD: interplay between clinical and animal studies. Eur Neuropsychopharmacol. 2011;21:796–809.

Chapter 2
Regulation of Excitatory Synapses by Stress Hormones

Marian Joëls, Harm Krugers and Henk Karst

Abstract Shortly after stress, brain levels of many transmitters and hormones such as corticosterone are elevated. In the brain, corticosterone affects those cells that express high-affinity mineralocorticoid receptors (MRs) and/or lower-affinity glucocorticoid receptors (GRs). Principal neurons in the hippocampal cornus ammoni 1 (CA1) area and dentate gyrus abundantly express both MR and GR, while principal cells in the basolateral amygdala have high GR but relatively low MR levels. Neurons in all three areas quickly respond to corticosterone with an enhancement in spontaneous glutamatergic transmission, an effect that is nongenomic and involves MR. This rapid effect is transient in hippocampal cells but sustained in amygdala neurons. The areas differ in their slow gene-mediated response to corticosterone. Hippocampal CA1 cells show an increased current amplitude in response to spontaneously released glutamate-containing vesicles; synaptically evoked responses are generally unaffected. The number of action potentials during a period of depolarization is attenuated, via a slow GR-dependent pathway. By contrast, basolateral amygdala neurons show higher excitability and more efficient transfer of action potentials several hours after corticosteroid exposure. The dichotomy between the two areas could explain why emotional aspects of stressful events are generally better retained than neutral aspects.

Abbreviations

AHP	Afterhyperpolarization
AMPA	α-Amino-3-hydroxy-5-methyl-4-isoxazolepropionic acid
BLA	Basolateral amygdala
BSA	Bovine serum albumin
CA1	Cornus ammoni 1
(m)EPSC/EPSP	(Miniature) Excitatory postsynaptic current/potential

M. Joëls (✉) · H. Karst
Department of Translational Neuroscience, Brain Center Rudolf Magnus, University Medical Center, 3584 Utrecht, The Netherlands
e-mail: m.joels@umcutrecht.nl

H. Krugers
Swammerdam Institute for Life Sciences—Center for Neuroscience, University of Amsterdam, Amsterdam, The Netherlands

M. Popoli et al. (eds.), *Synaptic Stress and Pathogenesis of Neuropsychiatric Disorders*,
DOI 10.1007/978-1-4939-1056-4_2,

ERK	Extracellular signal-regulated kinase
GR	Glucocorticoid receptor
LTD	Long-term depression
LTP	Long-term potentiation
MEK	Mitogen-activated protein kinase kinase
MR	Mineralocorticoid receptor
NMDA	N-methyl-D-aspartate

2.1 Introduction

When an organism encounters a situation that could (potentially) perturb its homeostatis, this is subjectively experienced as "stress." Two systems are activated upon stress exposure: (1) the autonomic nervous system, which quickly results in release of (nor)adrenaline from the adrenal medulla, but also from neurons in the locus coeruleus and nucleus tractus solitaries (for reviews see Valentino and Von Bockstaehle 2008; McIntyre et al. 2012) and (2) the hypothalamo-pituitary-adrenal axis, which eventually causes synthesis and secretion of corticosteroid hormones from the adrenal cortex (for reviews see De Kloet et al. 2005; Ulrich-Lai and Herman 2009). In humans, cortisol is the primary circulating corticosteroid, while in rodents corticosterone prevails. The stress-induced secretion of corticosteroid hormones occurs on top of ultradian pulses with a 1-h inter-pulse interval (Lightman and Conway-Campbell 2010). The peak of these ultradian pulses varies: low-amplitude pulses are seen at the start of the inactive period, and the amplitude of pulses gradually rises towards the start of the active period. Overall, the pulses give rise to a circadian release pattern of corticosteroid hormones.

Corticosterone easily enters the brain due to its lipophilic character. It reaches every cell in the brain but is only active in those cells that express receptors. Two corticosteroid receptors have been recognized, based on their molecular properties and pharmacological profile (Reul and de Kloet 1985; Evans and Arriza 1989). Low levels of corticosteroid hormones first bind to the mineralocorticoid receptor (MR), which has a Kd of approximately 0.5 nM. Expression levels of MR are high in all hippocampal neurons, as well as neurons in the lateral septum and some motor nuclei in the brain stem. In cortical cells and most of the amygdalar nuclei, MR expression is much lower. The brain MR is structurally similar to MRs in epithelial cells, such as in the kidney (see for review Funder 2010). However, in these cells cortisol and corticosterone are converted by the 11-β-hydroxysteroid dehydrogenase isoform 2 into metabolites with extremely low affinity for the MR, so that MRs become available for binding by the less prevalent hormone aldosterone (Wyrwoll et al. 2011). In most cells in the brain however, the 11-β-hydroxysteroid dehydrogenase isoform 2 is not highly expressed, explaining why corticosterone and cortisol are the main ligands of the brain MR.

With higher concentrations of corticosterone or cortisol, the hormones also bind to the glucocorticoid receptor (GR). This receptor has a Kd of 2–5 nM and is much more ubiquitously expressed (Reul and de Kloet 1985; Weinberger et al. 1985). The corticosteroid concentration reached at the trough of ultradian pulses is lower than the Kd of the GR; therefore, this receptor only becomes substantially occupied at the peak of high-amplitude ultradian pulses and after stress. The difference in Kd of the two receptor types is very relevant for neurons that express MR as well as GR, e.g., pyramidal neurons in the CA1 hippocampal area and granule cells in the dentate gyrus. These cells shuttle between on the one hand a condition of predominant MR activation during the circadian trough, and on the other hand concurrent MR and GR activation after stress or at the peak of high-amplitude ultradian pulses.

MR and GR reside in the cytoplasm when unbound to corticosteroids, in a complex with chaperone molecules such as heat shock proteins (Biddie and Hager 2009). When corticosteroids bind the receptor, the chaperones dissociate and the activated receptors move to the nucleus. There, they either homodimerize and directly bind to glucocorticoid response elements in the DNA; or they bind as monomers to other transcription factors, thus interfering with the efficacy of the latter. Through both pathways, corticosteroid receptors slowly and persistently change the expression of responsive genes, an approximate 2 % of the total (Datson et al. 2008). Potentially, this will alter neuronal function in many ways and for a prolonged period of time.

More recently, though, it has become evident that corticosteroid hormones are also active within minutes, via nongenomic signalling. This was first described extensively for parvocellular neurons in the hypothalamic paraventricular nucleus (Di et al. 2003, 2005). Rapid corticosteroid effects are probably mediated by MRs and GRs located on the plasma membrane rather than in the cytoplasm or nucleus. Although specific receptor molecules mediating fast effects by corticosteroids have been identified in nonmammalian vertebrates (Orchinik et al. 1991), convincing evidence for the existence of receptors exclusively mediating rapid actions was never obtained in rodents. In addition to corticosteroid actions developing over the course of minutes or hours, these hormones also seem to be able to change neuronal function in a third, intermediate time-domain which may depend on posttranslational modifications. For instance, recent evidence supports that GRs change Histone 3 methylation (Roozendaal et al. 2010; Gutièrrez-Mecinas et al. 2011; Hunter et al. 2012), which causes functional effects with a delay of approximately 20 min.

Evidently, variations in corticosteroid level will change the function of many neurons, over a wide range of time, starting directly after stress and lasting for hours to even days (for details see Joëls et al. 2012). In this chapter, we will particularly highlight rapid and slow cellular actions by corticosterone on glutamatergic transmission in three parts of the brain that are important for (emotional) memory formation, i.e., the hippocampal CA1 area, the dentate gyrus, and the basolateral amygdala.

2.2 Rapid effects

2.2.1 Hippocampus

Glutamate is the main excitatory transmitter in the brain. It mainly acts through AMPA and NMDA receptors. Upon arrival of action potentials in the presynaptic terminal, intracellular calcium levels are raised, which in turn promotes the release of glutamate. However, glutamate is to a limited extent also spontaneously released, i.e., in the absence of action potentials. This spontaneous activity can be detected postsynaptically through the recording of so-called miniature excitatory postsynaptic currents (mEPSCs), each of which represents the response to a spontaneously released synaptic vesicle containing glutamate.

CA1 hippocampal pyramidal cells show an enhanced mEPSC frequency during the application of corticosterone, while mEPSC amplitude, rise time, and decay remain unaffected by the hormone (Karst et al. 2005). Corticosterone diminished the second relative to the first evoked response in a paired pulse stimulation paradigm, supporting that the hormone increases the release probability of glutamate-containing vesicles, instead of increasing the number of synaptic contacts. The corticosterone-induced increase in mEPSC frequency is short-lived; when the hormone application is terminated, mEPSC frequency quickly returns to the pretreatment level. Corticosterone was found to exert very similar effects in the presence of a protein synthesis inhibitor, which argues against involvement of a genomic pathway. Corticosterone conjugated to bovine serum albumin (BSA), which does not pass the plasma membrane, caused very similar effects on mEPSC frequency; intracellular administration of corticosterone was ineffective (Karst et al. 2005; Olijslagers et al. 2008). These findings suggest that corticosterone binds to a receptor molecule located on (or close to) the membrane. Based on the just-effective concentration (10 nM), it was thought that these rapid actions of corticosterone involve GRs rather than MRs, similar to what had been reported for hypothalamic neurons (Di et al. 2003). Yet, the selective GR agonist RU 28386 was entirely ineffective, and effects of corticosterone were not blocked by the GR antagonist RU 38486 (Karst et al. 2005). Conversely, 10 nM of the MR agonist aldosterone in the presence of RU 38486 highly effectively increased mEPSC frequency, an effect that was completely blocked by the MR-antagonist spironolactone, indicating that the rapid effects are mediated by MR rather than GR. In agreement, the increased mEPSC frequency by corticosterone was not observed in forebrain specific MR knockouts, but remained intact in GR knockout mice. Recently, it was reported that the MR-mediated increase in mEPSC frequency depends on the expression of limbic system-associated membrane protein, Lsamp (Qiu et al. 2010). Using pharmacological tools it was shown that granule cells in the dentate also display an MR-dependent raise in mEPSC frequency, very similar to that seen in CA1 pyramidal cells (Pasricha et al. 2011).

The pathway through which corticosterone rapidly affects release probability has to some extent been resolved. Rapid effects are blocked by MEK inhibitors, pointing to involvement of ERK (Olijslagers et al. 2008). ERK activation is known to induce phosphorylation of Synapsin-I which promotes neurotransmitter release (Hilfiker et al. 1998). Interestingly, ERK activation and Synapsin-I were also proposed to be involved in slow GR-dependent modulation of glutamatergic transmission in the hippocampus (Revest et al. 2010). In agreement, ERK is important for stress-induced effects on hippocampus-dependent learning (Reul et al. 2009).

Corticosterone also rapidly changes postsynaptic properties of hippocampal cells, including aspects of glutamatergic transmission. Thus, in the postsynaptic membrane, lateral movement of GluA2 subunits of the AMPA receptor is rapidly increased by corticosterone and is linked to a long-lasting higher dwell-time in the postsynaptic density (Groc et al. 2008). This postsynaptic effect—like the presynaptic effect of corticosterone—involves MRs, is induced by the membrane-impermeable corticosterone-BSA conjugate and is not affected by a protein synthesis inhibitor. Both actions on glutamate transmission are expected to increase the (spontaneous) activity of hippocampal CA1 neurons. Since corticosterone also rapidly reduces the voltage dependent and transient A-current in CA1 neurons (Olijslagers et al. 2008), the changes in glutamatergic transmission are probably accompanied by more sustained firing. Overall, excitatory transmission is thought to be increased shortly after corticosterone reaches the brain.

Findings with regard to a slightly more delayed time-domain (approximately 20–60 min after stress) are more equivocal. One study (Tse et al. 2011) reported that CA1 cells respond more strongly to excitatory input 20–30 min after the start of corticosterone administration. At the single cell level, the NMDA/AMPA ratio was increased, most likely via GR. Extracellularly, an increase in the field excitatory postsynaptic potential evoked via NMDA—but not AMPA—receptors was found. However, most studies report *reduced* responses to synaptic input in this time-domain. For instance, spontaneous firing of hippocampal cells was reduced 20 min after peripheral injection of corticosterone (Pfaff et al. 1971). Various types of stress impaired the stability or reduced the firing rate of hippocampal place cells in this intermediate time-domain (Kim et al. 2007; Passecker et al. 2011). In vitro administered corticosterone (at a very high dose) was found to reduce the population spike amplitude in the CA1 area, reaching a plateau 20–40 min after corticosterone administration was started (Vidal et al. 1986). Also, the ability to evoke an action potential through synaptic stimulation and the amplitude of the EPSP in CA1 neurons declined with repeated stimulation of the afferents (Joëls and de Kloet 1993); these effects became evident approximately 20 min after the start of corticosterone administration. In neonatal cultured hippocampal neurons, corticosterone was found to reduce NMDA-evoked currents, through a membrane-bound receptor not blocked by classical MR- or GR-antagonists (Liu et al. 2007; Zhang et al. 2012).

The overall impact of corticosterone on CA1 pyramidal cell activity not only depends on its effect on excitatory transmission but also on inhibitory transmission. Corticosterone does change GABAergic inhibitory transmission in the intermediate time-domain, but the effects are variable and seem to depend on the recording method (Zeise et al. 1992; Teschemacher et al. 1996; Hu et al. 2010). Interestingly, inhibitory signals, i.e., spontaneous inhibitory postsynaptic current amplitude, were reported to be enhanced in the dorsal hippocampus via GRs (Maggio and Segal 2009), while in the ventral hippocampus an MR-dependent reduction in spontaneous inhibitory postsynaptic current frequency was reported. These effects were seen >25 min after onset of corticosterone administration and peaked at 55 min.

All in all, most studies agree that corticosterone quickly increases spontaneous glutamatergic transmission. Synaptically evoked field potentials, however, were mostly not rapidly altered by corticosterone administration (e.g., Wiegert et al. 2006; Pu et al. 2007). Of course, it should be realized that in this rapid time-window other transmitters and hormones released by stress are also active and will affect the excitability. For instance, noradrenaline acting via β-adrenoceptors (but not α-adrenoceptors) increases excitatory transmission (see, e.g., Gereau and Conn 1994; Croce et al. 2003; Zhou et al. 2012), a phenomenon that in the dentate gyrus was shown to involve Synapsin-I phosphorylation (Parfitt et al. 1991). The neuropeptide *corticotropin-releasing hormone* (CRH) is known to quickly potentiate population spikes in the CA1 hippocampal area evoked by Schaffer collateral stimulation (Blank et al. 2002). Therefore, enhanced hippocampal activity during this phase directly after stress probably prevails. After this initial phase, that is 20–60 min after corticosterone reaches hippocampal cells, mostly inhibitory actions have been reported.

2.2.2 Basolateral Amygdala

Corticosterone rapidly increases mEPSC frequency of principal cells in the BLA, similar to what has been described for neurons in the CA1 area and dentate gyrus (Karst et al. 2010). However, in contrast to the latter regions, mEPSC frequency in BLA cells remains high, even after washout of the hormone. While the onset is clearly too fast to involve genomic signalling, the persistence of the response was found to depend on protein synthesis and requires expression of both MR and GR (Karst et al. 2010). The sustained response to a first pulse of corticosterone changes BLA cell properties such that they show a *reduced* mEPSC frequency in response to a second pulse of corticosterone. In contrast to the rapid response to the first corticosteroid exposure, the response to a second pulse was shown to involve GRs and the cannabinoid receptor-1. These rapid inhibitory responses were also seen when animals were first exposed to stress and subsequently to a pulse of corticosterone in vitro. The reversal in response depending on the recent stress history of the organism was called metaplasticity (Karst et al. 2010; see Fig. 2.1). One explanation for the shift in responsiveness after the second exposure to corticosterone is a change in the number of MR and/or GR located on the membrane, e.g., caused by internalization of MRs after the first pulse of corticosterone. Obviously, this needs further investigation.

The functional relevance of the quick increase in spontaneous excitatory transmission induced by corticosterone in BLA neurons is still unclear. Both at the single cell and the field potential level, corticosterone did not quickly change AMPA- or NMDA-R mediated synaptic responses (Liebmann et al. 2009; Pu et al. 2009).

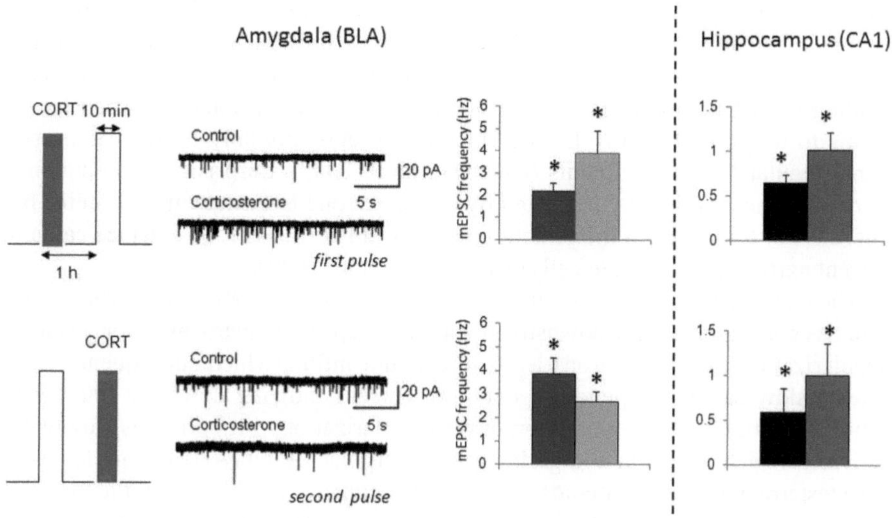

Fig. 2.1 In principal neurons of the *BLA*, a single pulse of corticosterone (*CORT*) (10 min, 100 nM) causes an increase in mEPSC frequency (*top left; dark bar:* mean mEPSC frequency prior to corticosterone application; *light grey bar:* mean mEPSC frequency during corticosterone application). The mEPSC frequency remains high even after wash-out (*bottom left*). Against this background, a second pulse of corticosterone decreases mEPSC frequency. By contrast, the response of hippocampal *CA1* neurons to a second pulse (*bottom right*) is highly comparable to the response to the first pulse (*top right*)
* indicates $p < 0.05$

However, as in the hippocampus, the rapid effect of corticosterone on spontaneous glutamatergic transmission may add to the overall change in excitability caused by other stress mediators like noradrenaline or CRH.

2.3 Delayed effects

2.3.1 Hippocampus

Passive or active membrane properties of dorsal CA1 pyramidal neurons, such as resting membrane potential, input resistance or characteristics of the action potential, are generally not much affected over the course of time after application of corticosterone (e.g., Joëls and de Kloet 1989; Kerr et al. 1989; but see Beck et al. 1994). However, neurons in the ventral-most (20 %) part of the hippocampus gradually become more excitable after corticosterone application (Maggio and Segal 2009).

Excitability could also be affected via corticosteroid actions on voltage-dependent ion channels. In the dorsal CA1 area corticosterone most prominently changes voltage-dependent calcium currents, much more so than sodium or potassium currents. A series of experiments showed that corticosterone or stress enhances the amplitude of sustained high-voltage-activated calcium currents in a slow manner, i.e., with a delay of > 1 h (Kerr et al. 1992; Karst et al. 1994; Joëls et al. 2003). The enhancement in calcium current amplitude requires protein synthesis and DNA-binding of GR homodimers (Kerr et al. 1992; Karst et al. 2000). Corticosterone seems to target particularly L-type calcium currents, possibly through transcriptional regulation of β4 subunits (Chameau et al. 2007). Surprisingly, β4 subunits were also transcriptionally regulated by corticosterone in dentate granule cells, but this was not translated to the protein level, nor did corticosterone enhance calcium current amplitude in granule cells (Van Gemert et al. 2009).

The increased calcium influx in CA1 neurons after stress or corticosterone exposure has consequences for downstream calcium-dependent pathways. For instance, depolarization of CA1 neurons leads to calcium influx, which subsequently activates a slow calcium-dependent potassium current, slowing down the transfer of action potentials. Upon termination of the depolarization, this current is slowly deactivated which results in a lingering afterhyperpolarization (AHP). High levels of corticosterone or glucocorticoids were found to enhance the AHP amplitude in CA1 pyramidal neurons recorded 1–4 h later and attenuated the transfer of action potentials during a period of depolarization (Joëls and de Kloet 1989; Kerr et al. 1989; Liebmann et al. 2008). Transfer of longer periods of excitatory information through the CA1 area is thus hampered 1–4 h after corticosterone levels are elevated. As with the passive and active membrane properties, the ventral-most part of the hippocampus reacted oppositely to the dorsal part after corticosterone application, showing *reduced* firing frequency accommodation and more spikes upon depolarization, i.e., higher excitability (Maggio and Segal 2009).

A third pathway through which corticosterone slowly changes excitability in the CA1 hippocampal area concerns its actions on spontaneous glutamatergic transmission. In both CA1 and (unidentified) cultured hippocampal neurons, a pulse of corticosteroids or of selective GR agonists increases the amplitude but not frequency of mEPSCs recorded several hours after corticosteroid exposure (Karst and Joëls 2005; Martin et al. 2009). This increase in amplitude is associated with a slow GR-dependent increase in surface expression of GluA2 subunits (Fig. 2.2) via a process requiring protein synthesis. At the same time the mobility of GluA2 subunits is enhanced (Groc et al. 2008; Martin et al. 2009). These effects on mEPSC amplitude develop > 1 h after corticosterone application and reach a maximal value between 150 and 200 min after onset of hormone treatment (Karst and Joëls 2005; Groc et al. 2008; Martin et al. 2009). Functionally, the increase in mEPSC amplitude occludes chemically induced LTP (Groc et al. 2008; Martin et al. 2009; Xiong et al., unpublished observations), while activity-dependent decreases in synaptic AMPA receptors (long-term depression, LTD) are facilitated.

How these effects on spontaneous glutamatergic transmission impact on synaptically evoked responses is presently unclear. Extracellularly recorded field potentials in the various hippocampal areas were in most studies not reported to be altered by corticosterone or stress (e.g., Pavlides et al. 1996; Bramham et al. 1998; Zhou et al. 2000; Yamada et al. 2003; Chen et al. 2010), although occasionally enhanced (Kavushansky et al. 2006; Avital et al. 2006) or reduced activity was observed

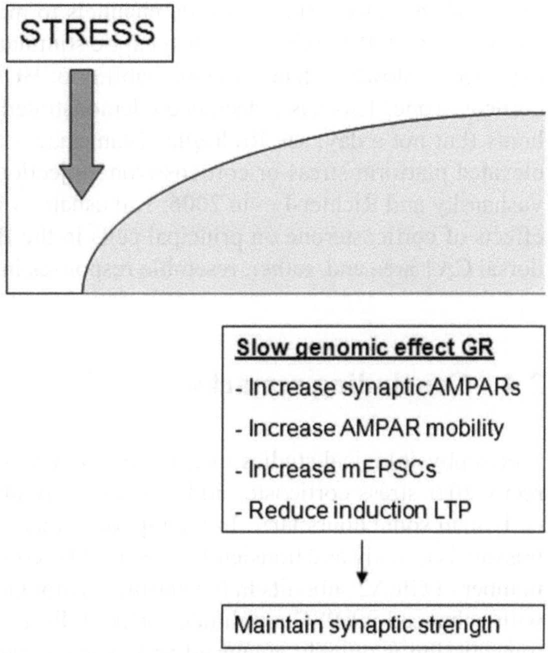

Fig. 2.2 Corticosterone slowly, via a genomic and glucocorticoid receptor (*GR*)-mediated action, enhances α-Amino-3-hydroxy-5-methyl-4-isoxazolepropionic acid receptor (*AMPAR*) synaptic transmission. At the same time—possibly via occlusion—corticosterone hampers the ability to strengthen synapses, which may promote the consolidation of relevant information. *mEPSC* miniature excitatory postsynaptic current, *LTP* long-term potentiation

(Hirata et al. 2008). Most likely, corticosteroid actions on excitatory (or inhibitory) transmission are restricted to a limited number of synapses and thus not discerned at a more global level, similar to what has been found after learning (Whitlock et al. 2006). It may also relate to the dose of corticosterone that was used or the intensity of the stressor. This is suggested by a study of Rey et al. (1987), showing that low doses of corticosterone enhance the amplitude of the population spike evoked by synaptic stimulation in the CA1 area, while high doses decrease the population spike.

2.3.2 Basolateral Amygdala

In the BLA, administration of a brief pulse of corticosterone increased input resistance and resulted in a more depolarized membrane potential of principal neurons some hours later (Duvarci and Pare 2007). This was only seen in a subpopulation of neurons with a very high input resistance (Duvarci and Pare 2007; Liebmann et al. 2008). In contrast to the CA1 area, firing frequency accommodation and AHP amplitude in the BLA were unaffected or even reduced by corticosterone (Duvarci and Pare 2007; Liebmann et al. 2008). Possibly, the low expression of alpha1.3 calcium channel subunits in the BLA contributes to this lack of modulation in firing frequency accommodation and AHP amplitude (Liebmann et al. 2008), despite a clear GR-dependent increase in sustained high-voltage-activated calcium currents (Karst et al. 2002). Corticosterone furthermore shifted the reversal potential of GABA-receptor linked chloride channels to more depolarized potentials, causing reduced IPSP amplitude upon synaptic stimulation. Altogether, these effects are expected to slowly enhance the excitability of BLA neurons after a single pulse of corticosterone. This has indeed been demonstrated at the field potential level some hours (but not a day; see Rodriguez Manzanaers et al. 2005) after restraint stress, elevated platform stress or corticosterone injection, both in vivo and in vitro (Kavushansky and Richter-Levin 2006; Kavushansky et al. 2006). In conclusion, slow effects of corticosterone on principal cells in the BLA differ from responses in the dorsal CA1 area and, rather, resemble responses in the ventral-most CA1 region.

2.4 Concluding remarks

Electrophysiological studies over the past decades have supplied evidence that directly after stress corticosteroid hormones may affect neuronal excitability differently than some hours later. In the hippocampus, spontaneous glutamatergic transmission is quickly and transiently enhanced by corticosterone. Some hours later, the number of GluA2 subunits in the plasma membrane is increased which is associated with enhanced mEPSC amplitudes. This indicates that spontaneous glutamatergic transmission is quickly enhanced and through another mechanism may remain elevated in those synapses that were involved in the initial response to stress. How these changes in spontaneous glutamatergic transmission translate to the transfer of information through the CA1 area or dentate gyrus is presently hard to predict. It is conceivable that glutamatergic transmission in some synapses is considerably facilitated in a similar manner as seen after high-frequency stimulation. However, when a series of (glutamate-mediated) signals reaches CA1 neurons >20 min after corticosterone exposure, the transfer of excitatory transmission is suppressed rather than enhanced. This may explain why it is difficult to induce long-term potentiation (LTP) at that time (see for review Kim and Diamond 2002). All of these actions may serve to enhance the signal-to-noise ratio and preserve stress-related information earlier encoded in the hippocampus.

Corticosteroid hormones affect transmission in the ventral-most part of the hippocampus and the BLA in a different manner than in the dorsal hippocampus. Neurons in, e.g., the BLA are quickly excited when corticosterone levels rise, but (different from dorsal hippocampal cells) both the spontaneous glutamatergic transmission and the response to multiple action potentials remain enhanced, also hours after the first exposure to corticosterone. This suggests that in the ventral-most part of the hippocampus and BLA the window for encoding of stress-related information is more prolonged than in the dorsal hippocampus. Given the role of the BLA in the processing of emotional information, this observation may explain why emotional aspects of a stressful situation are strongly retained, much more so than neutral aspects which particularly involve the dorsal hippocampus (Buchanan and Lovallo 2001; Kuhlmann and Wolf 2006; Van Stegeren et al. 2010), although not all studies report this (Abercrombie et al. 2003; Rimmele et al. 2003).

How the metaplasticity in responses to corticosterone, as described for the BLA, will impact on the signal transfer through this area after stress, is at this time hard to predict. To really appreciate the functional relevance of all of these changes recorded in brain slices, it will be necessary to focus on in vivo recordings, preferably in freely moving animals. This is technically demanding, certainly if one wants to correlate firing patterns of many cells in multiple regions of the brain. Nevertheless, such in vivo recordings will be necessary to understand how corticosteroid modulation of glutamatergic transmission can alter behavior in the aftermath of stress.

References

Abercrombie HC, Kalin NH, Thurow ME, Rosenkranz MA, Davidson RJ. Cortisol variation in humans affects memory for emotionally laden and neutral information. Behav Neurosci. 2003;117:505–16.

Avital A, Segal M, Richter-Levin G. Contrasting roles of corticosteroid receptors in hippocampal plasticity. J Neurosci. 2006;26:9130–4.

Beck SG, List TJ, Choi KC. Long- and short-term administration of corticosterone alters CA1 hippocampal neuronal properties. Neuroendocrinology. 1994;60:261–72.

Biddie SC, Hager GL. Glucocorticoid receptor dynamics and gene regulation. Stress. 2009;12:193–205.

Blank T, Nijholt I, Eckart K, Spiess J. Priming of long-term potentiation in mouse hippocampus by corticotropin-releasing factor and acute stress: implications for hippocampus-dependent learning. J Neurosci. 2002;22:3788–94.

Bramham CR, Southard T, Ahlers ST, Sarvey JM. Acute cold stress leading to elevated corticosterone neither enhances synaptic efficacy nor impairs LTP in the dentate gyrus of freely moving rats. Brain Res. 1998;789:245–55.

Buchanan TW, Lovallo WR. Enhanced memory for emotional material following stress-level cortisol treatment in humans. Psychoneuroendocrinology. 2001;26:307–17.

Chameau P, Qin Y, Spijker S, Smit AB, Joëls M. Glucocorticoids specifically enhance L-type calcium current amplitude and affect calcium channel subunit expression in the mouse hippocampus. J Neurophysiol. 2007;97:5–14.

Chen CC, Yang CH, Huang CC, Hsu KS. Acute stress impairs hippocampal mossy fiber-CA3 long-term potentiation by enhancing cAMP-specific phosphodiesterase 4 activity. Neuropsychopharmacology. 2010;35:1605–17.

Croce A, Astier H, Récasens M, Vignes M. Opposite effects of alpha 1- and beta-adrenoceptor stimulation on both glutamate- and gamma-aminobutyric acid-mediated spontaneous transmission in cultured rat hippocampal neurons. J Neurosci Res. 2003;71:516–25.

Datson NA, Morsink MC, Meijer OC, de Kloet ER. Central corticosteroid actions: search for gene targets. Eur J Pharmacol. 2008;583:272–89.

De Kloet ER, Joels M, Holsboer F. Stress and the brain: from adaptation to disease. Nat Rev Neurosci. 2005;6:463–75.

Di S, Malcher-Lopes R, Halmos KC, Tasker JG. Nongenomic glucocorticoid inhibition via endocannabinoid release in the hypothalamus: a fast feedback mechanism. J Neurosci. 2003;23:4850–7.

Di S, Malcher-Lopes R, Marcheselli VL, Bazan NG, Tasker JG. Rapid glucocorticoid-mediated endocannabinoid release and opposing regulation of glutamate and gamma-aminobutyric acid inputs to hypothalamic magnocellular neurons. Endocrinology. 2005;146:4292–301.

Duvarci S, Pare D. Glucocorticoids enhance the excitability of principal basolateral amygdala neurons. J Neurosci. 2007;27:4482–91.

Evans RM, Arriza JL. A molecular framework for the actions of glucocorticoid hormones in the nervous system. Neuron. 1989;2:1105–12.
Funder JW. Minireview: Aldosterone and mineralocorticoid receptors: past, present, and future. Endocrinology. 2010;151:5098–102.
Gereau RW 4th, Conn PJ. Presynaptic enhancement of excitatory synaptic transmission by beta-adrenergic receptor activation. J Neurophysiol 1994;72:1438–42.
Groc L, Choquet D, Chaouloff F. The stress hormone corticosterone conditions AMPAR surface trafficking and synaptic potentiation. Nat Neurosci. 2008;11:868–70.
Gutièrrez-Mecinas M, Trollope AF, Collins A, Morfett H, Hesketh SA, Kersanté F, Reul JM. Long-lasting behavioral responses to stress involve a direct interaction of glucocorticoid receptors with ERK1/2-MSK1-Elk-1 signaling. Proc Natl Acad Sci U S A. 2011;108:13806–11.
Hilfiker S, Schweizer FE, Kao HT, Czernik AJ, Greengard P, Augustine GJ. Two sites of action for synapsin domain E in regulating neurotransmitter release. Nat Neurosci. 1998;1(2):9–35.
Hirata R, Togashi H, Matsumoto M, Yamaguchi T, Izumi T, Yoshioka M. Characterization of stress-induced suppression of long-term potentiation in the hippocampal CA1 field of freely moving rats. Brain Res. 2008;1226:27–32.
Hu W, Zhang M, Czeh B, Flugge G, Zhang W. Stress impairs GABAergic network function in the hippocampus by activating nongenomic glucocorticoid receptors and affecting the integrity of the parvalbumin-expressing neuronal network. Neuropsychopharmacology. 2010;35:1693–707.
Hunter RG, Murakami G, Dewell S, Seligsohn M, Baker ME, Datson NA, McEwen BS, Pfaff DW. Acute stress and hippocampal histone H3 lysine 9 trimethylation, a retrotransposon silencing response. Proc Natl Acad Sci U S A. 2012;109:17657–62.
Joels M, de Kloet ER. Effects of glucocorticoids and norepinephrine on the excitability in the hippocampus. Science. 1989;245:1502–5.
Joels M, de Kloet ER. Corticosteroid actions on amino acid-mediated transmission in rat CA1 hippocampal cells. J Neurosci. 1993;13:4082–90.
Joels M, Velzing E, Nair S, Verkuyl JM, Karst H. Acute stress increases calcium current amplitude in rat hippocampus: temporal changes in physiology and gene expression. Eur J Neurosci. 2003;18:1315–24.
Joëls M, Sarabdjitsingh RA, Karst H. Unraveling the time domains of corticosteroid hormone influences on brain activity: rapid, slow, and chronic modes. Pharmacol Rev. 2012;64:901–38.
Karst H, Joels M. Corticosterone slowly enhances miniature excitatory postsynaptic current amplitude in mice CA1 hippocampal cells. J Neurophysiol. 2005;94:3479–86.
Karst H, Wadman WJ, Joëls M. Corticosteroid receptor-dependent modulation of calcium currents in rat hippocampal CA1 neurons. Brain Res. 1994;649:234–42.
Karst H, Karten YJ, Reichardt HM, de Kloet ER, Schutz G, Joels M. Corticosteroid actions in hippocampus require DNA binding of glucocorticoid receptor homodimers. Nat Neurosci. 2000;3:977–8.
Karst H, Nair S, Velzing E, Rumpff-van Essen L, Slagter E, Shinnick-Gallagher P, Joels M. Glucocorticoids alter calcium conductances and calcium channel subunit expression in basolateral amygdala neurons. Eur J Neurosci. 2002;16:1083–9.
Karst H, Berger S, Turiault M, Tronche F, Schutz G, Joels M. Mineralocorticoid receptors are indispensable for nongenomic modulation of hippocampal glutamate transmission by corticosterone. Proc Natl Acad Sci U S A. 2005;102:19204–7.
Karst H, Berger S, Erdmann G, Schutz G, Joels M. Metaplasticity of amygdalar responses to the stress hormone corticosterone. Proc Natl Acad Sci U S A. 2010;107:14449–54.
Kavushansky A, Richter-Levin G. Effects of stress and corticosterone on activity and plasticity in the amygdala. J Neurosci Res. 2006;84:1580–7.
Kavushansky A, Vouimba RM, Cohen H, Richter-Levin G. Activity and plasticity in the CA1, the dentate gyrus, and the amygdala following controllable vs. uncontrollable water stress. Hippocampus. 2006;16:35–42.
Kerr DS, Campbell LW, Hao SY, Landfield PW. Corticosteroid modulation of hippocampal potentials: increased effect with aging. Science. 1989;245:1505–9.

Kerr DS, Campbell LW, Thibault O, Landfield PW. Hippocampal glucocorticoid receptor activation enhances voltage-dependent Ca2+conductances: relevance to brain aging. Proc Natl Acad Sci U S A. 1992;89:8527–31.
Kim JJ, Diamond DM. The stressed hippocampus, synaptic plasticity and lost memories. Nat Rev Neurosci. 2002;3:453–62.
Kim JJ, Lee HJ, Welday AC, Song E, Cho J, Sharp PE, Jung MW, Blair HT. Stress-induced alterations in hippocampal plasticity, place cells, and spatial memory. Proc Natl Acad Sci U S A. 2007;104:18297–302.
Kuhlmann S, Wolf OT. Arousal and cortisol interact in modulating memory consolidation in healthy young men. Behav Neurosci. 2006;120:217–23.
Liebmann L, Karst H, Sidiropoulou K, van Gemert N, Meijer OC, Poirazi P, Joels M. Differential effects of corticosterone on the slow afterhyperpolarization in the basolateral amygdala and CA1 region: possible role of calcium channel subunits. J Neurophysiol. 2008;99:958–68.
Liebmann L, Karst H, Joels M. Effects of corticosterone and the beta-agonist isoproterenol on glutamate receptor-mediated synaptic currents in the rat basolateral amygdala. Eur J Neurosci. 2009;30:800–7.
Lightman SL, Conway-Campbell BL. The crucial role of pulsatile activity of the HPA axis for continuous dynamic equilibration. Nat Rev Neurosci. 2010;11:710–8.
Liu L, Wang C, Ni X, Sun J. A rapid inhibition of NMDA receptor current by corticosterone in cultured hippocampal neurons. Neurosci Lett. 2007;420:245–50.
Maggio N, Segal M. Differential corticosteroid modulation of inhibitory synaptic currents in the dorsal and ventral hippocampus. J Neurosci. 2009;29:2857–66.
Martin S, Henley JM, Holman D, Zhou M, Wiegert O, van Spronsen M, Joels M, Hoogenraad CC, Krugers HJ. Corticosterone alters AMPAR mobility and facilitates bidirectional synaptic plasticity. PLoS ONE. 2009;4:e4714.
McIntyre CK, McGaugh JL, Williams CL. Interacting brain systems modulate memory consolidation. Neurosci Biobehav Rev. 2012;36:1750–62.
Olijslagers JE, de Kloet ER, Elgersma Y, van Woerden GM, Joels M, Karst H. Rapid changes in hippocampal CA1 pyramidal cell function via pre- as well as postsynaptic membrane mineralocorticoid receptors. Eur J Neurosci. 2008;27:2542–50.
Orchinik M, Murray TF, Moore FL. A corticosteroid receptor in neuronal membranes. Science. 1991;252:1848–51.
Parfitt KD, Hoffer BJ, Browning MD. Norepinephrine and isoproterenol increase the phosphorylation of synapsin I and synapsin II in dentate slices of young but not aged Fisher 344 rats. Proc Natl Acad Sci U S A. 1991;88:2361–5.
Pasricha N, Joels M, Karst H. Rapid effects of corticosterone in the mouse dentate gyrus via a nongenomic pathway. J Neuroendocrinol. 2011;23:143–7.
Passecker J, Hok V, Della-Chiesa A, Chah E, O'Mara SM. Dissociation of dorsal hippocampal regional activation under the influence of stress in freely behaving rats. Front Behav Neurosci. 2011;5:66.
Pavlides C, Ogawa S, Kimura A, McEwen BS. Role of adrenal steroid mineralocorticoid and glucocorticoid receptors in long-term potentiation in the CA1 field of hippocampal slices. Brain Res. 1996;738:229–35.
Pfaff DW, Silva MT, Weiss JM. Telemetered recording of hormone effects on hippocampal neurons. Science. 1971;172:394–5.
Pu Z, Krugers HJ, Joels M. Corticosterone time-dependently modulates beta-adrenergic effects on long-term potentiation in the hippocampal dentate gyrus. Learn Mem. 2007;14:359–67.
Pu Z, Krugers HJ, Joels M. Beta-adrenergic facilitation of synaptic plasticity in the rat basolateral amygdala in vitro is gradually reversed by corticosterone. Learn Mem. 2009;16:155–60.
Qiu S, Champagne DL, Peters M, Catania EH, Weeber EJ, Levitt P, Pimenta AF. Loss of limbic system-associated membrane protein leads to reduced hippocampal mineralocorticoid receptor expression, impaired synaptic plasticity, and spatial memory deficit. Biol Psychiatry. 2010;68:197–204.
Reul JM, de Kloet ER. Two receptor systems for corticosterone in rat brain: microdistribution and differential occupation. Endocrinology. 1985;117:2505–11.

Reul JM, Hesketh SA, Collins A, Mecinas MG. Epigenetic mechanisms in the dentate gyrus act as a molecular switch in hippocampus-associated memory formation. Epigenetics. 2009;4:434–9.

Revest JM, Kaouane N, Mondin M, Le Roux A, Rougé-Pont F, Vallée M, Barik J, Tronche F, Desmedt A, Piazza PV. The enhancement of stress-related memory by glucocorticoids depends on synapsin-Ia/Ib. Mol Psychiatry. 2010;15:1140–51.

Rey M, Carlier E, Soumireu-Mourat B. Effects of corticosterone on hippocampal slice electrophysiology in normal and adrenalectomized BALB/c mice. Neuroendocrinology. 1987;46:424–9.

Rimmele U, Domes G, Mathiak K, Hautzinger M. Cortisol has different effects on human memory for emotional and neutral stimuli. Neuroreport. 2003;14:2485–8.

Rodriguez Manzanares PA, Isoardi NA, Carrer HF, Molina VA. Previous stress facilitates fear memory, attenuates GABAergic inhibition, and increases synaptic plasticity in the rat basolateral amygdala. J Neurosci. 2005;25:8725–34.

Roozendaal B, Hernandez A, Cabrera SM, Hagewoud R, Malvaez M, Stefanko DP, Haettig J, Wood MA. Membrane-associated glucocorticoid activity is necessary for modulation of long-term memory via chromatin modification. J Neurosci. 2010;30:5037–46.

Teschemacher A, Zeise ML, Zieglgansberger W. Corticosterone-induced decrease of inhibitory postsynaptic potentials in rat hippocampal pyramidal neurons in vitro depends on cytosolic factors. Neurosci Lett. 1996;215:83–6.

Tse YC, Bagot RC, Hutter JA, Wong AS, Wong TP. Modulation of synaptic plasticity by stress hormone associates with plastic alteration of synaptic NMDA receptor in the adult hippocampus. PLoS ONE. 2011;6:e27215.

Ulrich-Lai YM, Herman JP. Neural regulation of endocrine and autonomic stress responses. Nat Rev Neurosci. 2009;10:397–409.

Valentino RJ, Van Bockstaele E. Convergent regulation of locus coeruleus activity as an adaptive response to stress. Eur J Pharmacol. 2008;583:194–203.

Van Gemert NG, Carvalho DM, Karst H, van der Laan S, Zhang M, Meijer OC, Hell JW, Joels M. Dissociation between rat hippocampal CA1 and dentate gyrus cells in their response to corticosterone: effects on calcium channel protein and current. Endocrinology. 2009;150:4615–24.

Van Stegeren AH, Roozendaal B, Kindt M, Wolf OT, Joëls M. Interacting noradrenergic and corticosteroid systems shift human brain activation patterns during encoding. Neurobiol Learn Mem. 2010;93:56–65.

Vidal C, Jordan W, Zieglgansberger W. Corticosterone reduces the excitability of hippocampal pyramidal cells in vitro. Brain Res. 1986;383:54–9.

Weinberger C, Hollenberg SM, Ong ES, Harmon JM, Brower ST, Cidlowski J, Thompson EB, Rosenfeld MG, Evans RM. Identification of human glucocorticoid receptor complementary DNA clones by epitope selection. Science. 1985;228:740–2.

Whitlock JR, Heynen AJ, Shuler MG, Bear MF. Learning induces long-term potentiation in the hippocampus. Science. 2006;313:1093–7.

Wiegert O, Joels M, Krugers H. Timing is essential for rapid effects of corticosterone on synaptic potentiation in the mouse hippocampus. Learn Mem. 2006;13:110–3.

Wyrwoll CS, Holmes MC, Seckl JR. 11beta-hydroxysteroid dehydrogenases and the brain: from zero to hero, a decade of progress. Front Neuroendocrinol. 2011;32:265–86.

Yamada K, McEwen BS, Pavlides C. Site and time dependent effects of acute stress on hippocampal long-term potentiation in freely behaving rats. Exp Brain Res. 2003;152:52–9.

Zeise ML, Teschemacher A, Arriagada J, Zieglgansberger W. Corticosterone reduces synaptic inhibition in rat hippocampal and neocortical neurons in vitro. J Neuroendocrinol. 1992;4:107–12.

Zhang Y, Sheng H, Qi J, Ma B, Sun J, Li S, Ni X. Glucocorticoid acts on a putative G-protein coupled receptor to rapidly regulate the activity of NMDA receptors in hippocampal neurons. Am J Physiol Endocrinol Metab. 2012;302:E747–58.

Zhou J, Zhang F, Zhang Y. Corticosterone inhibits generation of long-term potentiation in rat hippocampal slice: involvement of brain-derived neurotrophic factor. Brain Res. 2000;885:182–91.

Zhou M, Hoogenraad CC, Joëls M, Krugers HJ. Combined β-adrenergic and corticosteroid receptor activation regulates AMPA receptor function in hippocampal neurons. J Psychopharmacol. 2012;26:516–24.

Chapter 3
Synaptic Stress, Changes in Glutamate Transmission and Circuitry, and Psychopathology

Laura Musazzi, Giulia Treccani, Carla Perego, Nicoletta Nava, Jens R Nyengaard and Maurizio Popoli

Abstract Dysfunction of the glutamate system is increasingly considered a core feature of stress-dependent neuropsychiatric disorders. Clinical neuroimaging studies have shown consistent volumetric changes in limbic and cortical areas, while preclinical studies with stress protocols in rodents found dendritic remodeling and reduction of synapses in the same areas, suggesting that destabilization of glutamate release/transmission, in turn induced by stress and glucocorticoids, is crucial for cognitive function and neural architecture. We found that acute stress rapidly enhances depolarization-evoked glutamate release/transmission in prefrontal and frontal cortex (PFC/FC), an effect mediated by stimulation of synaptic corticosterone receptors. Corticosterone rapidly increases the readily releasable pool of glutamate vesicles, through activation of synaptic receptor-mediated nongenomic mechanisms in PFC/FC. Moreover, we have shown that chronic antidepressants are able to prevent the enhancement of glutamate release induced by acute stressors in these areas.

While the predominant effect of acute stress is an enhancement of synaptic transmission, repeated exposure to stress brings about atrophy and remodeling of dendrites, loss of synapses, and reduction of synaptic transmission (except perhaps in amygdala). Understanding the mechanisms and effectors involved in this biphasic action of stress is essential to the development of new diagnostic and therapeutic means for stress-related disorders. Select brain-derived neurotrophic factor (BDNF) transcripts and their translation at synapses could be among these key effectors.

M. Popoli (✉) · L. Musazzi · G. Treccani
Laboratory of Neuropsychopharmacology and Functional Neurogenomics—Dipartimento di Scienze Farmacologiche e Biomolecolari and CEND, Università di Milano, 20133 Milano, Italy
Tel.: +39 02 5031 8361
e-mail: maurizio.popoli@unimi.it

C. Perego
Laboratory of Cell Physiology—Dipartimento di Scienze Farmacologiche e Biomolecolari, Università di Milano, 20133 Milano, Italy

N. Nava · J. R. Nyengaard
Stereology and Electron Microscopy Laboratory, Centre for Stochastic Geometry and Advanced Bioimaging, Aarhus University Hospital, 8000 Aarhus C, Denmark

M. Popoli et al. (eds.), *Synaptic Stress and Pathogenesis of Neuropsychiatric Disorders*,
DOI 10.1007/978-1-4939-1056-4_3,

Abbreviations

AMPA	α–Amino-3-hydroxy-methyl-4-isoxazole propionic acid
BDNF	Brain-derived neurotrophic factor
EM	Electron microscopy
FS	Footshock
GR	Glucocorticoid receptors
LTP	Long-term potentiation
MR	Mineralocorticoid receptors
MRI	Magnetic resonance imaging
mPFC	Medial prefrontal cortex
PFC/FC	Prefrontal and frontal cortex
PPF	Paired-pulse facilitation
RRP	Readily releasable pool
SNARE	Soluble N-ethylmaleimide-sensitive fusion protein attachment protein receptor
TIRF	Total internal reflection fluorescence
NMDA	N-Methyl-D-aspartate

3.1 Introduction: The Role of Synaptic Stress in Pathophysiology of Stress-Related Neuropsychiatric Disorders

A stressor is an event or experience that threatens the ability of an individual to adapt and cope. The impact of different behavioral stressors on cognitive/affective functions may vary depending on type, intensity, and duration of stress, and is influenced by genetic background. The outcome of stress may range from plasticity enhancing effects and improved cognition, when stress response is efficiently turned on and shut off, to noxious effects, when the response is dysregulated. A maladaptive stress response can be associated with impaired function and triggering of brain, systemic, and metabolic disorders. Recent lines of evidence have shown how tracing the effects of stress at synaptic level and on neurochemical mechanisms may supply essential information as to how different stressors affect the brain, induce adaptive or maladaptive changes, and trigger brain and metabolic disorders (Fig. 3.1; McEwen 2005; Gorman and Docherty 2010; Sanacora et al. 2012; Popoli et al. 2012; Sousa and Almeida 2012; Tokita et al. 2012; Gray et al. 2013).

3.1.1 Changes in Volume of Brain Areas and Neuronal Architecture in Stress-Related Neuropsychiatric Disorders

Stressful life events impact on memory and cognition and are known to precipitate neuropsychiatric diseases, in particular mood and anxiety disorders. Half a century after the monoamine hypothesis, which assigned a key role in pathophysiology

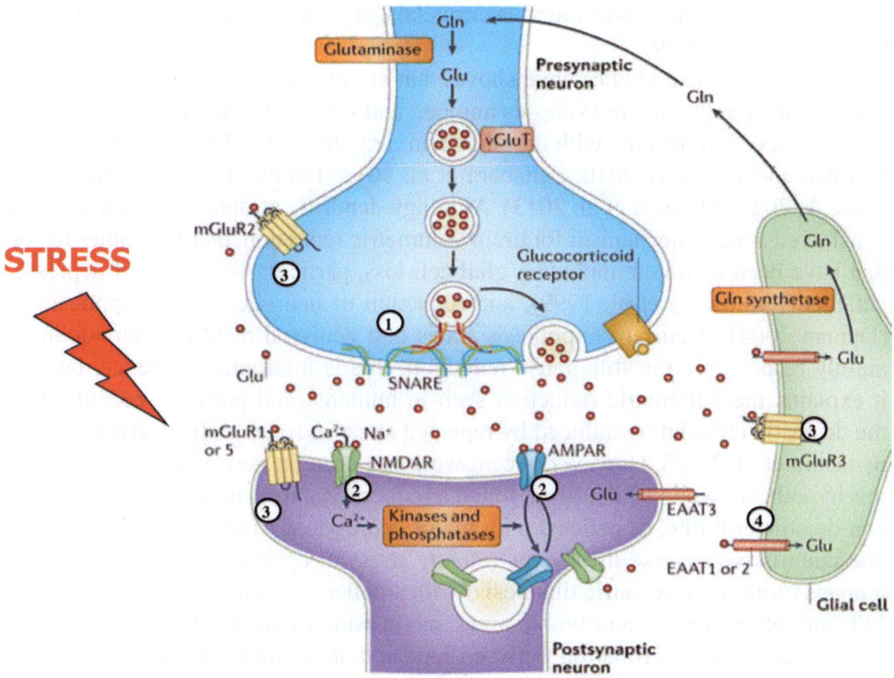

Fig. 3.1 Sites of action (targets) of stress in the glutamate synapse. Several sites/mechanisms of regulation of the glutamate synapse have been shown to be targets of stress: *1* presynaptic release of glutamate; *2* postsynaptic ionotropic receptors for glutamate (N-methyl-D-aspartate receptors (*NMDARs*) and α-amino-3-hydroxy-methyl-4-isoxazole propionic acid receptors (*AMPARs*); *3* metabotropic glutamate receptors (*mGluR*); and *4* glial-specific glutamate transporters. See text for details. Ca^{2+} calcium ions, *EAAT* excitatory amino-acid transporter, *Gln* glutamine, Na^{+} sodium ions, *SNARE* soluble N-ethylmaleimide-sensitive factor attachment protein receptor, *vGluT* vesicular glutamate transporter, *Glu* glutamate. (Adapted from Popoli et al. 2012)

to reduced availability of monoamine transmitters (Heninger et al. 1996), it has become increasingly acknowledged that maladaptive changes in the structure and function of excitatory/inhibitory circuitry (representing the vast majority of neurons and synapses in the brain) have a primary role in the pathophysiology of mood and anxiety disorders, particularly major depression (McEwen 2005; Gorman and Docherty 2010; Sousa and Almedia 2012; Tokita et al. 2012; Musazzi et al. 2013).

Clinical neuroimaging studies have shown consistent volumetric changes in brain areas where glutamate neurons and synapses predominate. Significant volumetric reduction of hippocampus and prefrontal cortex has been shown in MRI studies of patients with mood and anxiety disorders (Campbell and MacQueen 2006; Konarski et al. 2008; Koolschijn et al. 2009; Lorenzetti et al. 2009; Woon et al. 2010). For hippocampus, correlation of volume reduction with length of depressive episodes was found (Frodl et al. 2006; MacQueen et al. 2003; Sheline et al. 2003). Conversely volumetric enlargement was found in amygdala, at least in the early course of illness (Lorenzetti et al. 2009) .

In parallel, rodent studies have shown that chronic stress protocols induce dendritic atrophy, reduction of synapses number, and volumetric reductions resembling those observed in patients with mood and anxiety disorders (McEwen 2005, 2010; Gorman and Docherty 2010; Sanacora et al. 2012; Popoli et al. 2012; Sousa and Almeida 2012; Musazzi et al. 2013). Although dendritic atrophy and remodeling is considered a main mechanism for brain volumetric reduction, there are other factors that have been involved, including glial cell loss, particularly in PFC of depressed patients (Rajkowska et al. 1999), and reduction of neurogenesis in hippocampus (Duman 2004). A current hypothesis states that neuronal dendritic remodeling is mainly responsible for volumetric reduction; this is inferential evidence, because it explains the volumetric reduction seen in humans (and partly in rodents) with the dendritic remodeling induced by repeated stress in rodents (for a discussion see Sanacora et al. 2012). However, recent work has brought new evidence in favor of this hypothesis. Kang et al. (2012) found a reduced synapse number in the postmortem dorsolateral PFC of patients with major depression. Ansell et al. (2012) found that cumulative adverse life events (mostly stressful episodes) were associated in humans (with no psychiatric diagnosis) with smaller gray matter volume of medial PFC and other cortical and limbic areas, as measured with MRI. This last finding established a clear correlation between repeated stress and volumetric reduction. Finally, Kassem et al. (2013) reported that mice subjected to chronic (21 days) restraint stress showed significant gray matter loss in anterior cingulate cortex and hippocampus, measured with MRI. All these lines of evidence clearly support a correlation between stress, dendritic remodeling, and volumetric reduction.

A major role in this process is attributed to elevation of glucocorticoid hormones by stressors, which enhance glutamate release/transmission, in turn purported to induce retraction of dendrites. Converging evidence from various groups has shown that enhancement of glutamate release/transmission in cortical/limbic areas, in turn induced by stress and glucocorticoids, is crucial for these structural/functional changes (Fig. 3.2; Musazzi et al. 2013).

3.2 Acute Stress Enhances Glutamate Release and Transmission in Cortical and Limbic Brain Areas

Several studies have shown in the past that exposure of rodents to acute stressors or administration of glucocorticoids rapidly and transiently increase the level of extracellular glutamate, measured with microdialysis in vivo, in cortical and limbic brain areas (Bagley and Moghaddam 1997; Lowy et al. 1993; Moghaddam 1993; Venero and Borrell 1999). However, the nature and origin of extracellular glutamate measured by microdialysis has been questioned, because the pool of glutamate released at presynaptic terminals is only a small part of the total glutamate pool, which is largely made up of metabolic glutamate (for a discussion see: van der

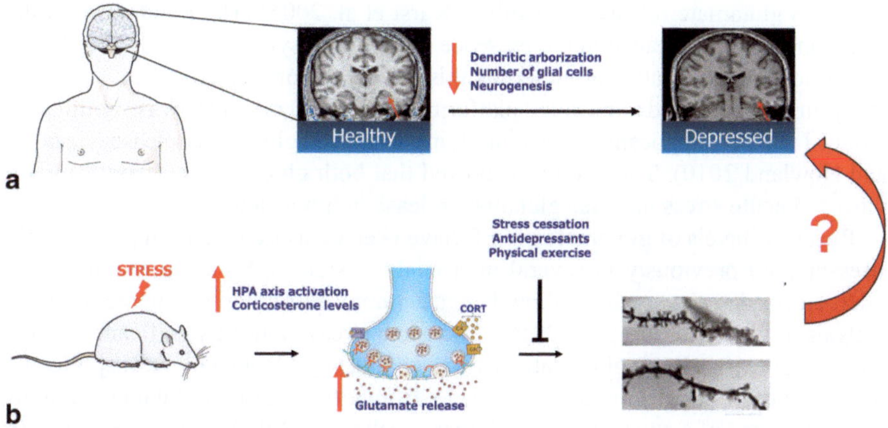

Fig. 3.2 Relationship between stress, glutamate system dysfunction and structural brain changes in stress-related neuropsychiatric disorders: a theoretical model. **a** Neuroimaging studies have consistently shown volumetric reduction of cortical and limbic areas in the brain of patients with mood and anxiety disorders. This has been attributed to several factors, including dendritic atrophy/remodeling, loss of glial cells and reduction of neurogenesis (in hippocampus). **b** Preclinical studies in rodents have shown that stress, through the action of glucocorticoids and other neurochemical/neuroendocrine mediators, may induce abnormal enhancement of glutamate release and excitatory transmission in select brain areas, including amygdala, hippocampus, and prefrontal cortex. If repeated or protracted, stress induces atrophy and remodeling of dendritic arbor at various locations in these areas (except for the amygdala), with loss of dendritic spines and synapses. In turn, dendritic/circuitry remodeling is envisaged as a major causal factor for volumetric changes, observed with magnetic resonance imaging in patients. The evidence that stress and consequent changes in excitatory transmission cause dendritic remodeling comes mostly from rodent studies, while volumetric changes have been observed in humans. The structural/functional changes induced by stress protocols in rodents are reversible with cessation of stress, and are prevented or reversed by pro-adaptive treatments, such as chronic antidepressant treatments and voluntary physical exercise. *HPA* hypothalamic-pituitary-adrenal, *CORT* corticosterone. (Adapted from Musazzi et al. 2013)

Zeyden et al. 2008; Musazzi et al. 2011). The effects of stress and glucocorticoids on glutamate release have been substantially confirmed and shown in details more recently by works employing different methodologies, including: (1) electrophysiological recordings; (2) measurement of endogenous glutamate release from synaptic terminals (synaptosomes) in superfusion; (3) measurement from synaptosomes in bulk by enzyme-linked fluorometric assay; and (4) measurement of resting glutamate levels in vivo by enzyme-based microelectrode arrays (Cazakoff and Howland 2010; Hascup et al. 2010; Karst et al. 2005; Musazzi et al. 2010; Reznikov et al. 2007; Satoh and Shimeki 2010; Wang and Wang 2009).

In vitro application of 100 nM corticosterone (the major stress hormone in rodents) to hippocampal slices was shown to enhance rapidly the frequency of miniature excitatory postsynaptic potentials in CA1 pyramidal neurons and reduce paired-pulse facilitation, a measure of synaptic facilitation induced by pairs of stimuli applied at increasing interpulse intervals, suggesting that the stress hormone increases glutamate release probability (Karst et al. 2005). The rapid onset of this effect and its maintenance in the presence of a protein synthesis inhibitor indicated that a nongenomic pathway underlies this action of corticosterone. In a different study, in rats subjected to elevated platform stress for 30 min, PPF was significantly reduced in CA1 hippocampal area, implying increased glutamate release (Cazakoff and Howland 2010). These studies showed that both glucocorticoid application in vitro and acute stress increase glutamate release in hippocampus.

Recently, levels of glutamate in PFC have been measured after tail pinch, a mild stressor used previously in several microdialysis studies, by using enzyme-based microelectrode arrays, which allow better temporal resolution compared to microdialysis in vivo (Hascup et al. 2010). The acute stress induced significant transient increase of extracellular glutamate that was completely blocked by local application of tetrodotoxin, suggesting this was due to exocytotic release of glutamate. Finally, different forms of acute stress were shown to enhance NMDA- and AMPA-receptor mediated synaptic currents in PFC of juvenile rats > 1 h after stress, that were sustained for 24 h after cessation of stress. The enhancement was mimicked by short-term treatment of slices with corticosterone and mediated by increased membrane expression of ionotropic glutamate receptors (Yuen et al. 2009, 2011).

3.2.1 Acute Stress Enhances Depolarization-Evoked Glutamate Release in Prefrontal and Frontal Cortex

In rat prefrontal and frontal cortex (PFC/FC), we have shown that acute stress rapidly enhances glutamate release and transmission, an effect mediated by glucocorticoid/mineralocorticoid receptors (GR/MR). We applied inescapable random foot-shock (FS)-stress to rats for 40 min, and then quickly purified synaptic terminals (synaptosomes) from PFC/FC with Percoll gradients (Musazzi et al. 2010). Basal and depolarization-evoked glutamate release was measured by using the technique of isolated synaptosomes in superfusion. This method is performed by applying a thin layer of purified synaptosomes on a microporous filter and a constant up-down superfusion to the samples (Popoli et al. 2012; Bonanno et al. 2005). By this way, endogenous transmitters/modulators released are immediately removed by the superfusion medium before they can be taken up by transporters, or activate autoreceptors/heteroreceptors present on synaptic terminals. Therefore, any indirect effects are minimized or prevented, and the release of a single amino acid transmitter can be measured precisely. We found that acute FS-stress markedly and significantly enhanced depolarization-evoked release of endogenous glutamate, with no changes in stimulated release of gamma-aminobutyric acid (GABA) or in the basal release of the two amino acids (Fig. 3.3). A selective antagonist of GR, injected systemically prior to stress application completely blocked the stress-induced enhancement of glutamate release (not shown).

Looking at the presynaptic machinery of PFC/FC in stressed rats we found a significant increase in the presynaptic membranes of the soluble N-ethylmaleimide-sensitive fusion protein attachment protein receptor (SNARE) protein complexes

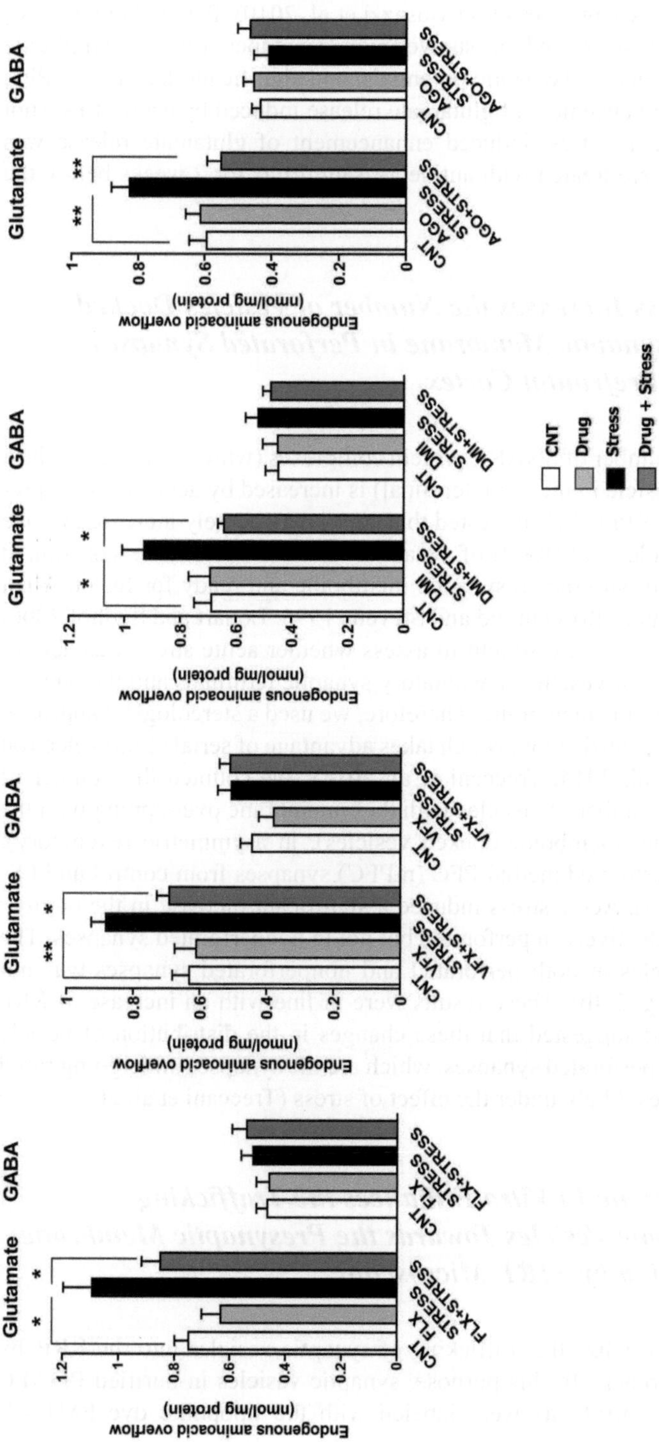

Fig. 3.3 Acute footshock stress enhances depolarization-evoked glutamate release in prefrontal and frontal cortex. Chronic antidepressants prevent the enhancement of release. The bar graphs show glutamate and gamma-aminobutyric acid (*GABA*) release from prefrontal and frontal cortex of superfused synaptosomes, evoked by 15 mM KCl. The rats were treated chronically (2 weeks) with vehicle (*CNT*), fluoxetine (*FLX*), venlafaxine (*VFX*), desipramine (*DMI*), agomelatine (*AGO*), or subjected to acute footshock stress, or treated chronically with either of the four drugs and then subjected to acute footshock stress. Data are expressed as mean ±SEM. *$p<0.05$, **$p<0.01$; Newman–Keuls post hoc tests following one-way ANOVA (n=6–9 rats/group). (From Musazzi et al. 2010)

that mediate fusion of synaptic vesicles (Musazzi et al. 2010). Patch-clamp recordings in slices of PFC from stressed rats showed significant increase in the amplitude of spontaneous excitatory postsynaptic potentials, and significant decrease of PPF, which confirmed an enhancement of glutamate release induced by acute stress (not shown). Interestingly, the stress-induced enhancement of glutamate release was abolished if the rats were treated with antidepressant drugs for 2 weeks before the acute stress (Fig. 3.3).

3.2.2 Acute Stress Increases the Number of Vesicles Docked to the Presynaptic Membrane in Perforated Synapses of Medial Prefrontal Cortex

The finding that the number of SNARE protein complexes (which is thought to be a constant number X vesicle in the same terminal) is increased by acute stress in presynaptic membranes of PFC/FC suggested that stress may acutely increase the size of the readily releasable pool (RRP) of synaptic vesicles. The RRP is constituted by the vesicles docked onto the presynaptic membrane and ready for fusion when the terminal is stimulated (Rosenmund and Stevens 1996; Lonart and Sudhof 2000; Milanese et al. 2011). First, we sought to assess whether acute stress was able to change the distribution of vesicles in excitatory synaptic terminals and the number of vesicles docked onto the membrane. Therefore, we used a stereological approach for synaptic vesicles quantification, which takes advantage of serial section electron microscopy (Nava et al. 2014; Treccani et al. 2014). We counted the number of total vesicles and the number of vesicles with their membrane overlapping with the presynaptic membrane (membrane-docked vesicles), in asymmetric (excitatory), perforated, and nonperforated medial PFC (mPFC) synapses from control and FS-stressed rats (Fig. 3.4a). Acute stress induced a significant increase in the number of docked vesicles, selectively in perforated but not in nonperforated synapses. The total number of vesicles in both perforated and nonperforated synapses was not changed by stress (Fig. 3.4b). These results were in line with an increase of RRP induced by stress, and suggested that these changes in the distribution of vesicle pools are localized to perforated synapses, which are the synapses undergoing rapid morphological changes, likely under the effect of stress (Treccani et al. 2014).

3.2.3 Corticosterone In Vitro Enhances the Trafficking of Glutamate Vesicles Towards the Presynaptic Membrane: Visualization by TIRF Microscopy

Next, we sought to visualize the trafficking of synaptic vesicles into the RRP, by using an in vitro approach. To this purpose, synaptic vesicles in purified PFC/FC synaptosomes from control rats were labeled with the lipophilic dye FM1–43,

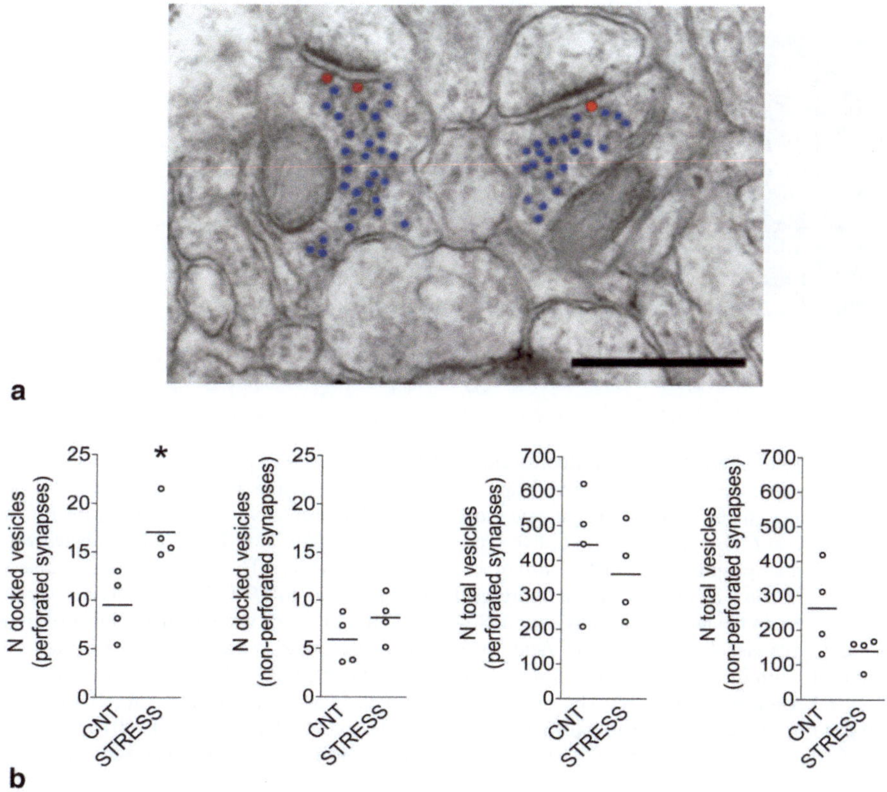

Fig. 3.4 Electron microscopy stereology: Acute stress increases the number of docked vesicles in perforated synapses of medial prefrontal cortex. **a** Representative transmission electron micrograph (EM) of medial prefrontal cortex non-perforated asymmetric synapse, showing docked (*red*) and total (*blue*) vesicles, used for serial reconstruction (28,000X, scale bar 500 nm). **b** Number of docked and total vesicles in perforated and non-perforated synapses of control (*CNT*) and acutely stressed (STRESS) rats. Data and statistics as in Fig. 3.3; $^{*}p<0.05$, $n=4$ rats/group

which intercalates with plasma membranes and allows monitoring vesicle trafficking (Cochilla et al. 1999). After labeling with FM1–43, live synaptosomes were incubated for 10 min with different concentrations of corticosterone (100 nM, 10 μM) to assess whether the hormone was able, by local action, to change vesicle trafficking. We used total internal reflection fluorescence (TIRF) microscopy, an imaging technique that allows the study of events occurring in or immediately beneath the plasma membrane (about 100 nm; Groves et al. 2008; Perego et al. 2012). Corticosterone application caused a time-dependent increase in the number of fluorescent spots in the TIRF field, which indicates a time-dependent accumulation of fluorescent vesicles in close proximity to the membrane (Fig. 3.5a). This increase was significant for both corticosterone concentrations, and started immediately after its

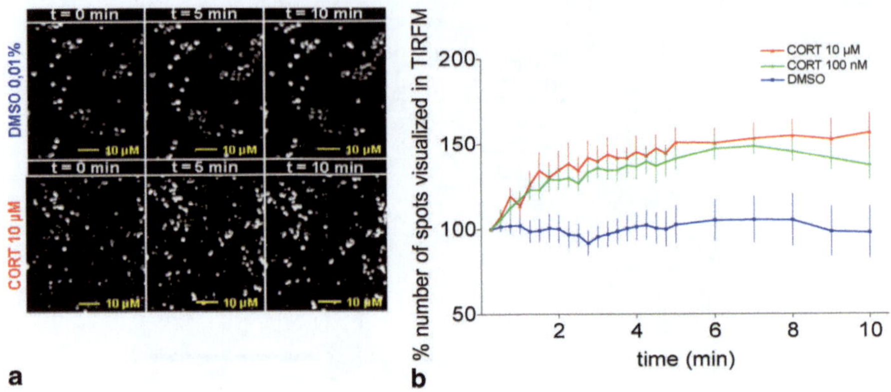

Fig. 3.5 Total internal reflection fluorescence (*TIRF*) microscopy of prefrontal and frontal cortex synaptosomes. Corticosterone (*CORT*) application in vitro increases the trafficking of FM1–43 fluorescent vesicles near the presynaptic membrane. **a** Representative *TIRF* images (magnification 100X) at $t=0$, $t=5$, $t=10$ min of live synaptosomes stained with FM1–43 and incubated with 0.01 % dimethyl sulfoxide (*DMSO, top*) or 10 μM *CORT* (*bottom*). The synaptosomes incubated with *CORT* show an increase in the fluorescent spots appearing in *TIRF* field after 5 and 10 min, compared with the number of spots at $t=0$. Scale bar: 2 μm. **b** Graph showing the number of fluorescent spots visualized in the *TIRF* field (expressed as percentage vs the number of spots at $t=0$) during 10 min of in vitro incubation with *DMSO*, 100 nM or 10 μM *CORT*. $n=6$–11 recordings, four independent experiments. The area under the curve of the recording curves for 100 nM and 10 μM *CORT* were significantly different from *DMSO* (control), for time ($p<0.0001$), treatment ($p<0.01$) and interaction ($p<0.0001$). One-way ANOVA followed by Newman–Keuls post hoc test

application; the fluorescent spots were maximal after 3–5 min incubation and remained constant for up to 10 min of recording (Fig. 3.5b). These results showed that in vitro incubation of PFC/FC synaptosomes with corticosterone induces rapid mobilization of vesicles towards the presynaptic membrane, consistent with an increase in the RRP size.

In separate experiments with purified synaptosomes in superfusion we found that both acute stress (ex vivo) and corticosterone application in vitro increased the RRP of glutamate (Treccani et al. 2014), measured after pulse stimulation with hypertonic sucrose, which is the standard method to measure RRP (Rosenmund and Stevens 1996; Lonart and Sudhof 2000; Milanese et al. 2011). We also observed that the rapid action of corticosterone on the trafficking of synaptic vesicles was mediated by local MR/GR, present on isolated synaptosomes, because the concomitant application with corticosterone of either MR/GR selective antagonist (spironolactone or RU486) blocked the translocation of FM1–43 fluorescent vesicles near the presynaptic membrane, observed with TIRF microscopy (not shown).

However, corticosterone application together with (15 mM KCl)-containing depolarizing buffer did not enhance glutamate release (not shown), as observed ex vivo with synaptosomes freshly isolated from acutely stressed rats (see Fig. 3.3). This finding clearly showed that, although corticosterone mediates an early effect of stress (e.g., the translocation of vesicles into the RRP), it is not able to fully replicate the effect of acute stress inducing the release of glutamate. This was confirmed by electrophysiological recordings in mPFC, which showed that application of 1 or 10 μM corticosterone to brain slices did not change synaptic transmission for up to 20 min (Treccani et al. 2014) .

3.2.4 Both Acute Stress and Corticosterone In Vitro Rapidly Increase the Readily Releasable Pool of Vesicles but only Stress Rapidly Increases Glutamate Release. Implications for the Mechanism of Acute Stress in Prefrontal and Frontal Cortex

Overall, our results showed that both acute stress and application of corticosterone in vitro to synaptosomes rapidly increase the RRP, but only stress rapidly increases glutamate release in PFC/FC. This effect of stress is seemingly mediated by a nongenomic action of the hormone, through the activation of synaptic MR/GR (Karst et al. 2005; Musazzi et al. 2010; Joëls et al. 2012). The presence of membrane-associated receptors for corticosterone has been shown in amygdala and PFC/FC (Treccani et al. 2014; Prager et al. 2010). However, the rapid synaptic action of corticosterone is necessary, but not sufficient, to increase glutamate release/transmission in PFC/FC, which likely requires activation of delayed, genomic, mechanisms. This is consistent with a previous work, which found that brief application of corticosterone to rat brain slices enhanced synaptic transmission only after 1 h (Yuen et al. 2011). The mechanism of stress seems to be different in PFC/FC compared to hippocampus, where local application of corticosterone is sufficient to induce rapid enhancement of glutamate release and synaptic transmission (Karst et al. 2005; Joëls et al. 2012) .

Therefore, although the enhancement of glutamate release induced by acute stress in PFC/FC appears to be mediated by corticosterone, the hormone seems necessary but not sufficient for this effect (Fig. 3.6). Corticosterone binds local MR/GR located at or near presynaptic terminals, and rapidly, by nongenomic action, increases the trafficking of synaptic vesicles into the RRP. The RRP increase is localized mostly to perforated synapses, which are the synapses undergoing rapid plastic changes under the effect of stress. This buildup of RRP primes the terminals for the enhancement of glutamate release, which may be delayed by about 1 h, for the subsequent involvement of unknown genomic effects of corticosterone and possibly additional effectors, including retrograde messenger(s) (Popoli et al. 2012; Treccani et al. 2014). Chronic antidepressant treatments mostly or completely prevent the stress-induced enhancement of glutamate release (see Fig. 3.3), suggesting that stabilization of glutamate release/transmission is a relevant part of their therapeutic action (Musazzi et al. 2013). This drug action may protect from buildup of dangerous concentrations of synaptic (or extrasynaptic) glutamate and contribute to preventing or reversing the dendritic remodeling and synaptic disconnection

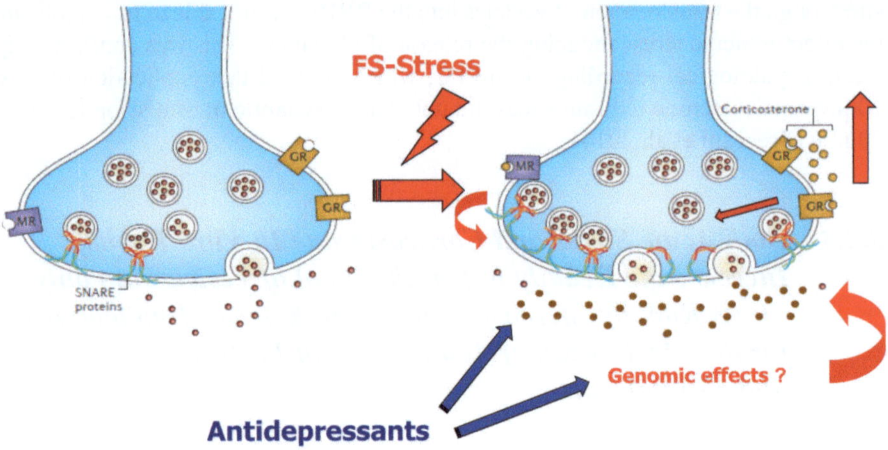

Fig. 3.6 Acute stress enhances glutamate release in prefrontal and frontal cortex. Corticosterone increases the readily releasable pool of glutamate vesicles by local synaptic action, and is necessary but not sufficient for stress-induced enhancement of glutamate release in cortical areas. Acute footshock stress enhances depolarization-evoked release of glutamate from presynaptic terminals of prefrontal and frontal cortex. The acute stress response involves a rapid increase of circulating levels of corticosterone, which binds to putative presynaptic membrane-associated receptor (*GR* and *MR*), in turn inducing increased trafficking of vesicles into the readily releasable pool (RRP). Corticosterone in vitro increases the RRP in purified synaptosomes, but does not enhance glutamate release for up to 20 min; evidence obtained by: purified synaptosomes in superfusion, EM-stereology of asymmetric synapses, TIRF microscopy of purified synaptosomes (see Musazzi et al. 2010; Treccani et al. 2014). In prefrontal and frontal cortex, different from hippocampus, corticosterone seems necessary but not sufficient to induce enhancement of glutamate release (at least in the first 20 min). The effect of corticosterone on RRP is likely a rapid non-genomic effect. Delayed, perhaps genomic, effects of corticosterone are necessary for completion of corticosterone action and enhancement of glutamate release and transmission (see Yuen et al. 2011). Previous chronic antidepressant treatments block the stress-induced enhancement of glutamate release. The mechanism of this drug action is not clear yet, but could be related to the delayed genomic effects of corticosterone on glutamate synapses. *SNARE*: N-ethylmaleimide-sensitive fusion protein attachment protein receptor.

which is thought to be a major factor in stress-related neuropsychiatric disorders (Sanacora et al. 2012; Musazzi et al. 2013). The mechanism of this anti-stress action of antidepressants is not clearly understood at present. The drugs prevent the enhancement of glutamate release in PFC/FC, but not the rise of corticosterone levels and the increase of number of SNARE complexes in presynaptic membranes in the same areas (Musazzi et al. 2010). Therefore, the action of these drugs seems to be downstream of the early action of corticosterone and could be related to the delayed, genomically mediated, effects that bring about the enhancement of glutamate release (Fig. 3.6). Further research is under way to dissect this mechanism, which may serve for the identification of new drug targets.

3.3 Structural/Functional Changes Induced by Chronic Stress

Acute stress protocols, as shown above (Sect. 3.2), may allow a careful dissection of the mechanisms whereby stress triggers the modifications in the glutamate synapses. However, experimental protocols employing repeated stress are more often used for animal models of neuropsychiatric pathology, mainly because stress is considered a major predisposing factor in psychopathology, such as for mood and anxiety disorders. There is a considerable literature which analyzed in rodents the effects of stress on: (1) structural features of synapses and circuitry and (2) synaptic transmission and plasticity. We have addressed the first issue in Sect. 3.1.1. For a detailed discussion see: McEwen 2005, 2010; Sanacora et al. 2012; Sousa and Almeida 2012.

Regarding the second issue, the outcome of acute stress episodes on the plasticity of glutamate synapses has been thoroughly analyzed, particularly in hippocampus, by measuring Long-Term Potentiation (LTP), an activity-dependent enhancement of synaptic strength that represents the most studied cellular process linked to memory and learning (Citri and Malenka 2008). Briefly, current hypotheses suggest that stress initially induces activation of synaptic transmission in the forebrain and facilitation of LTP, a phase that is partly coincident with rapid nongenomic action of corticosteroid hormones. The early enhancement of synaptic plasticity in hippocampus may have a role in the formation of traumatic memories that are saved in an individual's experience. The early enhancement is followed by a phase of inhibition of synaptic plasticity, in which the threshold for induction of LTP is raised, corresponding to delayed genomic-mediated action of corticosteroids, probably to consolidate the memory related to stressful events and avoid interference of subsequently formed memories (Kim and Diamond 2002; Diamond et al. 2007; Joëls 2008; Krugers et al. 2010). However, a high level of emotional arousal may impair proper evaluation and processing of experience by interfering with hippocampal function. Detailed information on the effects of stress on synaptic plasticity can be found in Sect. 2 of this chapter.

Regarding the effects of chronic stress on synaptic function, i.e., presynaptic glutamate release and function or membrane insertion of postsynaptic glutamate receptors, less information is available. Early evidence was provided by mycrodialysis studies, which found selective changes in the adaptation of glutamate release in hippocampus and PFC after application of a few consecutive stressors (Moghaddam 2002). Little or no evidence, obtained with later technologies (see above, Sect. 3.2), is available for glutamate release in rodents subjected to chronic stress protocols. However, recently it was found that depolarization-evoked glutamate release, measured in superfused synaptosomes from ventral hippocampus, was reduced in rats subjected to prenatal stress, which showed an anxious behavioral phenotype (Marrocco et al. 2012). In a different study, it was found that glutamate release induced by BDNF in slices of the PFC was attenuated in rats subjected to chronic restraint stress, coupled with anxious/depressive phenotype and downregulation of

GR (Chiba et al. 2012). These works suggest that the consequences of chronic stress protocols on glutamate release may be different from acute stress.

Moreover, recent work has shown that the outcome of chronic stress on the function and membrane expression of ionotropic glutamate receptors may be the opposite of the action of acute stress (Yuen et al. 2009, 2011, 2012). While acute stress was shown to enhance NMDA- and AMPA-receptor mediated synaptic currents in PFC of juvenile rats, repeated restraint or unpredictable stress caused marked reduction of ionotropic glutamate receptors mediated currents, due to ubiquitin/proteasome-mediated degradation of GluR1 and NR1 subunits. This effect of repeated stress was linked to GR activation and concomitant to significant impairment of temporal order recognition memory, a cognitive process controlled by the PFC. For more details and discussion see Chap. 4, this volume.

3.3.1 Structural/Functional Changes Induced by Stress in Glutamate Synapses and Circuitry: A Biphasic Process?

A remarkable feature of excitatory and inhibitory synapses is their continuous reorganization, with changes in morphology and stability, as well as the birth of new synapses and elimination of old ones (Holtmaat and Svoboda 2009; Yoshihara et al. 2009; Caroni et al. 2012). This phenomenon is regulated by synaptic activity, and the size of spine heads has been shown to be correlated with synaptic strength, such as in LTP and learning. Some studies have tried to correlate the time-dependent enhancement of transmission with spine growth and synaptogenesis. Thus, it has been suggested that synapses involving new spines are assembled within 12–18 h. However, recent evidence suggest that activity-dependent formation of new synaptic spines could be much faster. It has been shown in hippocampal slice cultures that new spines stimulated by glutamate uncaging may become functional within 10 min and show features of morphologically mature synapses already after 1.5 h (Nägerl et al. 2007; Zito et al. 2009).

As shown above, although the evidence is far from conclusive, the outcome of acute and repeated (chronic) stress on structural and functional features of the glutamate system could be different and often opposite. While acute stressors enhance glutamate release and excitatory transmission in select areas of the forebrain, chronic stress has been shown to reduce excitatory transmission and to induce consistently atrophy/remodeling of dendrites and loss of synapses, in line with reduction of excitatory transmission in the same areas. It is not known whether acute stress induces rapid morphological changes in synapses and circuitry, although hippocampal spinogenesis induced by stress has been reported (Shors et al. 2001; Diamond et al. 2006). Recently, it was shown that infradian corticosterone peaks promote learning-dependent formation of new spines in motor cortex, and application of corticosterone in vitro to hippocampal slices increased the density of spines in CA1 area after 1 h (Komatsuzaki et al. 2012; Liston et al. 2013). On the other hand, even single stress episodes, if measurements are performed at least 24 h later, induce loss of spines or dendritic retraction (Izquierdo et al. 2006; Hajszan et al. 2009). Currently, more work is under way to understand whether and how acute stress induces rapid changes in synaptic morphology.

Considering the different effects of acute and chronic stress in hippocampus and PFC on synaptic function and plasticity, and on synaptic spines reorganization, it may be conceived that, during the stress response, the early enhancement of glutamate transmission (perhaps coupled with and early increase of spines and synapses number) can be turned with time into its opposite. Repeated stressors, or the delayed consequences of acute stressors, seem to bring about destabilization of neuronal architecture and loss of synaptic connections in some pathways, with diffuse alterations in areas and circuits mediating cognitive and emotional behaviors (e.g., hippocampus, PFC). Therefore, the structural and functional changes in excitatory circuitry may follow a biphasic process (Fig. 3.7; see Popoli et al. 2012), with the exception of basolateral amygdala, where enhancement of excitatory transmission seems to prevail for a longer time. This is also mirrored by the effects of stress and physiological levels of glucocorticoids on related cognitive functions, which may be enhanced by acute stress and impaired by repeated stress (see Chap. 4, this volume). The stress response is a complex physiological process involving hormonal, neurochemical, and metabolic mechanisms. As it is often observed in pathophysiology, it can be envisaged that a continuum exists between physiological mechanisms of plasticity, allowing adaptation to the environmental stimuli, and pathological mechanisms, turning a normal response into maladaptive structural/functional changes. In this framework, the main target for research should be the identification of cellular effectors mediating the crucial passage from the early effects into later effects of stress (also linked with the action of repeated stress) on glutamate synapses and circuitry. If early enhancement of excitatory transmission (perhaps coupled with early increase of spines and synapses number) is recorded in the first few minutes and hours from the beginning of stress episodes, and delayed inhibitory effects (with atrophy and remodelling of dendrites) can be detected as early as 24 h after the stress episode (Izquierdo et al. 2006; Hajszan et al. 2009), the turning point in plasticity must be somewhere along this time frame. A better knowledge of the cellular effectors involved in this biphasic effect would be quite useful for a better understanding of stress-related pathophysiology. If early activity-dependent morphological changes at synapses can be observed in a matter of minutes (see above), it is possible that they are carried out by changes in local protein translation at dendrites, which is also consistent with early effects being linked to nongenomic effects of glucocorticoids (Joels et al. 2012, see also above). Later changes, particularly atrophy and remodelling of dendrites, must instead be linked to more robust changes in gene expression, involving both transcription and translation, as well as trafficking of signalling molecules.

We are currently investigating molecular effectors, which may be responsible for rapid changes induced in synaptic morphology by acute stress. The main target here may be the neurotrophin BDNF, which is encoded by different splice variant mRNAs, assembling together the mRNA transcribed from the 3′ coding exon, with one of the transcripts of at least eight 5′ noncoding exons (Aid et al. 2007). The protein product is the same, but it has been shown that different splice variants code for

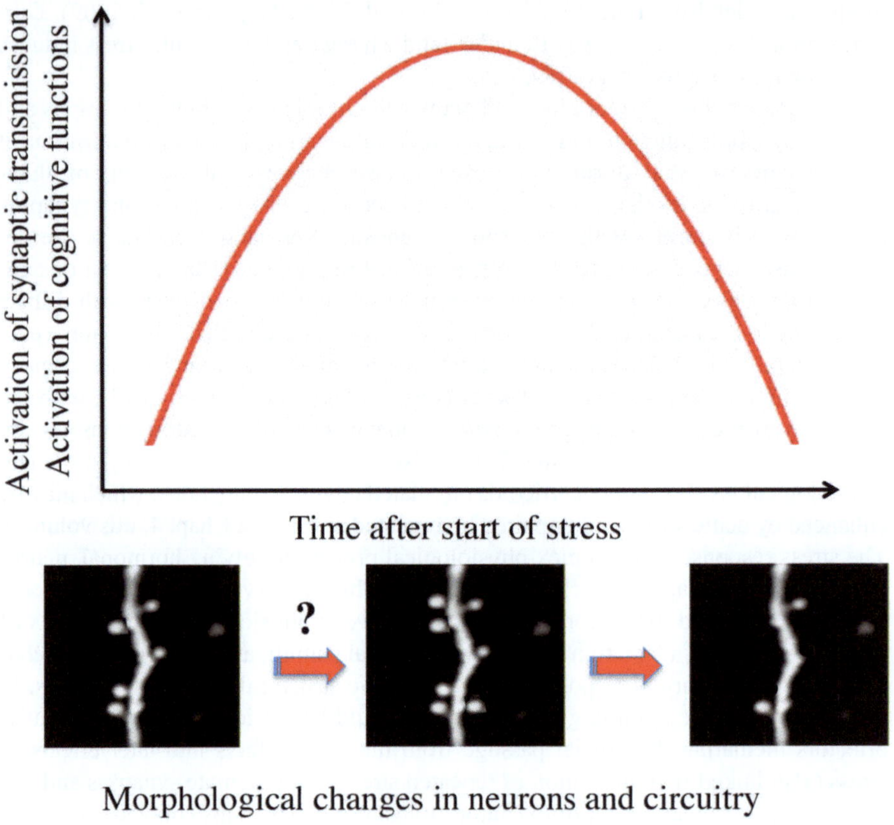

Fig. 3.7 Hypothetical scheme of structural/functional changes induced by stress in the glutamate system: a biphasic process. Acutely, e.g., in the first several minutes and hours, stress induces enhancement of excitatory synaptic transmission (often accompanied by cognitive enhancement). It is not clear yet if this is accompanied by an increase in the number of spines and synapses, although corticosterone was shown to exert rapid effect on spines morphology (Komatsuzaki et al. 2012; Liston et al. 2013). Later on, at least 24 h after application of the stressor (Izquierdo et al. 2006; Hajszan et al. 2009), a phase of inhibition follows, with reduction of synaptic transmission and related structural changes: dendritic atrophy and remodeling, loss of spines and synapses. This phase brings about destabilization of the glutamate system, with negative effects on cognitive functions. It is conceivable that, to a certain point, this is a compensatory physiologically adaptive process during the stress response. However, if the stress response is not correctly turned on and then shut off, because the stressor overcomes the coping capability of the system (stress is uncontrollable, too long, repeated, hits individual vulnerability, etc.) the structural/functional changes may become more stable or permanent, with the possible development of stress-related pathology (see text for explanation; McEwen 2005, 2010; Popoli et al. 2012). The rendering of dendrites at bottom emphasizes the biphasic changes in morphology over time

cellular localization of mRNAs, with a few of them targeted in activity-dependent fashion to dendrites, to subserve local dendritic translation of BDNF, and synaptic plasticity (Chiaruttini et al. 2008; Baj et al. 2011). We have recently shown that both voluntary physical exercise and chronic antidepressant treatments, two types of environmental factors that enhance adaptive neuroplasticity, increase expression and trafficking of select BDNF transcripts to hippocampal dendrites in rodents (Baj et al. 2012). In addition, we found that acute stress blocks the increase of total and dendritic BDNF expression induced by physical exercise, as well as the positive effect of physical exercise on dendritic trafficking of BDNF (not shown). Overall, these results point to BDNF dendritic transcripts as crucial mediators of adaptive/maladaptive changes in activity-dependent synaptic plasticity. In turn, BDNF regulates the trafficking of additional dendritic mRNAs and their translation at synapses, by selectively promoting the translation of a subset of dendritic mRNAs, including cytoskeletal proteins involved in synaptic rearrangement (Gray et al. 2013; Leal et al. 2013; Ruiz et al. 2013). Therefore, the investigation of BDNF and related pathways will supply essential information as to the nature of rapid versus delayed changes induced by stress in excitatory synapses and circuitry.

References

Aid T, Kazantseva A, Piirsoo M, Palm K, Timmusk T. Mouse and rat BDNF gene structure and expression revisited. J Neurosci Res. 2007;85:525–35.

Ansell EB, Rando K, Tuit K, Guarnaccia J, Sinha R. Cumulative adversity and smaller gray matter volume in medial prefrontal, anterior cingulate, and insula regions. Biol Psychiatry. 2012;72:57–64.

Bagley J, Moghaddam B. Temporal dynamics of glutamate efflux in the prefrontal cortex and in the hippocampus following repeated stress: effects of pretreatment with saline or diazepam. Neuroscience. 1997;77:65–73.

Baj G, Leone E, Chao MV, Tongiorgi E. Spatial segregation of BDNF transcripts enables BDNF to differentially shape distinct dendritic compartments. Proc Natl Acad Sci U S A. 2011;108:16813–8.

Baj G, D'Alessandro V, Musazzi L, Mallei A, Sartori CR, Sciancalepore M, Tardito D, Langone F, Popoli M, Tongiorgi E. Physical exercise and antidepressants enhance BDNF targeting in hippocampal CA3 dendrites: further evidence of a spatial code for BDNF splice variants. Neuropsychopharmacology. 2012;37:1600–11.

Bonanno G, Giambelli R, Raiteri L, Tiraboschi E, Zappettini S, Musazzi L, et al. Chronic antidepressants reduce depolarization-evoked glutamate release and protein interactions favoring formation of SNARE complex in hippocampus. J Neurosci. 2005;25:3270–9.

Campbell S, MacQueen G. An update on regional brain volume differences associated with mood disorders. Curr Opin Psychiatry. 2006;19:25–33.

Caroni P, Donato F, Muller D. Structural plasticity upon learning: regulation and functions. Nat Rev Neurosci. 2012;13:478–90.

Cazakoff BN, Howland JG. Acute stress disrupts paired pulse facilitation and long-term potentiation in rat dorsal hippocampus through activation of glucocorticoid receptors. Hippocampus. 2010;20:1327–31.

Chiaruttini C, Sonego M, Baj G, Simonato M, Tongiorgi E. BDNF mRNA splice variants display activity- dependent targeting to distinct hippocampal laminae. Mol Cell Neurosci. 2008;37:11–9.

Chiba S, Numakawa T, Ninomiya M, Richards MC, Wakabayashi C, Kunugi H. Chronic restraint stress causes anxiety- and depression-like behaviors, downregulates glucocorticoid receptor expression, and attenuates glutamate release induced by brain-derived neurotrophic factor in the prefrontal cortex. Prog Neuropsychopharmacol Biol Psychiatry. 2012;39:112–9.

Citri A, Malenka RC. Synaptic plasticity: multiple forms, functions, and mechanisms. Neuropsychopharmacology. 2008;33:18–41.

Cochilla AJ, Angleson JK, Betz WJ. Monitoring secretory membrane with FM1-43 fluorescence. Annu Rev Neurosci. 1999;22:1–10.

Diamond DM, Campbell AM, Park CR, Woodson JC, Conrad CD, Bachstetter AD, Mervis RF. Influence of predator stress on the consolidation versus retrieval of long-term spatial memory and hippocampal spinogenesis. Hippocampus. 2006;16:571–6.

Diamond DM, Campbell AM, Park CR, Halonen J, Zoladz PR. The temporal dynamics model of emotional memory processing: a synthesis on the neurobiological basis of stress-induced amnesia, flashbulb and traumatic memories, and the Yerkes-Dodson law. Neural. Plast. 2007;2007:60803.

Duman RS. Depression: a case of neuronal life and death? Biol Psychiatry. 2004;56:140–5.

Frodl T, Schaub A, Banac S, Charypar M, Jäger M, Kümmler P, Bottlender R, Zetzsche T, Born C, Leinsinger G, Reiser M, Möller HJ, Meisenzahl EM. Reduced hippocampal volume correlates with executive dysfunctioning in major depression. J Psychiatry Neurosci. 2006;31:316–23.

Gorman JM, Docherty JP. A hypothesized role for dendritic remodeling in the etiology of mood and anxiety disorders. J Neuropsychiatry Clin Neurosci. 2010;22:256–64.

Gray JD, Milner TA, McEwen BS. Dynamic plasticity: the role of glucocorticoids, brain-derived neurotrophic factor and other trophic factors. Neuroscience. 2013;239:214–27.

Groves JT, Parthasarathy R, Forstner MB. Fluorescence imaging of membrane dynamics. Annu Rev Biomed Eng. 2008;10:311–38.

Hajszan T, Dow A, Warner-Schmidt JL, Szigeti-Buck K, Sallam NL, Parducz A, Leranth C, Duman RS. Remodeling of hippocampal spine synapses in the rat learned helplessness model of depression. Biol Psychiatry. 2009;65:392–400.

Hascup ER, Hascup KN, Stephens M, Pomerleau F, Huettl P, Gratton A, Gerhardt GA. Rapid microelectrode measurements and the origin and regulation of extracellular glutamate in rat prefrontal cortex. J Neurochem. 2010;115:1608–20.

Heninger GR, Delgado PL, Charney DS. The revised monoamine theory of depression: a modulatory role for monoamines, based on new findings from monoamine depletion experiments in humans. Pharmacopsychiatry. 1996;29:2–11.

Holtmaat A, Svoboda K Experience-dependent structural synaptic plasticity in the mammalian brain. Nat Rev Neurosci. 2009;10:647–58. (Erratum in: Nat Rev Neurosci. 2009).

Izquierdo A, Wellman CL, Holmes A. Brief uncontrollable stress causes dendritic retraction in infralimbic cortex and resistance to fear extinction in mice. J Neurosci. 2006;26:5733–8.

Joëls M. Functional actions of corticosteroids in the hippocampus. Eur J Pharmacol. 2008;583:312–21.

Joëls M, Sarabdjitsingh A, Karst H. Unraveling the time domains of corticosteroid hormone influences on brain activity: rapid, slow, and chronic modes. Pharmacol Rev. 2012;64:901–38.

Kang HJ, Voleti B, Hajszan T, Rajkowska G, Stockmeier CA, Licznerski P, Lepack A, Majik MS, Jeong LS, Banasr M, Son H, Duman RS. Decreased expression of synapse-related genes and loss of synapses in major depressive disorder. Nat Med. 2012;18:1413–7.

Karst H, Berger S, Turiault M, Tronche F, Schütz G, Joëls M. Mineralocorticoid receptors are indispensable for nongenomic modulation of hippocampal glutamate transmission by corticosterone. Proc Natl Acad Sci U S A. 2005;102:19204–7.

Kassem MS, Lagopoulos J, Stait-Gardner T, Price WS, Chohan TW, Arnold JC, Hatton SN, Bennett MR. Stress-induced grey matter loss determined by MRI is primarily due to loss of dendrites and their synapses. Mol Neurobiol. 2013;47:645–61.

Kim JJ, Diamond DM. The stressed hippocampus, synaptic plasticity and lost memories. Nat Rev Neurosci. 2002;3:453–62.

Komatsuzaki Y, Hatanaka Y, Murakami G, Mukai H, Hojo Y, Saito M, Kimoto T, Kawato S. Corticosterone induces rapid spinogenesis via synaptic glucocorticoid receptors and kinase networks in hippocampus. PLoS ONE. 2012;7:e34124.

Konarski JZ, McIntyre RS, Kennedy SH, Rafi-Tari S, Soczynska JK, Ketter TA. Volumetric neuroimaging investigations in mood disorders: bipolar disorder versus major depressive disorder. Bipolar Disord. 2008;10:1–37.

Koolschijn PC, van Haren NE, Lensvelt-Mulders GJ, Hulshoff Pol HE, Kahn RS. Brain volume abnormalities in major depressive disorder: a meta-analysis of magnetic resonance imaging studies. Hum Brain Mapp. 2009;30:3719–35.

Krugers HJ, Hoogenraad CC, Groc L. Stress hormones and AMPA receptor trafficking in synaptic plasticity and memory. Nat Rev Neurosci. 2010;11:675–81.

Leal G, Comprido D, Duarte CB BDNF-induced local protein synthesis and synaptic plasticity. Neuropharmacology. 2013. [Epub ahead of print].

Liston C, Cichon JM, Jeanneteau F, Jia Z, Chao MV, Gan WB. Circadian glucocorticoid oscillations promote learning-dependent synapse formation and maintenance. Nat Neurosci. 2013;16:698–705.

Lonart G, Sudhof TC. Assembly of SNARE core complexes prior to neurotransmitter release sets the readily releasable pool of synaptic vesicles. J Biol Chem. 2000;275:27703–7.

Lorenzetti V, Allen NB, Fornito A, Yücel M. Structural brain abnormalities in major depressive disorder: a selective review of recent MRI studies. J Affect Disord. 2009;117:1–17.

Lowy M, Gault L, Yamamoto B. Adrenalectomy attenuates stress induced elevation in extracellular glutamate concentration in hippocampus. J Neurochem. 1993;61:1957–60.

MacQueen GM, Campbell S, McEwen BS, Macdonald K, Amano S, Joffe RT, Nahmias C, Young LT. Course of illness, hippocampal function, and hippocampal volume in major depression. Proc Natl Acad Sci U S A. 2003;100:1387–92.

Marrocco J, Mairesse J, Ngomba RT, Silletti V, Van Camp G, Bouwalerh H, Summa M, Pittaluga A, Nicoletti F, Maccari S, Morley-Fletcher S. Anxiety-like behavior of prenatally stressed rats is associated with a selective reduction of glutamate release in the ventral hippocampus. J Neurosci. 2012;32:17143–54.

McEwen BS. Stress, sex and neural adaptation to a changing environment: mechanisms of neuronal remodeling. Ann N Y Acad Sci. 2010;1204:E38–E59.

McEwen BS. Glucocorticoids, depression, and mood disorders structural remodeling in the brain. Metabolism. 2005;54:20–3.

Milanese M, Zappettini S, Onofri F, Musazzi L, Tardito D, Bonifacino T, Messa M, Racagni G, Usai C, Benfenati F, Popoli M, Bonanno G. Abnormal exocytotic release of glutamate in a mouse model of amyotrophic lateral sclerosis. J Neurochem. 2011;116:1028–42.

Moghaddam B. Stress preferentially increases extraneuronal levels of excitatory amino acids in the prefrontal cortex: comparison to hippocampus and basal ganglia. J Neurochem. 1993;60:1650–7.

Moghaddam B. Stress activation of glutamate neurotransmission in the prefrontal cortex: implications for dopamine-associated psychiatric disorders. Biol Psychiatry. 2002;51:775–87.

Musazzi L, Milanese M, Farisello P, Zappettini S, Tardito D, Barbiero VS, Bonifacino T, Mallei A, Baldelli P, Racagni G, Raiteri M, Benfenati F, Bonanno G, Popoli M. Acute stress increases depolarization- evoked glutamate release in the rat prefrontal/frontal cortex: the dampening action of antidepressants. PLoS ONE. 2010;5:e8566.

Musazzi L, Racagni G, Popoli M. Stress, glucocorticoids and glutamate release: effects of antidepressant drugs. Neurochem Int. 2011;59:138–49.

Musazzi L, Treccani G, Mallei A, Popoli M. The action of antidepressants on the glutamate system: regulation of glutamate release and glutamate receptors. Biol Psychiatry. 2013;73:1180–8.

Nägerl UV, Köstinger G, Anderson JC, Martin KA, Bonhoeffer T. Protracted synaptogenesis after activity- dependent spinogenesis in hippocampal neurons. J Neurosci. 2007;27:8149–56.

Nava N, Chen F, Wegener G, Popoli M, Nyengaard JR. A new efficient method for synaptic vesicles quantification reveals differences between medial prefrontal cortex perforated and nonperforated synapses. J Comp Neurol. 2014;522:284–97.

Perego C, Cairano ES, Ballabio M, Magnaghi V. Neurosteroid allopregnanolone regulates EAAC1-mediated glutamate uptake and triggers actin changes in Schwann cells. J Cell Physiol. 2012;227:1740–51.

Popoli M, Yan Z, McEwen BS, Sanacora G. The stressed synapse: the impact of stress and glucocorticoids on glutamate transmission. Nat Rev Neurosci. 2012;13:22–37.

Prager EM, Brielmaier J, Bergstrom HC, McGuire J, Johnson LR. Localization of mineralocorticoid receptors at mammalian synapses. PLoS ONE. 2010;5:e14344.

Rajkowska G, Miguel-Hidalgo JJ, Wei J, Dilley G, Pittman SD, Meltzer HY, Overholser JC, Roth BL, Stockmeier CA. Morphometric evidence for neuronal and glial prefrontal cell pathology in major depression. Biol Psychiatry. 1999;45:1085–98.

Reznikov LR, Grillo CA, Piroli GG, Pasumarthi RK, Reagan LP, Fadel J. Acute stress-mediated increases in extracellular glutamate levels in the rat amygdala: differential effects of antidepressant treatment. Eur J Neurosci. 2007;25:3109–14.

Rosenmund C, Stevens CF. Definition of the readily releasable pool of vesicles at hippocampal synapses. Neuron. 1996;16:1197–207.

Ruiz CR, Shi J, Meffert MK. Transcript specificity in BDNF-regulated protein synthesis. Neuropharmacology. 2013. [Epub ahead of print].

Sanacora G, Treccani G, Popoli M. Towards a glutamate hypothesis of depression: an emerging frontier of neuropsychopharmacology for mood disorders. Neuropharmacol. 2012;62:63–77.

Satoh E, Shimeki S. Acute restraint stress enhances calcium mobilization and glutamate exocytosis in cerebrocortical synaptosomes from mice. Neurochem Res. 2010;35:693–701.

Sheline YI, Gado MH, Kraemer HC. Untreated depression and hippocampal volume loss. Am J Psychiatry. 2003;160:1516–8.

Shors TJ, Chua C, Falduto J. Sex differences and opposite effects of stress on dendritic spine density in the male versus female hippocampus. J Neurosci. 2001;21:6292–7.

Sousa N, Almeida OF. Disconnection and reconnection: the morphological basis of (mal)adaptation to stress. Trends Neurosci. 2012;35:742–51.

Tokita K, Yamaji T, Hashimoto K. Roles of glutamate signaling in preclinical and/or mechanistic models of depression. Pharmacol Biochem Behav. 2012;100:688–704.

Treccani G, Musazzi L, Perego C, Milanese M, Nava N, Bonifacino T, Lamanna J, Malgaroli A, Drago F, Racagni G, Nyengaard JR, Wegener G, Bonanno G, Popoli M. Stress and corticosterone increase the readily releasable pool of glutamate vesicles in synaptic terminals of prefrontal and frontal cortex. Mol Psychiatry. 2014;19:433–43.

van derZM, Oldenziel WH, Rea K, Cremers TI, Westerink BH. Microdialysis of GABA and glutamate: analysis, interpretation and comparison with microsensors. Pharmacol Biochem Behav. 2008;90:135–47.

Venero C, Borrell J. Rapid glucocorticoid effects on excitatory amino acid levels in the hippocampus: a microdialysis study in freely moving rats. Eur J Neurosci. 1999;11:2465–73.

Wang CC, Wang SJ. Modulation of presynaptic glucocorticoid receptors on glutamate release from rat hippocampal nerve terminals. Synapse. 2009;63:745–51.

Woon FL, Sood S, Hedges DW. Hippocampal volume deficits associated with exposure to psychological trauma and posttraumatic stress disorder in adults: a meta-analysis. Prog Neuropsychopharmacol Biol Psychiatry. 2010;34:1181–8.

Yoshihara Y, De Roo M, Muller D. Dendritic spine formation and stabilization. Curr Opin Neurobiol. 2009;19:146–53.

Yuen EY, Liu W, Karatsoreos IN, Feng J, McEwen BS, Yan Z. Acute stress enhances glutamatergic transmission in prefrontal cortex and facilitates working memory. Proc Natl Acad Sci U S A. 2009;106:14075–9.

Yuen EY, Liu W, Karatsoreos IN, Ren Y, Feng J, McEwen BS, et al. Mechanisms for acute stress-induced enhancement of glutamatergic transmission and working memory. Mol Psychiatry. 2011;16:156–70.

Yuen EY, Wei J, Liu W, Zhong P, Li X, Yan Z. Repeated stress causes cognitive impairment by suppressing glutamate receptor expression and function in prefrontal cortex. Neuron. 2012;73:962–77.

Zito K, Scheuss V, Knott G, Hill T, Svoboda K. Rapid functional maturation of nascent dendritic spines. Neuron. 2009;61:247–58.

Chapter 4
Dual Regulation of Glutamatergic Transmission and Cognition by Stress in Prefrontal Cortex

Yan Zhen

Abstract Corticosterone, the major stress hormone, serves as a key controller for neuronal responses that underlie behavioral adaptation, as well as maladaptive changes that lead to cognitive and emotional disturbances in stress-related mental disorders. The molecular and cellular mechanisms underlying the complex actions of corticosteroid stress hormones are largely unknown. Here we demonstrate that acute versus chronic stress exerts opposite effects on glutamatergic transmission in prefrontal cortex (PFC), which leads to opposing effects on PFC-dependent cognitive functions. Acute stress induces synaptic potentiation by increasing surface delivery of N-methyl-D-aspartate (NMDA)-type and α-amino-3-hydroxy-methyl-4-isoxazole propionic acid (AMPA)-type glutamate receptor channels via glucocorticoid/serum- and glucocorticoid-inducible kinase (SGK)/Rab4 signaling, resulting in enhanced working memory performance. In contrast, repeated stress induces synaptic depression by increasing the ubiquitin/proteasome-mediated degradation of NMDA and AMPA receptor subunits, resulting in impaired recognition memory.

Abbereviation

AMPA	α-Amino-3-hydroxy-methyl-4-isoxazole propionic acid
NMDA	N-Methyl-D-aspartate
PFC	Prefrontal cortex
GR	Glucocorticoid receptor
MR	Mineralocorticoid receptors
ESPC	Excitatory postsynaptic current
SGK	Serum- and glucocorticoid-inducible kinase
WM	Working memory
TOR	Temporal order recognition
DR	Discrimination ratio

Y. Zhen (✉)
Department of Physiology and Biophysics, School of Medicine and Biomedical Sciences, State University of New York at Buffalo, Buffalo, NY 14214, USA
e-mail: zhenyan@buffalo.edu

M. Popoli et al. (eds.), *Synaptic Stress and Pathogenesis of Neuropsychiatric Disorders*,
DOI 10.1007/978-1-4939-1056-4_4,

4.1 Introduction

In response to stress, the brain recruits many neuronal circuits to adapt to the demand, leading to the activation of hypothalamic-pituitary-adrenocortical axis, and the production of adrenal corticosterone (cortisol in humans), the major stress hormone (de Kloet et al. 2005). Corticosterone exerts its cellular effects by acting on mineralocorticoid receptors (MRs) and glucocorticoid receptors (GRs). Importantly, stress hormones have both protective and damaging effects on the body (McEwen 1998). In situations of acute stress, they are essential for adaptation and maintenance of homeostasis, while in response to chronic and repeated stress, they can produce wear and tear on the body (McEwen 2007). Consistently, behavioral studies have found that moderate acute stress facilitates classical conditioning, associative learning, and working memory (WM) (Shors et al. 1992; Henckens et al. 2011), in contrast to the chronic stress-induced deficits in spatial and contextual memory performance and attentional control (McEwen 1999; Liston et al. 2006; Cerqueira et al. 2007). Thus, it has been proposed that the opposing effects that stress has on learning depend on the relative timing of the events (Joëls et al. 2006). Specifically, stress within the context of a learning situation leads to the release of corticosteroids, resulting in focused attention and improvements in memory (Joëls et al. 2006). It has also been suggested that there exists an "inverted U" relationship of stress to cognitive function (Diamond et al. 1992; Joëls 2006), such that a moderate level of glucocorticoids has pro-cognitive effects, while too low or too high glucocorticoid levels are detrimental to cognitive processing.

Given the strong impact of stress hormones on cognition and emotion, it is important to understand the neuronal basis underlying their actions in the brain. One of the primary targets of stress hormones is the prefrontal cortex (PFC), a brain region critical for WM, executive function, and extinction of learning. It has been proposed that glutamate receptor-mediated synaptic transmission that controls recurrent excitation within networks of PFC neurons is crucial for WM (Goldman-Rakic 1995; Lisman et al. 1998). Dysfunction of glutamatergic transmission is considered the core feature and fundamental pathology of stress-related mental disorders with impaired WM (Tsai and Coyle 2002; Moghaddam 2003). Thus, we speculate that NMDA receptors (NMDARs) and AMPA receptors (AMPARs) are potential targets of stress hormones critically involved in the regulation of PFC functions.

Our recent studies have found that acute stress induces a robust and sustained potentiation of glutamate receptor surface expression and excitatory synaptic currents in PFC pyramidal neurons, as well as a significant facilitation of PFC-mediated WM, via a mechanism dependent on serum- and glucocorticoid-inducible kinase (SGK) and the Rab family small guanosine triphosphatases (GTPases) (Yuen et al. 2009, 2011; Liu et al. 2010; Lee et al. 2012). On the other hand, we have found that repeated (subchronic) stress dampens PFC glutamatergic transmission by facilitating glutamate receptor turnover, which causes the detrimental effect on PFC-dependent cognitive processes (Yuen et al. 2012).

4.2 Methods

4.2.1 Stress Paradigm

Prepubertal (25–28 days of age) Sprague Dawley (SD) male rats were exposed to acute stressors of diverse types. For the forced-swim stress, rats were placed in a cylindrical glass tank (24.5 cm high × 18.5 cm diameter) filled with water to a depth of 20 cm. Rats were forced to swim in warm water (23–25 °C) for 20 min. For the acute restraint stress, rats were placed in air-assessable cylinders for 2 h. The size of the container was similar to the size of the animal, which made the animal almost immobile in the container. For the elevated-platform stress, rats were placed on an elevated platform (20 × 20 cm) for 20 min. For repeated unpredictable stress (7 days), rats were subjected each day to two stressors that were randomly chosen from six different stressors, forced swim (room temperature (RT), 30 min), elevated platform (30 min), cage movement (30 min), lights on overnight, immobilization (RT, 1 h), and food and water deprivation overnight.

4.2.2 Electrophysiological, Biochemical and Behavioral Experiments

Details can be found in our previous publications (Yuen et al. 2009, 2011, 2012; Liu et al. 2010; Lee et al. 2012; Wei et al. 2013).

4.3 Results

4.3.1 Differential Effects of Acute Versus Repeated Stress on Glutamate Transmission and Glutamate Receptors in PFC

To study the impact of stress on glutamate transmission, we examined synaptic strength by measuring input-output curves of evoked synaptic responses, such as NMDAR- and AMPAR-mediated excitatory postsynaptic current (EPSC), in PFC pyramidal neurons. Young male rats (4-week-old) were exposed to either a 20-min forced-swim acute stress paradigm, or repeated (7-day) restraint stress or unpredictable stress. As shown in Fig. 4.1a–d, AMPAR- or NMDAR-mediated excitatory synaptic responses were markedly potentiated in neurons from acutely stressed animals at 1–4 or 24 h post stress. No significant difference was found at 5 days post stress. In contrast, AMPAR-EPSC and NMDAR-EPSC amplitudes were markedly reduced in neurons from animals exposed to repeated stress (restraint or

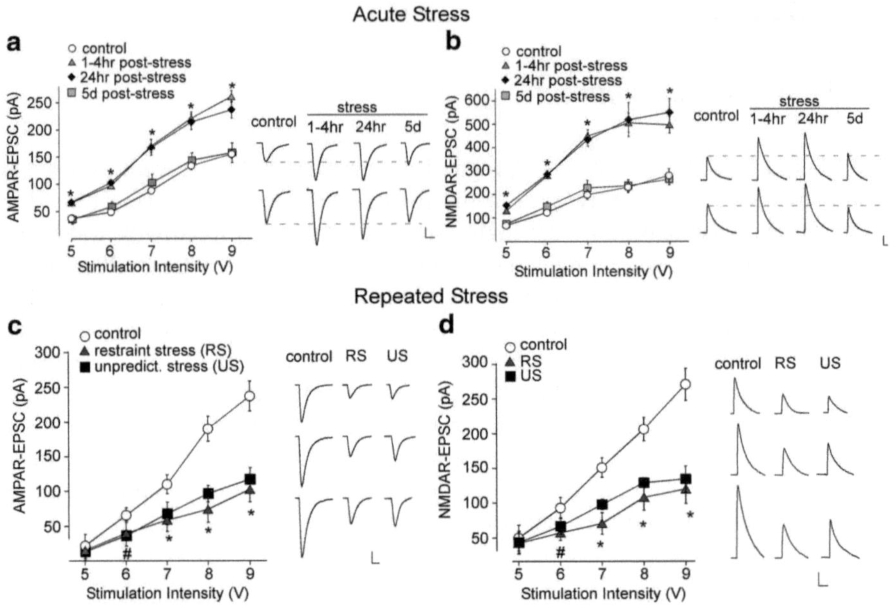

Fig. 4.1 Glutamatergic transmission in PFC pyramidal neurons is enhanced by acute stress, and impaired by repeated stress. **a, b** Summarized input-output curves of α-amino-3-hydroxy-methyl-4-isoxazole propionic acid receptor (*AMPAR*)-excitatory postsynaptic current (*EPSC*) (**a**) or N-methyl-D-aspartate receptor (*NMDAR*)-EPSC (**b**) evoked by a series of stimulus intensities in PFC pyramidal neurons taken from control or animals exposed to acute forced-swim stress (examined at 1–4, 24 h, and 5 days post stress). *Inset:* representative synaptic current traces. *Scale bars:* 100 pA, 100 ms (**a**); 50 pA, 20 ms (**b**). *$p<0.001$. **c, d** Summarized input-output curves of AMPAR-EPSC (**c**) or NMDAR-EPSC (**d**) in response to a series of stimulation intensity in control versus animals exposed to 7 days repeated restraint stress (*RS*) or unpredictable stress (*US*). *$p<0.01$, **$p<0.05$, ANOVA. *Inset*: representative EPSC traces. *Scale bars:* 50 pA, 20 ms (**c**) or 100 ms (**d**). (Adapted from Yuen et al. 2011, 2012)

unpredictable). Injection of the GR antagonist RU486 blocked both the enhancing effect of acute stress and the suppressing effect of repeated stress on glutamatergic responses (data not shown). These results suggest that stress exerts a bi-phasic effect on PFC glutamatergic transmission depending on the duration of stressor.

The alteration of glutamatergic transmission by stress could result from the changed number of glutamate receptors. To test this, we performed Western blotting and surface biotinylation experiments to detect the total and surface level of AMPAR and NMDAR subunits in PFC slices from stressed young male rats. As shown in Fig. 4.2a–d, animals exposed to acute restraint stress (single time, 2 h) showed a significant increase in surface AMPAR and NMDAR subunits, while the total proteins remained unchanged. Animals exposed to 5 or 7-day restraint stress showed a significant decrease in the amount of GluR1 and NR1 subunits. Moreover, repeated stress did not affect the total level of other glutamate receptor subunits, such as GluR2, NR2A, and NR2B, nor the expression of MAP2 (a dendritic

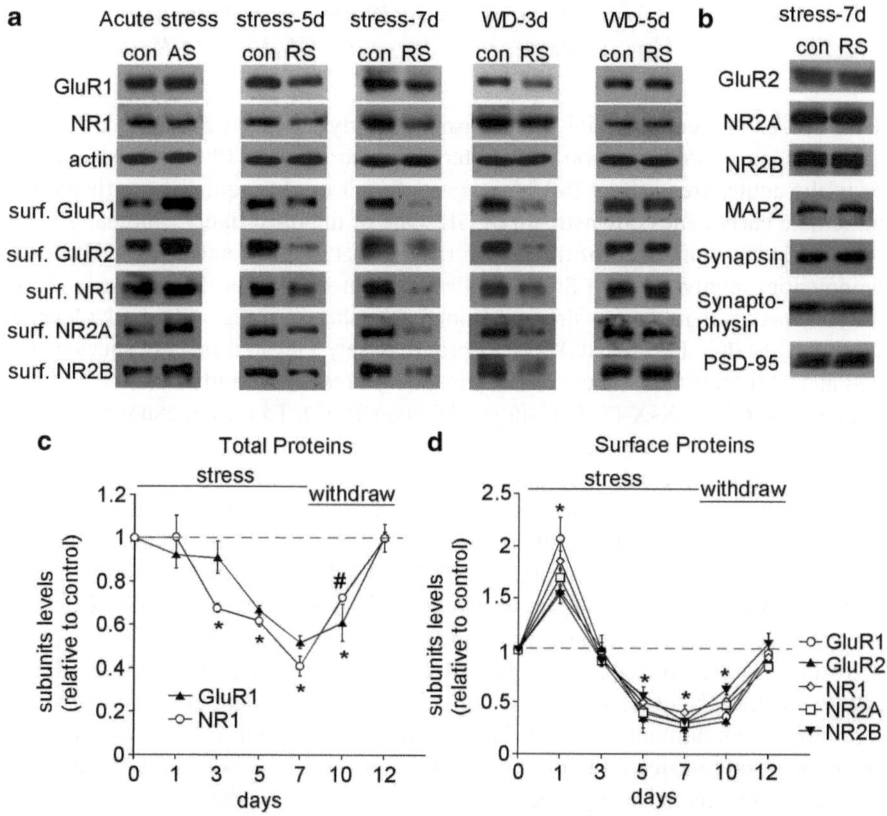

Fig. 4.2 The surface and total levels of AMPAR and NMDAR subunits in PFC are differentially altered by acute versus chronic stress. **a, c, d** Immunoblots (**a**) and quantification analysis (**c, d**) of the total and surface AMPAR and NMDAR subunits in PFC from control (*con*) versus rats exposed to acute restraint stress (*AS*, 1 day, single time of 2 h) or 5–7-day (2 h/day) repeated restraint stress (*RS*). Some animals were withdrawn (*WD*) for different durations (3 or 5 days) after being exposed to 7-day *RS*. $*p<0.01$; $**p<0.05$, *t* test. **b** Immunoblots of the total proteins in PFC from control versus repeatedly stressed (7-day restraint) rats. (Adapted from Yuen et al. 2012)

marker), synapsin, synaptophysin (presynaptic markers) or PSD-95 (a postsynaptic marker), suggesting that no general dendritic or synaptic loss has occurred under such conditions. The amount of AMPAR and NMDAR subunits in the surface pool was all significantly decreased by repeated stress, indicating the loss of glutamate receptors at the plasma membrane. To find out how long the effect of repeated stress can last, we exposed animals to 7-day restraint stress, and examined at 3–5 days after stressor cessation. After 3-day withdrawal of stress, the expression of total and surface AMPARs and NMDARs was still at a partially reduced level, but returned to the control level after 5-day withdrawal. These results suggest that stress-induced changes in glutamatergic transmission likely occur through GR-induced modification of postsynaptic NMDA and AMPA receptors in PFC pyramidal neurons.

4.3.2 Molecular Mechanisms Underlying the Differential Effects of Acute Versus Repeated Stress on Glutamate Receptors

Next, we examined potential mechanisms underlying the differential effects of acute versus repeated stress on glutamatergic transmission in PFC. The onset kinetics of the acute stress effect (>1 h) suggests that it might require the activation of immediate early genes downstream of GR. One of the most likely candidates is the SGK, which is composed of three isoforms, SGK1, SGK2, and SGK3. To assess the potential involvement of SGK, we first examined whether the expression level of SGK was up-regulated in stressed animals. As shown in Fig. 4.3a, b, the level of SGK1 and SGK3, but not SGK2, was progressively elevated in PFC slices examined at 1–2 h after acute stress. SGK phosphorylates serine and threonine residues in the motif R-X-R-X-X-(S/T) (Lang and Cohen 2001). To further examine the role of SGK in corticosterone regulation of NMDARs and AMPARs, we pretreated PFC neurons with a SGK substrate peptide (RPRAATF), which should competitively block the interaction of all SGK isoforms with their endogenous substrates. This peptide was coupled to the protein transduction domain of the human immunodeficiency virus (HIV) TAT protein (YGRKKRRQRRR), which rendered it cell-permeant. As shown in Fig. 4.3c, intravenous (i.v.) injection of TAT-SGK peptide prevented acute stress from increasing the amplitude of NMDAR-EPSC.

To identify which SGK is involved, we knocked down SGK isoforms in PFC cultures with siRNA transfection. We found that the enhancing effect of short-term corticosterone treatment (100 nM, 20 min) on NMDAR and AMPAR currents was lost in neurons transfected with SGK1 siRNA or SGK3 siRNA, but was unaltered in neurons transfected with SGK2 siRNA. Taken together, these data suggest that the regulation of glutamatergic signaling by acute stress requires the activation of SGK1/3 downstream of GRs.

The acute stress-induced potentiation of NMDA and AMPA responses is accompanied by increased surface NMDAR and AMPAR clusters, suggesting that GR activation might influence the membrane trafficking of glutamate receptors. It is known that the Rab family of small GTPases functions as specific regulators of vesicle transport between organelles, and different Rab members control vesicular fusion at different stages in the exocytic/endocytic cycle (Zerial and McBride 2001). Among them, the most likely candidates are: Rab5, which controls the transport from plasma membrane to early endosomes; Rab4, which controls a rapid direct recycling route from early endosomes to cell surface; and Rab11, which mediates recycling from recycling endosomes to plasma membrane. As demonstrated in Fig. 4.3d, knockdown of Rab4 blocked the increase of NMDAR or AMPAR current density by corticosterone treatment (100 nM, 20 min). In contrast, the enhancing effect of corticosterone was not altered by Rab5 siRNA or Rab11 siRNA. These results suggest that the corticosterone-induced increase in functional glutamate receptors is through a mechanism depending on Rab4-mediated receptor recycling.

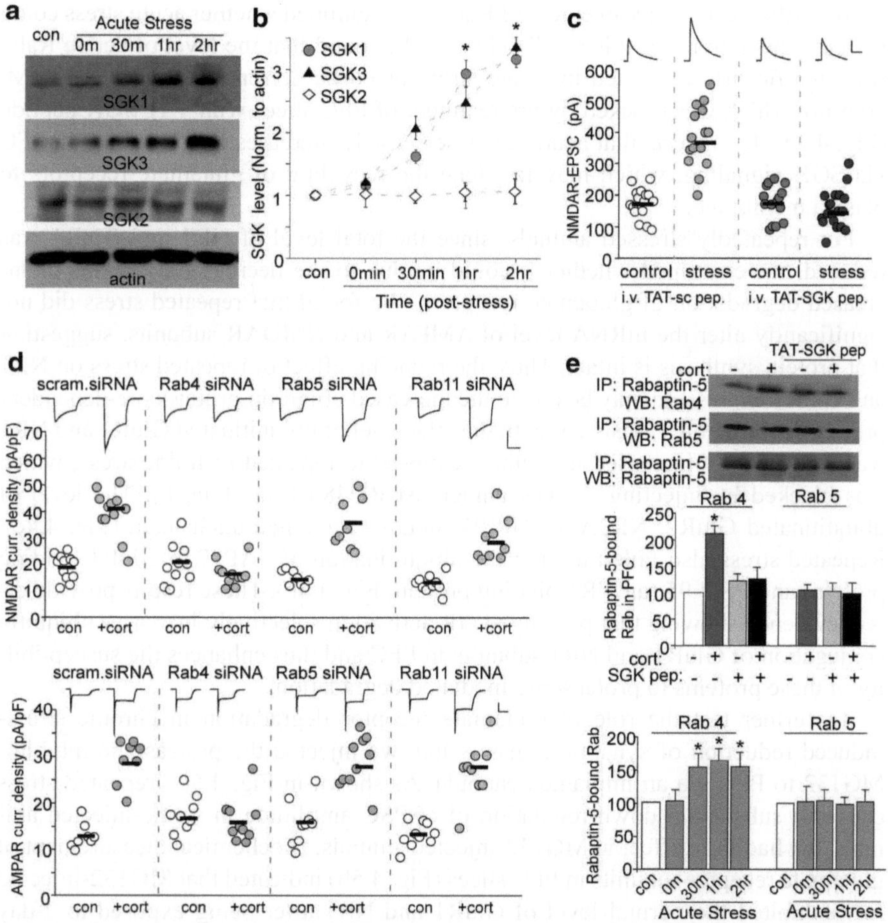

Fig. 4.3 Serum- and glucocorticoid-inducible kinase (*SGK*)/Rab4 signaling is required for acute stress-induced potentiation of glutamatergic transmission. **a, b** Western blots (**a**) and quantification (**b**) of SGKs in lysates of PFC slices taken from control or acutely stressed animals at various post stress time points (0 min, 30 min, 1 and 2 h). $^*p < 0.001$. **c** Dot plots of N-methyl-D-aspartate receptor (*NMDAR*)-excitatory postsynaptic current (*EPSC*) recorded in prefrontal cortex (*PFC*) slices from control versus acutely stressed animals i.v. injected with TAT-SGK peptide (0.6 pmol/g) or a scrambled control peptide (TAT-sc, 0.6 pmol/g). Peptides were administered 30 min prior to stress, and recordings were performed at 1–4 h post stress. *Inset*: Representative NMDAR-EPSC traces. *Scale bars:* 100 pA, 100 ms. **d** Dot plots showing the effect of corticosterone (*CORT*) treatment (100 nM, 20 min) on NMDAR or α-amino-3-hydroxy-methyl-4-isoxazole propionic acid receptor (*AMPAR*) current density in PFC cultures transfected with a scrambled siRNA or siRNA against Rab4, Rab5, or Rab11. Recordings were obtained 1–4 h after the treatment. *Inset*: Representative current traces. *Scale bars*: 200 pA, 1 s. **e** Co-immuoprecipitation blots and analysis showing the level of active (Rabaptin-5-bound) Rab4 or Rab5 in PFC slices without or with corticosterone treatment (100 nM, 20 min, collected 1 h after treatment) in the absence or presence of TAT-SGK peptide (50 µM, 30 min prior to corticosterone, *top*), or in PFC slices from control versus swim-stressed animals examined at various post stress time points (*bottom*). *IP* immunoprecipitation, *WB* Western blot. (Adapted from Yuen et al. 2011)

To further test the involvement of Rab4, we examined whether acute stress could increase the activity of this small GTPase. We found that the level of active Rab4 was significantly increased by acute stress or corticosterone treatment (100 nM, 20 min), which was blocked by pretreatment of PFC slices with TAT-SGK peptide (Fig. 4.3e). It suggests that acute stress selectively increases Rab4 activity in PFC via SGK signaling, which may facilitate the recycling of glutamate receptors to plasma membrane.

For repeatedly stressed animals, since the total level of NR1 and GluR1 was reduced, we examined whether it could be due to the decreased synthesis or increased degradation of glutamate receptors. We found that repeated stress did not significantly alter the mRNA level of AMPAR and NMDAR subunits, suggesting that protein synthesis is intact. Thus, the reducing effect of repeated stress on NR1 and GluR1 expression may be due to the increased ubiquitin/proteasome-dependent protein degradation. Consistent with this, the level of ubiquitinated GluR1 and NR1 was significantly increased in animals exposed to repeated restraint stress, which was blocked by injecting the GR antagonist RU486 (Fig. 4.4a, b). The level of ubiquitinated GluR2, NR2A, or NR2B subunits remained unchanged (Fig. 4.4c). Repeated stress also failed to alter the ubiquitination of SAP97 (a GluR1 binding protein) and PSD-95 (an NR1 binding protein, Fig. 4.4c). These results provide direct evidence showing that prolonged GR activation selectively increases ubiquitin conjugation of GluR1 and NR1 subunits in PFC and thus enhances the susceptibility of these proteins to proteasome-mediated degradation.

To further test the role of glutamate receptor degradation in chronic stress-induced reduction of synaptic transmission, we injected the proteasome inhibitor MG132 to PFC via an implanted cannula. As shown in Fig. 4.5a, repeated stress caused a substantial down-regulation of eEPSC amplitude in saline-injected animals, but had little effect in MG132-injected animals. Biochemical measurement of glutamate receptor subunits in PFC slices (Fig. 4.5b) indicated that MG132-injected rats exhibited the normal level of GluR1 and NR1 after being exposed to 7-day restraint stress, which was in sharp contrast to the reduced expression of GluR1 and NR1 in saline-injected rats after repeated stress. Taken together, these results suggest that repeated behavioral stress induces the ubiquitin/proteasome-dependent degradation of GluR1 and NR1, leading to the depression of glutamatergic transmission in PFC.

To find out which E3 ubiquitin ligases are potentially involved in the repeated stress-induced ubiquitination of GluR1 and NR1 subunits in PFC, we focused on two possible candidates, Nedd4–1 (neural-precursor cell-expressed developmentally downregulated gene 4-1), an E3 ligase necessary for GluR1 ubiquitination in response to the agonist AMPA (Schwarz et al. 2010; Lin et al. 2011), and Fbx2, an E3 ligase in the ER that ubiquitinates NR1 subunits (Kato et al. 2005). Nedd4-1 or Fbx2 shRNA lentivirus was delivered to rat frontal cortex via a stereotaxic injection to knockdown these E3 ligases in vivo. As illustrated in Fig. 4.5c, d, repeated stress caused a substantial down-regulation of the eEPSC amplitude in green fluorescent protein (GFP) lentivirus-injected animals, but had little effect on AMPAR-EPSC in Nedd4 shRNA lentivirus-injected animals or on NMDAR-EPSC in Fbx2 shRNA

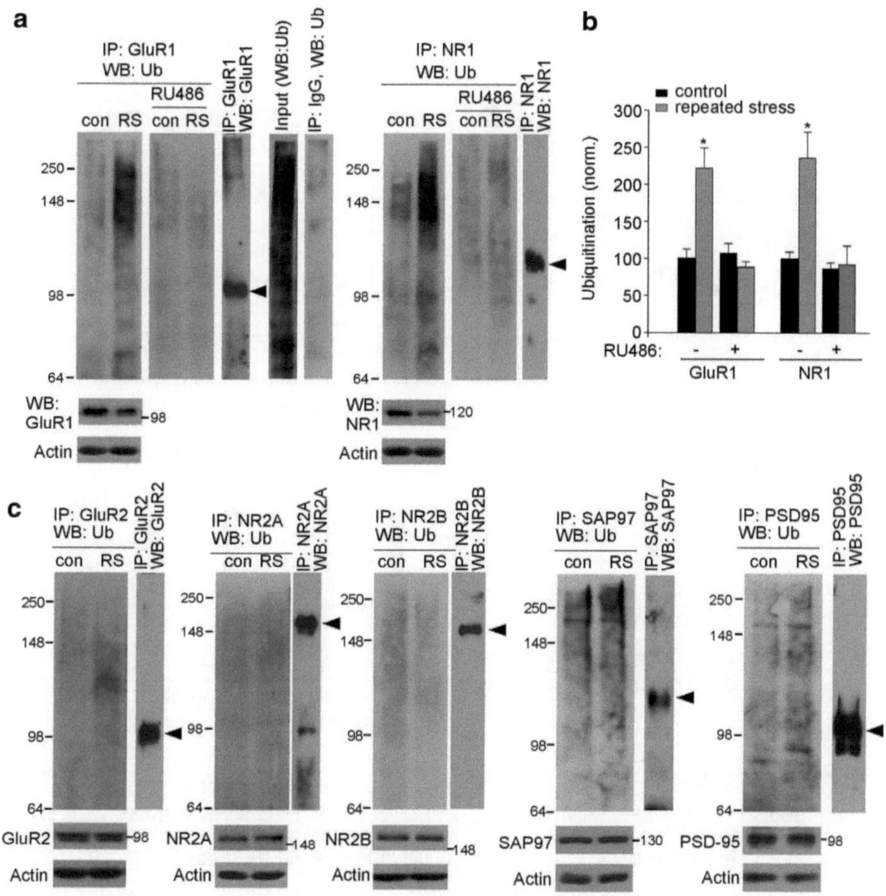

Fig. 4.4 Repeated stress increases the ubiquitination level of GluR1 and NR1 subunits. **a, b** Representative blots (**a**) and quantification (**b**) showing the ubiquitination of GluR1 and NR1 subunits in control versus stressed (7-day restraint) animals without or with RU486 injection (10 mg/kg). $^*p<0.01$, t test. Lysates of PFC slices were immunoprecipitated with an antibody against GluR1 or NR1, and then blotted with a ubiquitin (*Ub*) antibody. Also shown are the input control, the immunoprecipitation control, and the immunoblots of total proteins in control versus stressed animals. Note, in stressed rats, the immunoprecipitated GluR1 or NR1 showed ubiquitin staining at a molecular mass heavier than the unmodified protein itself. The ladder of ubiquitinated GluR1 or NR1 is typical of proteins that are polyubiquitinated to signal their degradation. (**c**) Ubiquitination of GluR2, NR2A, NR2B, SAP97, and PSD-95 in control versus stressed (7-day restraint) animals. *IP* immunoprecipitation, *WB* Western blot, *RS* restraint stress, *IgG immunoglobulin G*. (Adapted from Yuen et al. 2012)

lentivirus-injected animals. Nedd4-1 shRNA or Fbx2 shRNA lentivirus-injected rats also failed to exhibit the increased level of ubiquitinated GluR1 or NR1 after being exposed to 7-day restraint stress (data not shown). These results suggest that Nedd4-1 and Fbx2 mediate the repeated stress-induced downregulation of AMPAR and NMDAR responses in PFC, respectively.

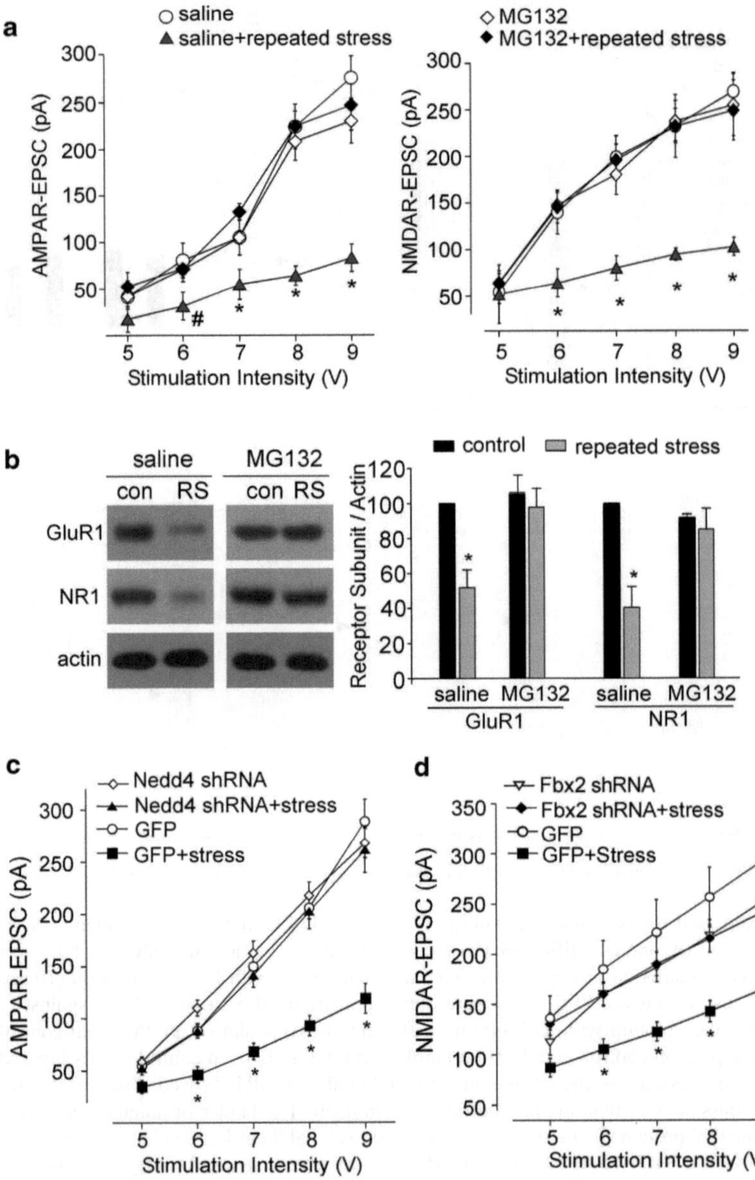

Fig. 4.5 PFC infusion of a proteasome inhibitor or knockdown of the E3 ubiquitin ligases Nedd4-1 and Fbx2 prevents the loss of glutamate receptors by repeated stress. **a** Summarized input-output curves of α-amino-3-hydroxy-methyl-4-isoxazole propionic acid receptor (*AMPAR*)-EPSC or N-methyl-D-aspartate receptor (*NMDAR*)- excitatory postsynaptic current (*EPSC*) in control versus repeatedly stressed (7-day restraint) animals with local injection of the proteasome inhibitor MG132 or saline control. *$p<0.01$, #$p<0.05$, ANOVA. **b** Immunoblots and quantification analysis of GluR1 and NR1 expression in control versus repeatedly stressed animals with PFC infusion of MG132 or saline. *$p<0.01$, *t* test. **c, d** Summarized input-output curves of AMPAR-EPSC (**c**) or NMDAR-EPSC (**d**) in control versus repeatedly stressed (7-day restraint) rats with the PFC injection of Nedd4-1 shRNA lentivirus (**c**), Fbx2 shRNA lentivirus (**d**), or GFP lentivirus control. *$p<0.01$, ANOVA. *RS* restraint stress, *GFP* green fluorescent protein. (Adapted from Yuen et al. 2012)

4.3.3 Behavioral Consequences of the Dual Effects of Stress on Glutamate Transmission

Since AMPAR- and NMDAR-mediated synaptic transmission at recurrent synapses in PFC networks is crucial for WM, the acute stress-induced enhancement of glutamatergic responses could be linked to improved WM in animals exposed to acute stress. Thus, we performed behavioral tests using the delayed alteration task in the T-maze, a well-established protocol for PFC-mediated WM. As shown in Fig. 4.6a, animals exposed to the acute forced-swim stress performed significantly better when examined at 4 h post stress or 1 day post stress. This difference disappeared at 2 day post stress. Except for the correctness, other parameters, such as the completion time and locomotor activity, were not significantly different between control versus stressed groups. These results indicate that acute stress facilitates WM within the time frame of a few hours to 1 day.

To test whether acute stress enhances WM via GR signaling, we injected (i.p.) animals with RU486 at 30 min prior to the stress procedure, and compared behavioral performance at 4 h or 1 day post stress. As shown in Fig. 4.6b, acutely stressed animals injected with saline showed better performance in the delayed alternation task. Injection of RU486 abolished the enhancing effect of acute stress on WM. These data suggest that the acute stress-induced enhancement of WM is mediated by GR activation.

To provide a "causal link" between stress-induced changes in glutamatergic transmission and WM, we tested whether TAT-SGK peptide, which blocked the effect of acute stress on glutamatergic transmission in vitro, could influence the effect of acute stress on WM. TAT-SGK peptide was stereotaxically injected into PFC prelimbic regions bilaterally via an implanted guide cannula. As shown in Fig. 4.6c, the enhancing effect of acute stress on WM was blocked by TAT-SGK peptide completely. These data suggest that GR/SGK-mediated enhancement of glutamatergic transmission within PFC may underlie the positive effect of acute stress on WM.

To test the impact of repeated stress on cognitive functions, we measured the recognition memory task, a fundamental explicit memory process requiring judgments of the prior occurrence of stimuli based on the relative familiarity of individual objects, the association of objects and places, or the recency information (Ennaceur and Delacour 1988; Dix and Aggleton 1999). Lesion studies have shown that medial PFC plays an obligatory role in the temporal order recognition (TOR) memory (Barker et al. 2007), so this behavioral task was used. Young (4-week-old) male rats, which had been exposed to 7-day repeated behavioral stressors, were examined at 24 h after stressor cessation.

The control groups spent much more time exploring the novel (less recent) object in the test trial, while the repeatedly stressed rats (restraint, 2 h/day, 7 days) lost the preference to the novel object. The discrimination ratio (DR), an index of the object recognition memory, indicated a profound impairment of TOR memory by repeated stress, which was blocked by injection of the GR antagonist RU486 (Fig. 4.7a). In contrast to the impaired TOR memory, rats exposed to repeated restraint stress showed no changes in anxiety-related behavior or locomotive activity (data not shown).

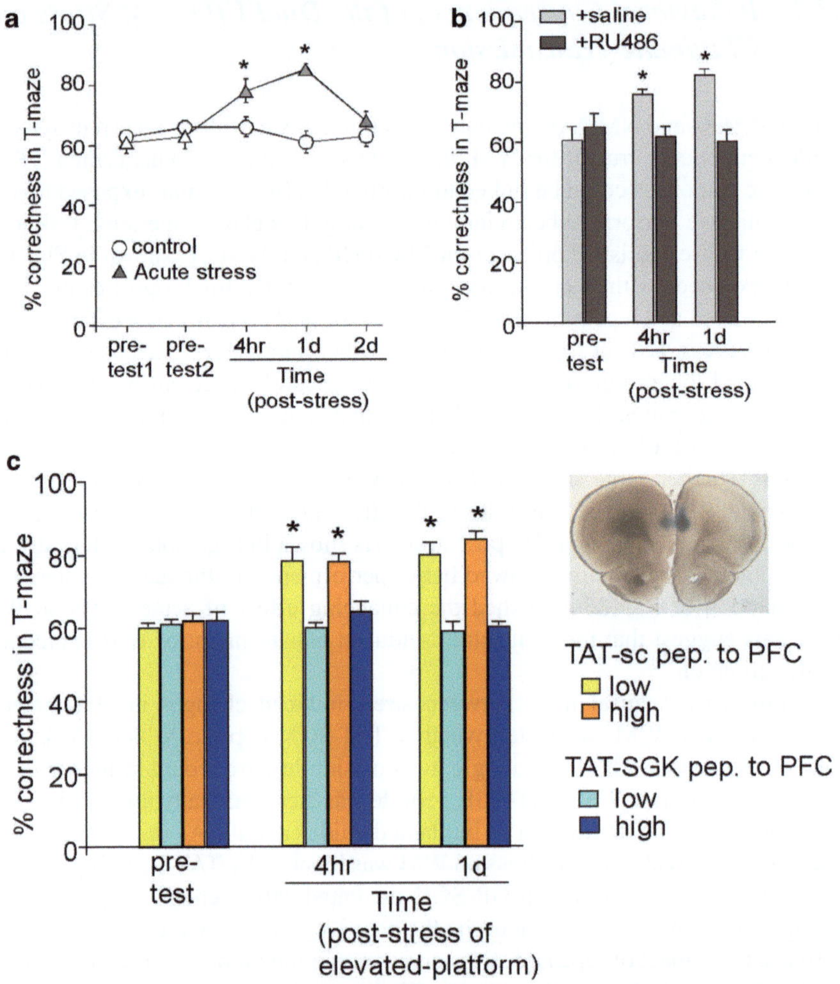

Fig. 4.6. Acute stress enhances working memory via a GR/serum- and glucocorticoid-inducible kinase (*SGK*)-dependent mechanism. **a** Cumulative data (mean ± SEM) showing percentage correct of responses in T-maze tests in control versus stressed (forced-swim) rats examined at various pre and post stress time points. $*p<0.01$, ANOVA. **b** Cumulative data (mean ± SEM) showing percentage correct in T-maze tests before and after forced-swim stress in rats injected with saline versus RU486. $*p<0.01$, ANOVA. **c** Cumulative data (mean ± SEM) showing the percentage correctness in T-maze tests before and after elevated platform stress in rats locally injected to prefrontal cortex (*PFC*) with TAT-SGK peptide versus scrambled TAT-sc peptide (high dose: 40 pmol/g; low dose: 0.12 pmol/g). *Inset:* A photograph showing the slice with a local injection of ink to PFC prelimbic regions to confirm the appropriate location of the cannula. $*p<0.01$, ANOVA. (Adapted from Yuen et al. 2009, 2011)

To test whether glutamatergic transmission in PFC is critical for the object recognition memory, we gave animals a stereotaxic injection of the NMDAR antagonist 2-amino-5-phosphonopentanoic acid (APV) and AMPAR antagonist 6-cyano-7-nitroquinoxaline-2,3-dione (CNQX) to PFC prelimbic regions bilaterally. As

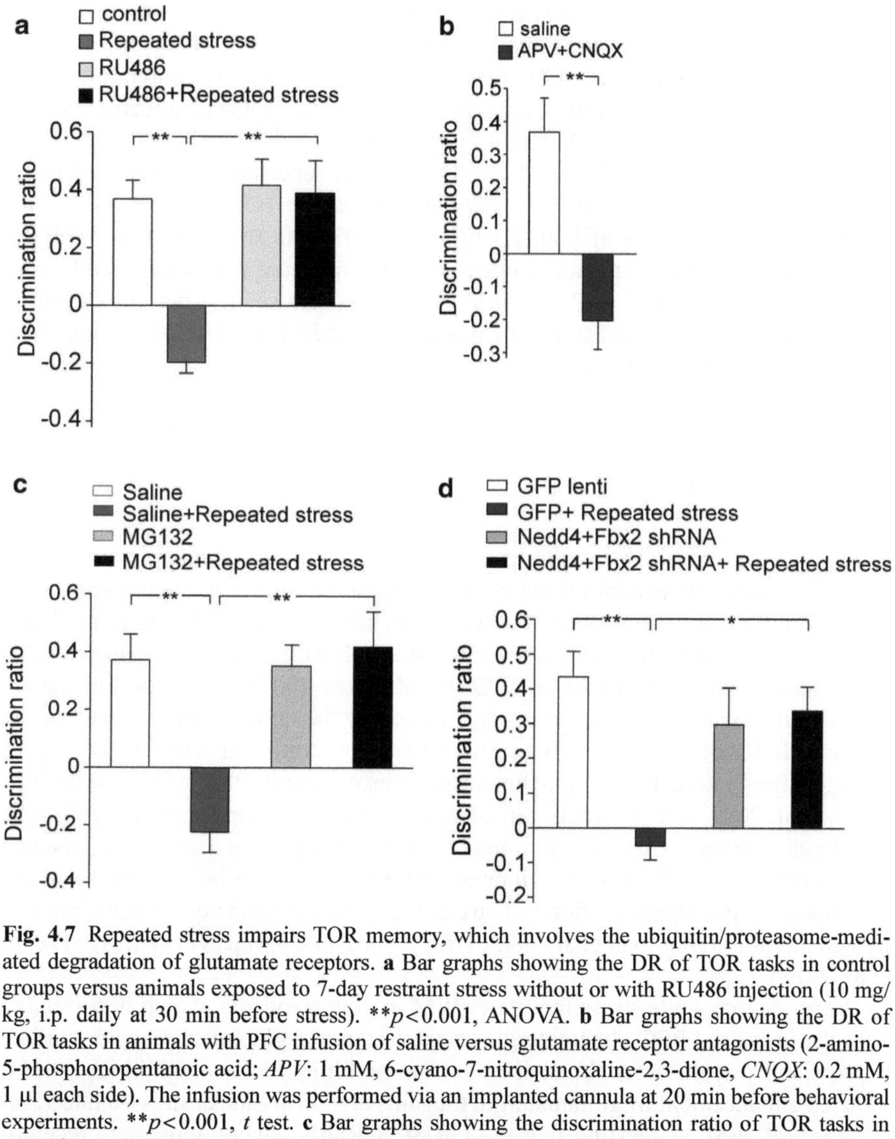

Fig. 4.7 Repeated stress impairs TOR memory, which involves the ubiquitin/proteasome-mediated degradation of glutamate receptors. **a** Bar graphs showing the DR of TOR tasks in control groups versus animals exposed to 7-day restraint stress without or with RU486 injection (10 mg/kg, i.p. daily at 30 min before stress). $^{**}p<0.001$, ANOVA. **b** Bar graphs showing the DR of TOR tasks in animals with PFC infusion of saline versus glutamate receptor antagonists (2-amino-5-phosphonopentanoic acid; *APV*: 1 mM, 6-cyano-7-nitroquinoxaline-2,3-dione, *CNQX*: 0.2 mM, 1 μl each side). The infusion was performed via an implanted cannula at 20 min before behavioral experiments. $^{**}p<0.001$, *t* test. **c** Bar graphs showing the discrimination ratio of TOR tasks in control groups versus repeatedly stressed animals (7-day restraint) with stereotaxic injections of saline or MG132 (0.5 μg each side; 21 pmol/g b.w., daily at 1 h before stress) into PFC via an implanted cannula. $^{**}p<0.001$, ANOVA. **d** Bar graphs showing the discrimination ratio of TOR tasks in control groups versus repeatedly stressed animals (7-day restraint) with PFC injection of GFP lentivirus or Nedd4-1 shRNA+Fbx2 shRNA lentiviruses. $^{**}p<0.001$, $^{*}p<0.01$, ANOVA. *GFP* green fluorescent protein. (Adapted from Yuen et al. 2012)

shown in Fig. 4.7b, APV+CNQX-injected animals lost the normal preference to the novel (less recent) object, similar to the animals exposed to repeated stress. Taken together, it suggests that repeated stress has a detrimental effect on recognition memory, which may be due to the loss of glutamatergic transmission in PFC.

To find out whether the proteasome-dependent degradation of glutamate receptors induced by repeated stress may underlie its detrimental effect on cognitive processes, we examined the TOR memory in animals with stereotaxic injections of MG132 into PFC bilaterally. As shown in Fig. 4.7c, MG132-injected animals exposed to repeated stress had normal TOR memory.

To find out the role of Nedd4-1 and Fbx2 in the repeated stress-induced detrimental effect on cognitive processes, we examined the TOR memory in animals with in vivo knockdown of both E3 ligases in PFC. As shown in Fig. 4.7d, the stress-induced TOR deficit was blocked in animals injected with both Nedd4-1 and Fbx2 shRNA lentiviruses to PFC. These behavioral data suggest that the cognitive impairment by repeated stress may be due to the Nedd4-1 and Fbx2-dependent loss of glutamate receptors in PFC.

4.4 Discussion

It is known that stress exerts complex influence on learning and memory processes, which is usually dependent on the action of stress hormones in combination with neuronal activities within the key target areas (Shors 2006). Mounting evidence has suggested that corticosteroid stress hormones induce divergent changes in different brain regions (de Kloet et al. 2005; McEwen 2007). In addition to the region specificity, the outcome is also determined by the duration and severity of the stressor (de Kloet et al. 2005; Joëls 2006). We have found that acute stress induces a long-lasting potentiation of glutamatergic transmission in PFC and facilitate WM (Yuen et al. 2009, 2011), which is in contrast to the strong suppression of PFC glutamatergic transmission and impairment of object recognition memory by repeated stress (Yuen et al. 2012). Thus, glutamate receptors seem to be a neural substrate that underlies the biphasic effects of stress and glucocorticoids on synaptic plasticity and memory (Diamond et al. 1992; Groc et al. 2008; Krugers et al. 2010; Popoli et al. 2012).

We show that acute stress facilitates WM in young rodents, which is correlated with the increased PFC glutamatergic transmission and glutamate receptor surface expression by acute stress (Yuen et al. 2009). Inhibiting SGK, which blocks stress-induced enhancement of glutamatergic transmission, also blocks stress-induced facilitation of WM, suggesting that the GR/SGK/Rab4-induced glutamate receptor trafficking in PFC may underlie the WM improvement by acute stress (Yuen et al. 2011). These results (Fig. 4.8a) have identified a form of long-term potentiation of synaptic transmission induced by natural stimuli in vivo, providing a potential molecular and cellular mechanism for the beneficial effects of acute stress on cognitive processes subserved by PFC.

On the other hand, the loss of glutamate receptors after repeated stress may involve the increased ubiquitin/proteasome-mediated degradation of GluR1 and NR1

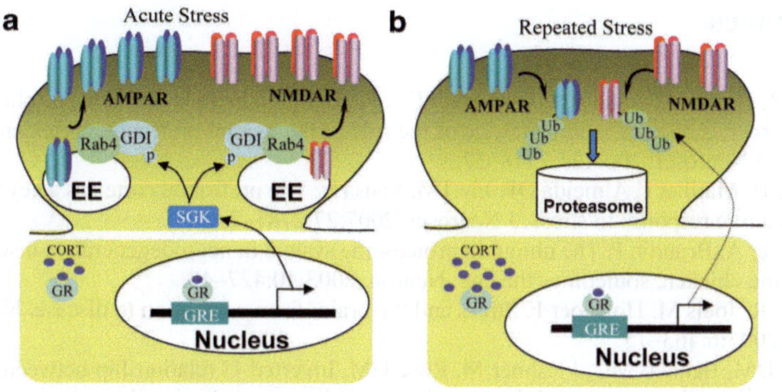

Fig. 4.8 A diagram illustrating the stress-induced changes in glutamate receptor trafficking and function in PFC. **a** In response to acute stress, glucocorticoid receptor (*GR*) activation triggers the upregulation of SGK1/3 (Yuen et al. 2011), leading to the phosphorylation of GTP dissociation inhibitor (GDP dissociation inhibitor *(GDI)*) and increased formation of GDI-Rab4 complex (Liu et al. 2010). Consequently, the functional cycle of Rab4 is facilitated and the Rab4-mediated recycling of N-methyl-D-aspartate receptors (*NMDARs*) and α-amino-3-hydroxy-methyl-4-isoxazole propionic acid receptors (*AMPARs*) from early endosome to plasma membrane is enhanced, resulting in the increased glutamate receptors at the synaptic membrane and potentiated glutamatergic transmission (Yuen et al. 2009, 2011). **b** In response to chronic stress, GR activation leads to the increased ubiquitination of NR1 and GluR1 subunits, probably via activating the E3 ubiquitin ligase Fbx2 and Nedd4, respectively. Consequently, the proteasome-mediated degradation of NMDARs and AMPARs is enhanced, leading to the loss of glutamate receptors from the synaptic membrane and depressed glutamatergic transmission (Yuen et al. 2012). *CORT* corticosterone, *EE* early endosome, *GRE* glucocorticoid response element. (Adapted from Popoli et al. 2012)

subunits. Posttranslational modification through the ubiquitin pathway at the postsynaptic membrane has emerged as a key mechanism for remodeling synaptic networks and altering synaptic transmission (Mabb and Ehlers 2010). Abnormalities in the brain ubiquitin/proteasome system have been implied in a variety of neurodegenerative and mental disorders (Ciechanover and Brundin 2003; Middleton et al. 2002), however little is known about the circumstances under which AMPAR and NMDAR ubiquitination occurs under normal and disease conditions. We demonstrate that the ubiquitination of GluR1 and NR1 subunits, but not their anchoring proteins, is specifically increased in PFC slices upon GR activation following repeated stress. The effect of repeated stress on glutamatergic responses and GluR1/NR1 expression is blocked by the specific inhibitors of proteasomes. This suggests that GR-induced ubiquitination of GluR1 and NR1 subunits tags them for degradation by proteasomes in the cytoplasm, therefore fewer heteromeric AMPARs and NMDARs channels are assembled and delivered to the synaptic membrane (Fig. 4.8b). Interestingly, infusion of a proteasome inhibitor into PFC prevents the loss of recognition memory in stressed animals, providing a potential approach to block the detrimental effects of repeated stress. The identification of E3 ligases involved in the effects of repeated stress provides drug targets for preventing chronic stress-induced impairment of cognitive processes.

References

Barker GR, Bird F, Alexander V, Warburton EC. Recognition memory for objects, place, and temporal order: a disconnection analysis of the role of the medial prefrontal cortex and perirhinal cortex. J Neurosci. 2007;27:2948–57.
Cerqueira JJ, Mailliet F, Almeida OF, Jay TM, Sousa N. The prefrontal cortex as a key target of the maladaptive response to stress. J Neurosci. 2007;27:2781–7.
Ciechanover A, Brundin P. The ubiquitin proteasome system in neurodegenerative diseases: sometimes the chicken, sometimes the egg. Neuron. 2003;40:427–46.
de Kloet ER, Joëls M, Holsboer F. Stress and the brain: from adaptation to disease. Nat Rev Neurosci. 2005;6:463–75.
Diamond DM, Bennett MC, Fleshner M, Rose GM. Inverted-U relationship between the level of peripheral corticosterone and the magnitude of hippocampal primed burst potentiation. Hippocampus. 1992;2:421–30.
Dix S, Aggleton J. Extending the spontaneous preference test of recognition: evidence of object-location and object-context recognition. Behav Brain Res. 1999;99:191–200.
Ennaceur A, Delacour J. A new one-trial test for neurobiological studies of memory in rats. 1: behavioral data. Behav Brain Res. 1988;31:47–59.
Goldman-Rakic PS. Cellular basis of working memory. Neuron. 1995;14:477–85.
Groc L, Choquet D, Chaouloff F. The stress hormone corticosterone conditions AMPAR surface trafficking and synaptic potentiation. Nat Neurosci. 2008;11:868–70.
Henckens MJ, van Wingen GA, Joëls M, Fernández G. Time-dependent corticosteroid modulation of prefrontal working memory processing. Proc Natl Acad Sci U S A. 2011;108:5801–6.
Joëls M. Corticosteroid effects in the brain: U-shape it. Trends Pharmacol Sci. 2006;27:244–50.
Joëls M, Pu Z, Wiegert O, Oitzl MS, Krugers HJ. Learning under stress: how does it work? Trends Cogn Sci. 2006;10:152–8.
Kato A, Rouach N, Nicoll RA, Bredt DS. Activity-dependent NMDA receptor degradation mediated by retrotranslocation and ubiquitination. Proc Natl Acad Sci U S A. 2005;102:5600–5.
Krugers HJ, Hoogenraad CC, Groc L. Stress hormones and AMPA receptor trafficking in synaptic plasticity and memory. Nat Rev Neurosci. 2010;11:675–81.
Lang F, Cohen P. Regulation and physiological roles of serum- and glucocorticoid-induced protein kinase isoforms. Sci STKE. 2001;2001(108):re17.
Lee JB, Wei J, Liu W, Cheng J, Feng J, Yan Z. Histone Deacetylase 6 gates the synaptic action of acute stress in prefrontal cortex. J Physiol. 2012;90:1535–46.
Lin A, Hou Q, Jarzylo L, Amato S, Gilbert J, Shang F, Man HY. Nedd4-mediated AMPA receptor ubiquitination regulates receptor turnover and trafficking. J Neurochem. 2011;119:27–39.
Lisman JE, Fellous JM, Wang XJ. A role for NMDA-receptor channels in working memory. Nat Neurosci. 1998;1:273–5.
Liston C, Miller MM, Goldwater DS, Radley JJ, Rocher AB, Hof PR, et al. Stress-induced alterations in prefrontal cortical dendritic morphology predict selective impairments in perceptual attentional set-shifting. J Neurosci. 2006;26:7870–4.
Liu W. Yuen EY, Yan Z. The stress hormone corticosterone increases synaptic AMPA receptors via SGK regulation of the GDI-Rab4 complex. J Biol Chem. 2010;285:6101–8.
Mabb AM, Ehlers MD. Ubiquitination in postsynaptic function and plasticity. Annu Rev Cell Dev Biol. 2010;26:179–210.
McEwen BS. Protective and damaging effects of stress mediators. N Engl J Med. 1998;338:171–9.
McEwen BS. Stress and hippocampal plasticity. Annu Rev Neurosci. 1999;22:105–22.
McEwen BS. Physiology and neurobiology of stress and adaptation: central role of the brain. Physiol Rev. 2007;87:873–904.
Middleton FA, Mirnics K, Pierri JN, Lewis DA, Levitt P. Gene expression profiling reveals alterations of specific metabolic pathways in schizophrenia. J Neurosci. 2002;22:2718–29.
Moghaddam B. Bringing order to the glutamate chaos in schizophrenia. Neuron. 2003;40:881–4.

Popoli M, Yan Z, McEwen BS, Sanacora G. The stressed synapse: the impact of stress and glucocorticoids on glutamate transmission. Nat Rev Neurosci. 2012;13:22–37.

Schwarz LA, Hall BJ, Patrick GN. Activity-dependent ubiquitination of GluA1 mediates a distinct AMPA receptor endocytosis and sorting pathway. J Neurosci. 2010;30:16718–29.

Shors TJ. Stressful experience and learning across the lifespan. Annu Rev Psychol. 2006;57:55–85.

Shors TJ, Weiss C, Thompson RF. Stress-induced facilitation of classical conditioning. Science. 1992;257:537–9.

Tsai G, Coyle JT. Glutamatergic mechanisms in schizophrenia. Annu Rev Pharm Toxicol. 2002;42:165–79.

Wei J, Yuen EY, Liu W, Li X, Zhong P, Karatsoreos IN, McEwen BS, Yan Z Estrogen protects against the detrimental effects of repeated stress on glutamatergic transmission and cognition. Mol Psychiatry. 2013 (Epub ahead of print).

Yuen EY, Liu W, Karatsoreos IN, Feng J, McEwen BS, Yan Z. Acute stress enhances glutamatergic transmission in prefrontal cortex and facilitates working memory. Proc Natl Acad Sci U S A. 2009;106:14075–9.

Yuen EY, Liu W, Karatsoreos IN, Ren Y, Feng J, McEwen BS, Yan Z. Mechanisms for acute stress-induced enhancement of glutamatergic transmission and working memory. Mol Psychiatry. 2011;16:156–70.

Yuen EY, Wei J, Liu W, Zhong P, Li X, Yan Z. Repeated stress causes cognitive impairment by suppressing glutamate receptor expression and function in prefrontal cortex. Neuron. 2012;73:962–77.

Zerial M, McBride H. Rab proteins as membrane organizers. Nat Rev Mol Cell Biol. 2001;2:107–17.

Chapter 5
Role of Endocannabinoids in Regulating Glucocorticoid Effects on Memory for Emotionally Arousing Experiences

Piray Atsak, Benno Roozendaal and Patrizia Campolongo

Abstract There is extensive evidence that glucocorticoid hormones, normally released from the adrenal cortex during stressful events, enhance the consolidation of long-term memory of emotionally arousing training experiences, yet impair the retrieval of previously acquired information during emotionally arousing test situations. In contrast, glucocorticoids have little effect on the consolidation or retrieval of memory of low-arousing or neutral information. Although it is now well established that glucocorticoid effects on these two memory functions depend on rapid interactions with arousal-induced noradrenergic activity within the basolateral amygdala and several other brain regions, the exact neurobiological mechanism underlying this presumably nongenomically mediated glucocorticoid action remained to be elucidated. In this chapter, we present compelling evidence indicating that the endocannabinoid system, a rapid lipid signaling system in the brain, plays an essential role in regulating glucocorticoid effects on different memory processes via actions through a membrane-associated glucocorticoid receptor.

Abbreviations

AEA	Anandamide
AMPA	α-amino-3-hydroxy-5-methyl-4-isoxazolepropionic acid receptor
BLA	Basolateral complex of the amygdala
CB1	Cannabinoid receptor type 1
CB2	Cannabinoid receptor type 2
cort-BSA	Corticosterone conjugated to a bovine serum albumin molecule
CREB	cAMP response-element binding protein

P. Atsak (✉) · B. Roozendaal
Department of Cognitive Neuroscience, Radboud University Medical Centre, Nijmegen, The Netherlands
e-mail: pirayatsak@radboudumc.nl

Donders Institute for Brain, Cognition and Behaviour, Radboud University Nijmegen, Nijmegen, The Netherlands

P. Campolongo
Department of Physiology and Pharmacology, Sapienza University of Rome, P.le A. Moro 5, Rome, Italy

M. Popoli et al. (eds.), *Synaptic Stress and Pathogenesis of Neuropsychiatric Disorders*,
DOI 10.1007/978-1-4939-1056-4_5,

pCREB	Phosphorylated CREB
HDAC	Histone deacetylase
HPA-axis	Hypothalamus-pituitary-adrenocortical-axis
GABA	Gamma Amino Butyric Acid
GR	Glucocorticoid receptor
PKA	cAMP-dependent protein kinase
PTSD	Posttraumatic stress disorder
THC	Tetrahydrocannabinol
2-AG	2-arachidonoylglycerol

5.1 Introduction

Stress is defined as any stimulus that represents a perceived or actual threat to the psychological and physiological equilibrium of an organism (Selye 1976). As a response to stress, the organism strives to reinstate homeostasis by activating several autonomic and humoral stress-response systems. Typically, stress leads to an activation of the sympathetic nervous system and HPA-axis, culminating in the release of catecholamines and glucocorticoids from the adrenal medulla and cortex, respectively (McCarty and Gold 1981; de Boer et al. 1990; Roozendaal et al. 1996b). These hormones promote the organism's ability to cope with stress by acting on target systems in the periphery but also inducing a myriad of effects on the brain. In addition to preparing an individual for the acute consequences of dangerous or threatening situations and the return to homeostasis, an important function of the stress response is to induce long-term adaptive changes (McEwen 1998, 2001). For instance, stressful or emotionally arousing life events typically leave lasting and vivid memories. Extensive evidence indicates that glucocorticoid hormones, in concert with several other stress-activated systems, mediate the selective better storage of memory of emotionally significant experiences (Oitzl and de Kloet 1992; Sandi and Rose 1994; de Kloet et al. 1999; Roozendaal 2000; Joëls and Baram 2009; Roozendaal et al. 2009a). While this is considered to be a highly adaptive survival mechanism that enables the organism to retain lasting memories of biologically significant life events, intense emotional experiences such as automobile accidents, fires, muggings, rapes, wartime battles, or terrorists' bombings can also create maladaptive traumatic memories and result in the development of psychiatric disorders such as (PTSD).

It is now well established that stress and glucocorticoid hormones do not only influence the formation and long-term storage of new memories, but also affect the remembrance of previously acquired information. In contrast to the enhancing effects of glucocorticoids on memory consolidation, these hormones typically impair the retrieval of memory processing (de Quervain et al. 1998; Het et al. 2005). However, glucocorticoids do not modulate memory of all experiences alike; rather, they appear to preferentially influence the consolidation and retrieval of memory of emotionally arousing experiences. Extensive evidence from our as well as other laboratories indicates that this selectivity derives from a critical dependence of glucocorticoid actions on concomitant arousal-induced activation of noradrenergic transmission within the (BLA) as well as several other brain regions (Roozendaal et al. 2006a, 2009a). Despite the different time courses of these hormones, i.e., norepinephrine is rapidly released within the brain followed several minutes later by a rise of glucocorticoid levels in general circulation, there appears to be an overlapping presence of norepinephrine and glucocorticoids in time and space that allows the stage for interactions (Joëls et al. 2011). Importantly, recent evidence indicates that such interplay between glucocorticoids and the noradrenergic system is not mediated via the classical genomic action of glucocorticoids but likely to involve fast actions through an activation of membrane-associated steroid receptors.

The scope of this chapter is to summarize recent findings on some novel mechanisms underlying the acute effects of glucocorticoid hormones on memory. We will first summarize the opposing effects of glucocorticoids on memory consolidation and memory retrieval. Then, we will describe how glucocorticoids interact with noradrenergic activity within the BLA to selectively modulate memory of emotionally arousing experiences. Finally, we will present evidence indicating a critical involvement of the endocannabinoid system, a fast-acting lipid system in the brain, in mediating such rapid effects of glucocorticoids onto the noradrenergic system in influencing both the consolidation and retrieval of memory of emotionally significant experiences.

5.2 Acute Glucocorticoid Effects on Memory Consolidation and Retrieval

Over the last decades, considerable evidence has accumulated indicating that glucocorticoids (cortisol in humans, corticosterone in rodents) are crucially involved in modulating memory processes. Early reports on both enhancing and impairing properties of glucocorticoids on memory have revealed that these hormones have complex effects on cognitive functions (Bohus and Lissak 1968; Flood et al. 1978; Beckwith et al. 1986; Luine et al. 1993; Arbel et al. 1994). However, more recent studies investigating glucocorticoid effects on distinct memory phases allowed for a disentangling of the multifaceted actions of these stress hormones. Glucocorticoids are now known to enhance the consolidation of memory of emotionally arousing experiences, but to impair memory retrieval and working memory during emotionally arousing test situations (de Quervain et al. 1998; Roozendaal 2000; Roozendaal et al. 2004b; de Quervain et al. 2009).

There is extensive evidence from animal studies that glucocorticoids are critically involved in regulating the consolidation of memory processing (Flood et al. 1978; de Kloet 2000; Roozendaal 2000; McGaugh and Roozendaal 2002). Acute administration of corticosterone or a specific GR agonist typically enhances long-term memory consolidation when given either before or shortly after a training experience (Flood et al. 1978; Sandi and Rose 1994; Pugh et al. 1997; Roozendaal et al. 1999a; Cordero et al. 2002). In contrast, a blockade of glucocorticoid production with the synthesis inhibitor metyrapone impairs memory consolidation (Roozendaal et al. 1996a; Maheu et al. 2004) and prevents stress-induced memory enhancement (Roozendaal et al. 1996b; Liu et al. 1999). Such glucocorticoid effects on memory consolidation follow an inverted U-shape dose–response relationship. Moderate doses enhance memory, whereas higher doses are typically less effective or may even impair memory consolidation (Roozendaal et al. 1999b). In rodents, enhancing effects of glucocorticoids on memory consolidation have been observed in many different kinds of learning tasks, including inhibitory avoidance, contextual and cued fear conditioning, water-maze spatial and cued training, object recognition, and conditioned taste aversion (Roozendaal et al. 2006a). These findings indicate that, in animals, glucocorticoids not only enhance memory of training on hippocampus-dependent tasks that have a strong spatial/contextual component, but also memory of recognition- and procedural training that are known to depend on other brain systems. In humans, glucocorticoid effects on consolidation have mostly been investigated with respect to declarative memory(Het et al. 2005; Wolf 2008).

Recent findings indicate that glucocorticoids enhance memory consolidation of emotionally arousing training experiences but do not affect the consolidation of emotionally neutral information. Learning tasks in animal experiments are usually emotionally arousing because of the motivational stimulation necessary to elicit changes in behavior. We investigated the importance of emotional arousal in mediating glucocorticoid effects on memory consolidation in rats trained on an object recognition task (Okuda et al. 2004). Although no rewarding or aversive stimulation is used during this learning paradigm, training on this task induces modest novelty-induced stress or arousal (de Boer et al. 1990). However, extensive habituation of rats to the experimental context (in the absence of any objects) reduces the arousal component of the task during the training. We found that corticosterone, administered systemically immediately after training, enhanced 24-h retention of rats that were not previously habituated to the experimental context. In contrast, posttraining corticosterone administration did not affect 24-h retention of rats that had received extensive prior habituation to the experimental context and, thus, had decreased novelty-induced emotional arousal during training (Okuda et al. 2004). Human studies support the hypothesis that learning-associated arousal is a prerequisite for the enhancing effects of glucocorticoids on memory consolidation (Abercrombie et al. 2006; Wolf 2008; de Quervain et al. 2009; van Stegeren et al. 2010).

In contrast to the enhancing effects of glucocorticoids on memory consolidation, these hormones typically impair memory retrieval. In the first study investigating the effects of stress and glucocorticoids on retrieval processes, de Quervain et al. (1998) reported that 30 min after exposure to footshock stress, rats had impaired retrieval of spatial memory on a water-maze task they had acquired 24 h earlier. Interestingly, memory performance was not impaired when rats were tested either 2 min or 4 h after the stress exposure. These time-dependent effects of stress exposure on retrieval processes corresponded to the circulating corticosterone levels at the time of retention testing, which suggested that the retrieval impairment might be directly related to stress-induced increases in adrenocortical function. In support of this idea, we found that suppression of corticosterone synthesis with metyrapone blocked the stress-induced impairment in memory retrieval. Moreover, systemic corticosterone administered to nonstressed rats 30 min before retention testing induced dose-dependent retrieval impairment (de Quervain et al. 1998). In the next step, we translated these findings to healthy humans and found that a single administration of cortisone shortly before retention testing impaired free recall of words learned 24 h earlier (de Quervain et al. 2000). Several further studies from different laboratories have indicated that stress exposure, glucocorticoids or selective GR agonists (such as dexamethasone and RU 28362) impair the retrieval of hippocampus-dependent spatial or contextual memory in rats and declarative (mostly episodic) memory in humans (Wolf et al. 2001; Roozendaal et al. 2003; Buss et al. 2004; Rashidy-Pour et al. 2004; Roozendaal et al. 2004b; Het et al. 2005; Kuhlmann et al. 2005a; Sajadi et al. 2007; Coluccia et al. 2008; Wolf 2008), yet few studies revealed that the impairing effects of stress and glucocorticoids extend to hippocampus-independent memory tasks (Guenzel et al. 2013). Highly comparable to the previously described effects of glucocorticoids on memory consolidation, these hormones selectively impair the retrieval of memory of emotionally arousing information or during emotionally arousing test situations (Kuhlmann et al. 2005a; Kuhlmann et al. 2005b; de Quervain et al. 2007; Smeets et al. 2008).

5.3 Glucocorticoids Interact with Noradrenergic Mechanisms Within the Basolateral Amygdala

As summarized up to this point, glucocorticoids selectively modulate the consolidation and retrieval of memory of emotionally arousing, but not of emotionally neutral, information. An apparent question is what neurobiological mechanism might underlie this selectivity? Our findings indicate that interactions between glucocorticoids and arousal-induced noradrenergic activity within the BLA may be key in determining this selectivity. It is well established that emotionally arousing training experiences that induce the release of adrenal stress hormones also increase BLA neuronal activity (Pelletier et al. 2005). Norepinephrine is also released into the amygdala during emotionally arousing training (Galvez et al. 1996; Quirarte et al. 1998; McIntyre et al. 2002), whereas posttraining infusion of norepinephrine or a β-adrenoceptor agonist into the BLA enhances memory of training on several learning tasks (Ferry and McGaugh 1999; Hatfield et al. 1999; LaLumiere et al. 2003; Roozendaal et al. 2008). Considerable evidence indicates that glucocorticoids interact with this training-associated noradrenergic activation within the amygdala in enhancing the consolidation of memory of emotionally arousing training experiences (Roozendaal et al. 2009a). For example, as shown in Fig. 5.1, an in vivo microdialysis study reported that the administration of a memory-enhancing dose of corticosterone after inhibitory avoidance training rapidly augmented norepinephrine levels within the amygdala (McReynolds et al. 2010). In contrast,

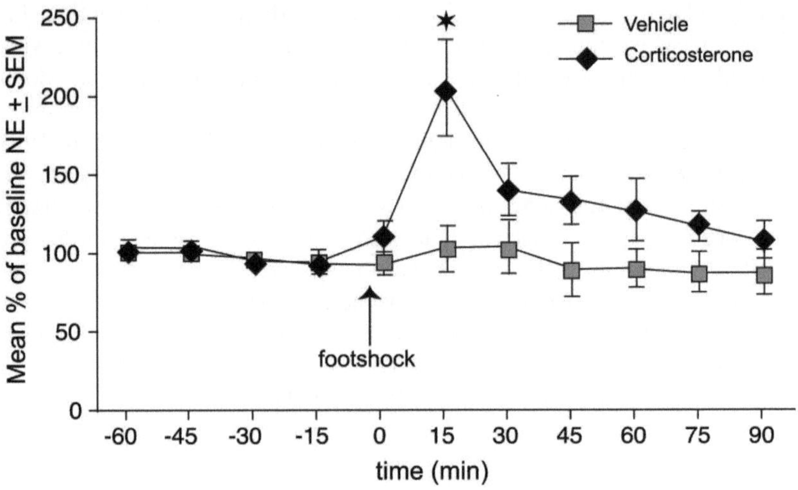

Fig. 5.1 Effect of immediate posttraining corticosterone treatment on norepinephrine (NE) levels in the basolateral complex of the amygdala (BLA). Microdialysis samples were collected every 15 min. Norepinephrine levels (mean ± SEM) are expressed as a percentage change from average baseline levels. Corticosterone treatment (3 mg/kg, i.p.) significantly increased norepinephrine release in the amygdala of animals trained on an inhibitory avoidance task compared with vehicle-injected animals. *$p < 0.05$ versus vehicle. (Adapted from McReynolds et al. 2010, with permission)

the same dose of corticosterone, administered to nontrained control rats did not modify amygdala norepinephrine levels. Moreover, attenuation of noradrenergic signaling with the β-adrenoceptor antagonists propranolol or atenolol infused into the BLA, but not into the neighboring central amygdala, blocked the memory enhancement induced by a glucocorticoid administered either systemically or directly into the BLA (Quirarte et al. 1997; Roozendaal et al. 2002). In subsequent studies we showed that glucocorticoids enhance memory consolidation, in a permissive fashion, by potentiating β-adrenoceptor-PKA efficacy and downstream phosphorylation of CREB protein (Roozendaal 2002; Roozendaal et al. 2002; Roozendaal et al. 2006a; Roozendaal et al. 2010). Importantly, a β-adrenoceptor antagonist infused into the BLA also prevented memory consolidation enhancement induced by a glucocorticoid administered into other brain regions, including the hippocampus (Roozendaal et al. 1999a), supporting the general hypothesis that norepinephrine-induced BLA activity is required for regulating neural plasticity and information storage processes in its many efferent brain regions (McGaugh 2004).

Based on the evidence summarized above, it may be hypothesized that an arousal-induced increase in noradrenergic activity within the BLA is essential in enabling glucocorticoid effects on memory consolidation. Such a mechanism may then provide a direct explanation for the finding that glucocorticoids selectively enhance memory consolidation of emotionally arousing experiences. We investigated this issue in rats trained on an object recognition task. As already mentioned, corticosterone enhances memory of object recognition training when administered to

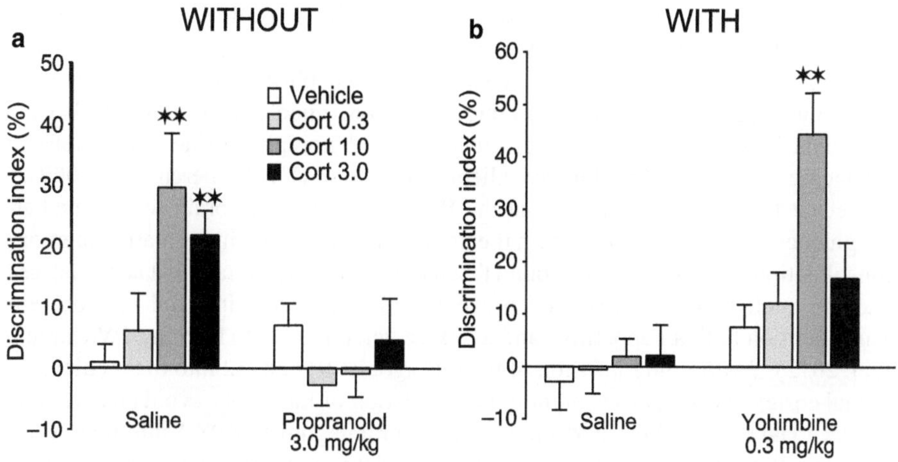

Fig. 5.2 Glucocorticoid effects on memory consolidation for object recognition training require arousal-induced noradrenergic activation. Rats were either habituated to the training context for 7 days (*WITH*) or not habituated (*WITHOUT*). On day 8, they were given a 3-min training trial during which they could freely explore two identical objects, training was followed by systemic drug administration. Retention was tested 24 h later by placing the rats back into the apparatus for 3 min; in this trial, one object was similar to the training objects whereas the other was novel. Data represent discrimination index (%) on a 24-h retention trial, expressed as mean ± SEM. The discrimination index was calculated as the difference in the time spent exploring the novel and the familiar object, expressed as the ratio of the total time spent exploring both objects. **a** Effects of immediate posttraining administration of the β-adrenoceptor antagonist propranolol (3 mg/kg, s.c.) on corticosterone (0.3, 1.0, 3.0 mg/kg, s.c)-induced enhancement of object recognition memory in naïve (emotionally aroused) rats. **b** Effect of co-administration of the α_2-adrenoceptor antagonist yohimbine (0.3 mg/kg, s.c.) with corticosterone on object recognition memory in habituated (emotionally nonaroused) rats. $^{**}p < 0.0001$ versus vehicle. (Adapted from Roozendaal et al. 2006b)

naïve rats, but is ineffective when training-associated arousal levels are reduced by extensive prior habituation (Okuda et al. 2004). In a follow-up study we found that, in nonhabituated (i.e., emotionally aroused) rats, the β-adrenoceptor antagonist propranolol administered systemically after training blocked the corticosterone-induced memory enhancement (Roozendaal et al. 2006b). Propranolol infused directly into the BLA also blocked the enhancing effects of corticosterone on object recognition memory. To determine whether the failure of corticosterone to enhance memory consolidation under low-arousing conditions was due to insufficient training-induced noradrenergic activation, low doses of the α_2-adrenoceptor antagonist yohimbine, which increases norepinephrine levels in the brain, were co-administered with the corticosterone to well-habituated rats immediately after object recognition training. As shown in Fig. 5.2, the critical finding of this latter experiment is that such an augmented noradrenergic tone was sufficient to mimic the effects of emotional arousal in that simultaneously administered corticosterone now enhanced memory consolidation (Roozendaal et al. 2006b). Further, in habituated rats, corticosterone increased the activity of BLA neurons, as assessed by pCREB immunoreactivity levels, only in animals also given yohimbine. Such observations strongly suggest that because glucocorticoid effects on memory consolidation require noradrenergic activation within the BLA, they only modulate memory under emotionally arousing conditions that induce the release of norepinephrine. Interestingly, a recent functional magnetic resonance imaging study confirmed that in humans also the amygdala is an important locus of glucocorticoid–norepinephrine interactions in enhancing memory of emotionally salient information (Van Stegeren et al. 2007).

Recent findings have shown that the BLA is not the only brain region mediating glucocorticoid interactions with the noradrenergic system in regulating memory consolidation. For example, we found that a β-adrenoceptor antagonist administered into the nucleus accumbens shell prevented glucocorticoid-induced memory enhancement on both an appetitive and aversive version of taste learning (Wichmann et al. 2012). Posttraining infusion of the GR agonist RU 28362 into the medial prefrontal cortex also enhances memory consolidation of inhibitory avoidance training (Roozendaal et al. 2009b), and a β-adrenoceptor antagonist or PKA inhibitor co-infused into the medial prefrontal cortex prevented this memory enhancement (Barsegyan et al. 2010). Moreover, corticosterone administered systemically immediately after inhibitory avoidance training increased PKA activity in the medial prefrontal cortex within 30 min. These findings suggested that glucocorticoid effects on noradrenergic signaling might have an onset that is too fast to be mediated via transcriptional regulation in the nucleus and likely involve a rapid, nongenomic mode of action. In support of the view that these glucocorticoid effects might require a GR that is located in or near the cell membrane, we found that posttraining infusion of corticosterone conjugated to a bovine serum albumin molecule (i.e., cort:BSA), a ligand that selectively activates adrenal steroid receptors on the cell surface, into the insular cortex enhanced memory consolidation, and that this enhancing effect was blocked by co-administration of a GR, but not mineralocorticoid receptor, antagonist (Roozendaal et al. 2010). In an entirely new line of research, we found that glucocorticoid effects on norepinephrine signaling and downstream pCREB activation in the insular cortex might enhance memory consolidation via chromatin modification (Roozendaal et al. 2010). Systemic corticosterone increased histone acetylation, a form of chromatin modification, in the insular cortex as assessed 1 h after training on an object recognition task. Furthermore, infusion of the HDAC inhibitor sodium butyrate administered into the insular cortex enhanced memory consolidation of this training. Inducing a histone hyperacetylated state via HDAC inhibition appears to facilitate transcription by relaxing chromatin structure, resulting in enhanced synaptic plasticity, and long-term memory processes (Barrett and Wood 2008). However, the effect of the HDAC inhibitor on memory enhancement was completely abolished by blocking GR activity. Additionally, a PKA inhibitor also blocked the ability of HDAC inhibition to enhance memory in the insular cortex. Thus, these findings indicate that inducing a histone hyperacetylated state via HDAC inhibition is not sufficient to enhance long-term memory. It is still necessary to have upstream signaling via GR and PKA activity. Presumably, these signaling events are triggering steps necessary to activate transcription factors and co-activators such as CREB and CREB binding protein.

Glucocorticoid effects on memory retrieval are highly comparable to the effects on memory consolidation in that emotionally arousing information or an emotionally arousing test situation, both inducing the release of norepinephrine, is required for enabling glucocorticoid effects on memory retrieval (Smeets et al. 2008; Wolf 2008; de Quervain et al. 2009; Roozendaal et al. 2009a). Systemic administration of the β-adrenoceptor antagonist propranolol blocked the memory retrieval impairment of spatial/contextual information induced by a concurrent injection of corticosterone (Roozendaal et al. 2004a). Extensive evidence from studies in amnesic patients, human imaging studies, and lesion studies in animals indicates that the medial temporal lobe (hippocampus and parahippocampal gyrus) is crucially involved in the retrieval of spatial and contextual memory in animals and declarative memory in humans (Squire 1992; Moser and Moser 1998; Cabeza and Nyberg 2000). We found that local infusions of a GR agonist into the hippocampus of rats induce retrieval impairment on a water-maze spatial task comparable to that seen after systemic administration (Roozendaal et al. 2003) and that a β-adrenoceptor antagonist co-infused into the hippocampus prevented the retrieval-impairing effect of the GR agonist (Roozendaal et al. 2004b). As stimulation of β_1-adrenoceptors with systemic injections of the selective agonist xamoterol induces memory retrieval impairment comparable to that seen after corticosterone administration (Roozendaal et al. 2004b), the findings suggest that glucocorticoid effects on memory retrieval impairment involve a facilitation of noradrenergic mechanisms. Further studies in animals have indicated that the BLA interacts with the hippocampus in mediating glucocorticoid effects on memory retrieval of emotionally arousing information (Roozendaal et al. 2003, 2004b). We found that the administration of a β-adrenoceptor antagonist into the BLA blocks the impairing effect of a GR agonist infused into the hippocampus on retrieval of spatial memory (Roozendaal et al. 2004b). Findings of animal studies addressing the importance of interactions between the amygdala and the hippocampus during retrieval of emotionally arousing information are corroborated by human imaging studies indicating that the degree of interaction between these two brain regions is greater during the retrieval of emotionally arousing declarative information as compared to neutral information (Dolcos et al. 2005; Smith and Vale 2006).

Collectively, these findings indicate that glucocorticoids interact with the noradrenergic system in strengthening the consolidation of long-term memory of emotionally significant events, while at the same time inducing temporary impairment of the recall of previously acquired information. Figure 5.3 summarizes these findings. Given that the onset of such glucocorticoid interactions with the noradrenergic system is rapid and likely involves binding to a membrane-associated receptor for corticosterone, it is highly plausible that these glucocorticoid effects are mediated through a nongenomic mode of action. Therefore, in the next section we will first briefly discuss some general mechanisms that have been described in the literature that might regulate such rapid, nongenomic effects of glucocorticoids on physiology and behavior, followed by a more extensive discussion of the possible involvement of the endocannabinoid system in mediating such rapid glucocorticoid effects.

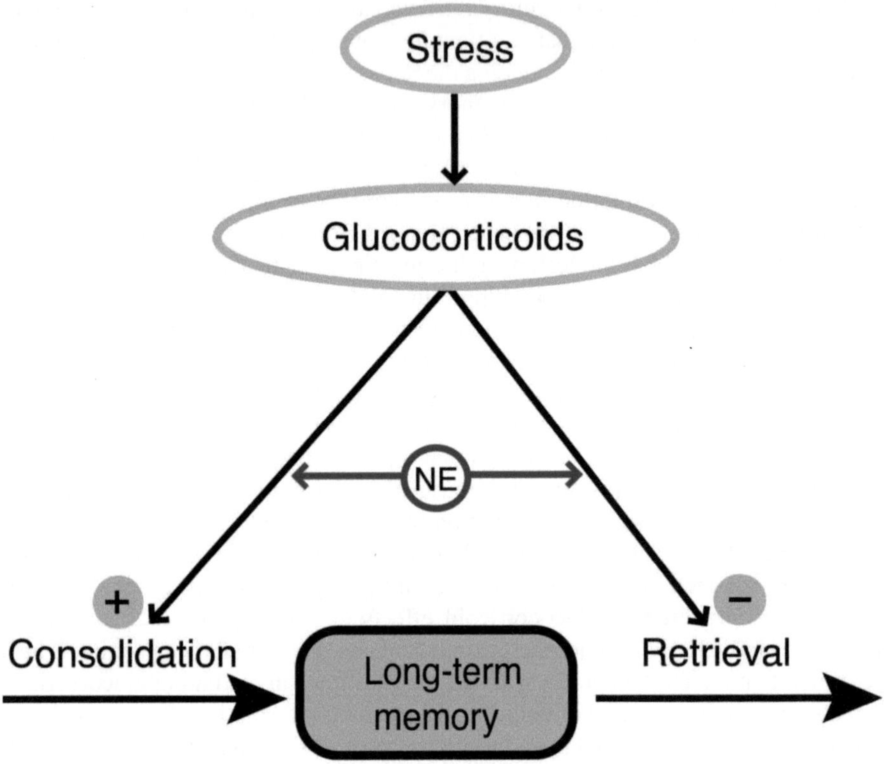

Fig. 5.3 Effects of stress and glucocorticoids on memory functions. Glucocorticoids enhance memory consolidation, whereas they impair memory retrieval. Both of these glucocorticoid effects depend on emotional arousal-induced noradrenergic activity. *NE* norepinephrine. (Adapted from de Quervain et al. 2009, with permission)

5.4 Nongenomic Glucocorticoid Actions

Glucocorticoids are known to modulate cellular function, including learning and memory, through both genomic (slow) and nongenomic (rapid) pathways (de Kloet 2000; Dallman 2005; Popoli et al. 2011). Genomic glucocorticoid effects are mediated by classical steroid mechanisms involving transcriptional regulation. Glucocorticoids can influence transcription through both DNA-binding–dependent and DNA-binding–independent mechanisms (de Kloet 2000). Although many glucocorticoid actions suit the time frame for a genomic mechanism, some behavioral and physiological effects of glucocorticoids, for example, the previously described effects on the noradrenergic system, have a rapid onset, occurring in seconds to minutes, that is not readily compatible with transcriptional regulation. Rapid glucocorticoid actions have been reported in different limbic and brainstem structures, where they control functions ranging from learning and memory to neuroendocrine functions (Dallman 2005; Tasker et al. 2006; Haller et al. 2008; Riedemann et al.

2010). It is important to note that glucocorticoid effects on the consolidation of long-term memory might depend on an interplay between genomic and nongenomic actions (Falkenstein et al. 2000), whereas glucocorticoids' ability to temporarily impair memory retrieval might depend solely on nongenomic glucocorticoid actions. In support of this view, it has been reported that protein synthesis inhibitors fail to prevent glucocorticoid effects on memory retrieval (Sajadi et al. 2006).

Nongenomic glucocorticoid actions likely involve the activation of a membrane-associated variant(s) of the steroid receptor (Losel et al. 2003; Dallman 2005; Tasker et al. 2006; Riedemann et al. 2010). Orchinik and colleagues (Orchinik et al. 1991; Rose et al. 1993) were the first to provide evidence that glucocorticoids exert behavioral effects through the activation of a corticosteroid receptor on the neuronal membrane. In this series of experiments, glucocorticoids rapidly suppressed mating behavior in the amphibian Taricha granulosa (rough-skinned newt) by binding to a receptor on neuronal membranes. As mentioned, recent findings indicate that the administration of the membrane-impermeable glucocorticoid ligand cort:BSA into a variety of brain regions of the rat is sufficient to enhance the consolidation of long-term memory of emotionally arousing training experiences (Roozendaal et al. 2010; Lee et al. 2011). As these cort:BSA effects are blocked by co-administration of a GR antagonist (Barsegyan et al. 2010; Roozendaal et al. 2010), these findings suggest a role for a membrane-associated GR in mediating rapid glucocorticoid effects on memory. Studies employing GR immunoreactivity, at both the light and the electron microscopic level, provided anatomical evidence for the existence of membrane-associated GRs in neurons of the hippocampus, hypothalamus (Liposits and Bohn 1993), and postsynaptic membranes of lateral amygdala neurons (Johnson et al. 2005).

Current evidence indicates a variety of nongenomic glucocorticoid actions on neuroplasticity and memory, ranging from a rapid increase in glutamate-release probability from presynaptic sites (Karst et al. 2005) to a rapid insertion of AMPA receptor subunits in postsynaptic membranes (Groc et al. 2008; Pasricha et al. 2011). Recently, the endocannabinoid system emerged as an important mediator of some of the rapid effects of glucocorticoids. The first evidence derived from in vitro studies indicating an involvement of endocannabinoids in mediating glucocorticoid-induced rapid inhibition of the HPA-axis within the hypothalamus (Di et al. 2003; Di et al. 2005a; Evanson et al. 2010; Hill and Tasker 2012). Consistently, later studies pointed out that both stress and glucocorticoids significantly alter endocannabinoid content in limbic brain regions that can function to both mount and terminate the stress response (Hill and McEwen 2010). Although the interest in endocannabinoid signaling as a candidate for mediating fast glucocorticoid effects has been quickly growing, it is noteworthy to also mention the existence of other candidate systems that might regulate rapid glucocorticoid actions. For instance, an activation of membrane GRs evokes the release of nitric oxide from pyramidal cells in the hippocampus (Hu et al. 2010) that acts as a retrograde messenger and induces the release of GABA from hippocampal interneurons and hypothalamic magnocellular neurons (Di et al. 2009; Hu et al. 2010). Glucocortiocoids also enhance glutamate transmission in hippocampal CA1 pyramidal neurons in the rat by a mineralocorticoid receptor-dependent mechanism. Although the mechanism underlying this fast mineralocorticoid receptor-mediated effect on glutamatergic transmission is not known, it has been shown not to rely on endocannabinoid signaling (Karst et al. 2005; Olijslagers et al. 2008).

5.5 Role of the Endocannabinoid System in Mediating Glucocorticoid Effects on Memory Consolidation and Retrieval

In the previous sections we have shown that glucocorticoids, because of critical interactions with arousal-activated noradrenergic mechanisms, selectively influence the consolidation and retrieval of emotionally arousing learning experiences or under emotionally arousing test situations. However, the onset of these glucocorticoid effects on the noradrenergic system is, at least in part, not readily compatible with its classical action of inducing transcriptional regulation in the nucleus. We have subsequently described several novel mechanisms by which glucocorticoids might be able to induce rapid and nongenomically mediated effects on physiology and behavior. In this section, we will first introduce the endocannabinoid system and give a brief overview of its general role in neuronal plasticity and learning and memory, and then we focus on recent findings indicating that the endocannabinoid system might be essentially involved in mediating the rapid effects of glucocorticoids onto the noradrenergic system in regulating both the consolidation and retrieval of memory.

5.5.1 The Endocannabinoid System in the Brain

The endocannabinoid system, a fast lipid system in the brain, recently emerged as an important stress-response system (Hill and Tasker 2012). It is composed of two G protein-coupled receptors, the CB1 and the CB2, and two endogenous cannabinoid ligands such as *N*-arachidonylethanolamine (AEA) and (2-AG). Endocannabinoids are produced upon activation by both neurons and glia cells and operate primarily as interneuronal signaling molecules (Freund et al. 2003; Kano et al. 2009). Cannabinoid receptors are also activated by external ligands such as plant-derived cannabinoids (e.g., THC, produced by the cannabis plant) and synthetic cannabinoids (e.g., WIN55,212-2). CB1 receptors are expressed almost ubiquitously throughout the brain (Katona et al. 1999, 2001), whereas CB2 receptors are mostly present in peripheral immunological tissues, but they have also been found within the central nervous system (Onaivi et al. 2006) Postsynaptic depolarization induces an elevation of intracellular Ca^{2+} concentrations that triggers the release of endocannabinoids into the synapse. Once released, endocannabinoids contribute to several forms of short-term and long-term synaptic plasticity by acting as a retrograde messenger and binding to CB1 receptors at the presynaptic membrane, eventually suppressing neurotransmitter release either transiently or persistently (Hashimotodani et al. 2007; Kano et al. 2009). A vast number of studies demonstrated that CB1 receptor activation influences the release of various neurotransmitters, including glutamate, GABA, glycine, acetylcholine, norepinephrine, dopamine, serotonin, and cholecystokinin (Kano et al. 2009).

5.5.2 Cannabinoid Effects on Learning and Memory

The cannabinoid system emerged as an important modulator of different learning and memory processes (Wotjak 2005; Kano et al. 2009; Marsicano and Lafenetre 2009; Akirav 2011). Early studies, examining the effects of pretraining administration of cannabinoid agonists, in particular THC or WIN55212-2, reported impairing effects on the acquisition of water maze, contextual fear memory, and object recognition training in rodents (Lichtman et al. 1995; Da and Takahashi 2002; Pamplona and Takahashi 2006). Moreover, concurrent administration of the CB1 receptor antagonist/inverse agonist SR141716 (rimonabant) blocked these impairments (Lichtman et al. 1995; Da and Takahashi 2002; Pamplona and Takahashi 2006). More recent studies employing targeted pharmacological manipulations of the cannabinoid system by local infusions into the brain have illustrated more consistent results with regard to their wide-ranging effects on different memory phases. Pretraining administration of a CB1 receptor agonist into the hippocampus has consistently been shown to impair spatial learning (Lichtman et al. 1995; Egashira et al. 2002; Wegener et al. 2008; Abush and Akirav 2010). However, drug treatment given before a learning experience could affect performance by influencing nonspecific attentional, locomotor, and motivational processes during acquisition. To address whether cannabinoid drugs directly modulate the consolidation of memory, we investigated the effect of the CB receptor agonist WIN55,212-2 on long-term retention when infused into the BLA immediately after training on an inhibitory avoidance task. As shown in Fig. 5.4a and b, we found that WIN55,212-2 dose-dependently enhanced 48-h retention of this training, whereas the CB1 receptor antagonist AM251 administered posttraining into the BLA impaired memory consolidation (Campolongo et al. 2009b). Consistent with these findings, others have reported that infusion of the CB1 receptor antagonist AM251 into the amygdala (Bucherelli et al. 2006) or hippocampus (de Oliveira Alvares et al. 2005) disrupts the consolidation of long-term memory, possibly by inhibiting long-term potentiation (de Oliveira Alvares et al. 2006). More recently, similar to the effects of glucocorticoids on memory consolidation, we found that endocannabinoid effects on the consolidation of long-term memory of inhibitory avoidance training follow an inverted-U shaped dose–response relationship. Moderate doses enhanced memory whereas both lower and higher doses were less effective (P. Atsak et al. unpublished observation).

Recent studies indicated that baseline arousal levels can influence the sensitivity to cannabinoid drugs in influencing memory processes. For instance, it has been reported that cannabinoid receptor activation differently influences neural processes

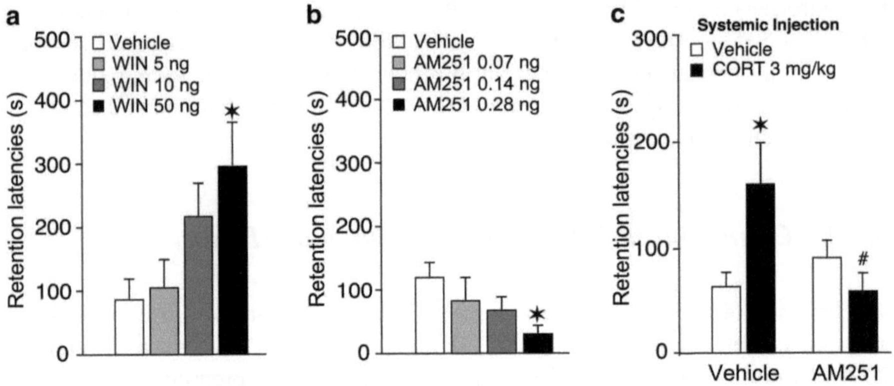

Fig. 5.4 Endocannabinoids in the basolateral complex of the amygdala (BLA) enhance memory consolidation and enable glucocorticoid modulation of memory. **a** Immediately posttraining bilateral intra-BLA infusions of the CB1 receptor agonist WIN55,212-2 (5, 10, 50 ng in 0.2 μL) enhance 48-h inhibitory avoidance retention. **b** Immediate posttraining intra-BLA infusions of the CB1 receptor antagonist AM251 (0.07, 0.14, 0.28 ng in 0.2 μL) impair inhibitory avoidance retention. **c** Immediate posttraining bilateral infusions of AM251 (0.14 ng in 0.2 μL) into the BLA block retention enhancement induced by subcutaneous injections of corticosterone (3 mg/kg, s.c.). Data represent step-through latencies (mean + SEM) in seconds on the 48-h inhibitory avoidance retention test. *$p<0.05$ versus vehicle; #$p<0.05$ versus corticosterone group. (Adapted from Campolongo et al. 2009b)

underlying the formation of emotional memory as compared to nonemotional memory (Chhatwal and Ressler 2007; Akirav 2011). We further demonstrated that the endocannabinoid-uptake inhibitor AM404, which enhances endocannabinoid tone, induces different effects on recognition memory performance in rats subjected to different levels of emotional arousal induced by the changes in environmental condition (Campolongo et al. 2012). In agreement with these findings, a recent experiment in humans reported that cannabinoid drugs such as THC also preferentially modulate memory for emotionally arousing, and not mundane, experiences (Ballard et al. 2012). Recently, we investigated cannabinoid effects on both short- and long-term memory of object recognition training under two conditions that differed in their training-associated level of emotional arousal (Campolongo et al. 2013). As shown in Fig. 5.5a, WIN55,212-2 administered immediately after object recognition training to rats that were not previously habituated to the experimental context induced impairment of short-term retention performance. In contrast, the same dose of WIN55,212-2 enhanced short-term memory of rats that had received extensive prior habituation to the experimental context (Campolongo et al. 2013). The effects of posttraining WIN55,212-2 administration on long-term memory of the object recognition training were different. WIN55,212-2 enhanced long-term retention of object recognition memory in nonhabituated rats, but had no effect on long-term memory of extensively habituated rats (Fig. 5.5c and d). This arousal-dependent cannabinoid effect on memory is thus highly comparable to the glucocorticoid effects described earlier and lend support for the idea that the origin of

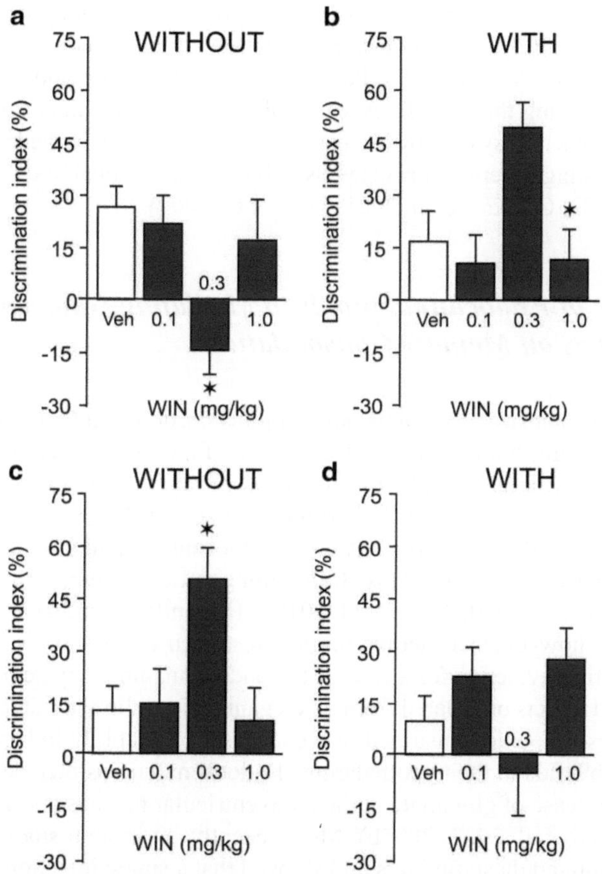

Fig. 5.5 Effects of the CB receptor agonist WIN55,212-2 (*WIN*) on both short- and long-term retention of object recognition are influenced by training-associated emotional arousal. For both experiments, rats were either habituated to the training context for 7 days (*WITH*) or not habituated (*WITHOUT*). On day 8, they were given a 3-min training trial during which they could freely explore two identical objects, training was followed by a systemic administration of WIN 0.1, 0.3, 1.0 i.p. Retention was tested either 1 or 24 h later by placing the rats back into the apparatus for 3 min; in this trial, one object was similar to the training objects whereas the other was novel. Data represent discrimination index (%) on the retention trial, expressed as mean ± SEM. The discrimination index was calculated as the difference in the time spent exploring the novel and the familiar object, expressed as the ratio of the total time spent exploring both objects. Posttraining WIN dose-dependently impaired 1-h object recognition performance of nonhabituated rats **a**, but enhanced object recognition performance of extensively habituated rats **b**. In contrast, posttraining administration of WIN, in a dose that impaired 1-h performance, enhanced 24-h object recognition performance of nonhabituated rats **c**, but not of well-habituated rats **d**. *$p < 0.05$ versus vehicle. (Adapted from Campolongo et al. 2013, with permission)

the altered sensitivity to cannabinoids results from a differential activation of the noradrenergic system during arousing versus low-arousing conditions (Patel and Hillard 2003; Oropeza et al. 2005; Page et al. 2007; Carvalho and Van Bockstaele 2012). Corroborating these findings, cannabinoid drugs have been shown to influence the noradrenergic system by increasing neuronal activity in the locus coeruleus or directly boosting norepinephrine levels in limbic and cortical brain regions (Patel and Hillard 2003; Oropeza et al. 2005; Page et al. 2007).

5.5.3 Role of Endocannabinoids in Mediating Glucocorticoid Effects on Memory Consolidation

Recent evidence consistently points out that glucocorticoids interact with the endocannabinoid system in influencing different brain functions, including learning and memory (Atsak et al. 2012b; Crosby and Bains 2012; Hill and Tasker 2012; Ramot and Akirav 2012; Riebe et al. 2012; de Bitencourt et al. 2013). Some of these studies clearly demonstrated an involvement of the endocannabinoid system in mediating the rapid effects of glucocorticoids (Campolongo et al. 2009b; Hill and McEwen 2009; Evanson et al. 2010; Atsak et al. 2012b; Hill and Tasker 2012). Although the mechanism of how glucocorticoids might exert such rapid actions remains to be clarified, the first evidence for a role of the endocannabinoid system in regulating glucocorticoid effects originated from an elegant series of in vitro studies by Tasker and colleagues. They demonstrated that corticosterone rapidly induces the release of endocannabinoids in the hypothalamus. Endocannabinoids then act retrogradely to inhibit the release of glutamate in the paraventricular nucleus and suppress HPA-axis activity (Di et al. 2003, 2005b). More recently, an in vivo study by Hill et al. (2010) corroborated these findings and showed that a single injection of corticosterone rapidly (within 10 min) elevated AEA levels in the hypothalamus, but also in the amygdala and hippocampus. Collectively, these and other data (Hill and Tasker 2012) suggested that the endocannabinoid system might play a critical role in mediating rapid glucocorticoid effects on the stress response.

In a series of experiments, we sought to examine whether endocannabinoid transmission might play a role in mediating glucocorticoid effects on memory consolidation. For this, rats were trained on an inhibitory avoidance task and received immediate posttraining infusions of the CB1 receptor antagonist AM251 into the BLA together with a systemic administration of corticosterone. As is shown in Fig. 5.4c, intra-BLA administration of the CB1 receptor antagonist blocked the ability of systemic corticosterone to facilitate memory consolidation of inhibitory avoidance training (Campolongo et al. 2009b). Similarly, other researchers found that a CB1 receptor antagonist infused into the hippocampus blocked memory enhancement induced by the synthetic glucocorticoid dexamethasone (de Oliveira Alvares et al. 2010). To investigate whether this glucocorticoid effect on the endocannabinoid system is dependent upon an adrenal steroid receptor on the cell surface, we performed an additional experiment. The CB1 receptor antagonist AM251 infused into the BLA blocked the memory-enhancing effects induced by concurrent infusions of either a specific GR agonist or the membrane-impermeable ligand cort:BSA (P. Atsak et al. unpublished observation). In contrast, the GR antagonist RU38486 infused into the BLA did not alter the memory-enhancing effects of WIN55,212-2. Therefore, these findings indicate that endocannabinoid transmission is required for mediating glucocorticoid effects on memory consolidation, presumably involving the activation of a GR on the cell surface and downstream endocannabinoid signaling. While these findings clearly indicate that endocannabinoids essentially mediate glucocorticoid effects on memory consolidation, they do not address whether the endocannabinoid system mediates the rapid effects of glucocorticoids onto the noradrenergic system. To investigate this issue, we examined whether endocannabinoid effects on memory consolidation might depend on concurrent noradrenergic activity within the BLA. Highly comparable to the above-described effects of glucocorticoids on memory consolidation, the β-adrenoceptor antagonist propranolol administered into the BLA prevented the memory enhancement induced by concurrent administration of the CB receptor agonist WIN55,212-2 (P. Atsak et al. unpublished observation). In an earlier study, we already reported that systemic administration of the endocannabinoid oleoylethanolamide enhances memory consolidation of inhibitory avoidance training. As the β-adrenoceptor antagonist propranolol infused into the BLA blocks this memory enhancement (Campolongo et al. 2009a), these findings indicate that also oleoylethanolamide enhances memory consolidation via a norepinephrine-dependent mechanism in the BLA. These findings are thus in line with previous evidence showing that systemic or local administration of a CB1 receptor agonist increases norepinephrine levels in cortical and limbic brain regions (Oropeza et al. 2005; Page et al. 2007). These findings might not only explain the observation that cannabinoids, like glucocorticoids, preferentially modulate memory of emotionally arousing information, but they also illustrate that the endocannabinoid is a likely target for glucocorticoids in influencing noradrenergic activity in the context of memory consolidation processes.

5.5.4 Role of Endocannabinoids in Mediating Glucocorticoid Effects on Memory Retrieval

As discussed, glucocorticoids induce temporary impairment of the retrieval of memory of previously acquired information (Wolf 2008; de Quervain et al. 2009; Roozendaal et al. 2009a). Importantly, these glucocorticoid effects on memory retrieval are mediated through GRs and, similar to the consolidation effects, essentially depend on arousal-induced noradrenergic activity (Roozendaal et al. 2006a). Highly comparable to glucocorticoid effects, cannabinoid drugs, including THC, induce impairment of memory retrieval (Castellano et al. 2003; Ranganathan and D'Souza 2006). We recently examined whether endocannabinoid signaling within

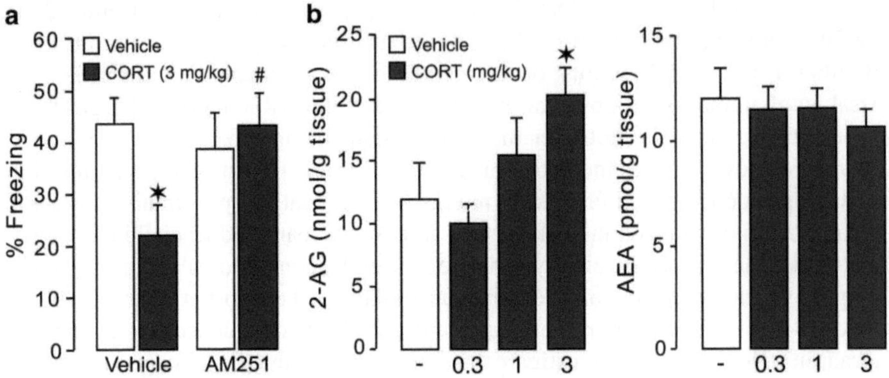

Fig. 5.6 Role of the endocannabinoid system in regulating glucocorticoid effects on retrieval of contextual fear memory. **a** Hippocampal infusion of the CB1 receptor antagonist AM251 (0.35 ng in 0.5 μL) administered 1 h before retention testing blocks the impairment of retrieval of contextual fear memory induced by concurrent systemic corticosterone (*CORT*; 3 mg/kg) treatment. Results represent mean ± SEM. *$p < 0.05$ versus vehicle; #$p < 0.05$ versus corticosterone alone. **b** Systemic corticosterone (0.3, 1, or 3 mg/kg) treatment dose-dependently increased hippocampal 2-AG, but not AEA, levels in the same time window of the retention test. All results represent mean ± SEM. *$p < 0.05$ versus vehicle. (Adapted from Atsak et al. 2012)

the hippocampus is involved in mediating glucocorticoid-induced impairment of retrieval of contextual fear memory. In this experiment, rats were trained on a contextual fear conditioning task and tested 24 h later for fear memory retention (Atsak et al. 2012a). As shown in Fig. 5.6a, we found that a blockade of hippocampal CB1 receptors by local infusions of AM251, 1 h before retention testing prevented the impairing effects of systemically co-administered glucocorticoids on retrieval of contextual fear memory. Moreover, we found that a retrieval-impairing dose of corticosterone elevated hippocampal levels of 2-AG, but not AEA (Fig. 5.6b). As mentioned before, glucocorticoid effects on memory retrieval highly depend on noradrenergic activity, thus in order to determine whether endocannabinoids mediate the effects of glucocorticoids on the noradrenergic system, we further examined possible interactions between the endocannabinoid and noradrenergic systems during retrieval processing of contextual fear memory. We found that the CB receptor agonist WIN55,212-2 infused into the hippocampus 1 h before retention testing impaired the retrieval of contextual fear memory; however, the β-adrenoceptor antagonist propranolol blocked the impairing effect of WIN55,212-2 on memory retrieval (Fig. 5.7a). Conversely, the CB1 receptor antagonist AM251 infused into hippocampus together with an impairing dose of norepinephrine failed to abolish the impairing effect of norepinephrine on memory retrieval (Fig. 5.7b). Collectively, these findings indicate that endocannabinoids interact with the noradrenergic system in inducing memory retrieval impairment and that the noradrenergic system appears to be located downstream, at least functionally, from the endocannabinoid system.

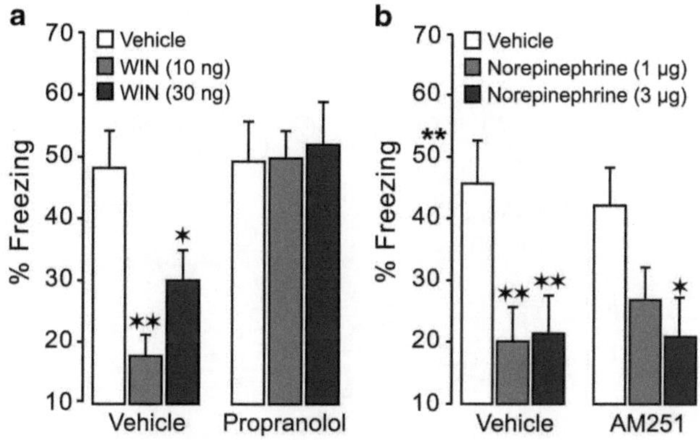

Fig. 5.7 Endocannabinoid and norepinephrine interactions in the dorsal hippocampus on retrieval of contextual fear memory. **a** The CB receptor agonist WIN55,212-2 (WIN, 10 or 30 ng in 0.5 µL) infused into the hippocampus 1 h before the retention test impaired retrieval of contextual fear memory. Concurrent infusion of the β-adrenoceptor antagonist propranolol (1.25 µg) blocked this WIN55,212-2-induced memory retrieval impairment. Results represent mean ± SEM. *$p<0.05$, **$p<0.001$ versus vehicle. **b** Intrahippocampal infusions of norepinephrine (1 or 3 µg in 0.5 µL) administered 1 h before the retention testing impaired retrieval of contextual fear memory. Concurrent infusion of the CB1 receptor antagonist AM251 (0.35 ng) did not block this impairment. Results represent mean ± SEM. *$p<0.05$; **$p<0.01$ versus vehicle. (Adapted from Atsak et al. 2012a)

5.5.5 The Model

In both the hippocampus and amygdala, CB1 receptors are expressed in GABAergic cells and to a minor extent in glutamatergic cells. Thus, an activation of CB1 receptors can modify the release of both neurotransmitters (Katona et al. 1999, 2001; Azad et al. 2003; Kawamura et al. 2006; Kano et al. 2009). Although our behavioral findings provide strong support for the view that the endocannabinoid system is crucially involved in mediating the fast effects of glucocorticoids on the noradrenergic system in modulating both the consolidation and the retrieval of memory, the underlying mechanism remains unknown. The endocannabinoid system might either directly influence noradrenergic activity or, alternatively, alter noradrenergic function indirectly via a modulation of GABAergic or glutamatergic activity. Within the BLA, CB1 receptors are in particular abundantly expressed in GABAergic interneurons (Katona et al. 2001) and activation of CB1 receptors has consistently been shown to suppress the release of GABA (Katona et al. 1999, 2001; Ohno-Shosaku et al. 2001) via a rapid inhibition of calcium entry into the terminals (Hoffman and Lupica 2000; Wilson et al. 2001). It is well established that the amygdala GABAergic system is involved in memory modulation such that posttraining infusions of GABA receptor antagonists into the BLA enhance memory consolidation, whereas posttraining infusions of GABA receptor agonists impair memory consolidation (McGaugh and Roozendaal 2002). Importantly, the modulatory effects of GABAergic transmission on memory crucially depend on an interaction with

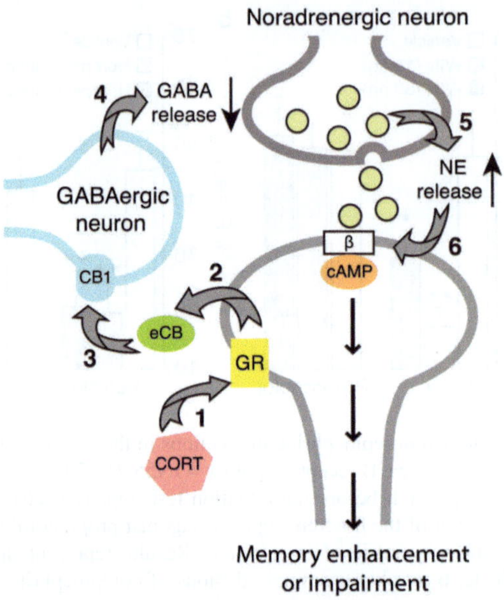

Fig. 5.8 Model on the role of the endocannabinoid system in the BLA in mediating glucocorticoid effects on norepinephrine release in regulating memory consolidation. Corticosterone (*CORT*) is released during training on an emotionally arousing tasks and binds to a membrane-bound glucocorticoid receptor (*GR*) **1**, that activates a pathway to induce endocannabinoid synthesis **2**. Endocannabinoids are then released into the synapse where they bind to CB1 receptors on GABAergic terminals **3** and thereby inhibit the release of GABA **4**. This suppression of GABA release subsequently disinhibits norepinephrine (*NE*) release **5** and this results in an activation of the postsynaptic β-adrenoceptor and the downstream cAMP/PKA/pCREB intracellular signaling pathway **6**. These stress hormone effects on noradrenergic activation in the BLA are required for enhancement of memory consolidation or impairment of memory retrieval. (Adapted from Atsak et al. 2012b, with permission)

the noradrenergic system. A *β*-adrenoceptor antagonist administered systemically or directly into the BLA prevents the modulatory effects of GABAergic drugs on memory consolidation (McGaugh 2004). Moreover, an in vivo microdialysis study indicated that the administration of a GABA receptor antagonist increases norepinephrine levels in the amygdala, whereas that of a GABA receptor agonist decreases norepinephrine levels (Hatfield et al. 1999). Thus, endocannabinoids might increase BLA neuronal activity by decreasing GABAergic neurotransmission, leading to increased noradrenergic activity within the BLA. Interestingly, a recent study indicated that glucocorticoids also increase the excitability of BLA neurons by decreasing the impact of GABAergic influences (Duvarci and Paré 2007).

As shown in Fig. 5.8, corticosterone binds to a membrane-associated GR and induces the release of endocannabinoids. Then, endocannabinoids bind to CB1 receptors and suppress GABAergic transmission that can then result in increased levels of norepinephrine. This increased norepinephrine level is associated with enhanced consolidation and temporary impairment of memory recall (McGaugh and Roozendaal 2002). Nevertheless, it is possible that glucocorticoid-induced memory effects might be also a result of endocannabinoid-mediated changes in glutamatergic signaling (Popoli et al. 2011).

5.6 Concluding Remarks

The evidence summarized in this chapter indicates that glucocorticoids enhance memory consolidation while impair memory retrieval in various animal and human memory tasks. Although glucocorticoids may act in many different brain regions to modulate these memory processes, the effects appear to depend critically on arousal-induced BLA activation and noradrenergic neurotransmission within the BLA. These findings may help to explain why glucocorticoids do not uniformly modulate memory for all kinds of information but, rather, preferentially influence the memory of emotionally arousing information. Furthermore, the findings indicate that glucocorticoids do not only modulate memory via their classically recognized genomic actions, but that glucocorticoid interactions with the noradrenergic arousal system depend critically on rapid, nongenomic actions via an activation of membrane-bound GRs and increased endocannabinoid signaling. Future studies will have to determine whether and how such rapid glucocorticoid effects on arousal mechanisms might cooperate with the slow actions in influencing gene transcription and the formation of strong and stabile memories of emotionally significant experiences.

References

Abercrombie HC, Speck NS, Monticelli RM. Endogenous cortisol elevations are related to memory facilitation only in individuals who are emotionally aroused. Psychoneuroendocrinology. 2006;31:187–96.

Abush H, Akirav I. Cannabinoids modulate hippocampal memory and plasticity. Hippocampus. 2010;20:1126–38.

Akirav I. The role of cannabinoids in modulating emotional and non-emotional memory processes in the hippocampus. Front Behav Neurosci. 2011;5:34.

Arbel I, Kadar T, Silbermann M, Levy A. The effects of long-term corticosterone administration on hippocampal morphology and cognitive performance of middle-aged rats. Brain Res. 1994;657:227–35.

Atsak P, Hauer D, Campolongo P, Schelling G, McGaugh JL, Roozendaal B. Glucocorticoids interact with the hippocampal endocannabinoid system in impairing retrieval of contextual fear memory. Proc Natl Acad Sci U S A. 2012a;109:3504–9.

Atsak P, Roozendaal B, Campolongo P. Role of the endocannabinoid system in regulating glucocorticoid effects on memory for emotional experiences. Neuroscience. 2012b;204:104–16.

Azad SC, Eder M, Marsicano G, Lutz B, Zieglgansberger W, Rammes G. Activation of the cannabinoid receptor type 1 decreases glutamatergic and GABAergic synaptic transmission in the lateral amygdala of the mouse. Learn Mem. 2003;10:116–28.

Ballard ME, Bedi G, de Wit H. Effects of delta-9-tetrahydrocannabinol on evaluation of emotional images. J Psychopharmacol. 2012;26:1289–98.
Barrett RM, Wood MA. Beyond transcription factors: the role of chromatin modifying enzymes in regulating transcription required for memory. Learn Mem. 2008;15:460–7.
Barsegyan A, Mackenzie SM, Kurose BD, McGaugh JL, Roozendaal B. Glucocorticoids in the prefrontal cortex enhance memory consolidation and impair working memory by a common neural mechanism. Proc Natl Acad Sci U S A. 2010;107:16655–60.
Beckwith BE, Petros TV, Scaglione C, Nelson J. Dose-dependent effects of hydrocortisone on memory in human males. Physiol Behav. 1986;36:283–6.
Bohus B, Lissak K. Adrenocortical hormones and avoidance behaviour of rats. Int J Neuropharmacol. 1968;7:301–6.
Bucherelli C, Baldi E, Mariottini C, Passani MB, Blandina P. Aversive memory reactivation engages in the amygdala only some neurotransmitters involved in consolidation. Learn Mem. 2006;13:426–30.
Buss C, Wolf OT, Witt J, Hellhammer DH. Autobiographic memory impairment following acute cortisol administration. Psychoneuroendocrinology. 2004;29:1093–6.
Cabeza R, Nyberg L. Neural bases of learning and memory: functional neuroimaging evidence. Curr Opin Neurol. 2000;13:415–21.
Campolongo P, Roozendaal B, Trezza V, Cuomo V, Astarita G, Fu J, et al. Fat-induced satiety factor oleoylethanolamide enhances memory consolidation. Proc Natl Acad Sci U S A. 2009a;106:8027–31.
Campolongo P, Roozendaal B, Trezza V, Hauer D, Schelling G, McGaugh JL, et al. Endocannabinoids in the rat basolateral amygdala enhance memory consolidation and enable glucocorticoid modulation of memory. Proc Natl Acad Sci USA. 2009b;106:4888–93.
Campolongo P, Ratano P, Manduca A, Scattoni ML, Palmery M, Trezza V, et al. The endocannabinoid transport inhibitor AM404 differentially modulates recognition memory in rats depending on environmental aversiveness. Front Behav Neurosci. 2012;6:11.
Campolongo P, Morena M, Scaccianoce S, Trezza V, Chiarotti F, Schelling G, et al. Novelty-induced emotional arousal modulates cannabinoid effects on recognition memory and adrenocortical activity. Neuropsychopharmacol. 2013;7:174.
Carvalho AF, Van Bockstaele EJ. Cannabinoid modulation of noradrenergic circuits: Implications for psychiatric disorders. Prog Neuropsychopharmacol Biol Psychiatry. 2012;38:59–67.
Castellano C, Rossi-Arnaud C, Cestari V, Costanzi M. Cannabinoids and memory: animal studies. Curr Drug Targets CNS Neurol Disord. 2003;2:389–402.
Chhatwal JP, Ressler KJ. Modulation of fear and anxiety by the endogenous cannabinoid system. CNS Spectr. 2007;12:211–20.
Coluccia D, Wolf OT, Kollias S, Roozendaal B, Forster A, de Quervain DJ. Glucocorticoid therapy-induced memory deficits: acute versus chronic effects. J Neurosci. 2008;28:3474–8.
Cordero MI, Kruyt ND, Merino JJ, Sandi C. Glucocorticoid involvement in memory formation in a rat model for traumatic memory. Stress. 2002;5:73–9.
Crosby KM, Bains JS. The intricate link between glucocorticoids and endocannabinoids at stress-relevant synapses in the hypothalamus. Neuroscience. 2012;204:31–7.
Da S, Takahashi RN. SR 141716A prevents delta 9-tetrahydrocannabinol-induced spatial learning deficit in a Morris-type water maze in mice. Prog Neuropsychopharmacol Biol Psychiatry. 2002;26:321–5.
Dallman M. Fast glucocorticoid actions on brain: back to the future. Front Neuroendocrinol. 2005;26:103–8.
de Bitencourt RM, Pamplona FA, Takahashi RN. A current overview of cannabinoids and glucocorticoids in facilitating extinction of aversive memories: potential extinction enhancers. Neuropharmacology. 2013;64:389–95.
de Boer SF, Koopmans SJ, Slangen JL, Van derGJ. Plasma catecholamine, corticosterone and glucose responses to repeated stress in rats: effect of interstressor interval length. Physiol Behav. 1990;47:1117–24.
de Kloet ER. Stress in the brain. Eur J Pharmacol. 2000;405:187–98.

de Kloet ER, Oitzl MS, Joëls M. Stress and cognition: are corticosteroids good or bad guys? Trends Neurosci. 1999;22:422–6.

de Oliveira Alvares L, de Oliveira LF, Camboim C, Diehl F, Genro BP, Lanziotti VB et al. Amnestic effect of intrahippocampal AM251, a CB1-selective blocker, in the inhibitory avoidance, but not in the open field habituation task, in rats. Neurobiol Learn Mem. 2005;83:119–24.

de Oliviera Alvares L, Genro BP, Vaz Breda R, Pedroso MF, Da Costa JC, Quillfeldt JA. AM251, a selective antagonist of the CB1 receptor, inhibits the induction of long-term potentiation and induces retrograde amnesia in rats. Brain Res. 2006;1075:60–7.

de Oliveira Alvares L, Engelke DS, Diehl F, Scheffer-Teixeira R, Haubrich J, de Freitas Cassini L, et al. Stress response recruits the hippocampal endocannabinoid system for the modulation of fear memory. Learn Mem. 2010;17:202–9.

de Quervain DJ, Roozendaal B, McGaugh JL. Stress and glucocorticoids impair retrieval of long-term spatial memory. Nature. 1998;394:787–90.

de Quervain DJ, Roozendaal B, Nitsch RM, McGaugh JL, Hock C. Acute cortisone administration impairs retrieval of long-term declarative memory in humans. Nat Neurosci. 2000;3:313–4.

de Quervain DJ, Aerni A, Roozendaal B. Preventive effect of beta-adrenoceptor blockade on glucocorticoid-induced memory retrieval deficits. Am J Psychiatry. 2007;164:967–9.

de Quervain DJ, Aerni A, Schelling G, Roozendaal B. Glucocorticoids and the regulation of memory in health and disease. Front Neuroendocrinol. 2009;30:358–70.

Di S, Malcher-Lopes R, Halmos KC, Tasker JG. Nongenomic glucocorticoid inhibition via endocannabinoid release in the hypothalamus: a fast feedback mechanism. J Neurosci. 2003;23:4850–7.

Di S, Boudaba C, Popescu IR, Weng FJ, Harris C, Marcheselli VL, et al. Activity-dependent release and actions of endocannabinoids in the rat hypothalamic supraoptic nucleus. J Physiol. 2005a;569:751–60.

Di S, Malcher-Lopes R, Marcheselli VL, Bazan NG, Tasker JG. Rapid glucocorticoid-mediated endocannabinoid release and opposing regulation of glutamate and gamma-aminobutyric acid inputs to hypothalamic magnocellular neurons. Endocrinology. 2005b;146:4292–301.

Di S, Maxson M, Franco A, Tasker J. Glucocorticoids regulate glutamate and GABA synapse-specific retrograde transmission via divergent nongenomic signaling pathways. J Neurosci. 2009;29:393–401.

Dolcos F, LaBar KS, Cabeza R. Remembering one year later: role of the amygdala and the medial temporal lobe memory system in retrieving emotional memories. Proc Natl Acad Sci U S A. 2005;102:2626–31.

Duvarci S, Paré D. Glucocorticoids enhance the excitability of principal basolateral amygdala neurons. J Neurosci. 2007;27:4482–91.

Egashira N, Mishima K, Iwasaki K, Fujiwara M. Intracerebral microinjections of delta 9-tetrahydrocannabinol: search for the impairment of spatial memory in the eight-arm radial maze in rats. Brain Res. 2002;952:239–45.

Evanson NK, Tasker JG, Hill MN, Hillard CJ, Herman JP. Fast feedback inhibition of the HPA-axis by glucocorticoids is mediated by endocannabinoid signaling. Endocrinology. 2010;151:4811–9.

Falkenstein E, Tillmann HC, Christ M, Feuring M, Wehling M. Multiple actions of steroid hormones-a focus on rapid, nongenomic effects. Pharmacol Rev. 2000;52:513–56.

Ferry B, McGaugh JL. Clenbuterol administration into the basolateral amygdala post-training enhances retention in an inhibitory avoidance task. Neurobiol Learn Mem. 1999;72:8–12.

Flood JF, Vidal D, Bennett EL, Orme AE, Vasquez S, Jarvik ME. Memory facilitating and anti-amnesic effects of corticosteroids. Pharmacol Biochem Behav. 1978;8:81–7.

Freund TF, Katona I, Piomelli D. Role of endogenous cannabinoids in synaptic signaling. Physiol Rev. 2003;83:1017–66.

Galvez R, Mesches MH, McGaugh JL. Norepinephrine release in the amygdala in response to footshock stimulation. Neurobiol Learn Mem. 1996;66:253–7.

Groc L, Choquet D, Chaouloff F. The stress hormone corticosterone conditions AMPAR surface trafficking and synaptic potentiation. Nat Neurosci. 2008;11:868–70.

Guenzel FM, Wolf OT, Schwabe L. Stress disrupts response memory retrieval. Psychoneuroendocrinology. 2013.
Haller J, Mikics É, Makara G. The effects of non-genomic glucocorticoid mechanisms on bodily functions and the central neural system: a critical evaluation of findings. Front Neuroendocrinol. 2008;29:273–91.
Hashimotodani Y, Ohno-Shosaku T, Kano M. Endocannabinoids and synaptic function in the CNS. The Neuroscientist. 2007;13:127–37.
Hatfield T, Spanis C, McGaugh JL. Response of amygdalar norepinephrine to footshock and GABAergic drugs using in vivo microdialysis and HPLC. Brain Res. 1999;835:340–5.
Het S, Ramlow G, Wolf OT. A meta-analytic review of the effects of acute cortisol administration on human memory. Psychoneuroendocrinology. 2005;30:771–84.
Hill M, McEwen B. Endocannabinoids: the silent partner of glucocorticoids in the synapse. Proc Natl Acad Sci USA. 2009;106:4579–80.
Hill M, McEwen B. Involvement of the endocannabinoid system in the neurobehavioural effects of stress and glucocorticoids. Prog Neuropsychopharmacol Biol Psychiatry. 2010;34:791–7.
Hill MN, Tasker JG. Endocannabinoid signaling, glucocorticoid-mediated negative feedback, and regulation of the hypothalamic-pituitary-adrenal axis. Neuroscience. 2012;204:5–16.
Hill M, Karatsoreos I, Hillard C, Mcewen B. Rapid elevations in limbic endocannabinoid content by glucocorticoid hormones in vivo. Psychoneuroendocrinology. 2010;35:1333–8.
Hoffman AF, Lupica CR. Mechanisms of cannabinoid inhibition of GABA(A) synaptic transmission in the hippocampus. J Neurosci. 2000;20:2470–9.
Hu W, Zhang M, Czéh B, Flügge G. Zhang W. Stress impairs GABAergic network function in the hippocampus by activating nongenomic glucocorticoid receptors and affecting the integrity of the parvalbumin-expressing neuronal network. Neuropsychopharmacology. 2010;35:1693–707.
Joëls M, Baram TZ. The neuro-symphony of stress. Nat Rev Neurosci. 2009;10:459–66.
Joëls M, Fernandez G, Roozendaal B. Stress and emotional memory: a matter of timing. Trends Cogn Sci. 2011;15:280–8.
Johnson L, Farb C, Morrison J, McEwen B, Ledoux J. Localization of glucocorticoid receptors at postsynaptic membranes in the lateral amygdala. Neuroscience. 2005;136:289–99.
Kano M, Ohno-Shosaku T, Hashimotodani Y, Uchigashima M, Watanabe M. Endocannabinoid-mediated control of synaptic transmission. Physiol Rev. 2009;89:309–80.
Karst H, Berger S, Turiault M, Tronche F, Schütz G, Joëls M. Mineralocorticoid receptors are indispensable for nongenomic modulation of hippocampal glutamate transmission by corticosterone. Proc Natl Acad Sci USA. 2005;102:19204–7.
Katona I, Sperlágh B, Sík A, Käfalvi A, Vizi ES, Mackie K, et al. Presynaptically located CB1 cannabinoid receptors regulate GABA release from axon terminals of specific hippocampal interneurons. J Neurosci. 1999;19:4544–58.
Katona I, Rancz EA, Acsady L, Ledent C, Mackie K, Hajos N, et al. Distribution of CB1 cannabinoid receptors in the amygdala and their role in the control of GABAergic transmission. J Neurosci. 2001;21:9506–18.
Kawamura Y, Fukaya M, Maejima T, Yoshida T, Miura E, Watanabe M, et al. The CB1 cannabinoid receptor is the major cannabinoid receptor at excitatory presynaptic sites in the hippocampus and cerebellum. J Neurosci. 2006;26:2991–3001.
Kuhlmann S, Kirschbaum C, Wolf OT. Effects of oral cortisol treatment in healthy young women on memory retrieval of negative and neutral words. Neurobiol Learn Mem. 2005a;83:158–62.
Kuhlmann S, Piel M, Wolf OT. Impaired memory retrieval after psychosocial stress in healthy young men. J Neurosci. 2005b;25:2977–82.
LaLumiere RT, Buen TV, McGaugh JL. Post-training intra-basolateral amygdala infusions of norepinephrine enhance consolidation of memory for contextual fear conditioning. J Neurosci. 2003;23:6754–8.
Lee E, Son G, Chung S, Lee S, Kim J, Choi S, et al. Impairment of fear memory consolidation in maternally stressed male mouse offspring: evidence for nongenomic glucocorticoid action on the amygdala. J Neurosci. 2011;31:7131–40.

Lichtman AH, Dimen KR, Martin BR. Systemic or intrahippocampal cannabinoid administration impairs spatial memory in rats. Psychopharmacology (Berl). 1995;119:282–90.

Liposits Z, Bohn MC. Association of glucocorticoid receptor immunoreactivity with cell membrane and transport vesicles in hippocampal and hypothalamic neurons of the rat. J Neurosci Res. 1993;35:14–9.

Liu L, Tsuji M, Takeda H, Takada K, Matsumiya T. Adrenocortical suppression blocks the enhancement of memory storage produced by exposure to psychological stress in rats. Brain Res. 1999;821:134–40.

Losel RM, Falkenstein E, Feuring M, Schultz A, Tillmann HC, Rossol-Haseroth K, et al. Nongenomic steroid action: controversies, questions, and answers. Physiol Rev. 2003;83:965–1016.

Luine VN, Spencer RL, McEwen BS. Effects of chronic corticosterone ingestion on spatial memory performance and hippocampal serotonergic function. Brain Res. 1993;616:65–70.

Maheu FS, Joober R, Beaulieu S, Lupien SJ. Differential effects of adrenergic and corticosteroid hormonal systems on human short- and long-term declarative memory for emotionally arousing material. Behav Neurosci. 2004;118:420–8.

Marsicano G, Lafenetre P. Roles of the endocannabinoid system in learning and memory. Curr Top Behav Neurosci. 2009;1:201–30.

McCarty R, Gold PE. Plasma catecholamines: effects of footshock level and hormonal modulators of memory storage. Horm Behav. 1981;15:168–82.

McEwen BS. Stress, adaptation, and disease. allostasis and allostatic load. Ann N Y Acad Sci. 1998;840:33–44.

McEwen BS. Plasticity of the hippocampus: adaptation to chronic stress and allostatic load. Ann N Y Acad Sci. 2001;933:265–77.

McGaugh JL. The amygdala modulates the consolidation of memories of emotionally arousing experiences. Annu Rev Neurosci. 2004;27:1–28.

McGaugh JL, Roozendaal B. Role of adrenal stress hormones in forming lasting memories in the brain. Curr Opin Neurobiol. 2002;12:205–10.

McIntyre C, Hatfield T, McGaugh JL. Amygdala norepinephrine levels after training predict inhibitory avoidance retention performance in rats. Eur J Neurosci. 2002;16:1223–6.

McReynolds JR, Donowho K, Abdi A, McGaugh JL, Roozendaal B, McIntyre CK. Memory-enhancing corticosterone treatment increases amygdala norepinephrine and Arc protein expression in hippocampal synaptic fractions. Neurobiol Learn Mem. 2010;93:312–21.

Moser MB, Moser EI. Distributed encoding and retrieval of spatial memory in the hippocampus. J Neurosci. 1998;18:7535–42.

Ohno-Shosaku T, Maejima T, Kano M. Endogenous cannabinoids mediate retrograde signals from depolarized postsynaptic neurons to presynaptic terminals. Neuron. 2001;29:729–38.

Oitzl MS, de Kloet ER. Selective corticosteroid antagonists modulate specific aspects of spatial orientation learning. Behav Neurosci. 1992;106:62–71.

Okuda S, Roozendaal B, McGaugh JL. Glucocorticoid effects on object recognition memory require training-associated emotional arousal. Proc Natl Acad Sci USA. 2004;101:853–8.

Olijslagers J, De Kloet E, Elgersma Y, Van Woerden G, Joëls M, Karst H. Rapid changes in hippocampal CA1 pyramidal cell function via pre- as well as postsynaptic membrane mineralocorticoid receptors. Eur J Neurosci. 2008;27:2542–50.

Onaivi ES, Ishiguro H, Gong JP, Patel S, Perchuk A, Meozzi PA, et al. Discovery of the presence and functional expression of cannabinoid CB2 receptors in brain. Ann N Y Acad Sci. 2006;1074:514–36.

Orchinik M, Murray TF, Moore FL. A corticosteroid receptor in neuronal membranes. Science. 1991;252:1848–51.

Oropeza VC, Page ME, Van Bockstaele EJ. Systemic administration of WIN 55,212-2 increases norepinephrine release in the rat frontal cortex. Brain Res. 2005;1046:45–54.

Page ME, Oropeza VC, Sparks SE, Qian Y, Menko AS, Van Bockstaele EJ. Repeated cannabinoid administration increases indices of noradrenergic activity in rats. Pharmacol Biochem Behav. 2007;86:162–8.

Pamplona FA, Takahashi RN. WIN 55212-2 impairs contextual fear conditioning through the activation of CB1 cannabinoid receptors. Neurosci Lett. 2006;397:88–92.

Pasricha N, Joels M, Karst H. Rapid effects of corticosterone in the mouse dentate gyrus via a nongenomic pathway. J Neuroendocrinol. 2011;23:143–7.

Patel S, Hillard CJ. Cannabinoid-induced Fos expression within A10 dopaminergic neurons. Brain Res. 2003;963:15–25.

Pelletier JG, Likhtik E, Filali M, Paré D. Lasting increases in basolateral amygdala activity after emotional arousal: implications for facilitated consolidation of emotional memories. Learn Mem. 2005;12:96–102.

Popoli M, Yan Z, McEwen BS, Sanacora G. The stressed synapse: the impact of stress and glucocorticoids on glutamate transmission. Nat Rev Neurosci. 2011;13:22–37.

Pugh CR, Tremblay D, Fleshner M, Rudy JW. A selective role for corticosterone in contextual-fear conditioning. Behav Neurosci. 1997;111:503–11.

Quirarte GL, Roozendaal B, McGaugh JL. Glucocorticoid enhancement of memory storage involves noradrenergic activation in the basolateral amygdala. Proc Natl Acad Sci U S A. 1997;94:14048–53.

Quirarte GL, Galvez R, Roozendaal B, McGaugh JL. Norepinephrine release in the amygdala in response to footshock and opioid peptidergic drugs. Brain Res. 1998;808:134–40.

Ramot A, Akirav I. Cannabinoid receptors activation and glucocorticoid receptors deactivation in the amygdala prevent the stress-induced enhancement of a negative learning experience. Neurobiol Learn Mem. 2012;97:393–401.

Ranganathan M, D'Souza DC. The acute effects of cannabinoids on memory in humans: a review. Psychopharmacology (Berl). 2006;188:425–44.

Rashidy-Pour A, Sadeghi H, Taherain AA, Vafaei AA, Fathollahi Y. The effects of acute restraint stress and dexamethasone on retrieval of long-term memory in rats: an interaction with opiate system. Behav Brain Res. 2004;154:193–8.

Riebe CJ, Pamplona FA, Kamprath K, Wotjak CT. Fear relief-toward a new conceptual frame work and what endocannabinoids gotta do with it. Neuroscience. 2012;204:159–85.

Riedemann T, Patchev A, Cho K, Almeida O. Corticosteroids: way upstream. Mol Brain. 2010;3:2.

Roozendaal B. Curt P. Richter award. Glucocorticoids and the regulation of memory consolidation. Psychoneuroendocrinology. 2000;25:213–38.

Roozendaal B. Stress and memory: opposing effects of glucocorticoids on memory consolidation and memory retrieval. Neurobiol Learn Mem. 2002;78:578–95.

Roozendaal B, Bohus B, McGaugh JL. Dose-dependent suppression of adrenocortical activity with metyrapone: effects on emotion and memory. Psychoneuroendocrinology. 1996a;21:681–93.

Roozendaal B, Carmi O, McGaugh JL. Adrenocortical suppression blocks the memory-enhancing effects of amphetamine and epinephrine. Proc Natl Acad Sci U S A. 1996b;93:1429–33.

Roozendaal B, Nguyen BT, Power AE, McGaugh JL. Basolateral amygdala noradrenergic influence enables enhancement of memory consolidation induced by hippocampal glucocorticoid receptor activation. Proc Natl Acad Sci U S A. 1999a;96:11642–7.

Roozendaal B, Williams CL, McGaugh JL. Glucocorticoid receptor activation in the rat nucleus of the solitary tract facilitates memory consolidation: involvement of the basolateral amygdala. Eur J Neurosci. 1999b;11:1317–23.

Roozendaal B, Quirarte GL, McGaugh JL. Glucocorticoids interact with the basolateral amygdala beta-adrenoceptor-cAMP/PKA system in influencing memory consolidation. Eur J Neurosci. 2002;15:553–60.

Roozendaal B, Griffith QK, Buranday J, de Quervain DJ, McGaugh JL. The hippocampus mediates glucocorticoid-induced impairment of spatial memory retrieval: dependence on the basolateral amygdala. Proc Natl Acad Sci USA. 2003;100:1328–33.

Roozendaal B, de Quervain DJ, Schelling G, McGaugh JL. A systemically administered beta-adrenoceptor antagonist blocks corticosterone-induced impairment of contextual memory retrieval in rats. Neurobiol Learn Mem. 2004a;81:150–4.

Roozendaal B, Hahn EL, Nathan SV, de Quervain DJ, McGaugh JL. Glucocorticoid effects on memory retrieval require concurrent noradrenergic activity in the hippocampus and basolateral amygdala. J Neurosci. 2004b;24:8161–9.

Roozendaal B, Okuda S, de Quervain DJ, McGaugh JL. Glucocorticoids interact with emotion-induced noradrenergic activation in influencing different memory functions. Neuroscience. 2006a;138:901–10.

Roozendaal B, Okuda S, Van der Zee EA, McGaugh JL. Glucocorticoid enhancement of memory requires arousal-induced noradrenergic activation in the basolateral amygdala. Proc Natl Acad Sci USA. 2006b;103:6741–6.

Roozendaal B, Castello NA, Vedana G, Barsegyan A, McGaugh JL. Noradrenergic activation of the basolateral amygdala modulates consolidation of object recognition memory. Neurobiol Learn Mem. 2008;90:576–9.

Roozendaal B, McEwen B, Chattarji S. Stress, memory and the amygdala. Nat Rev Neurosci. 2009a;10:423–33.

Roozendaal B, McReynolds JR, Van der Zee EA, Lee S, McGaugh JL, McIntyre CK. Glucocorticoid effects on memory consolidation depend on functional interactions between the medial prefrontal cortex and basolateral amygdala. J Neurosci. 2009b;29:14299–308.

Roozendaal B, Hernandez A, Cabrera SM, Hagewoud R, Malvaez M, Stefanko DP, et al. Membrane-associated glucocorticoid activity is necessary for modulation of long-term memory via chromatin modification. J Neurosci. 2010;30:5037–46.

Rose JD, Moore FL, Orchinik M. Rapid neurophysiological effects of corticosterone on medullary neurons: relationship to stress-induced suppression of courtship clasping in an amphibian. Neuroendocrinology. 1993;57:815–24.

Sajadi AA, Samaei SA, Rashidy-Pour A. Intra-hippocampal microinjections of anisomycin did not block glucocorticoid-induced impairment of memory retrieval in rats: an evidence for non-genomic effects of glucocorticoids. Behav Brain Res. 2006;173:158–62.

Sajadi AA, Samaei SA, Rashidy-Pour A. Blocking effects of intra-hippocampal naltrexone microinjections on glucocorticoid-induced impairment of spatial memory retrieval in rats. Neuropharmacology. 2007;52:347–54.

Sandi C, Rose SP. Corticosterone enhances long-term retention in one-day-old chicks trained in a weak passive avoidance learning paradigm. Brain Res. 1994;647:106–12.

Selye H. The stress of life by Hans Selye. New York:McGraw-Hill; 1976.

Smeets T, Otgaar H, Candel I, Wolf OT. True or false? Memory is differentially affected by stress-induced cortisol elevations and sympathetic activity at consolidation and retrieval. Psychoneuroendocrinology. 2008;33:1378–86.

Smith SM, Vale WW. The role of the hypothalamic-pituitary-adrenal axis in neuroendocrine responses to stress. Dialogues Clin Neurosci. 2006;8:383–95.

Squire LR. Memory and the hippocampus: a synthesis from findings with rats, monkeys, and humans. Psychol Rev. 1992;99:195–231.

Tasker JG, Di S, Malcher-Lopes R. Minireview: rapid glucocorticoid signaling via membrane-associated receptors. Endocrinology. 2006;147:5549–56.

Van Stegeren A, Wolf OT, Everaerd W, Scheltens P, Barkhof F, Rombouts S. Endogenous cortisol level interacts with noradrenergic activation in the human amygdala. Neurobiol Learn Mem. 2007;87:57–66.

van Stegeren AH, Roozendaal B, Kindt M, Wolf OT, Joels M. Interacting noradrenergic and corticosteroid systems shift human brain activation patterns during encoding. Neurobiol Learn Mem. 2010;93:56–65.

Wegener N, Kuhnert S, Thuns A, Roese R, Koch M. Effects of acute systemic and intra-cerebral stimulation of cannabinoid receptors on sensorimotor gating, locomotion and spatial memory in rats. Psychopharmacology (Berl). 2008;198:375–85.

Wichmann R, Fornari RV, Roozendaal B. Glucocorticoids interact with the noradrenergic arousal system in the nucleus accumbens shell to enhance memory consolidation of both appetitive and aversive taste learning. Neurobiol Learn Mem. 2012;98:197–205.

Wilson RI, Kunos G, Nicoll RA. Presynaptic specificity of endocannabinoid signaling in the hippocampus. Neuron. 2001;31:453–62.

Wolf OT. The influence of stress hormones on emotional memory: relevance for psychopathology. Acta Psychol. 2008;127:513–31.

Wolf OT, Schommer NC, Hellhammer DH, McEwen BS, Kirschbaum C. The relationship between stress induced cortisol levels and memory differs between men and women. Psychoneuroendocrinology. 2001;26:711–20.

Wotjak CT. Role of endogenous cannabinoids in cognition and emotionality. Mini Rev Med Chem. 2005;5:659–70.

Chapter 6
Endocannabinoid Signaling and Synaptic Plasticity During Stress

J. Megan Gray, Haley A. Vecchiarelli and Matthew N. Hill

Abstract This chapter summarizes and highlights advances from the last decade which have significantly contributed to our understanding of how endocannabinoid signaling is influenced during acute and chronic stress conditions, and in turn is able to importantly shape endocrine and behavioral stress responses through a variety of stress-responsive nuclei. The reviewed literature underscores a pivotal interaction of glucocorticoid-mediated changes during stress scenarios, and region-specific changes that display specialized responses depending on whether encountered stressors are experienced acutely or chronically. While the majority of reviewed content discusses our current understanding of *in vitro* and *in vivo* animal work, promising translational studies which have documented similar parallels in human literature are additionally spotlighted.

Abbreviations

2-AG	2-arachidonoylglycerol
ACTH	Adrenocorticotropin
AEA	Anandamide
CB_1R	Cannabinoid receptor 1
CB_2R	Cannabinoid receptor 2
CUS	Chronic unpredictable stress
CORT	Corticosterone
CRH	Corticotropin releasing hormone
THC	Delta9-tetrahydrocannabinol
DAG	Diacylglyerol
FAAH	Fatty acid amide hydrolase
FST	Forced swim test
GR	Glucocorticoid receptor

M. N. Hill (✉) · J. M. Gray · H. A. Vecchiarelli
Department of Cell Biology and Anatomy, Hotchkiss Brain Institute, University of Calgary, Calgary, Canada
e-mail: mnhill@ucalgary.ca

M. Popoli et al. (eds.), *Synaptic Stress and Pathogenesis of Neuropsychiatric Disorders*,
DOI 10.1007/978-1-4939-1056-4_6, © Springer Science+Business Media New York 2014

HPA axis	Hypothalamic-pituitary-adrenal axis
MAG lipase	Monoacylglyceride lipase
PVN	Paraventricular nucleus
PFC	Prefrontal cortex
DMH	Dorsomedial hypothalamus
BLA	Basolateral amygdala
GABA	Gamma-aminobutyric acid

6.1 Introduction

More than a decade ago, cannabinoids were shown to act as novel retrograde messengers capable of synaptic modulation, which prompted interest in a possible application to stress-neurocircuitry (Auclair et al. 2000; Wilson and Nicoll 2001; Ohno-Shosaku et al. 2001). Anecdotally, the stress-reducing effects of cannabinoids and cannabis usage are traced back to antiquity (Skaper and Di Marzo 2012). And yet the examination of cannabinoids in the regulation of stress only seriously emerged following the identification of cannabinoid receptors in the brain (Devane et al. 1988; Herkenham et al. 1991), and the ability to selectively stimulate or antagonize them through advances in genetics and pharmacology. These developments have since led to pivotal discoveries in the area of stress research and established that: (1) cannabinoids inhibit excitation of the hypothalamic-pituitary-adrenal (HPA) axis, which ultimately regulates endocrine stress responses and (2) this neurotransmitter system is activated by glucocorticoid elevations during stress, enabling cannabinoids to significantly shape the magnitude and duration of neural excitation imposed on the HPA axis. Thus, the cannabinoid system has quickly become a target of interest for stakeholders engaged in stress research including scientists, clinicians, and pharmaceutical corporations.

6.2 Endocannabinoid Basics

The endogenous cannabinoid system, denoted as the "*endo*cannabinoid system," is a neurotransmitter family composed of two lipid-based ligands and two G protein-coupled receptors. These receptors are activated by endogenous and exogenous cannabinoid molecules (i.e., THC or delta9-tetrahydrocannabinol) and are commonly referred to as cannabinoid receptors 1 and 2, or CB_1 and CB_2. CB_1 receptors (CB_1Rs) are widely distributed in the brain with notable distribution in stress-responsive regions like the hippocampus, amygdala, cortex, hypothalamus, septum, and brainstem (Herkenham et al. 1991; Marsicano and Lutz 1999; Egertova et al. 2003). CB_1Rs are coupled to G_i/G_o proteins and as their expression is almost exclusively confined to axon terminals, activation of this receptor results in a suppression of voltage-gated calcium channels, activation of outward rectifying potassium channels, and a net inhibition of synaptic release of neurotransmitters (Katona and Freund 2012). Initial perspectives thought that CB_1Rs were exclusively found in the brain and its counterpart CB_2R was isolated to peripheral immune-regulating cells or cells that had peripheral origins e.g., leukocytes, macrophages, microglia), and peripheral organs (e.g., the spleen) (Munro et al. 1993; Parolaro 1999; Cabral and Marciano-Cabral 2005; Atwood and Mackie 2010). However, although CB_1R and CB_2R distribution is still regarded as distinct and largely non-overlapping, views on the distribution of these receptors continues to change. CB_1R has also been found in the spine, vascular tissue, adipocytes, and on peripheral organs including all endocrine glands (Herkenham et al. 1991; Parolaro 1999; Cota et al. 2003; Bellocchio et al. 2008). Emerging evidence also indicates CB_2R is limitedly expressed within neural tissue (Nunez et al. 2004; Van Sickle et al. 2005; Gong et al. 2006; Palazuelos et al. 2006; Onaivi 2011; Xi et al. 2011). Based on the initial discoveries which suggested that CB_1Rs were exclusively found in the brain, the effects of endocannabinoid signaling on HPA axis activity has been entirely focused on CB_1R synaptic contributions. Therefore, the remainder of this chapter will discuss the effects of endocannabinoid signaling with attention specifically on the existing CB_1R-related evidence.

6.3 Endocannabinoid Synthesis and Metabolism

Just as the lipid structure of glucocorticoid steroids allows easy passage through cell membranes and penetration throughout the brain and body, the two endocannabinoid ligands N-arachidonoyl-ethanolamine (anandamide(AEA)) (Devane et al. 1992) and 2-arachidonoylglycerol (2-AG) (Mechoulam et al. 1995; Sugiura et al. 1995), are similarly composed of lipids, thus providing them ubiquitous systemic access. Contrary to typical neurotransmitters which usually move across synapses from a pre- to postsynaptic membrane surface, these modulators are instead made postsynaptically during neuronal activation through intracellular elevations in calcium and the activation of specific phospholipases in an "on demand" fashion, then released retrogradely, allowing them to act on presynaptic CB_1Rs (Wilson and Nicoll 2001; Alger 2002). Endocannabinoids are not packaged into synaptic vesicles like classic neurotransmitters, but are instantaneously released into the synaptic cleft following their membrane-based production. CB_1Rs are also found on the axon terminals of many different neural phenotypes including glutaminergic, GABAergic, and monoaminergic neurons (Schlicker and Kathmann 2001; Freund et al. 2003), thus it is not surprising that CB_1R activation has region-specific effects, which is dictated by the excitatory or inhibitory nature of the cell populations involved.

Synthesis of AEA and 2-AG during neuronal depolarization, or as a result of postsynaptic signaling cascades, is thought to occur through enzyme-mediated cleavage of membrane-associated phospholipids. Although production of these coordinating enzymes is believed to be triggered by changes in intracellular calcium, activation of metabotropic receptors is also a major factor for endocannabinoid mobilization (Freund et al. 2003). In the case of 2-AG, phospholipase C and D can both stimulate production of diacylglyerol (DAG), which is readily converted to 2-AG via enzymatic actions of DAG lipase (Hillard 2000; Sugiura et al. 2002; Di Marzo 2008). The pathway coordinating AEA production however is less clear as three independent mechanisms have been reported (Liu et al. 2006; Simon and Cravatt 2006; Okamoto et al. 2007). It also remains to be confirmed which possible pathways drive AEA synthesis in the brain (Ahn et al. 2008; Bisogno 2008).

Following postsynaptic release, endocannabinoids exhibit a very transient lifespan and are metabolized quickly, which allows for tight regulation of their temporal influence on synaptic transmission. However, AEA and 2-AG are not uniformly metabolized by the same enzyme. Fatty acid amide hydrolase (FAAH), which is a postsynaptically expressed enzyme found on the membrane of the endoplasmic reticulum, is the only known catabolic enzyme capable of hydrolyzing AEA into ethanolamine and arachidonic acid (Deutsch et al. 2002; Ueda 2002). 2-AG can be metabolized by FAAH, however this appears to be an artifact of *in vitro* preparations, as *in vivo* testing has shown it is primarily degraded (85 %) by presynaptic monoacylglyceride (MAG) lipase into glycerol and arachidonic acid, while the rest (15%) is degraded by the recently identified postsynaptic enzymes ABHD6 and ABHD12 (Ueda 2002; Dinh et al. 2002; Blankman et al. 2007; Marrs et al. 2010). The capacity that cells have to selectively metabolize 2-AG without altering AEA tone intriguingly suggests functional differences in these ligands—but the implications and the nature of these differences remain unresolved.

6.4 Current Trends in Endocannabinoid-Stress Research

Initially, AEA and 2-AG were thought to have similar physiological and behavioral effects; however there exists differences in binding affinity, pharmacokinetics, and ligand signaling efficacy (Sugiura et al. 2006), which has led researchers to suspect that AEA and 2-AG act during different temporal phases of neuronal activation and regulate different neuronal states. In applying this concept to activation of the HPA axis, an on-going hypothesis we and others are pursuing is the idea that constituent levels of AEA provide "tonic inhibition" on synaptic signaling allowing tight regulation of neurotransmitter release under normal basal conditions (Hill and Tasker 2012). Conversely, it appears 2-AG is produced "on demand" and is robustly increased during scenarios of sustained neuronal activation, contributing to the onset of adaptive forms of synaptic plasticity (Ahn et al. 2008; Gorzalka et al. 2008). This framework is importantly shaping how previous and emerging endocannabinoid research is being viewed. This categorization of roles for AEA and 2-AG also foreshadows the current trends in this field; which as discussed below, emphasizes a prominent role for increased 2-AG signaling during acute and mild repetitive stress conditions, whereby enhanced HPA axis inhibition could be adaptive and appropriate in the face of predictable, non-threatening scenarios to prevent HPA axis hyperactivation. Conversely, at the other end of the stress-scenario spectrum, when conditions involve chronic unpredictable physical and emotional stressors, the endocannabinoid system appears to respond with both ligand and receptor changes to promote HPA axis responsiveness downstream of the prefrontal cortex (PFC), while enhancing the inhibitory strength of the PFC via CB_1R upregulation. Although HPA axis sensitization provides certain survival advantages in the context of physical or predatory threats, it may be the case however, that chronic stress-induced adaptations to the central endocannabinoid system create a physiological state vulnerable to excitotoxicity, neuroinflammation, and stress-related disorders (Zoppi et al. 2011).

6.5 Origins of Endocannabinoid-Stress Research

The first characterizations of CB_1R expression revealed a wide distribution throughout the brain with notable expression in stress-sensitive regions communicating with the HPA axis, and low but detectable levels in the hypothalamus, median eminence, and anterior pituitary (Herkenham et al. 1991; Gonzalez et al. 1999; Marsicano and Lutz 1999; Egertova et al. 2003). With the advent of receptor-specific pharmacological drugs, and the ability to measure stress-induced changes in endocannabinoid content, this neurotransmitter system has been an exciting new target in the field of stress research. As previously mentioned, cannabinoids have long been perceived as having anxiolytic effects, however it has only been in the last decade that the underlying mechanisms explaining these effects have been explored. Initial studies administering THC intracerebroventricularly to rodents in tandem with a CB_1R antagonist, showed that CB_1R blockade at high concentrations increased basal levels of adrenocorticotrophin (ACTH) and corticosterone (CORT), suggesting an inhibitory role of the endocannabinoid system over the HPA axis (Manzanares et al. 1999).

In trying to further clarify the role of CB_1R in the stress response, it was work from Jeff Tasker and colleagues who used a more isolated and direct approach involving hypothalamic rat slices to show that endocannabinoids can modulate neurosecretory cells within the command center of the HPA axis, the paraventricular nucleus (PVN). This groundbreaking study was the first *in vitro* experiment to establish that endocannabinoids can inhibit HPA axis signaling, as they found that CB_1R activation decreases presynaptic glutamate release onto PVN parvocellular populations, which included corticotropin releasing hormone (CRH) positive cells, and other stress-regulating oxytocin-, vasopressin-, and thyrotrophin-releasing hormone-positive cells (Di et al. 2003). Continued work from Tasker's group has shown that endocannabinoid signaling in the PVN does not merely rely on postsynaptic activation, but is contingent on rapid non-genomic glucocorticoid signaling (Tasker 2006). This exciting work has contributed significantly to our understanding of glucocorticoid negative feedback by providing insight into how activation of the lower affinity glucocorticoid receptor (GR) actually coordinates an inhibitory influence on synaptic communication. These findings have also revealed that a downstream component of this long-established GR-mediated negative feedback cascade relies on endocannabinoids; opening up new and exciting avenues for investigating the etiology and treatment of diseases marked by glucocorticoid hypersecretion.

6.6 Early Studies in Acute Stress Literature

The seminal work of Di et al. (2003) have since set the stage for follow-up studies to confirm and further explore with *in vitro* and *in vivo* approaches how acute stress and glucocorticoids effect endocannabinoid synaptic transmission. These findings have also inspired the use of knockout approaches to examine the consequences of endocannabinoid dysregulation on stress-related endocrine and behavioral measures. Genetic deletion of CB_1R in knockout models has been found to enhance stress-induced peak responses of ACTH and CORT under a variety of stress conditions including restraint (Uriguen et al. 2004), tail suspension (Aso et al. 2008), forced swim (Steiner et al. 2008), and novel cage stress (Barna et al. 2004; Haller et al. 2004). CB_1R knockout mice (CB_1R-/-) also have enhanced HPA axis circadian peaks and impaired glucocorticoid feedback (Cota et al. 2007). Although knockout models are susceptible to possible compensatory changes, the knowledge generated using this approach has been consistent with experiments using pharmacological manipulations, which also have underscored that CB_1R antagonism potentiates peak ACTH, CORT, and cFos mRNA responses during noise stress (Newsom et al. 2012); potentiates CORT elevations during restraint recovery when administered locally into the PFC (Hill et al. 2011a); potentiates CORT responses during forced swim (Steiner et al. 2008) and social defeat (Steiner and Wotjak 2008); and increases basal circadian CORT levels (Atkinson et al. 2010). This work has led to the suggestion that CB_1Rs negatively influence activation of the HPA axis in two regards: (1) by dampening the initial activation of the HPA axis to attenuate peak increases and (2) by facilitating termination of HPA axis activity to reduce the overall duration that glucocorticoid elevations are experienced systemically (Barna et al. 2004; Haller et al. 2004; Uriguen et al. 2004; Steiner and Wotjak 2008; Hill et al. 2010a, 2011a).

6.7 Endocannabinoid Changes During Acute Stress

In vitro studies modeling acute stress conditions have shown that bath application of CORT and dexamethasone increases CB_1R-mediated inhibition of glutamate release in the PVN, supraoptic nucleus, basolateral amygdala, dorsal raphe, but not the cerebellum, suggesting a CORT-dependent relationship selective to stress-regulating circuits (Di et al. 2003, 2005; Malcher-Lopes et al. 2006; Karst et al. 2010; Wang et al. 2012a). These studies have confirmed that CB_1R-mediated inhibition of glutamate release occurs throughout the brain; and in examining the PVN specifically, that this effect is found in a variety of cell populations including parvo-, magno-, and pre-autonomic cells (Tasker 2006; Boychuk et al. 2013). In modeling hemorrhage-stress, CB_1R-mediated inhibition of PVN glutamate release has been shown to be activated by alpha-2-adrenergic receptors (Kuzmiski et al. 2009). Tasker and colleagues have also revealed that glucocorticoid-induced biosynthesis of endocannabinoids in the PVN is blocked by the satiety hormone leptin (Malcher-Lopes et al. 2006). It additionally appears that endocannabinoids do not only modulate glutamate release in the PVN, but display CORT-dependent CB_1R regulation of GABA synapses as well (Wamsteeker et al. 2010). A similar relationship is also found outside the hypothalamus, as CORT-dependent inhibition of GABA release has been documented in the hippocampus (Wang et al. 2012b) and PFC (Hill et al. 2011a). Taken together these studies have led to the consensus that the inhibitory effects of endocannabinoid signaling on stress responsivity show a prominent, although not exclusive, glucocorticoid dependence (Kuzmiski et al. 2009; Crosby et al. 2011), and underscore that CB_1R plays a prominent regulatory role on both glutamatergic and GABAergic neurons throughout the brain. Our knowledge of stress-induced CB_1R signaling also continues to expand as microdialysis studies have shown that stress-induced CB_1R activation in the hippocampus is able to limit acetylcholine transmission, in addition to GABA release (Degroot et al. 2006).

Having established that glucocorticoids can significantly alter the endocannabinoid system, many studies in the last decade have focused on determining if stress scenarios alter endocannabinoid tone by testing for possible stress-induced changes to the receptor, ligands, and the metabolic enzymes composing this neuromodulatory family. During acute *physical* stressors like foot shock, AEA and 2-AG increases have been demonstrated in the periaqueductal gray (Hohmann et al. 2005). However, when stressors are primarily *psychological,* such as, acute restraint, increases appear to be dominated by 2-AG rises in the PFC, hippocampus, and hypothalamus (Evanson et al. 2010; Hill et al. 2011a; Wang et al. 2012b), with no change in the amygdala (Hill et al. 2009a; Patel et al. 2009). 2-AG increases in the PFC, hippocampus, and hypothalamus are considered CORT-dependent (Hill et al. 2010b; Wang et al. 2012b)—unlike the rapid nongenomic effects observed in the hypothalamus (Di et al. 2003; Hill et al. 2010b)—as CORT application to the PFC elicits 2-AG rises with a slower onset (1 h) suggesting genomic actions (Hill et al. 2011a). Similarly, CORT application to the hippocampus also produces slower (30 min) 2-AG increases (Wang et al. 2012b). When further tested *in vivo,* CB_1R antagonist administered into the PFC does not alter restraint-induced CORT peak responses, but does potentiate post-stress recovery levels of CORT via a mechanism that is glucocorticoid-dependent (Hill et al. 2011a). These data suggest that CORT-initiated 2-AG increases in the PFC have a greater contribution to the termination of the stress response, as opposed to its initiation and maintenance. These findings also beg the question as to whether antagonism of hippocampal CB_1Rs would also have a greater influence during stress recovery, on the basis that lesion studies have revealed that its inhibitory HPA axis contribution is most apparent during the recovery phase (Herman et al. 2005). As yet, the mechanisms causing acute 2-AG increases is unknown, but preliminary indications point to a CORT-mediated decrease in

MAG lipase, which may have a facilitatory role by reducing 2-AG metabolism, herein enhancing its synaptic availability (Sumislawski et al. 2011).

In many cases, a corresponding rapid AEA decrease is found in the PFC, hippocampus, and amygdala following forced swim stress (McLaughlin et al. 2012) or restraint stress (Hill et al. 2009a; Wang et al. 2012b); which in the case of the amygdala appears to coincide with increases in FAAH-mediated AEA metabolism (Hill et al. 2009a). Given that CORT-dependent endocannabinoid mobilization and CB_1R activation has mostly been studied *in vitro,* our laboratory has made attempts to study the *in vivo* effects of CORT elevations on AEA and 2-AG regional levels. Acute intraperitoneal CORT injections have a stimulatory effect on AEA content in the amygdala, hippocampus, and hypothalamus, and elicit increases in 2-AG content within the hypothalamus (Hill et al. 2010b). These data would suggest that glucocorticoids on their own possess the ability to increase both AEA and 2-AG (consistent with *in vitro* studies) (Malcher-Lopes et al. 2006), but under conditions of stress, an additional stress-induced neural signal (possibly CRH or norepinephrine) seems to engage FAAH activity to instead reduce AEA content. Our working hypothesis is that CORT-mediated increases in AEA account for the recovery in AEA levels following cessation from stress, but that the reductions in AEA content following stress are through a CORT-independent mechanism.

With respect to CB_1R function, acute restraint exposure does not appear to alter CB_1R binding density (Rademacher et al. 2008; Hill et al. 2009a; Evanson et al. 2010), while acute social defeat stress has been found to blunt CB_1R-mediated inhibition of GABAergic transmission in the striatum (Rossi et al. 2008). Additionally, 24 h food deprivation stress extinguishes CB_1R-mediated inhibition of GABA synapses in the dorsomedial hypothalamus (DMH) in a manner that is CORT- and nitric oxide-dependent (Crosby et al. 2011). Given that the DMH, striatum, and limited brainstem regions have been found to be vulnerable to stress-induced endocannabinoids changes, future research examining ligand and receptor changes in these regions, in addition to, and in comparison to the more typical target structures for stress research (i.e. PFC, hippocampus, hypothalamus, amygdala), should aid in rounding out our understanding of the neuroanatomical impact of emotional and physical stressors. Recent work from our laboratory also suggests measurement of inducible serum endocannabinoid changes may be an area for bridging and comparing rodent and human studies. Using the Trier social stress test entailing a mock job interview, female participants were found to exhibit rapid increases in plasma 2-AG levels with no change in circulating AEA (Hill et al. 2009b). Together this literature has established that endocannabinoid levels do change in the brain and blood during acute stressors and indicate 2-AG rises during psychological stressors show a fair degree of consistency across rodents and humans thus far.

6.7.1 Circuit Implications

Based on our findings in the amygdala that AEA concentrations negatively correlate with stress-induced CORT (Hill et al. 2009a), the evolving model that our laboratory has proposed is that AEA in the amygdala serves as a gatekeeper—tonically inhibiting amygdalar glutamatergic projections to the PVN via both limited direct (Prewitt and Herman 1998; Csaki et al. 2000), and more prominent indirect routes (Dong et al. 2001). So far stress-induced FAAH increases have been localized to the amygdala, suggesting that FAAH-mediated hydrolysis of AEA may create a state of stress-hypersensitivity in the amygdala allowing it to play an enhanced role during the initial stages of stress detection and appraisal. In other regions like the hippocampus, PFC, and hypothalamus, where both AEA and 2-AG changes occur but in opposite directions (Hill et al. 2007; Rademacher et al. 2008; Evanson et al. 2010; Hill et al. 2011; McLaughlin et al. 2012; Wang et al. 2012b), there may be differences in the temporal onset of these changes allowing for CB_1R activation to be selectively decreased through rapid AEA reductions, but then later increased once HPA activation has been achieved, through CORT-dependent 2-AG rises (see Hill and McEwen 2010, for review). From stress onset, glucocorticoid increases typically take 2–3 min to become significantly elevated within plasma, and 10–15 min to become significantly increased centrally (Vahl et al. 2005; Droste et al. 2008). This suggests that the initial moments of HPA axis activation may favor early events coordinating FAAH-mediated AEA hydrolysis to facilitate HPA axis stimulation through disinhibition of the amygdala. Then following successful glucocorticoid mobilization, the effects of CORT-negative feedback likely initiate "on demand" 2-AG increases to inhibit glutamate release in the PVN and amygdala, while inhibiting GABA transmission in the PFC and hippocampus (Katona et al. 1999; Irving et al. 2000; Hill and Tasker 2012; Wang et al. 2012b), to enhance activation of glutamatergic projections to downstream inhibitory PVN relays such as the bed nucleus of the stria terminalis (Cullinan et al. 1993; Radley et al. 2006b; Choi et al. 2008; Radley et al. 2009) (Table 6.1, Fig. 6.1). Notably, certain aspects of this proposed cascade still need to be elucidated—the mechanisms driving stress-induced FAAH increases remain unknown, as well the developmental onset of these mechanisms. Additionally, limited studies have examined these processes in female rodents (Cota et al. 2007; Reich et al. 2009; Atkinson et al. 2010); or fully explored the contributions of the lower affinity, membrane-bound mineralocorticoid receptor that was recently uncovered (Karst et al. 2005; de Kloet et al. 2008; Olijslagers et al. 2008; Karst et al. 2010).

6.8 Endocannabinoid Changes During Repeated Homotypic Stress and Chronic Unpredictable Stress

The emerging pattern of endocannabinoid changes during repeated homotypic stress consistently shows 2-AG increases isolated to stress-sensitive relays like the hypothalamus, amygdala, and the PFC (Patel et al. 2004, 2005b; Rademacher et al. 2008; Patel et al. 2009). Although 2-AG increases are known to be CORT-dependent in many stress structures, the mechanisms involved remain unknown (Malcher-Lopes et al. 2006; Hill et al. 2010b; Bowles et al. 2012). While CORT-induced decreases in MAG lipase may contribute to acute stress 2-AG increases (Sumislawski et al.

Table 6.1 Summarization of the effects of acute stress or CORT exposure on tissue and serum levels of endocannabinoid ligands AEA and 2-AG, as well as the CB_1R and the maximal hydrolytic activity of FAAH

Species/Strain	Stress paradigm	Region/Sample	AEA	2-AG	CB_1R	FAAH	Reference
ICR mice	Restraint	Hypothalamus	NC	–	nd	nd	Patel et al. (2004)
ICR mice	Restraint	Amygdala/BLA	nd	NC	nd	nd	Patel et al. (2009)
ICR mice	Restraint	Prefrontal cortex	–	NC	NC[a]	NC	Rademacher et al. (2008)
		Amygdala	NC	NC	NC[a]	NC	
		Ventral striatum	NC	NC	NC[a]	NC	
Sprague Dawley rats	Restraint	Amygdala	–	NC	NC[a]	+	Hill et al. (2009a)
Sprague Dawley rats	Restraint	Hypothalamus	NC	+	NC[a]	nd	Evanson et al. (2010)
Sprague Dawley rats	Restraint	Prefrontal cortex	NC	+	nd	nd	Hill et al. (2011a)
Sprague Dawley rats	Restraint	Hippocampus	–	+	nd	nd	Wang et al. (2012b)
	Restraint- recovery	Hippocampus	–	+	nd	nd	
Sprague Dawley rats	Swim stress	Prefrontal cortex	–	NC	nd	nd	McLaughlin et al. (2012)
Sprague Dawley rats	Electroconvulsive shock	Prefrontal cortex	–	NC	–[a]	–	Hill et al. (2007)
		Hippocampus	–	NC	–[a]	NC	
		Hypothalamus	NC	NC	–[a]	NC	
		Amygdala	NC	NC	+[a]	NC	
Sprague Dawley rats	Foot shock	PAG	+	+	nd	nd	Hohmann et al. (2005)
Human females	Trier social stress	Serum	NC	+	nd	nd	Hill et al. (2009b)
Sprague Dawley rats	CORT-injection (10 min)	Amygdala	+	NC	nd	nd	Hill et al. (2010b)
		Hippocampus	+	NC	nd	nd	
		Hypothalamus	+	+	nd	nd	
		Prefrontal cortex	NC	NC	nd	nd	
	CORT-injection (1 h)	Amygdala	+	NC	nd	nd	
		Hippocampus	NC	NC	nd	nd	
		Hypothalamus	NC	NC	nd	nd	
		Prefrontal cortex	NC	NC	nd	nd	

Hill et al. (2007) reported widespread increases in CB_1R affinity following acute shock, but as illustrated in the table no change in binding density occurred, except for increases in the amygdala. *NC* no change, (−) significant decrease, (+) significant increase, *nd* not determined, *PAG* periaqueductal gray, *BLA* basolateral amygdala, *CORT* corticosterone, *AEA* anandamide, *2-AG* 2-arachidonylglycerol, *CB* cannabinoid receptor, *FAAH* fatty acid amide hydrolase, *ICR* imprinting control region

[a] Bmax

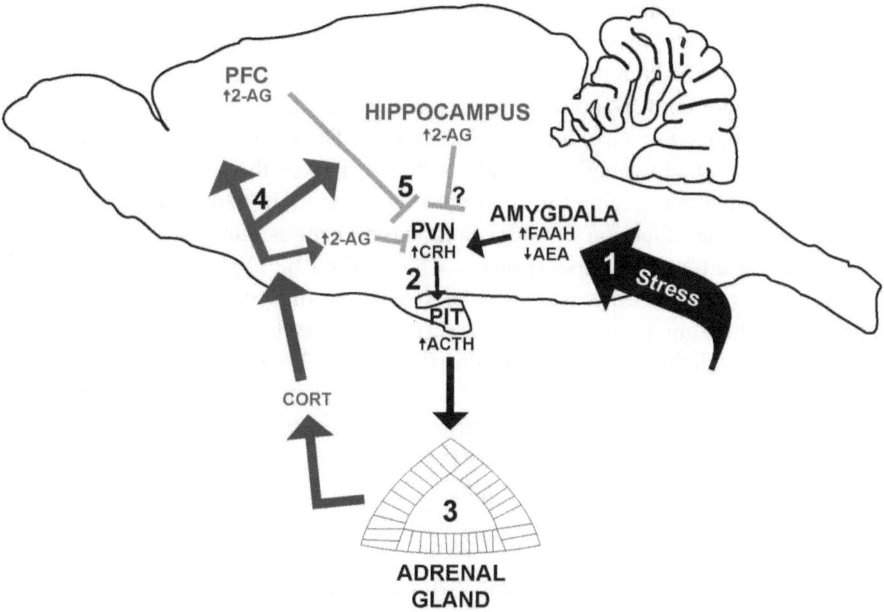

Fig. 6.1 Acute effects of stress- and glucocorticoid-mediated changes in endocannabinoids. *1.* Stress causes a decrease in anandamide (*AEA*) content in the BLA, through an increase in fatty acid amide hydrolase (*FAAH*) content within this region. This increase in FAAH and subsequent decrease in AEA content lessens the basal gate-keeping tone in the BLA—and through this excitatory facilitation of amygdalar projections, eventually their downstream projections lead to a removal of the GABAergic inhibition of the paraventricular nucleus (*PVN*) in the hypothalamus, thus driving the HPA response. *2.* Corticotropin releasing hormone (*CRH*) is released from the PVN into the anterior pituitary, causing the release of adrenocorticotropin (*ACTH*), which is then released into circulation. *3.* ACTH drives the release of corticosterone (*CORT*) from the adrenal cortex. CORT is released into circulation and exerts negative feedback on HPA axis signaling. There is direct negative feedback at the level of the pituitary and PVN and indirect feedback, both mediated by endocannabinoids at upstream limbic regions. *4.* Circulating CORT causes an increase in 2-arachidonoylglycerol (*2-AG*) in multiple regions, including the PVN, prefrontal cortex (*PFC*), and hippocampus. *5.* At the level of the PVN and amygdala, the rise in 2-AG content inhibits glutamate transmission, thus rapidly inhibiting the drive on the HPA axis. Additionally, the increase in 2-AG in the PFC and hippocampus, leads to a decrease in GABA transmission, which, in the case of the PFC and possibly in the case of the hippocampus, leads to an activation of glutamatergic projections to downstream inhibitory circuits on the PVN, thus providing a slower mechanism of shutting down the drive on the HPA axis. Finally, AEA content within the BLA is increased, thus restoring the basal inhibitory gate-keeping tone on the HPA axis

2011), upregulation of the 2-AG precursor DAG during repeated restraint appears to be an underlying contributing factor when looking in the BLA (Patel et al. 2009). Unlike 2-AG, repeated stress studies typically report stress-induced AEA reductions occurring in regions like the amygdala, PFC, hypothalamus, and hippocampus (Patel et al. 2004, 2005b; Hill et al. 2007, 2008a; Rademacher et al. 2008; Patel et al. 2009; Hill et al. 2010b). Based on the discriminative expression of CB_1R within the amygdala, such that it is predominately found in the basolateral aspect and less so in the medial and central divisions, it now appears that AEA and 2-AG induced changes, and their ensuing immediate effects on synaptic communication, have prominent effects in the BLA (Hill et al. 2009a; Patel et al. 2009). This is supported by antagonist work confirming that CB_1R blockade increases stress-induced CORT elevations when introduced locally into the BLA and not neighboring nuclei (Hill et al. 2009a). However, it should not be overlooked that CB_1R activation also has downstream consequences for neuronal signaling in the central amygdala (Patel et al. 2005a). The induction of endocannabinoid changes during repeated restraint also show variations in temporal onset, which might be aligned with species differences and regional differences in the sensitivity of synapses to initiate 2-AG increases. Following 5 days of repeated restraint, mice show 2-AG increases in the amygdala, hypothalamus, and forebrain (Patel et al. 2004, 2005b), although there are reports that the amygdala and PFC take 10 days, and not 7 to show increases in 2-AG (Rademacher et al. 2008). In contrast, rats show increases in amygdalar 2-AG following 9 days of repeated stress (Hill et al. 2010a), with no detectable increases elsewhere. Patel et al. (2009) have found 2-AG increases in the amygdala following repeated restraint at 20 min following stress onset, but are non-detectable at 60 min, suggesting possible discrepancies among studies may be due to the transient nature of 2-AG increases. Similarly in the rat, 2-AG levels return to normal, 24 h following the final stressor (Hill et al. 2009c), suggesting that the ability of repeated stress to increase 2-AG content is a transient response.

Few repeated stress studies have quantified changes in CB_1R binding or mRNA levels (Rademacher et al. 2008; Hill et al. 2012; Lee and Hill 2012); but *in vitro* tests indicate CB_1R function is downregulated in the hypothalamus (Wamsteeker et al. 2010), nucleus accumbens (Wang et al. 2010), BLA (Patel et al. 2009), and hippocampus (Hu et al. 2011). As stress paradigms shift from repeated homotypic stress to more intense chronic physical and emotional stressors, the resulting effects on the endocannabinoid system show a prominent shift, and a greater impact on CB_1R levels. When looking at the effects of chronic unpredictable stress (CUS), the net effect of CB_1R changes appears adaptive, in that it increases the efficiency by which the HPA axis is both activated and terminated, therein creating a faster "on" and "off" switch. Consistent across rodent studies CUS induces significant increases in PFC CB_1R binding density, but prevalent CB_1R decreases within downstream HPA axis relays including the hippocampus, amygdala, and hypothalamus (Hillard et al. 2006; Bortolato et al. 2007; Hill et al. 2008a; McLaughlin et al. 2013). Given that CORT-dependent downregulation of CB_1R has been reported in the hippocampus, amygdala, hypothalamus, and striatum (Hill et al. 2008b; Rossi et al. 2008; Wamsteeker et al. 2010; Bowles et al. 2012), it is likely CUS-induced CB_1R decreases are CORT-mediated, and quite possible that PFC CB_1Rs are exceptionally sensitive to CORT-upregulation as well. Consistent with this, postmortem tissue of individuals with major depression also present with PFC CB_1R elevations (Hungund et al. 2004), which has highlighted CB_1R forebrain increases as a potentially very important synaptic compensatory change during states of chronic stress. These findings are also complemented by evidence from selective knockout models generated by Beat Lutz and Giovanni Marsicano. The effects of CB_1R knockout on cortical glutamatergic (Glu-CB_1R-/-), just GABAergic (GAB-CB_1R-/-), and all principal forebrain neurons (CaMK-CB_1R-/-), have shown that removing CB_1R from cortical glutamate and GABA synapses has no effect on CORT release during the forced swim test (FST), whereas CB_1R deletion from principal forebrain neurons elevates FST endocrine stress response (Steiner et al. 2008). These findings suggest that abolishing CB_1R from cortical glutamatergic and CB_1R-GABAergic expression throughout the brain results in a net change that does not significantly alter CORT output, whereas CB_1Rs on principal neurons in the forebrain have the capacity to significantly inhibit stress-induced CORT responses (Steiner et al. 2008). The PFC has long been regarded as an important inhibitory influence on the PVN (Diorio et al. 1993; Radley et al. 2006a), however until now little has been known about the synaptic mechanisms coordinating this effect. Together, these data suggest CB_1Rs are differently regulated in a site-specific manner with glucocorticoids negatively regulating CB_1Rs in the hippocampus, amygdala, striatum, and hypothalamus, and possibly having an opposite effect on CB_1Rs in the PFC (McLaughlin et al. 2013). CUS may be associated with widespread AEA reductions across the hippocampus, hypothalamus, ventral striatum, amygdala, and midbrain (Hill et al. 2008a), although this possibility has yet to be consistently reported (Hill et al. 2005; Wang et al. 2010). Similar to repeated restraint, CUS also induces 2-AG increases; however these increases have only been reported in the hypothalamus, midbrain, and thalamus (Bortolato et al. 2007; Hill et al. 2008a). More studies are needed to confirm the effects of CUS on induced 2-AG levels, and particularly the temporal nature of these changes given that the effects of repeated stress seem to be temporally constrained to stress exposure.

In addition to stress-induced changes in endocannabinoid signaling, stress-induced structural changes also represent an important influence on synaptic transmission during chronic stress. FAAH-dependent amygdalar changes in excitability are associated with stress-induced increases in dendritic arborization, complexity, and spine density, which parallel increases in anxiety behavior (Hill et al. 2011b). These effects are abolished in FAAH-knockout mice—verifying that FAAH activity within the BLA increases amygdalar excitability and promotes a hyper-anxious state during chronic stress. Similarly CB_1R-/- mice are also vulnerable to stress-induced dendritic changes in the amygdala, and under nonstressed conditions show prelimbic structural changes which mirror the dendritic retraction and reductions in branch points typically induced by chronic stress (Hill et al. 2011b). Together these data suggest PFC CB_1Rs are critical for maintaining normal synaptic function and structure, and are an important point of comparison when investigating the hallmark changes of depression and chronic stress. It additionally appears that amygdalar synaptic changes induced by stress are multifaceted, entailing structural, ligand, and receptor changes, paired with altered endocannabinoid anabolic and catabolic capacities.

6.8.1 Circuit Implications

As neurons sense their external environment changing and consistently experience glucocorticoid elevations, repeated restraint appears to cause AEA reductions paired with 2-AG elevations throughout the limbic-HPA axis. Widespread AEA declines likely prime the HPA axis and its afferents for future anticipated stress by lowering the activation threshold of HPA axis relays to enhance synaptic communication. While at the same time "on demand" CORT-dependent increases in 2-AG become heightened to provide a more robust "brake" on activated stress-circuitry, leading to faster and efficient termination of behavioral and endocrine stress responses. In contrast to repeated restraint which favors an upregulation of ligands to enhance CB_1R-activated HPA inhibition, the utility of significantly reducing CB_1R expression during CUS in subcortical regions is likely necessary for maintaining HPA axis responsiveness. CORT-dependent CB_1R declines in the amygdala are poised to enhance glutamatergic amygdalar activation, thus promoting and maintaining HPA axis responsivity. Similarly, hippocampal CB_1R declines may promote HPA axis activation by enhancing hippocampal GABA release, thus silencing the hippocampus and reducing its capacity to provide indirect inhibition on the PVN (Sapolsky et al. 1984; Herman et al. 1992, 2005). Thus it appears that CB_1R is necessary for promoting adaptation during repeated homotypic stress conditions, but under chronic stress conditions, subcortical downregulation of CB_1R is more favorable. CB_1R decreases could be beneficial in the face of life-threatening physical stressors and especially adaptive when repeated stressors are unpredictable, but still highly anticipated. Based on the conditional knockout models which have shown that forebrain CB_1Rs are essential for dampening endocrine stress responses (Steiner and Wotjak 2008), the data seem to suggest that CUS-induced CB_1R increases in the PFC should protect individuals from HPA axis hyperactivation. In the PFC, CB_1Rs are almost entirely expressed on GABAergic terminals in the prelimbic division (Hill and Tasker 2012), indicating stress-induced CB_1R increases are positioned to promote activation of PFC projections to downstream inhibitory PVN afferents like the bed nucleus (Radley et al. 2006a, 2009). Based on the evidence that depressed, suicidal individuals show higher CB_1R levels in the PFC (Hungund et al. 2004), and that this is a similar hallmark of rodent CUS models, CB_1R PFC increases could be a compensatory change aimed at preventing hyper-glucocorticoid secretion and promoting termination of the stress response once the threatening stimulus is removed. This is consistent with a recent report which suggests that upregulation of prefrontal cortical CB_1R is an adaptive response aimed at limiting the adverse effects of stress (McLaughlin et al. 2013) (Table 6.2, Fig. 6.2).

Table 6.2 Summarization of the effects of RR, CUS, and CORT on tissue and serum levels of endocannabinoid ligands AEA and 2-AG, as well as the CB_1R and the maximal hydrolytic activity of FAAH

Species/Strain	Stress paradigm	Region/Sample	AEA	2-AG	CB_1R	FAAH	Reference
ICR mice	RR (5 days)	Hypothalamus	NC	+	nd	nd	Patel et al. (2004)
ICR mice	RR (5 days)	Forebrain	NC	+	nd	nd	Patel et al. (2005b)
		Amygdala	−	+	nd	nd	
		Cerebellum	NC	NC	nd	nd	
ICR mice	RR (7 days)	Prefrontal cortex	−	NC	nd	nd	Rademacher et al. (2008)
		Amygdala	−	NC	nd	nd	
		Ventral striatum	NC	−	nd	nd	
	RR (10 days)	Prefrontal cortex	−	+	NC[a]	+	
		Amygdala	−	+	NC[a]	+	
		Ventral striatum	+	NC	NC[a]	−	
ICR mice	RR (10 days) 20 min	Amygdala/BLA	nd	+	nd	nd	Patel et al. (2009)
	RR (10 days) 60 min	Amygdala/BLA	nd	NC	nd	nd	
C57/BL6 mice	RR (21 days)	Amygdala	−	nd	NC[a]	+	Hill et al. (2012)
Sprague Dawley rats	RR (9 days)	Amygdala	−	+	nd	nd	Hill et al. (2010a)
		Hypothalamus	−	NC	nd	nd	
		Prefrontal cortex	−	NC	nd	nd	
		Hippocampus	−	NC	nd	nd	
		Thalamus	NC	NC	nd	nd	
Sprague Dawley rats	RR (10 days) P75	Prefrontal cortex	nd	+	nd	nd	Lee and Hill (2012)
		Hippocampus	nd	nd	−[a]	nd	
		Amygdala	nd	nd	NC[a]	nd	
	P35	Prefrontal cortex	nd	nd	+[a]	nd	
		Hippocampus	nd	nd	NC[a]	nd	
		Amygdala	nd	nd	+[a]	nd	
Sprague Dawley rats	Electroconvulsive shock (10 days)	Prefrontal cortex	−	NC	−[a]	−	Hill et al. (2007)
		Hippocampus	NC	NC	NC[a]	NC	
		Hypothalamus	NC	NC	NC[a]	NC	
		Amygdala	NC	NC	NC[a]	NC	
C57BL/6J mice	Sub-CUS (1 wk)	Striatum	NC	NC	nd	nd	Wang et al. (2010)
	CUS (5–6 wk)	Striatum	NC	NC	nd	nd	
CB_1R-/- and WT mice	Sub-CUS (4 days)	Prefrontal cortex	nd	nd	+[b]	nd	Zoppi et al. (2011)
ICRS mice	CUS (21 days)	Prefrontal cortex	nd	nd	+[b]	nd	Hillard et al. (2006)

Table 6.2 (continued)

Species/ Strain	Stress paradigm	Region/Sample	AEA	2-AG	CB_1R	FAAH	Reference
		Hippocampus	nd	nd	–[b]	nd	
		Hypothalamus	nd	nd	–[b]	nd	
		Amygdala	nd	nd	–[b]	nd	
C57/BL6 mice	CORT-$H_2$0 (4 wk)	Hippocampus	–	+	–[a] NC[b]	+	Bowles et al. (2012)
			–	NC	–[a] NC[b]	+	
Long Evans rats	CUS (21 days)	Limbic forebrain	NC	NC	NC[a]	nd	Hill et al. (2005)
		Hippocampus	NC	–	–[a]	nd	
Long Evans rats	CUS (21 days)	Prefrontal cortex	–	NC	+[a]	NC	Hill et al. (2008a)
		Hippocampus	–	NC	–[a]	NC	
		Hypothalamus	–	+	–[a]	NC	
		Amygdala	–	NC	NC[a]	NC	
		Ventral striatum	–	NC	–[a]	NC	
		Midbrain	–	+	NC[a]	NC	
		Plasma	+	NC	nd	nd	
Sprague Dawley rats	CUS (21 days)	Prefrontal cortex	nd	nd	+[a]	nd	Hillard et al. (2006)
		Hippocampus	nd	nd	–[a]	nd	
		Amygdala	nd	nd	NC[a]	nd	
		Hypothalamus	nd	nd	–[a]	nd	
Sprague Dawley rats	CUS (21 days)	Cortex-vmPFC	nd	nd	+[a]	nd	McLaughlin et al. (2013)
		Cortex-dmPFC	nd	nd	–[a]	nd	
Sprague Dawley rats	CUS (21 days)	Hippocampus-CA1	nd	nd	NC[a]	nd	Hill et al. (2009c)
		Hippocampus-CA3	nd	nd	+[a]	nd	
		Hippocampus-dentate	nd	nd	–[a]	nd	
		Retrospinal ctx	nd	nd	NC[a]	nd	
		Laterodorsal thal	nd	nd	NC[a]		
Wistar rats	CUS (70 days)	Prefrontal cortex	NC	NC	+[b]	nd	Bortolato et al. (2007)
		Striatum	NC	NC	NC[b]	NC	
		Thalamus	NC	+	nd[b]	nd	
		Hippocampus	NC	NC	NC[b]	NC	
		Midbrain	NC	NC	–[b]	NC	
Long Evans rats	CORT-injection (21 days)	Hippocampus	NC	NC	–[a]	nd	Hill et al. (2008b)
		Amygdala	nd	+	NC[a]	nd	Hill et al. (2005)
Humans (post-mortem)	Major depression	Prefrontal cortex	nd	nd	+[a]	nd	Hungund et al. (2004)

Table 6.2 (continued)

Species/Strain	Stress paradigm	Region/Sample	AEA	2-AG	CB_1R	FAAH	Reference
Human female (medication-free)	Minor depression	Serum	+	NC	nd	nd	Hill et al. (2008c)
	Major depression	Serum	NC	−	nd	nd	
Human females	Depression	Serum	−	−	nd	nd	Hill et al. (2009b)

NC no change, (−) significant decrease, (+) significant increase, *nd* not determined, *vmPFC* ventromedial prefrontal cortex, *dmPFC* dorsomedial prefrontal cortex, *retrospinal ctx* retrospinal cortical gyrus, *laterodorsal thal* laterodorsal thalamus, *RR* repeated restraint, *CUS* chronic unpredictable stress, *CORT* corticosterone, *AEA* anandamide, *2-AG* 2-arachidonylglycerol, *CB1* cannabinoid receptor, *FAAH* fatty acid amide hydrolase, *ICR* imprinting control region
a Bmax
b mRNA

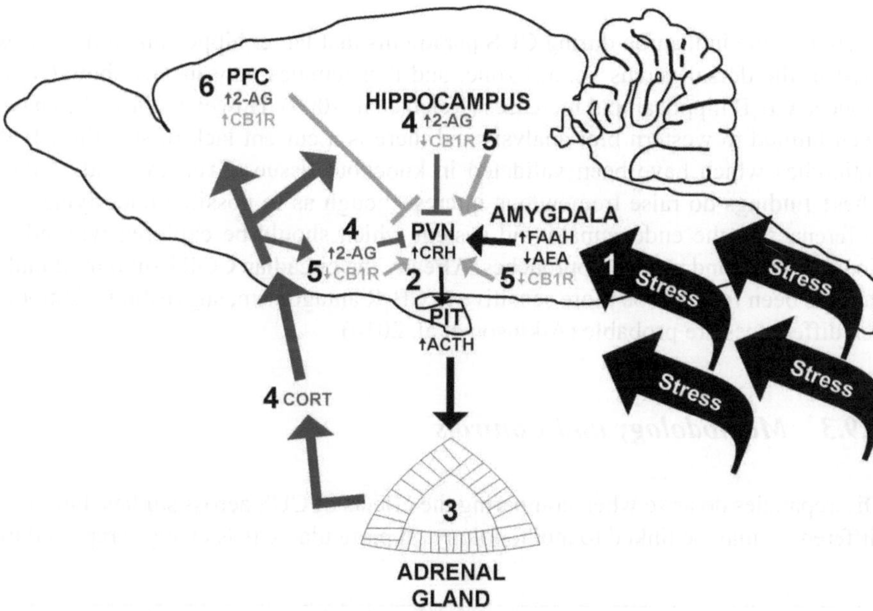

Fig. 6.2 Chronic effects of stress- and glucocorticoid-mediated changes in endocannabinoids. *1.* Repeated restraint leads to a decrease of the anandamide (*AEA*) tone in the BLA, through an increase in fatty acid amide hydrolase (FAAH) activity, which possibly lowers the activation threshold for HPA axis activation. *2.* Upon loss of the gate-keeping tone in the primed BLA, the paraventricular nucleus (*PVN*) is activated to release corticotropin releasing hormone (*CRH*), which is released into the anterior pituitary causing the release of adrenocorticotropin (*ACTH*). *3.* ACTH is released into circulation and causes the adrenal cortex to release corticosterone (*CORT*). In the case of repeated stress, there is a habituation in the amount of CORT released. *4.* CORT-induced 2-arachidonoylglycerol (*2-AG*) increases in the prefrontal cortex (*PFC*), hypothalamus, and hippocampus are elevated, which may be causing a more effective and quicker termination of

6.9 Future Considerations

6.9.1 Psychological Versus Physical Stress Circuits

Restraint is primarily a psychological stress, thus studies are currently needed to confirm that restraint induced 2-AG increases are indeed isolated to prominent limbic-HPA axis regions such as the hippocampus. It also has yet to be shown if physical and psychological stimuli induce similar or anatomically distinct endocannabinoid responses. Since limbic-PVN circuits are primarily recruited during psychological stress, and brainstem-PVN circuits are differently responsive to physical stress (Herman and Cullinan 1997; Dayas et al. 2001) it may be the case that physical stressors elicit distinct regional changes within the brainstem and spine that warrant more detailed investigation.

6.9.2 CB_1R Quantification Tools

There is some indication during CUS paradigms that larger hippocampal decreases exist in the dorsal versus ventral zone, and that females may in fact show CUS-induced CB_1R hippocampal increases (Reich et al. 2009). However, these data have been limited to western blot analysis and there is a current lack of specific CB_1R antibodies which have been validated in knockout tissue (Grimsey et al. 2008). These findings do raise tremendous interest though as to possible underlying sex differences in the endocannabinoid system which should be explored with additional binding and mRNA approaches. Already the circadian CORT rhythm of male rats has been found to be more sensitive to CB_1R antagonism, suggesting additional sex differences are probable (Atkinson et al. 2010).

6.9.3 Methodology and Controls

Discrepancies do arise when comparing the effects of CUS across studies, but these differences may be linked to methodology. In particular, CB_1R changes reported by

the HPA axis response to repeated homotypic stressors. *5.* This is in contrast to chronic unpredictable stressors. Animals exposed to CUS do not show CORT habituation. Furthermore, after CUS, there is a decrease in cannabinoid receptor 1 (CB_1R) in the amygdala and hippocampus. These declines could promote HPA axis signaling through different mechanisms. In the amygdala, a decrease in CB_1R would lead to an enhancement of glutamatergic amygdalar activation, which would promote HPA axis signaling. In the hippocampus, it is through enhancing GABA signaling on hippocampal interneurons, which silences the hippocampus and its inhibitory relays to the PVN. *6.* In the PFC, CB_1R is upregulated under chronic stress conditions. This is in contrast to the subcortical decreases in CB_1R, which facilitate HPA axis activation. CB_1R upregulation in the PFC could serve to protect against hyperactivation of the HPA axis and by terminating the stress response through downstream inhibitory projections to the PVN.

Bortolato et al. (2007) may be different compared to other reports since the control rats in this experiment were exposed to isolation as well as food and water deprivation stress which may have generated unintended stress-mediated CB_1R changes, making it difficult to separate out, and detect CUS-induced treatment effects. Studies which have been subsequent to Hungund et al. (2004) in examining CB_1R changes in depressed, suicidal individuals are also difficult to apply to existing rodent findings as these studies are usually restricted to alcoholic populations without the inclusion of nonalcoholic controls (Vinod et al. 2005, 2010).

6.9.4 Permanence and Plasticity

Proving that stress-induced changes display a great deal of plasticity, the permanence of stress-induced changes have been tested to a limited extent. Looking at repeated social defeat stress Rossi et al. (2008) have found that glucocorticoid-dependent CB_1R-mediated inhibition of GABAergic transmission in the striatum arises after 3 and 7 days of stress exposure, and that they were able to reverse these effects by providing rodents access to running wheels, sucrose, and cocaine. These data have importantly shown that changes to the efficacy of synaptic signaling can be recovered through physical and metabolic experiences which are known to activate central reward systems (Rossi et al. 2008). As well, simple cessation of repeated restraint for 1 week is also sufficient to reverse signs of long-term depression at inhibitory BLA synapses and behavioral changes in feeding latency (Sumislawski et al. 2011). Recently our laboratory has shown that repeated restraint results in a reduction in CB_1 receptor binding in the hippocampus and increased CB_1 receptor binding in the PFC, and that following a 4-week recovery period the PFC returns to normal, while in the hippocampus there is actually a surprising rebound effect where CB_1R densities increase significantly above what is seen in control animals (Lee and Hill 2012). These findings highlight the plasticity of synaptic changes, enabling neural systems to dynamically respond with reversible changes as situational changes arise. Although the structural consequences of CUS stress have yet to be examined, this synaptic flexibility may be compromised in chronic conditions creating a vulnerable state of hyper-excitable stress centers, exacerbating an individual's susceptibility to glucocorticoid hypersecretion.

6.10 Conclusion

In summary, the role of endocannabinoids within stress neural-circuitry aligns with the inhibitory and excitatory influences of each structure. Under acute conditions, HPA axis *stimulatory* regions such as the PVN and amygdala show CORT-mediated recruitment of endocannabinoids to inhibit presynaptic glutamate release, leading to reduced neural activation. Whereas in HPA axis *inhibitory* structures, like the

PFC and hippocampus, CORT-mediated recruitment of endocannabinoids inhibits GABA release to increase neural activation of glutamatergic projections which communicate with intermediate inhibitory PVN afferents (i.e. the bed nucleus of the stria terminalis and PVN surround). The effects of chronic stress on this neurotransmitter system lead to widespread receptor and ligand alterations whereby CB_1R activity is reduced throughout the brain, but selectively increased in the PFC to provide an increased descending inhibitory input, while enhancing the stress-sensitivity of subcortical relays. Evidently, endocannabinoid and glucocorticoid signaling robustly interact at the synaptic level to regulate endocrine stress responses; however the full breadth of this relationship and its application to stress-linked disorders remains to be elucidated.

References

Ahn K, McKinney MK, Cravatt BF. Enzymatic pathways that regulate endocannabinoid signaling in the nervous system. Chem Rev. 2008;108(5):1687–707.

Alger BE. Retrograde signaling in the regulation of synaptic transmission: focus on endocannabinoids. Prog Neurobiol. 2002;68(4):247–86.

Aso E, Ozaita A, Valdizan EM, Ledent C, Pazos A, Maldonado R, et al. BDNF impairment in the hippocampus is related to enhanced despair behavior in CB1 knockout mice. J Neurochem. 2008;105(2):565–72.

Atkinson HC, Leggett JD, Wood SA, Castrique ES, Kershaw YM, Lightman SL. Regulation of the hypothalamic-pituitary-adrenal axis circadian rhythm by endocannabinoids is sexually diergic. Endocrinology. 2010;151(8):3720–7.

Atwood BK, Mackie K. CB2: a cannabinoid receptor with an identity crisis. Br J Pharmacol. 2010;160(3):467–79.

Auclair N, Otani S, Soubrie P, Crepel F. Cannabinoids modulate synaptic strength and plasticity at glutamatergic synapses of rat prefrontal cortex pyramidal neurons. J Neurophysiol. 2000;83(6):3287–93.

Barna I, Zelena D, Arszovszki AC, Ledent C. The role of endogenous cannabinoids in the hypothalamo-pituitary-adrenal axis regulation: in vivo and in vitro studies in CB1 receptor knockout mice. Life Sci. 2004;75(24):2959–70.

Bellocchio L, Cervino C, Vicennati V, Pasquali R, Pagotto U. Cannabinoid type 1 receptor: another arrow in the adipocytes' bow. J Neuroendocrinol. 2008;20(Suppl 1):130–38.

Bisogno T. Endogenous cannabinoids: structure and metabolism. J Neuroendocrinol. 2008;20(Suppl 1):1–9.

Blankman JL, Simon GM, Cravatt BF. A comprehensive profile of brain enzymes that hydrolyze the endocannabinoid 2-arachidonoylglycerol. Chem Biol. 2007;14(12):1347–56.

Bortolato M, Mangieri RA, Fu J, Kim JH, Arguello O, Duranti A, et al. Antidepressant-like activity of the fatty acid amide hydrolase inhibitor URB597 in a rat model of chronic mild stress. Biol Psychiatry. 2007;62(10):1103–10.

Bowles NP, Hill MN, Bhagat SM, Karatsoreos IN, Hillard CJ, McEwen BS. Chronic, noninvasive glucocorticoid administration suppresses limbic endocannabinoid signaling in mice. Neuroscience. 2012;204:83–9.

Boychuk CR, Zsombok A, Tasker JG, Smith BN. Rapid glucocorticoid-induced activation of TRP and CB1 receptors causes biphasic modulation of Glutamate release in gastric-related hypothalamic preautonomic neurons. Front Neurosci. 2013;7:3.

Cabral GA, Marciano-Cabral F. Cannabinoid receptors in microglia of the central nervous system: immune functional relevance. J Leukoc Biol. 2005;78(6):1192–7.

Choi DC, Furay AR, Evanson NK, Ulrich-Lai YM, Nguyen MM, Ostrander MM, et al. The role of the posterior medial bed nucleus of the stria terminalis in modulating hypothalamic-pituitary-adrenocortical axis responsiveness to acute and chronic stress. Psychoneuroendocrinology. 2008;33(5):659–69.

Cota D, Marsicano G, Tschop M, Grubler Y, Flachskamm C, Schubert M, et al. The endogenous cannabinoid system affects energy balance via central orexigenic drive and peripheral lipogenesis. J Clin Invest. 2003;112(3):423–31.

Cota D, Steiner MA, Marsicano G, Cervino C, Herman JP, Grubler Y, et al. Requirement of cannabinoid receptor type 1 for the basal modulation of hypothalamic-pituitary-adrenal axis function. Endocrinology. 2007;148(4):1574–81.

Crosby KM, Inoue W, Pittman QJ, Bains JS. Endocannabinoids gate state-dependent plasticity of synaptic inhibition in feeding circuits. Neuron. 2011;71(3):529–41.

Csaki A, Kocsis K, Halasz B, Kiss J. Localization of glutamatergic/aspartatergic neurons projecting to the hypothalamic paraventricular nucleus studied by retrograde transport of [3H] D-aspartate autoradiography. Neuroscience. 2000;101(3):637–55.

Cullinan WE, Herman JP, Watson SJ. Ventral subicular interaction with the hypothalamic paraventricular nucleus: evidence for a relay in the bed nucleus of the stria terminalis. J Comp Neurol. 1993;332(1):1–20.

Dayas CV, Buller KM, Crane JW, Xu Y, Day TA. Stressor categorization: acute physical and psychological stressors elicit distinctive recruitment patterns in the amygdala and in medullary noradrenergic cell groups. Eur J Neurosci. 2001;14(7):1143–52.

de Kloet ER, Karst H, Joels M. Corticosteroid hormones in the central stress response: quick-and-slow. Front Neuroendocrinol. 2008;29(2):268–72.

Degroot A, Kofalvi A, Wade MR, Davis RJ, Rodrigues RJ, Rebola N, et al. CB1 receptor antagonism increases hippocampal acetylcholine release: site and mechanism of action. Mol Pharmacol. 2006;70(4):1236–45.

Deutsch DG, Ueda N, Yamamoto S. The fatty acid amide hydrolase (FAAH). Prostaglandins Leukot Essent Fatty Acids. 2002;66(2–3):201–10.

Devane WA, Hanus L, Breuer A, Pertwee RG, Stevenson LA, Griffin G, et al. Isolation and structure of a brain constituent that binds to the cannabinoid receptor. Science. 1992;258(5090):1946–9.

Devane WA, Dysarz FA 3rd, Johnson MR, Melvin LS, Howlett AC. Determination and characterization of a cannabinoid receptor in rat brain. Mol Pharmacol. 1988;34(5):605–13.

Di Marzo V. Endocannabinoids: synthesis and degradation. Rev Physiol Biochem Pharmacol. 2008;160:1–24.

Di S, Malcher-Lopes R, Halmos KC, Tasker JG. Nongenomic glucocorticoid inhibition via endocannabinoid release in the hypothalamus: a fast feedback mechanism. J Neurosci. 2003;23(12):4850–7.

Di S, Boudaba C, Popescu IR, Weng FJ, Harris C, Marcheselli VL, et al. Activity-dependent release and actions of endocannabinoids in the rat hypothalamic supraoptic nucleus. J Physiol. 2005;569(Pt 3):751–60.

Dinh TP, Freund TF, Piomelli D. A role for monoglyceride lipase in 2-arachidonoylglycerol inactivation. Chem Phys Lipids. 2002;121(1–2):149–58.

Diorio D, Viau V, Meaney MJ. The role of the medial prefrontal cortex (cingulate gyrus) in the regulation of hypothalamic-pituitary-adrenal responses to stress. J Neurosci. 1993;13(9):3839–47.

Dong HW, Petrovich GD, Swanson LW. Topography of projections from amygdala to bed nuclei of the stria terminalis. Brain Res Brain Res Rev. 2001;38(1–2):192–246.

Droste SK, de Groote L, Atkinson HC, Lightman SL, Reul JM, Linthorst AC. Corticosterone levels in the brain show a distinct ultradian rhythm but a delayed response to forced swim stress. Endocrinology. 2008;149(7):3244–53.

Egertova M, Cravatt BF, Elphick MR. Comparative analysis of fatty acid amide hydrolase and cb(1) cannabinoid receptor expression in the mouse brain: evidence of a widespread role for fatty acid amide hydrolase in regulation of endocannabinoid signaling. Neuroscience. 2003;119(2):481–96.

Evanson NK, Tasker JG, Hill MN, Hillard CJ, Herman JP. Fast feedback inhibition of the HPA axis by glucocorticoids is mediated by endocannabinoid signaling. Endocrinology. 2010;151(10):4811–9.

Freund TF, Katona I, Piomelli D. Role of endogenous cannabinoids in synaptic signaling. Physiol Rev. 2003;83(3):1017–66.

Gong JP, Onaivi ES, Ishiguro H, Liu QR, Tagliaferro PA, Brusco A, et al. Cannabinoid CB2 receptors: immunohistochemical localization in rat brain. Brain Res. 2006;1071(1):10–23.

Gonzalez S, Manzanares J, Berrendero F, Wenger T, Corchero J, Bisogno T, et al. Identification of endocannabinoids and cannabinoid CB(1) receptor mRNA in the pituitary gland. Neuroendocrinology. 1999;70(2):137–45.

Gorzalka BB, Hill MN, Hillard CJ. Regulation of endocannabinoid signaling by stress: implications for stress-related affective disorders. Neurosci Biobehav Rev. 2008;32(6):1152–60.

Grimsey NL, Goodfellow CE, Scotter EL, Dowie MJ, Glass M, Graham ES. Specific detection of CB1 receptors; cannabinoid CB1 receptor antibodies are not all created equal! J Neurosci Methods. 2008;171(1):78–86.

Haller J, Varga B, Ledent C, Barna I, Freund TF. Context-dependent effects of CB1 cannabinoid gene disruption on anxiety-like and social behaviour in mice. Eur J Neurosci. 2004;19(7):1906–12.

Herkenham M, Lynn AB, Johnson MR, Melvin LS, de Costa BR, Rice KC. Characterization and localization of cannabinoid receptors in rat brain: a quantitative in vitro autoradiographic study. J Neurosci. 1991;11(2):563–83.

Herman JP, Cullinan WE. Neurocircuitry of stress: central control of the hypothalamo-pituitary-adrenocortical axis. Trends Neurosci. 1997;20(2):78–84.

Herman JP, Cullinan WE, Young EA, Akil H, Watson SJ. Selective forebrain fiber tract lesions implicate ventral hippocampal structures in tonic regulation of paraventricular nucleus corticotropin-releasing hormone (CRH) and arginine vasopressin (AVP) mRNA expression. Brain Res. 1992;592(1–2):228–38.

Herman JP, Ostrander MM, Mueller NK, Figueiredo H. Limbic system mechanisms of stress regulation: hypothalamo-pituitary-adrenocortical axis. Prog Neuropsychopharmacol Biol Psychiatry. 2005;29(8):1201–13.

Hill MN, McEwen BS. Involvement of the endocannabinoid system in the neurobehavioural effects of stress and glucocorticoids. Prog Neuropsychopharmacol Biol Psychiatry. 2010;34(5):791–7.

Hill MN, Tasker JG. Endocannabinoid signaling, glucocorticoid-mediated negative feedback, and regulation of the hypothalamic-pituitary-adrenal axis. Neuroscience. 2012;204:5–16.

Hill MN, Patel S, Carrier EJ, Rademacher DJ, Ormerod BK, Hillard CJ, et al. Downregulation of endocannabinoid signaling in the hippocampus following chronic unpredictable stress. Neuropsychopharmacology. 2005;30(3):508–15.

Hill MN, Barr AM, Ho WS, Carrier EJ, Gorzalka BB, Hillard CJ. Electroconvulsive shock treatment differentially modulates cortical and subcortical endocannabinoid activity. J Neurochem. 2007;103(1):47–56.

Hill MN, Carrier EJ, Ho WS, Shi L, Patel S, Gorzalka BB, et al. Prolonged glucocorticoid treatment decreases cannabinoid CB1 receptor density in the hippocampus. Hippocampus. 2008b;18(2):221–6.

Hill MN, Carrier EJ, McLaughlin RJ, Morrish AC, Meier SE, Hillard CJ, et al. Regional alterations in the endocannabinoid system in an animal model of depression: effects of concurrent antidepressant treatment. J Neurochem. 2008a;106(6):2322–36.

Hill MN, McLaughlin RJ, Morrish AC, Viau V, Floresco SB, Hillard CJ, et al. Suppression of amygdalar endocannabinoid signaling by stress contributes to activation of the hypothalamic-pituitary-adrenal axis. Neuropsychopharmacology. 2009a;34(13):2733–45.

Hill MN, Miller GE, Carrier EJ, Gorzalka BB, Hillard CJ. Circulating endocannabinoids and N-acyl ethanolamines are differentially regulated in major depression and following exposure to social stress. Psychoneuroendocrinology. 2009b;34(8):1257–62.

Hill MN, Miller GE, Ho WSV, Gorzalka BB, Hillard CJ. Serum endocannabinoid content is altered in females with depressive disorders: a preliminary report. Pharmacopsychiatry. 2008c; 41(2):48-53.

Hill MN, Hunter RG, McEwen BS. Chronic stress differentially regulates cannabinoid CB1 receptor binding in distinct hippocampal subfields. Eur J Pharmacol. 2009c;614(1–3):66–9.

Hill MN, McLaughlin RJ, Bingham B, Shrestha L, Lee TT, Gray JM, et al. Endogenous cannabinoid signaling is essential for stress adaptation. Proc Natl Acad Sci U S A. 2010a;107(20):9406–11.

Hill MN, Karatsoreos IN, Hillard CJ, McEwen BS. Rapid elevations in limbic endocannabinoid content by glucocorticoid hormones in vivo. Psychoneuroendocrinology. 2010b;35(9):1333–8.

Hill MN, McLaughlin RJ, Pan B, Fitzgerald ML, Roberts CJ, Lee TT, et al. Recruitment of prefrontal cortical endocannabinoid signaling by glucocorticoids contributes to termination of the stress response. J Neurosci. 2011a;31(29):10506–15.

Hill MN, Hillard CJ, McEwen BS. Alterations in corticolimbic dendritic morphology and emotional behavior in cannabinoid CB1 receptor-deficient mice parallel the effects of chronic stress. Cereb Cortex. 2011b;21(9):2056–64.

Hill MN, Kumar SA, Filipski SB, Iverson M, Stuhr KL, Keith JM, et al. Disruption of fatty acid amide hydrolase activity prevents the effects of chronic stress on anxiety and amygdalar microstructure. Mol Psychiatry. 2012 July 10.

Hillard CJ. Biochemistry and pharmacology of the endocannabinoids arachidonylethanolamide and 2-arachidonylglycerol. Prostaglandins Other Lipid Mediat. 2000;61(1–2):3–18.

Hillard CJ, Hill MN, Carrier EJ, Shi L, Cullinan WE, Gorzalka BB. Regulation of cannabinoid receptor expression by chronic, unpredictable stress in rats and mice. Soc Neurosci Abstr. 2006;746:19.

Hohmann AG, Suplita RL, Bolton NM, Neely MH, Fegley D, Mangieri R, et al. An endocannabinoid mechanism for stress-induced analgesia. Nature. 2005;435(7045):1108–12.

Hu W, Zhang M, Czeh B, Zhang W, Flugge G. Chronic restraint stress impairs endocannabinoid mediated suppression of GABAergic signaling in the hippocampus of adult male rats. Brain Res Bull. 2011;85(6):374–9.

Hungund BL, Vinod KY, Kassir SA, Basavarajappa BS, Yalamanchili R, Cooper TB, et al. Upregulation of CB1 receptors and agonist-stimulated [35S]GTPgammaS binding in the prefrontal cortex of depressed suicide victims. Mol Psychiatry. 2004;9(2):184–90.

Irving AJ, Coutts AA, Harvey J, Rae MG, Mackie K, Bewick GS, et al. Functional expression of cell surface cannabinoid CB(1) receptors on presynaptic inhibitory terminals in cultured rat hippocampal neurons. Neuroscience. 2000;98(2):253–62.

Karst H, Berger S, Turiault M, Tronche F, Schutz G, Joels M. Mineralocorticoid receptors are indispensable for nongenomic modulation of hippocampal glutamate transmission by corticosterone. Proc Natl Acad Sci U S A. 2005;102(52):19204–7.

Karst H, Berger S, Erdmann G, Schutz G, Joels M. Metaplasticity of amygdalar responses to the stress hormone corticosterone. Proc Natl Acad Sci U S A. 2010;107(32):14449–54.

Katona I, Sperlagh B, Sik A, Kafalvi A, Vizi ES, Mackie K, et al. Presynaptically located CB1 cannabinoid receptors regulate GABA release from axon terminals of specific hippocampal interneurons. J Neurosci. 1999;19(11):4544–58.

Katona I and Freund TF. Multiple functions of endocannabinoid signaling in the brain. Annu Rev Neurosci. 2012;35:529-58.

Kuzmiski JB, Pittman QJ, Bains JS. Metaplasticity of hypothalamic synapses following in vivo challenge. Neuron. 2009;62(6):839–49.

Lee TT, Hill MN. Age of stress exposure modulates the immediate and sustained effects of repeated stress on corticolimbic cannabinoid CB(1) receptor binding in male rats. Neuroscience. 2012 Nov 27.

Liu J, Wang L, Harvey-White J, Osei-Hyiaman D, Razdan R, Gong Q, et al. A biosynthetic pathway for anandamide. Proc Natl Acad Sci U S A. 2006;103(36):13345–50.

Malcher-Lopes R, Di S, Marcheselli VS, Weng FJ, Stuart CT, Bazan NG, et al. Opposing crosstalk between leptin and glucocorticoids rapidly modulates synaptic excitation via endocannabinoid release. J Neurosci. 2006;26(24):6643–50.

Manzanares J, Corchero J, Fuentes JA. Opioid and cannabinoid receptor-mediated regulation of the increase in adrenocorticotropin hormone and corticosterone plasma concentrations induced by central administration of delta(9)-tetrahydrocannabinol in rats. Brain Res. 1999;839(1):173–9.

Marrs WR, Blankman JL, Horne EA, Thomazeau A, Lin YH, Coy J, et al. The serine hydrolase ABHD6 controls the accumulation and efficacy of 2-AG at cannabinoid receptors. Nat Neurosci. 2010;13(8):951–7.

Marsicano G, Lutz B. Expression of the cannabinoid receptor CB1 in distinct neuronal subpopulations in the adult mouse forebrain. Eur J Neurosci. 1999;11(12):4213–25.

McLaughlin RJ, Hill MN, Dang SS, Wainwright SR, Galea LA, Hillard CJ, et al. Upregulation of CB(1) receptor binding in the ventromedial prefrontal cortex promotes proactive stress-coping strategies following chronic stress exposure. Behav Brain Res. 2013;237:333–7.

McLaughlin RJ, Hill MN, Bambico FR, Stuhr KL, Gobbi G, Hillard CJ, et al. Prefrontal cortical anandamide signaling coordinates coping responses to stress through a serotonergic pathway. Eur Neuropsychopharmacol. 2012;22(9):664–71.

Mechoulam R, Ben-Shabat S, Hanus L, Ligumsky M, Kaminski NE, Schatz AR, et al. Identification of an endogenous 2-monoglyceride, present in canine gut, that binds to cannabinoid receptors. Biochem Pharmacol. 1995;50(1):83–90.

Munro S, Thomas KL, Abu-Shaar M. Molecular characterization of a peripheral receptor for cannabinoids. Nature. 1993;365(6441):61–5.

Newsom RJ, Osterlund C, Masini CV, Day HE, Spencer RL, Campeau S. Cannabinoid receptor type 1 antagonism significantly modulates basal and loud noise induced neural and hypothalamic-pituitary-adrenal axis responses in male Sprague-Dawley rats. Neuroscience. 2012;204:64–73.

Nunez E, Benito C, Pazos MR, Barbachano A, Fajardo O, Gonzalez S, et al. Cannabinoid CB2 receptors are expressed by perivascular microglial cells in the human brain: an immunohistochemical study. Synapse. 2004;53(4):208–13.

Ohno-Shosaku T, Maejima T, Kano M. Endogenous cannabinoids mediate retrograde signals from depolarized postsynaptic neurons to presynaptic terminals. Neuron. 2001;29(3):729–38.

Okamoto Y, Wang J, Morishita J, Ueda N. Biosynthetic pathways of the endocannabinoid anandamide. Chem Biodivers. 2007;4(8):1842–57.

Olijslagers JE, de Kloet ER, Elgersma Y, van Woerden GM, Joels M, Karst H. Rapid changes in hippocampal CA1 pyramidal cell function via pre- as well as postsynaptic membrane mineralocorticoid receptors. Eur J Neurosci. 2008;27(10):2542–50.

Onaivi ES. Commentary: functional neuronal CB2 cannabinoid receptors in the CNS. Curr Neuropharmacol. 2011;9(1):205–8.

Palazuelos J, Aguado T, Egia A, Mechoulam R, Guzman M, Galve-Roperh I. Non-psychoactive CB2 cannabinoid agonists stimulate neural progenitor proliferation. FASEB J. 2006;20(13):2405–7.

Parolaro D. Presence and functional regulation of cannabinoid receptors in immune cells. Life Sci. 1999;65(6–7):637–44.

Patel S, Roelke CT, Rademacher DJ, Cullinan WE, Hillard CJ. Endocannabinoid signaling negatively modulates stress-induced activation of the hypothalamic-pituitary-adrenal axis. Endocrinology. 2004;145(12):5431–8.

Patel S, Cravatt BF, Hillard CJ. Synergistic interactions between cannabinoids and environmental stress in the activation of the central amygdala. Neuropsychopharmacology. 2005a;30(3):497–507.

Patel S, Roelke CT, Rademacher DJ, Hillard CJ. Inhibition of restraint stress-induced neural and behavioural activation by endogenous cannabinoid signalling. Eur J Neurosci. 2005b;21(4):1057–69.

Patel S, Kingsley PJ, Mackie K, Marnett LJ, Winder DG. Repeated homotypic stress elevates 2-arachidonoylglycerol levels and enhances short-term endocannabinoid signaling at inhibitory synapses in basolateral amygdala. Neuropsychopharmacology. 2009;34(13):2699–709.

Prewitt CM, Herman JP. Anatomical interactions between the central amygdaloid nucleus and the hypothalamic paraventricular nucleus of the rat: a dual tract-tracing analysis. J Chem Neuroanat. 1998;15(3):173–85.

Rademacher DJ, Meier SE, Shi L, Ho WS, Jarrahian A, Hillard CJ. Effects of acute and repeated restraint stress on endocannabinoid content in the amygdala, ventral striatum, and medial prefrontal cortex in mice. Neuropharmacology. 2008;54(1):108–16.

Radley JJ, Arias CM, Sawchenko PE. Regional differentiation of the medial prefrontal cortex in regulating adaptive responses to acute emotional stress. J Neurosci. 2006a;26(50):12967–76.

Radley JJ, Rocher AB, Miller M, Janssen WG, Liston C, Hof PR, et al. Repeated stress induces dendritic spine loss in the rat medial prefrontal cortex. Cereb Cortex. 2006b;16(3):313–20.

Radley JJ, Gosselink KL, Sawchenko PE. A discrete GABAergic relay mediates medial prefrontal cortical inhibition of the neuroendocrine stress response. J Neurosci. 2009;29(22):7330–40.

Reich CG, Taylor ME, McCarthy MM. Differential effects of chronic unpredictable stress on hippocampal CB1 receptors in male and female rats. Behav Brain Res. 2009;203(2):264–9.

Rossi S, De Chiara V, Musella A, Kusayanagi H, Mataluni G, Bernardi G, et al. Chronic psychoemotional stress impairs cannabinoid-receptor-mediated control of GABA transmission in the striatum. J Neurosci. 2008;28(29):7284–92.

Sapolsky RM, Krey LC, McEwen BS. Glucocorticoid-sensitive hippocampal neurons are involved in terminating the adrenocortical stress response. Proc Natl Acad Sci U S A. 1984;81(19):6174–7.

Schlicker E, Kathmann M. Modulation of transmitter release via presynaptic cannabinoid receptors. Trends Pharmacol Sci. 2001;22(11):565–72.

Simon GM, Cravatt BF. Endocannabinoid biosynthesis proceeding through glycerophospho-N-acyl ethanolamine and a role for alpha/beta-hydrolase 4 in this pathway. J Biol Chem. 2006;281(36):26465–72.

Skaper SD, Di Marzo V. Endocannabinoids in nervous system health and disease: the big picture in a nutshell. Philos Trans R Soc Lond B Biol Sci. 2012;367(1607):3193–200.

Steiner MA, Wotjak CT. Role of the endocannabinoid system in regulation of the hypothalamic-pituitary-adrenocortical axis. Prog Brain Res. 2008;170:397–432.

Steiner MA, Marsicano G, Wotjak CT, Lutz B. Conditional cannabinoid receptor type 1 mutants reveal neuron subpopulation-specific effects on behavioral and neuroendocrine stress responses. Psychoneuroendocrinology. 2008;33(8):1165–70.

Sugiura T, Kondo S, Sukagawa A, Nakane S, Shinoda A, Itoh K, et al. 2-Arachidonoylglycerol: a possible endogenous cannabinoid receptor ligand in brain. Biochem Biophys Res Commun. 1995;215(1):89–97.

Sugiura T, Kobayashi Y, Oka S, Waku K. Biosynthesis and degradation of anandamide and 2-arachidonoylglycerol and their possible physiological significance. Prostaglandins Leukot Essent Fatty Acids. 2002;66(2–3):173–92.

Sugiura T, Kishimoto S, Oka S, Gokoh M. Biochemistry, pharmacology and physiology of 2-arachidonoylglycerol, an endogenous cannabinoid receptor ligand. Prog Lipid Res. 2006;45(5):405–46.

Sumislawski JJ, Ramikie TS, Patel S. Reversible gating of endocannabinoid plasticity in the amygdala by chronic stress: a potential role for monoacylglycerol lipase inhibition in the prevention of stress-induced behavioral adaptation. Neuropsychopharmacology. 2011;36(13):2750–61.

Tasker JG. Rapid glucocorticoid actions in the hypothalamus as a mechanism of homeostatic integration. Obesity (Silver Spring). 2006;14(Suppl 5):259S–65S.

Ueda N. Endocannabinoid hydrolases. Prostaglandins Other Lipid Mediat. 2002;68–69:521–34.

Uriguen L, Perez-Rial S, Ledent C, Palomo T, Manzanares J. Impaired action of anxiolytic drugs in mice deficient in cannabinoid CB1 receptors. Neuropharmacology. 2004;46(7):966–73.

Vahl TP, Ulrich-Lai YM, Ostrander MM, Dolgas CM, Elfers EE, Seeley RJ, et al. Comparative analysis of ACTH and corticosterone sampling methods in rats. Am J Physiol Endocrinol Metab. 2005;289(5):E823–8.

Van Sickle MD, Duncan M, Kingsley PJ, Mouihate A, Urbani P, Mackie K, et al. Identification and functional characterization of brainstem cannabinoid CB2 receptors. Science. 2005;310(5746):329–32.

Vinod KY, Arango V, Xie S, Kassir SA, Mann JJ, Cooper TB, et al. Elevated levels of endocannabinoids and CB1 receptor-mediated G-protein signaling in the prefrontal cortex of alcoholic suicide victims. Biol Psychiatry. 2005;57(5):480–6.

Vinod KY, Kassir SA, Hungund BL, Cooper TB, Mann JJ, Arango V. Selective alterations of the CB1 receptors and the fatty acid amide hydrolase in the ventral striatum of alcoholics and suicides. J Psychiatr Res. 2010;44(9):591–7.

Wamsteeker JI, Kuzmiski JB, Bains JS. Repeated stress impairs endocannabinoid signaling in the paraventricular nucleus of the hypothalamus. J Neurosci. 2010;30(33):11188–96.

Wang J, Shen RY, Haj-Dahmane S. Endocannabinoids mediate the glucocorticoid-induced inhibition of excitatory synaptic transmission to dorsal raphe serotonin neurons. J Physiol. 2012a;590(Pt 22):5795–808.

Wang M, Hill MN, Zhang L, Gorzalka BB, Hillard CJ, Alger BE. Acute restraint stress enhances hippocampal endocannabinoid function via glucocorticoid receptor activation. J Psychopharmacol. 2012b;26(1):56–70.

Wang W, Sun D, Pan B, Roberts CJ, Sun X, Hillard CJ, et al. Deficiency in endocannabinoid signaling in the nucleus accumbens induced by chronic unpredictable stress. Neuropsychopharmacology. 2010;35(11):2249–61.

Wilson RI, Nicoll RA. Endogenous cannabinoids mediate retrograde signalling at hippocampal synapses. Nature. 2001;410(6828):588–92.

Xi ZX, Peng XQ, Li X, Song R, Zhang HY, Liu QR, et al. Brain cannabinoid CB(2) receptors modulate cocaine's actions in mice. Nat Neurosci. 2011;14(9):1160–6.

Zoppi S, Perez Nievas BG, Madrigal JL, Manzanares J, Leza JC, Garcia-Bueno B. Regulatory role of cannabinoid receptor 1 in stress-induced excitotoxicity and neuroinflammation. Neuropsychopharmacology. 2011;36(4):805–18.

Chapter 7
Stress-Induced Metaplasticity at GABA Synapses

Jaideep S. Bains

Abstract Responding quickly and effectively to stress is necessary for the survival of any organism. Each response to stress relies on hard-wired, evolutionarily conserved neural circuitry, but importantly, is also shaped by previous experience. These modifications provide an adaptive advantage, but if left unchecked may result in inappropriate activation of the stress axis. Exposure to a single, severe stressful event can result in long-lasting psychopathological consequences such as post-traumatic stress disorder, which is characterized by a hyperreactivity to stressors not directly related to the traumatic situation. Understanding the neurobiological consequences of stress exposure will provide treatment targets for stress disorders. In our efforts to better understand stress physiology and plasticity, we have made the surprising finding that gamma-aminobutyric acid (GABA), which is an inhibitory neurotransmitter in the adult nervous system, becomes excitatory during an acute stress. More recently, we discovered that acute stress also causes a novel form of priming that increases the number of functional GABA synapses in the hypothalamus in response to a second stress. Here we discuss these findings along with new information about the specific intracellular pathways responsible for this plasticity that may be key determinants of plasticity and hyperactivity of the stress axis.

Abbreviations

ACTH	Adrenocorticotropin hormone
CAMKII	Ca^{2+}/calmodulin-dependent protein kinase II
Cl^-	Chloride
CORT	Corticosterone
CRH	Corticotropin releasing hormone
ECl^-	Cl^- reversal potential
HPA	Hypothalamic-pituitary-adrenal

J. S. Bains (✉)
Department of Physiology and Pharmacology, Hotchkiss Brain Institute, University of Calgary, Calgary, AB, Canada
Tel.: 403-220-7585
e-mail: jsbains@ucalgary.ca

M. Popoli et al. (eds.), *Synaptic Stress and Pathogenesis of Neuropsychiatric Disorders*,
DOI 10.1007/978-1-4939-1056-4_7,

LTD	Long-term depression
LTP	Long-term potentiation
NA	Noradrenaline
PVN	Paraventricular nucleus

7.1 Introduction

Stress signifies a potential or actual threat that requires immediate hormonal and behavioural responses, and necessitates a modification of future responses. Exposure to a stressor results in the activation of the hypothalamic-pituitary-adrenal (HPA) axis to meet the impending challenges (Joels and Baram 2009). Repeated exposure to the same stress results, over time, in diminished HPA output (Grissom and Bhatnagar 2009; Franklin et al. 2012). In contrast, exposure to a single severe stressor induces a long-lasting sensitization of neuroendocrine responsiveness to subsequent novel stressors (Bruijnzeel et al. 1999, 2001; Armario et al. 2008). This priming, which manifests as facilitated adrenocorticotropin hormone (ACTH) and corticosterone (CORT) responses to a subsequent challenge triggered by the release of corticotropin releasing hormone (CRH) from parvocellular neurosecretory cells in the paraventricular nucleus (PVN) of the hypothalamus. These neurons represent the final central integrative and output step of the HPA axis (Joels and Baram 2009). While sensitization of the stress response is complex and likely involves multiple central stress pathways, it appears that at least part of the plasticity in HPA axis regulation occurs at the level of the PVN (Grissom and Bhatnagar 2009; Franklin et al. 2012). The altered sensitivity to future stressors is both appropriate and necessary for promoting survival, yet persistent hyperactivity of the stress axis can be maladaptive (Sapolsky et al. 1985; McEwen and Sapolsky 1995), and has been implicated in contributing to a host of pathologies, including memory impairment (Lupien et al. 1998), anxiety disorders (Joëls 2011), depression (Krugers et al. 2010) and hypertension (De Kloet et al. 1998). The putative cellular mechanisms involved in the long-term effects of acute stress on the HPA axis, however, have remained largely unresolved.

Here we will discuss the underlying synaptic principles of synapses in the PVN, then examine the observations supporting a role for modifications of this microcircuitry following a single stress. What might such a modification require? At a minimum, the first stress must impart a signal, from which the neural network responding to stress extracts salient information (i.e. learns). This information is then stored (i.e. remembered) and then recalled during a subsequent stress to modify output. In many systems, neuromodulators function as 'associative' signals during a specific event to effectively embed information in a neural circuit that modifies synaptic function and network output in the future. Neuromodulators 'prime' the network through changes in intracellular machinery that may not impact ongoing synaptic transmission, but instead, alters how these synapse when recruited during the next behaviour. These experience-dependent changes in the rules for synaptic plasticity, known as metaplasticity, have been explored extensively at glutamate

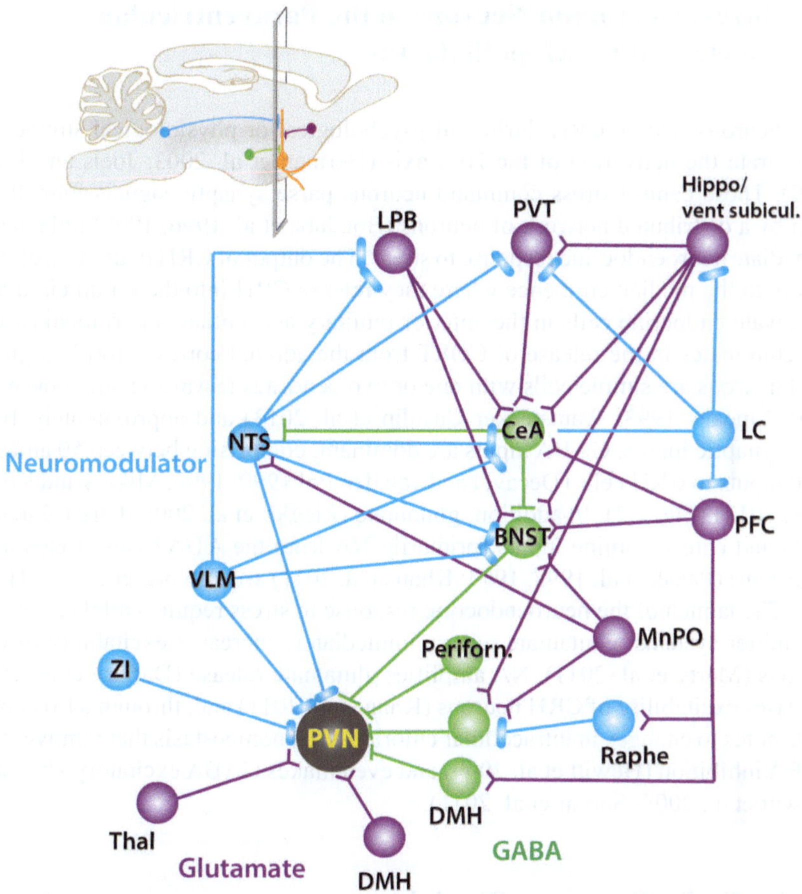

Fig. 7.1 Stress connectome. Compilation of the stress 'connectome' based on comprehensive analysis of anatomical and physiological literature examining recruitment of different brain nuclei in response to various stressors. Note convergence of inputs at the level of the paraventricular nucleus (*PVN*). Neuromodulators include noradrenaline, CRH, serotonin. The source of glutamate inputs to *PVN* remains poorly defined. The vast majority of gamma-aminobutyric acid (*GABA*) inputs originate in local hypothalamic subnuclei circumnavigating *PVN*. *PFC* prefrontal cortex, *LPB* lipopolysaccharide-binding protein, *PVT* paraventricular nucleus of the thalamus, *NTS* nucleus tractus solitarius, *VLM* ventrolateral medulla, *ZI* zona incerta, *DMH* dorsal medial hypothalamus, *BNST* bed nucleus of the stria terminalis, *LC* locus coeruleus

synapses (Perez-Otano and Ehlers 2005; Panatier et al. 2006; Kuzmiski et al. 2009). For stress-related behaviours, studies have examined synaptic function and information storage in the hippocampus, prefrontal cortex and amygdala (Sapolsky et al. 1985; Pavlides et al. 1993; Joels and Baram 2009). These structures are key components of many, but not every stress response. By comparison, every single stress activates CRH neurons in the PVN (Fig. 7.1); yet there are only a handful of studies that have attempted to link synaptic changes at this level to inappropriate activation of the stress axis (Hewitt et al. 2009; Wamsteeker et al. 2010; Kuzmiski et al. 2011).

7.2 Stress Command Neurons in the Paraventricular Nucleus of the Hypothalamus

PVN neurons release CRH during all psychological or physiological stresses and orchestrate the activation of the HPA axis (Herman et al. 2003; Joels and Baram 2009). These central stress command neurons parse synaptic signals funnelled to them by a distributed network of neurons (Boudaba et al. 1996, 1997) to launch an immediate neuroendocrine response to stress. The output of CRH neurons is directed largely to the median eminence where they release CRH into the portal circulation to activate endocrine cells in the anterior pituitary and initiate a hormonal cascade that culminates in the release of CORT from the adrenal cortex. Morphologically, CRH neurons are simple cells with one or two dendrites (Swanson and Sawchenko 1980; Liposits 1993; Wamsteeker Cusulin et al. 2013) and approximately 1000–3000 synaptic inputs. GABA inputs are dominant, comprising between 50 and 65 % of all inputs to CRH cells (Decavel and van den Pol 1990, 1992; Miklos and Kovacs 2002, 2012) (Fig. 7.2). In addition, glutamate (Ziegler et al. 2005; Ulrich-Lai et al. 2011) and catecholamine inputs (primarily NA from the A1/A2 cell groups in the brainstem) (Pacak et al. 1992, 1993; Khan et al. 2011) are also present on CRH neurons. The launch of the neuroendocrine response to stress requires all three of these transmitter systems. Glutamate release immediately increases excitability of CRH neurons (Marty et al. 2011), NA amplifies glutamate release (Daftary et al. 2000), increases excitability of CRH neurons (Khan et al. 2011) and, through α1 receptors, contributes to changes in intracellular chloride (Cl^-) homeostasis that removes tonic GABA inhibition (Hewitt et al. 2009) and even makes GABA excitatory after stress (Hewitt et al. 2009; Sarkar et al. 2011).

7.2.1 GABA Synapses: Physiology

GABA nerve terminals form dense clusters around CRH neurons (Miklos and Kovacs 2002; Fig. 7.2) to provide inhibitory tone that effectively restrains the activation of these cells under basal (non-stressed) conditions. This tone is the result of both spontaneous GABA release that acts on postsynaptic $GABA_A$ receptors (Hewitt et al. 2009) as well as high ambient extracellular levels of GABA that recruit extrasynaptic $GABA_A$ receptors (Sarkar et al. 2011). Relief of CRH neurons from tonic inhibition is necessary to launch the neuroendocrine response to stress. We have shown that this relief occurs when NA, released in the PVN at the onset of an acute stress, recruits α1 adrenoceptors and downregulates the K-Cl co-transporter, KCC2 (Hewitt et al. 2009). This causes an increase in intracellular Cl^- and results in a depolarizing shift in the Cl^- reversal potential (ECl^-) at the onset of stress (Fig. 7.3). Since $GABA_A$ inhibition relies on an electrochemical gradient that drives Cl^- into the cell at resting membrane potential, this shift in E_{GABA} collapses the Cl^- gradient resulting in anion efflux upon activation of $GABA_A$ receptors. This effectively converts GABA-mediated inhibitory synapses to excitatory synapses following a single stress (Hewitt et al. 2009). GABA excitations following KCC2

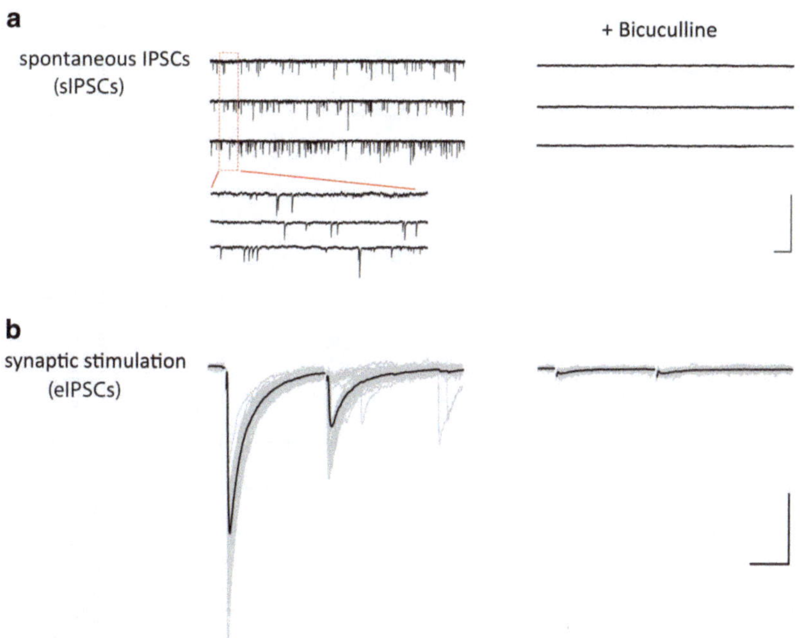

Fig. 7.2 GABA synapses on CRH neurons—physiology. **a** *Left panel* shows spontaneous *IPSCs* in a CRH neuron. *Right panel* shows complete block by 30 µM bicuculline methiodide, confirming they are $GABA_A$ receptor mediated (scale bars = 100 pA, 2 s). **b** *IPSCs* evoked by electrical stimulation of fibres immediately adjacent to CRH neuron. *Grey traces* are individual *IPSCs*. *Black trace* is averaged. *Right panel* shows block by bicuculline methiodide (scale bars = 50 pA, 20 ms). *sIPSC* spontaneous inhibitory synaptic current, *eIPSC* evoked inhibitory synaptic current

downregulation have been reported in other systems (Coull et al. 2003, 2005). Our findings were recently confirmed by others in the field who also went on to show a key contribution of extrasynaptic $GABA_A$ receptors in providing a tonic excitation of CRH neurons immediately after stress (Sarkar et al. 2011).

7.2.2 GABA Synapses: Plasticity

GABA synapses exhibit classical forms of plasticity including long-term potentiation (LTP) and long-term depression (LTD). In most instances, glutamate acting on postsynaptic N-methyl-D-aspartate receptors (NMDARs) or mGluRs induces changes in GABA efficacy (Chevaleyre and Castillo 2003; Marsden et al. 2007). In many cases, this results in the release of a retrograde signal from the postsynaptic neuron—nitric oxide for LTP (Bains and Ferguson 1997; Nugent et al. 2007; Crosby et al. 2011) and endocannabinoids for LTD (Gerdeman et al. 2002; Chevaleyre and Castillo 2003; Gerdeman and Lovinger 2003; Lovinger 2007). Although less frequently described, GABA synapses also undergo enduring postsynaptic changes (Jacob et al. 2008; Tyagarajan and Fritschy 2009; Castillo et al. 2011; Saliba et al.

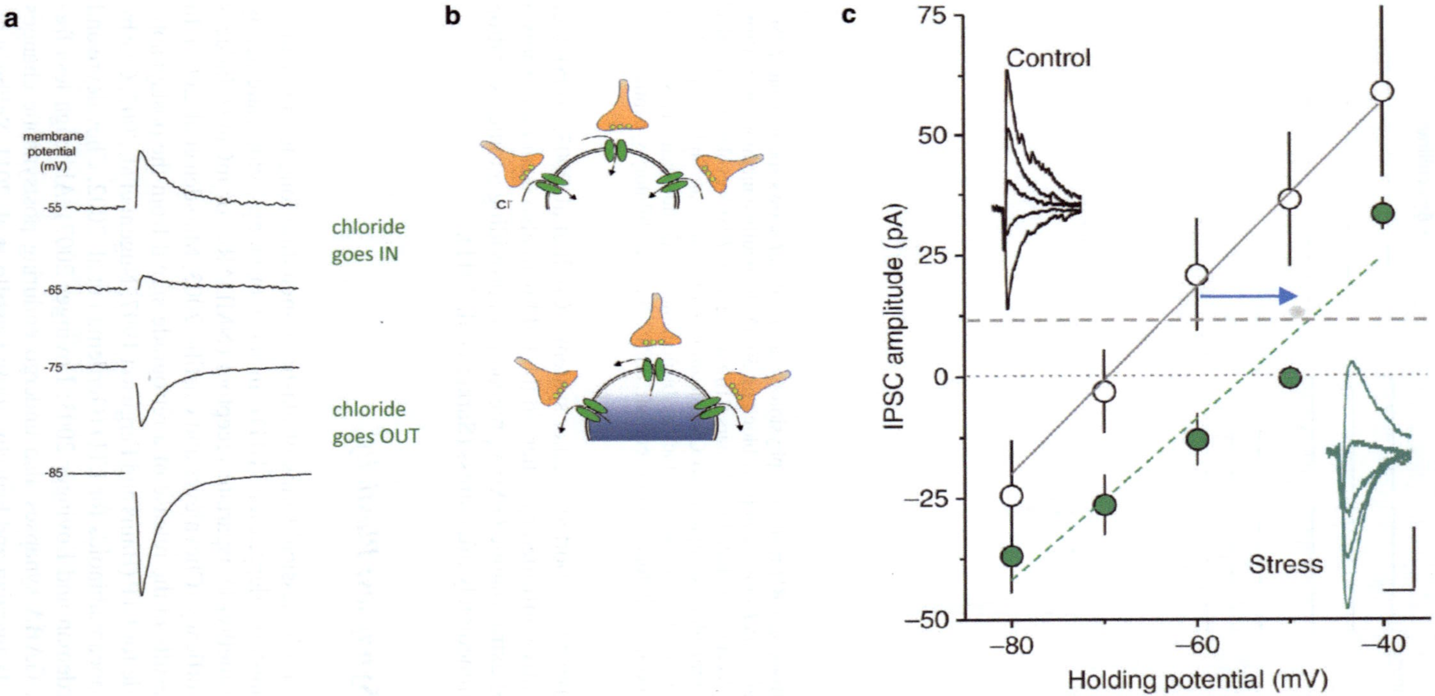

Fig. 7.3 Stress causes depolarizing shift in E_{GABA}. **a** Averaged *IPSCs* at different holding potentials. **b** Cartoon depicting chloride flux through GABAA receptor under conditions of low (*top*) and high (*bottom*) intracellular chloride. **c** I–V curve shows shift in the reversal potential for GABA-mediated synaptic currents. Data obtained using gramicidin perforated patch clamp recordings from naïve animals (*black*) and animals subjected to 30 min immobilization stress (*green*, scale bars = 50 pA, 20 ms). (Adapted from Hewitt et al. 2009)

2012), including a rapid insertion of $GABA_A$ receptors. Seminal work in the hippocampus (Patenaude et al. 2003) and cerebellum (Ouardouz and Sastry 2000; Sugiyama et al. 2008; Hirano and Kawaguchi 2012) makes a compelling link between mGluRs and postsynaptic changes in the strength of GABA synapses and implicates postsynaptic intracellular Ca^{2+} stores (Ouardouz and Sastry 2000), Ca^{2+}/calmodulin-dependent protein kinase II (CAMKII) and vesicular fusion (Kawaguchi and Hirano 2007). Increasing evidence in reduced preparations indicates that $GABA_A$ receptors in postsynaptic membranes are dynamic, with continuous turnover between synaptic and extrasynaptic pools. This regulated trafficking effectively controls the strength of synaptic inhibition and may be a precursor for the formation of new synapses (Wierenga et al. 2008; Dobie and Craig 2011) or a functional re-organization of synaptic input. Interestingly, consensus is building around the idea that $GABA_A$ receptors are inserted at extrasynaptic sites and then trafficked to synaptic targets (Bogdanov et al. 2006; Luscher et al. 2011). While the formation of new functional GABA synapses is well described in the developmental literature (Elmariah et al. 2005; Wierenga et al. 2008; Dobie and Craig n.d.), there are, few examples supporting the unmasking of new functional GABA synapses following behavioural manipulations in the adult animal.

7.3 Glutamate and mGluR1s for the Induction of LTP_{GABA}

We have described the basic properties of glutamate synaptic transmission onto putative CRH neurons in the PVN (Marty et al. 2011). Consistent with observations throughout the vertebrate nervous system, these synapses use α-amino-3-hydroxy-methyl-4-isoxazole propionic acid (AMPA) and NMDA receptors for fast transmission. Additionally, a number of mGluR subtypes have been identified in the PVN (Kiss et al. 1996; Kocsis et al. 1998). Below, I will describe the essential role of mGluR1 in the induction of LTP at GABA synapses. When bound by glutamate, these Gq-coupled receptors signal by increasing intracellular levels of PKC and IP3. The latter binds to IP3 receptors to liberate Ca^{2+} from intracellular stores.

7.4 NA and β-Adrenergic Receptors for Priming the System

Histological and electrophysiological studies provide evidence for the expression of both α- and β- adrenergic receptors (α-ARs and β-ARs) in PVN (Day et al. 1999). As noted above, α-ARs are necessary for driving the depolarizing shift in E_{GABA} at the onset of stress (Hewitt et al. 2009) and contribute to the immediate excitatory effects of NA on CRH neurons (Pacak et al. 1992, 1993, 1995). The role of β-ARs in this system remains unclear. In other systems, activation of β-ARs causes lasting downstream biochemical changes that position synapses and neural networks in a 'learning-ready' or labile state (Gelinas and Nguyen, 2005; Gelinas et al. 2008; O'Dell et al. 2010; Tenorio et al. 2010). One consequence of this labile state is that the threshold for the induction of activity-dependent plasticity is lowered for prolonged periods of time after exposure of synapses to NA. This has been documented extensively at glutamate synapses in numerous brain regions. For example, when released during emotional arousal and stress experience, NA enhances LTP induction and persistence at glutamate synapses in the amygdala (Hu et al. 2007). In brain slices, NA primes glutamate synapses in the hippocampus allowing them to undergo LTP in response to stimuli that are 'subthreshold' in the absence of NA (O'Dell et al. 2010; Tenorio et al. 2010). As noted above, stress is accompanied by activation of NA cell populations in the brainstem that project directly to the PVN.

7.5 LTP_{GABA} Following a Single Stress

A stereotyped recruitment of CRH stress command neurons is vital for managing the impending challenges of stress. We used an experimental approach in which rats or mice were exposed to acute behavioural stress and then asked questions about changes in synaptic function/plasticity using in vitro electrophysiological approaches. As noted above, we have shown that GABA synapses, which are critical for regulating the output of stress command neurons in the hypothalamus, are excitatory (not inhibitory) after stress. We have recently discovered that these synapses are also a key site for stress information processing during stress (Inoue et al. 2013). Specifically, we have observed that NA, released in the PVN during a single episode of stress, is sufficient to induce metaplasticity at GABA synapses on CRH neurons. This means that GABA synapses undergo activity-dependent, long-term potentiation (LTP_{GABA}) after stress, but not prior to stress (Inoue et al. 2013). Using a number of techniques, including electrophysiology and optogenetics, we show that the manifestation of this plasticity requires three essential steps:

1. During stress, NA primes intracellular pathways in CRH neurons. This relies on β-AR-mediated up-regulation of PKA and provides a necessary target for the induction LTP_{GABA}.
2. Following this stress, glutamate, released during subsequent bursts of synaptic activity activates mGluR1 to rapidly target primed pathways in CRH neurons and induce LTP_{GABA}.
3. LTP_{GABA} is expressed as a rapid insertion of GABAA receptors at previously silent GABA synapses in CRH neurons.

In combination with excitatory GABA, this potentiation may be particularly important in sensitizing this system to future stressors. Importantly, this metaplasticity is generalizable to other species (mice) and other intense stressors; in addition to immobilization stress, exposure to predator odour (30-min exposure to a chemical component of fox faeces in a fresh cage) is also sufficient to induce metaplasticity.

7.6 Summary

The observations described above provide new information about the synaptic regulation of PVN CRH neurons. These cells are key integration points for the neuroendocrine response to stress and our observations here indicate they may also contribute to stress sensitization. This is a key building block to better understand how HPA axis hyperactivity contributes to multiple stress-related disorders. Importantly, the changes in signalling at GABA synapses likely act in concert with impaired glucocorticoid negative feedback and/or other neuromodulators such as CRH and serotonin. It is important to note that the effects described above are only observed if hypothalamic slices are prepared immediately after exposure to acute stress. Extending the temporal window following stress, but prior to the preparation of slices results in a loss of the potentiation. It is intriguing that abnormalities in NA and, in particular, β adrenoceptors are thought to play a key role in PTSD and other stress related affective disorders, such as anxiety and depression. Our observations indicate that CRH neurons may be a key and under-investigated site for the emergence of these disorders.

References

Armario A, Escorihuela RM, Nadal R. Long-term neuroendocrine and behavioural effects of a single exposure to stress in adult animals. Neurosci Biobehav Rev. 2008;32:1121–35.

Bains JS, Ferguson AV. Nitric oxide regulates NMDA-driven GABAergic inputs to type I neurones of the rat paraventricular nucleus. J Physiol. 1997;499(Pt 3):733–46.

Bogdanov Y, Michels G, Armstrong-Gold C, Haydon PG, Lindstrom J, Pangalos M, Moss SJ. Synaptic GABAA receptors are directly recruited from their extrasynaptic counterparts. EMBO J. 2006;25:4381–9.

Boudaba C, Szabo K, Tasker JG. Physiological mapping of local inhibitory inputs to the hypothalamic paraventricular nucleus. J Neurosci. 1996;16:7151–60.

Boudaba C, Schrader LA, Tasker JG. Physiological evidence for local excitatory synaptic circuits in the rat hypothalamus. J Neurophysiol. 1997;77:3396–400.

Bruijnzeel AW, Stam R, Compaan JC, Croiset G, Akkermans LM, Olivier B, Wiegant VM. Long-term sensitization of Fos-responsivity in the rat central nervous system after a single stressful experience. Brain Res. 1999;819:15–22.

Bruijnzeel AW, Stam R, Compaan JC, Wiegant VM. Stress-induced sensitization of CRH-ir but not P-CREB-ir responsivity in the rat central nervous system. Brain Res. 2001;908:187–96.

Castillo PE, Chiu CQ, Carroll RC. Long-term plasticity at inhibitory synapses. Curr Opin Neurobiol. 2011;21:328–38.

Chevaleyre V, Castillo PE. Heterosynaptic LTD of hippocampal GABAergic synapses: a novel role of endocannabinoids in regulating excitability. Neuron. 2003;38:461–72.

Coull JA, Boudreau D, Bachand K, Prescott SA, Nault F, Sik A, De Koninck P, De Koninck Y. Trans-synaptic shift in anion gradient in spinal lamina I neurons as a mechanism of neuropathic pain. Nature. 2003;424:938–42.

Coull JA, Beggs S, Boudreau D, Boivin D, Tsuda M, Inoue K, Gravel C, Salter MW, De Koninck Y. BDNF from microglia causes the shift in neuronal anion gradient underlying neuropathic pain. Nature. 2005;438:1017–21.

Crosby KM, Inoue W, Pittman QJ, Bains JS. Endocannabinoids gate state-dependent plasticity of synaptic inhibition in feeding circuits. Neuron. 2011;71:529–41.

Daftary SS, Boudaba C, Tasker JG. Noradrenergic regulation of parvocellular neurons in the rat hypothalamic paraventricular nucleus. Neuroscience. 2000;96:743–51.

Day HE, Campeau S, Watson SJJ, Akil H. Expression of alpha(1b) adrenoceptor mRNA in corticotropin-releasing hormone-containing cells of the rat hypothalamus and its regulation by corticosterone. J Neurosci. 1999;19:10098–106.

De Kloet ER, Vreugdenhil E, Oitzl MS, Joels M. Brain corticosteroid receptor balance in health and disease. Endocr Rev. 1998;19:269–301.

Decavel C, van den Pol AN. GABA: a dominant neurotransmitter in the hypothalamus. J Comp Neurol. 1990;302:1019–37.

Decavel C, van den Pol AN. Converging GABA- and glutamate-immunoreactive axons make synaptic contact with identified hypothalamic neurosecretory neurons. J Comp Neurol. 1992;316:104–16.

Dobie FA, Craig AM. Inhibitory synapse dynamics: coordinated presynaptic and postsynaptic mobility and the major contribution of recycled vesicles to new synapse formation. J Neurosci. 2011;31:10481–93.

Elmariah SB, Oh EJ, Hughes EG, Balice-Gordon RJ. Astrocytes regulate inhibitory synapse formation via Trk-mediated modulation of postsynaptic GABAA receptors. J Neurosci. 2005;25:3638–50.

Franklin TB, Saab BJ, Mansuy IM. Neural mechanisms of stress resilience and vulnerability. Neuron. 2012;75:747–61.

Gelinas JN, Nguyen PV. Beta-adrenergic receptor activation facilitates induction of a protein synthesis-dependent late phase of long-term potentiation. J Neurosci. 2005;25:3294–303.

Gelinas JN, Tenorio G, Lemon N, Abel T, Nguyen PV. Beta-adrenergic receptor activation during distinct patterns of stimulation critically modulates the PKA-dependence of LTP in the mouse hippocampus. Learn Mem. 2008;15:281–9.

Gerdeman GL, Lovinger DM. Emerging roles for endocannabinoids in long-term synaptic plasticity. Br J Pharmacol. 2003;140:781–9.

Gerdeman GL, Ronesi J, Lovinger DM. Postsynaptic endocannabinoid release is critical to long-term depression in the striatum. Nat Neurosci. 2002;5:446–51.

Grissom N, Bhatnagar S. Habituation to repeated stress: get used to it. Neurobiol Learn Mem. 2009;92:215–24.

Herman JP, Figueiredo H, Mueller NK, Ulrich-Lai Y, Ostrander MM, Choi DC, Cullinan WE. Central mechanisms of stress integration: hierarchical circuitry controlling hypothalamo-pituitary-adrenocortical responsiveness. Front Neuroendocrinol. 2003;24:151–80.

Hewitt SA, Wamsteeker JI, Kurz EU, Bains JS. Altered chloride homeostasis removes synaptic inhibitory constraint of the stress axis. Nat Neurosci. 2009;12:438–43.

Hirano T, Kawaguchi S-Y. Regulation of inhibitory synaptic plasticity in a Purkinje neuron. Cerebellum. 2012;11:453–4.

Hu H, Real E, Takamiya K, Kang M-G, Ledoux J, Huganir RL, Malinow R. Emotion enhances learning via norepinephrine regulation of AMPA-receptor trafficking. Cell. 2007;131:160–73.

Inoue W, Baimoukhametova DV, Füzesi T, Cusulin JIW, Koblinger K, Whelan PJ, Pittman QJ, Bains JS. Noradrenaline is a stress-associated metaplastic signal at GABA synapses. Nat Neurosci. 2013;16:605–12.

Jacob TC, Moss SJ, Jurd R. GABA(A) receptor trafficking and its role in the dynamic modulation of neuronal inhibition. Nat Rev Neurosci. 2008;9:331–43.

Joëls M. Impact of glucocorticoids on brain function: relevance for mood disorders. Psychoneuroendocrinology. 2011;36:406–14.

Joels M, Baram TZ. The neuro-symphony of stress. Nat Rev Neurosci. 2009;10:459–66.

Kawaguchi SY, Hirano T. Sustained structural change of GABA(A) receptor-associated protein underlies long-term potentiation at inhibitory synapses on a cerebellar Purkinje neuron. J Neurosci. 2007;27:6788–99.
Khan AM, Kaminski KL, Sanchez-Watts G, Ponzio TA, Kuzmiski JB, Bains JS, Watts AG. MAP kinases couple hindbrain-derived catecholamine signals to hypothalamic adrenocortical control mechanisms during glycemia-related challenges. J Neurosci. 2011;31:18479–91.
Kiss J, Gorcs TJ, Kuhn R, Knopfel T, Csaky A, Halasz B. Distribution of metabotropic glutamate receptor 1a in the rat hypothalamus: an immunocytochemical study using monoclonal and polyclonal antibody. Acta Biol Hung. 1996;47:221–37.
Kocsis K, Kiss J, Gorcs T, Halasz B. Metabotropic glutamate receptor in vasopressin, CRF and VIP hypothalamic neurones. Neuroreport. 1998;9:4029–33.
Krugers HJ, Lucassen PJ, Karst H, Joëls M. Chronic stress effects on hippocampal structure and synaptic function: relevance for depression and normalization by anti-glucocorticoid treatment. Front Synaptic Neurosci. 2010;2:24.
Kuzmiski JB, Pittman QJ, Bains JS. Metaplasticity of hypothalamic synapses following in vivo challenge. Neuron. 2009;62:839–49.
Kuzmiski JB, Marty V, Baimoukhametova DV, Bains JS. Stress-induced priming of glutamate synapses unmasks associative short-term plasticity. Nat Neurosci. 2011;13:1257–64.
Liposits Z. Ultrastructure of hypothalamic paraventricular neurons. Crit Rev Neurobiol. 1993;7:89–162.
Lovinger DM. Endocannabinoid liberation from neurons in transsynaptic signaling. J Mol Neurosci. 2007;33:87–93.
Lupien SJ, de Leon M, de Santi S, Convit A, Tarshish C, Nair NP, Thakur M, McEwen BS, Hauger RL, Meaney MJ. Cortisol levels during human aging predict hippocampal atrophy and memory deficits. Nat Neurosci. 1998;1:69–73.
Luscher B, Fuchs T, Kilpatrick CL. GABAA receptor trafficking-mediated plasticity of inhibitory synapses. Neuron. 2011;70:385–409.
Marsden KC, Beattie JB, Friedenthal J, Carroll RC. NMDA receptor activation potentiates inhibitory transmission through GABA receptor-associated protein-dependent exocytosis of GABA(A) receptors. J Neurosci. 2007;27:14326–37.
Marty V, Kuzmiski JB, Baimoukhametova DV, Bains JS. Short-term plasticity impacts information transfer at glutamate synapses onto parvocellular neuroendocrine cells in the paraventricular nucleus of the hypothalamus. J Physiol. 2011;589:4259–70.
McEwen BS, Sapolsky RM. Stress and cognitive function. Curr Opin Neurobiol. 1995;5:205–16.
Miklos IH, Kovacs KJ. GABAergic innervation of corticotropin-releasing hormone (CRH)-secreting parvocellular neurons and its plasticity as demonstrated by quantitative immunoelectron microscopy. Neuroscience. 2002;113:581–92.
Miklos IH, Kovacs KJ. Reorganization of synaptic inputs to the hypothalamic paraventricular nucleus during chronic psychogenic stress in rats. Biol Psychiatry. 2012;71:301–8.
Nugent FS, Penick EC, Kauer JA. Opioids block long-term potentiation of inhibitory synapses. Nature. 2007;446:1086–90.
O'Dell TJ, Connor SA, Gelinas JN, Nguyen PV. Viagra for your synapses: enhancement of hippocampal long-term potentiation by activation of beta-adrenergic receptors. Cell Signal. 2010;22:728–36.
Ouardouz M, Sastry BR. Mechanisms underlying LTP of inhibitory synaptic transmission in the deep cerebellar nuclei. J Neurophysiol. 2000;84:1414–21.
Pacak K, Armando I, Fukuhara K, Kvetnansky R, Palkovits M, Kopin IJ, Goldstein DS. Noradrenergic activation in the paraventricular nucleus during acute and chronic immobilization stress in rats: an in vivo microdialysis study. Brain Res. 1992;589:91–6.
Pacak K, Palkovits M, Kvetnansky R, Kopin IJ, Goldstein DS. Stress-induced norepinephrine release in the paraventricular nucleus of rats with brainstem hemisections: a microdialysis study. Neuroendocrinology. 1993;58:196–201.
Pacak K, Palkovits M, Kopin IJ, Goldstein DS. Stress-induced norepinephrine release in the hypothalamic paraventricular nucleus and pituitary-adrenocortical and sympathoadrenal activity: in vivo microdialysis studies. Front Neuroendocrinol. 1995;16:89–150.

Panatier A, Gentles SJ, Bourque CW, Oliet SH. Activity-dependent synaptic plasticity in the supraoptic nucleus of the rat hypothalamus. J Physiol. 2006;573:711–21.

Patenaude C, Chapman CA, Bertrand S, Congar P, Lacaille J-C. GABAB receptor- and metabotropic glutamate receptor-dependent cooperative long-term potentiation of rat hippocampal GABAA synaptic transmission. J Physiol. 2003;553:155–67.

Pavlides C, Watanabe Y, McEwen BS. Effects of glucocorticoids on hippocampal long-term potentiation. Hippocampus. 1993;3:183–92.

Perez-Otano I, Ehlers MD. Homeostatic plasticity and NMDA receptor trafficking. Trends Neurosci. 2005;28:229–38.

Saliba RS, Kretschmannova K, Moss SJ. Activity-dependent phosphorylation of GABAA receptors regulates receptor insertion and tonic current. EMBO J. 2012;31:2937–51.

Sapolsky RM, Krey LC, McEwen BS. Prolonged glucocorticoid exposure reduces hippocampal neuron number: implications for aging. J Neurosci. 1985;5:1222–7.

Sarkar J, Wakefield S, MacKenzie G, Moss SJ, Maguire J. Neurosteroidogenesis is required for the physiological response to stress: role of neurosteroid-sensitive GABAA receptors. J Neurosci. 2011;31:18198–210.

Sugiyama Y, Kawaguchi SY, Hirano T. mGluR1-mediated facilitation of long-term potentiation at inhibitory synapses on a cerebellar Purkinje neuron. Eur J Neurosci. 2008;27:884–96.

Swanson LW, Sawchenko PE. Paraventricular nucleus: a site for the integration of neuroendocrine and autonomic mechanisms. Neuroendocrinology. 1980;31:410–7.

Tenorio G, Connor SA, Guévremont D, Abraham WC, Williams J, O'Dell TJ, Nguyen PV. "Silent" priming of translation-dependent LTP by ß-adrenergic receptors involves phosphorylation and recruitment of AMPA receptors. Learn Mem. 2010;17:627–38.

Tyagarajan SK, Fritschy JM. GABA(A) receptors, gephyrin and homeostatic synaptic plasticity. J Physiol. 2009;588:101–6.

Ulrich-Lai YM, Jones KR, Ziegler DR, Cullinan WE, Herman JP. Forebrain origins of glutamatergic innervation to the rat paraventricular nucleus of the hypothalamus: differential inputs to the anterior versus posterior subregions. J Comp Neurol. 2011;519:1301–19.

Wamsteeker JI, Kuzmiski JB, Bains JS. Repeated stress impairs endocannabinoid signaling in the paraventricular nucleus of the hypothalamus. J Neurosci. 2010;30:11188–96.

Wamsteeker Cusulin JI, Füzesi T, Watts AG, Bains JS. Characterization of corticotropin-releasing hormone neurons in the paraventricular nucleus of the hypothalamus of Crh-IRES-Cre mutant mice. PLoS ONE. 2013;8:e64943.

Wierenga CJ, Becker N, Bonhoeffer T. GABAergic synapses are formed without the involvement of dendritic protrusions. Nat Neurosci. 2008;11:1044–52.

Ziegler DR, Cullinan WE, Herman JP. Organization and regulation of paraventricular nucleus glutamate signaling systems: N-methyl-D-aspartate receptors. J Comp Neurol. 2005;484:43–56.

Chapter 8
Stress Modulation of Synaptic Plasticity in the Hippocampus

Menahem Segal and Nicola Maggio

Abstract Despite its homogeneous, highly ordered structure, the hippocampus serves very different functions along its septo-temporal axis; while the dorsal (septal) end is associated with cognition, its ventral (temporal) region regulates emotion and anxiety. As stress has been known to affect cognitive functions in the brain, it is of prime interest to try and understand how the hippocampus assumes its cognitive roles under stressful conditions. We hypothesize that stress switches the focus of control of hippocampal functions by differential modulation of synaptic plasticity in the dorsal and ventral sectors of the hippocampus through the activation/suppression of steroid hormones and monoamine neurotransmission. Herein, we will review recent studies on the effects of stress on synaptic plasticity in the dorsal and ventral hippocampus and outline the outcomes of this modulation on stress-related global functions of the temporal lobe, which hosts the hippocampus. We propose that steroid hormones act as molecular switches to change the strength of synaptic connectivity in the hippocampus following stress, thus regulating the routes by which the hippocampus is functionally linked to the rest of the brain. This role has profound implications for the pathophysiology of psychiatric disorders.

Abbreviations

CRH	Corticotropin releasing hormone
DH	Dorsal hippocampus
iGR	Intracellular glucocorticosterone receptors
iMR	Intracellular mineralocorticosterone receptors
LTP	Long term potentiation
mGluR	Metabotropic glutamate receptor
mGR	Membrane glucocorticosterone receptor
mIPSC	Miniature inhibitory post synaptic currents

M. Segal (✉)
Department of Neurobiology, Weizmann Institute of Science, Rehovot, Israel
e-mail: menahem.segal@weizmann.ac.il

N. Maggio
Department of Neurology, Talpiot Medical Leadership Program, J. Sagol Neuroscience Center, The Chaim Sheba Medical Center, Ramat Gan, Israel

M. Popoli et al. (eds.), *Synaptic Stress and Pathogenesis of Neuropsychiatric Disorders*,
DOI 10.1007/978-1-4939-1056-4_8,

mMR	Membrane mineralocorticosterone receptors
VGCC	Voltage gated calcium current
VH	Ventral Hippocampus

8.1 Introduction: The Hippocampus, more than One?

The view of the hippocampus, the most intensively studied brain structure, has changed drastically over the past century. Considered part of the Papez circuit, the hippocampus was originally related to affective circuits in the brain. The striking observation of loss of short-term memory in epileptic patients undergoing hippocampectomy, led to a series of animal studies testing the hypothesis that the hippocampus is a locus of short-term storage of memories. This enthusiasm about the role of the hippocampus in memory neglected the fact that some of these patients had little cognitive deficits, but suffered from severe emotional problems following the operation. Only more recent studies began to appreciate the significant role of the ventral hippocampus (VH) in emotion and anxiety, distinctly different from the more traditional role of the dorsal hippocampus (DH) in cognitive functions. This assertion is based on lesion and stimulation studies as well as recording of single neurons in freely moving animals, and on studies in hippocampal slice preparations. Indeed, there are distinct differences in the distribution of synaptic proteins between the two poles of the hippocampus. While the roles of the two regions of the hippocampus in cognitive versus affective functions becomes evident (see below), there are still unsolved issues related to this distinction. First, why is it so important to have two major brain functions in one rather small structure. Second, while the hippocampus has a lamellar organization, meaning that the entire input/output pathway is embedded in parallel lamella along the septo-temporal axis, it does contain extensive longitudinal fiber systems that unite the entire hippocampus into one apparent functional unit. Since the VH and the DH have different connections with the rest of the brain, with the DH projecting mainly to cortical structures, whereas the VH mainly to the amygdala and hypothalamus, it is apparent that the weight of connectivity of the hippocampus may switch between the dorsal and ventral poles, in relation to the ambient state of the animal. The factors that determine this switch and the rules that govern them will be discussed below.

8.2 Corticosteroid receptors in the brain

Steroid hormones have been traditionally associated with regulation of peripheral organs, associated with stress (corticosterone) or with gonadal function (estrogen and androgens). Over the years, it became evident that these hormones also act within the hypothalamus, in a feedback regulatory loop, to affect the release of the neural factors that modulate production of the steroid hormones. More recently, several observations have elucidated new roles of steroid hormones in modulating higher CNS functions. Specifically, both stress and steroid hormones have been shown to affect synaptic receptors and ion channels and therefore regulate synaptic transmission and neuronal plasticity in several different ways. Furthermore, corticosterone is not the only player in the control of stress responses, and the central factor that regulates it, corticotropin releasing hormone (CRH) has been described to exert an important role in modulating neuronal plasticity in the hippocampus and elsewhere (Joels and Baram 2009). Consequently, stress hormones have been implicated in processes ranging from homeostatic to cognitive functions. Likewise, in some disorders of the nervous system, hormones have been shown to play critical roles: favoring or halting the disease process. Thus, the interaction between peripheral hormones and central networks seem to be more intense than ever before.

In the present study, we review current knowledge on the effects of steroid hormones on synaptic plasticity and define their influence on cognitive and emotional functions of the DH and VH.

Following the exposure to stressful stimuli, the steroid hormone corticosterone (cortisol in humans) is released from the adrenal glands in order to set up the best response to the challenge by acting on steroid receptors (de Kloet et al. 2005). These are distributed throughout the body and have a particularly dense distribution in the CNS (de Kloet et al. 2005). In the brain, the cellular and molecular targets for the action of corticosterone include, in addition to basic metabolic processes, an effect on excitatory (Karst and Joels 2005) and inhibitory (Maggio and Segal 2009a, b) synaptic transmission, as well as an effect on voltage-gated calcium channels (VGCC) (Karst et al. 2000; Chameau et al. 2007). These effects are mediated by the activation of mineralocorticoid receptors (MRs) and glucocorticoid receptors (GRs) (Joels 1999; de Kloet et al. 2005; Joels 2008, Joels et al. 2008). Initially, it was suggested that both receptors act as nuclear transcription factors that modify protein synthesis and produce a slow, persistent change in the function of the cell (de Kloet et al. 1993, 2008). More recently, the existence of a new family of membrane-bound MR and GR (mMR and mGR, respectively), which act through novel nongenomic pathways, has been reported (Karst et al. 2005; de Kloet et al. 2008). In this route, mMR and mGR can rapidly affect ionic conductances and thereby modify cell excitability and function (Karst et al. 2005; de Kloet et al. 2008). These membrane-bound receptors appear to differ from their intracellular cognates, not only in their location on the cell membrane, but also in their molecular structures (Joels et al. 2008), in their affinities for corticosterone, and in their downstream mechanisms of action which involve activation of G proteins (Joels et al. 2008). Specifically, intracellular MR (iMR) have a very high affinity for corticosterone and are highly expressed in all hippocampal subfields, as well as in cells of the central amygdala, lateral septum, and some motor nuclei in the brainstem (Joels 2006). Intracellular GR (iGR) have a relatively low affinity, are widely distributed throughout the brain, and are expressed both in neurons and in glia (Joels 2006). Consequently, it has been proposed that iMR hardly participates, if at all, in the fast response to stressful stimuli, due to their characteristic of being already saturated by the low ambi-

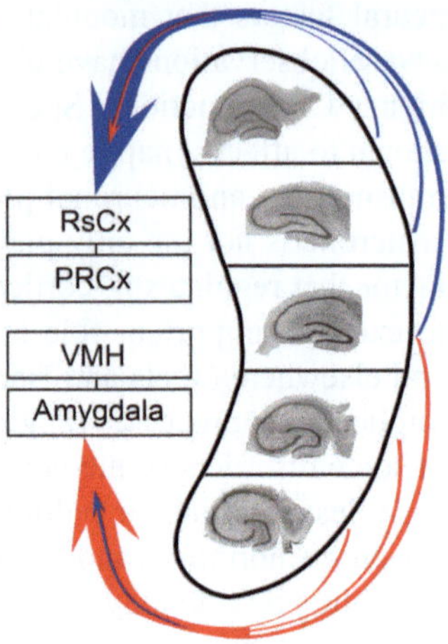

Fig. 8.1 Schematic of the hippocampus and its connections with the main efferent systems. While the dorsal hippocampus is connected with cortical structures, e.g., prefrontal and retrosplenial cortex (PRCx and RsCx, respectively), the VH is linked to the amygdala and ventromedial nucleus of the hypothalamus (VMH). *Blue* indicates the main functional connections at rest and *red* the main functional connections during stress.

ent levels of corticosterone at rest (Joels 2006, 2008). Conversely, iGR have been reported to become gradually activated by rising levels of corticosterone following a stressful event (Joels 2006, 2008) (Fig. 8.1). Therefore, under physiological conditions, cells that co-express both receptor types, such as principal cells in the CA1 region, the dentate gyrus (DG), and the central amygdala, will shift between predominant iMR activation and concurrent mMR and iGR activation (Joels and Krugers 2007).

8.3 The Hippocampus: One structure, two functions?

The realization that there might be intrinsic differences between CA1 neurons of the DH and VH, which may underlie the differences in their firing properties as well as their ability to undergo plastic changes, led to several attempts to characterize the biophysical properties of CA1 neurons in the two sectors. The first study, by Maggio and Segal (2009a) reported that neurons in the two sectors had similar resting potentials, input resistance and membrane time constant, but the VH neurons generated fewer action potentials to a depolarizing current pulse than DH neurons. A more recent study (Dougherty et al. 2012) reported opposite results, with the VH neurons being more depolarized by 7 mV than the DH neurons, a difference that resulted in more action potential discharges to the same depolarizing current pulse. In a more recent study, they propose (Dougherty et al. 2013, Marcelin et al. 2012) that VH neurons have different compositions of HCN channels, responsible for Ih in these neurons. This might underlie the 7 mV depolarization and the higher input resistance of their VH CA1 neurons and eventually explain the difference in excitability between the two studies. Notably, if the VH neurons would be more excitable, then a larger amplitude theta rhythm is expected to be generated in this area. However, theta rhythm of smaller amplitude compared to their cognates in

DH has been reported (Patel et al. 2012). In addition, a different excitability of the CA1 neurons in the two regions should generate larger response amplitudes to Schaffer's collateral stimulation in VH cells compared to DH neurons. However, several experiments have shown that both DH and VH have similar input/output relations (Papatheodoropoulos and Kostopoulos 2000; Grigoryan et al. 2012). Altogether these studies suggest that there is a genuine difference in synaptic plasticity between DH and VH.

Several other hippocampal features are affected by stress differently in the DH and VH. For example, neurogenesis is one of the unique properties of the DG of the hippocampus, one of two locations in the brain where adult neurogenesis was characterized (O'Leary et al. 2012). VH neurogenesis is more affected by stress than DH, and drugs that reduce the effects of stress are active primarily in the VH (Felice et al. 2012, Xia et al. 2012, Tanti et al. 2012, Hawley and Leasure 2012, Hawley et al. 2012). This difference may be related to different regulation of brain derived neurotrophic factor (BDNF), which has been linked to neurogenesis, to depression and to the DH/VH disparity (Roth et al. 2011). Furthermore, stress-induced memory impairments involve different steroid receptors in DH and VH (Dorey et al. 2012), and stimulation of the VH ameliorates fear memory (Cleren et al. 2013). Also, exercise facilitates recovery from stress-induced protein synthesis decline in VH (Daniels et al. 2012). Finally, the VH is more sensitive to redox dysregulation than the DH, and the difference is reflected in GABAergic interneurons as well as electrical activity (Steullet et al. 2010).

These and other studies indicate that the DH and VH may react to stressful stimulation in a different manner, and thus, a careful analysis of the direct effect of stress and corticosterone in the VH and DH is justified.

8.4 Corticosteroid receptors in the regulation of hippocampal LTP

The identification of the molecular cascades of corticosteroids actions in the brain resulted in a series of studies examining the role of corticosterone in neuronal plasticity as well as in the cellular mechanisms underlying learning and memory such as long-term potentiation (LTP) and long-term depression (LTD) (Bliss and Collingridge 1993). Initial studies indicated that induction of LTP in the hippocampal area CA1 is impaired in a rat exposed to behavioral stress, such as inescapable shock (Foy et al. 1987; Shors et al. 1989). Administration of high doses of corticosterone either in vivo (Diamond et al. 1992) or in vitro (Pavlides et al. 1996; Alfarez et al. 2002) mimicked this effect, indicating that corticosterone is likely to mediate this action of stress. Specifically, corticosterone-induced impairment of LTP seems to be due to the activation of iGR, which depresses NMDA receptors (Calabrese et al. 2012) and NMDA-dependent LTP (Krugers et al. 2005) (Fig. 8.1b). Conversely, it was also shown that LTP could be enhanced in the presence of low-to-moderate concentrations of corticosterone, while in absence of corticosterone LTP induction was impaired (Diamond et al. 1992). These studies show that the effects of corticosteroids on LTP induction are dose-dependent and follow an inverted U-shaped curve (Fig. 8.1) (Diamond et al. 1992; Joels 2006).

Further studies, however, have presented a more complex view on the effects of steroid hormones on synaptic plasticity. Specifically, it seems that the same dosage of corticosterone that impairs NMDA-dependent LTP can in fact enhance VGCC-dependent LTP (Krugers et al. 2005). This species of LTP is found in the amygdale where it is believed to underlie the formation of fear memories (Blair et al. 2001; Bauer et al. 2002) and can be evoked in the hippocampus as well (Borroni et al. 2000) (Fig. 8.1b). Interestingly, in the hippocampus, corticosterone appears to enhance VGCC LTP through an iGR-dependent mechanism (Krugers et al. 2005). It has been proposed that this effect requires a genomic pathway, as it occurs after a long delay between the exposure to stress and/or corticosterone and the recordings (Krugers et al. 2005), thus probably depending on the binding of GR homodimers to DNA that causes an increase in calcium currents (Karst and Joels 2005; Chameau et al. 2007). Recent data from our group have shown that MRs are also able to enhance VGCC LTP (Maggio and Segal 2007b): either stress or physiological concentrations of corticosterone can enhance LTP in the VH, while inhibiting it in the DH (Maggio and Segal 2007b). In particular, corticosterone enhances LTP through MRs since a selective MR agonist, aldosterone, shares the same effect in the VH (Maggio and Segal 2007b). The proposed mechanism excludes an interaction between MR and NMDA receptors, as aldosterone by itself does not increase NMDA-dependent synaptic potentials (Maggio and Segal 2007b). Conversely, MR-induced LTP can be blocked by nifedipine, suggesting that VGCCs are likely responsible for this effect (Maggio and Segal 2007b) (Fig. 8.1b). It is likely that MR activates VGCC by modulating ionic conductances or changing VGCC activation kinetics. In vivo experiments have shown that MR activation is able to increase LTP in the DH as well (Avital et al. 2006). Specifically, animals which were injected with a GR antagonist prior to the stressful exposure, such that only MR could be activated by stress, show a much larger LTP than controls. In contrast, those animals previously injected with an MR antagonist and then exposed to stress, allowing only GR activation, show a much lower LTP than controls (Avital et al. 2006). These recordings were performed in the DG and even though there could be differences in the effects of stress and steroids between the DG and CA1 (Joels and Krugers 2007), MRs were still shown to mediate an enhancement of LTP.

These experiments raise several issues. It could be argued that the experiments in the VH were conducted using an in vitro preparation where ambient corticosterone maintained normally through the circulation is washed out. Consequently, MRs are not occupied in the slice, and are ready to be activated by the superfused drug and produce LTP enhancement in the VH. This might not reflect the situation in the intact animal, where the brain is constantly exposed to fluctuating concentrations of corticosterone. In fact, MR should be already saturated by the resting concentration of corticosterone and should not respond to the stress-induced rise of corticosterone in the presence of a GR blockade. This, however, does not seem to be the case (Avital et al. 2006). Furthermore, even though both MRs and GRs are expressed in the VH, corticosterone action is mediated by activation of MR rather than GR. This reflects the observation that in the VH, MR concentration is double that of GR (Robertson et al. 2005). If so, according to the U-shaped curve model of corticosterone effects, MR should be saturated faster by the rising concentration of corticosterone and their effect should fade away faster in favor of the slower GR activation. This is in contrast with the experimental evidence. Altogether, it seems that the simple, dose-dependent, inverted U-shaped curve does not fully explain the modulatory functions of MR and GR on LTP in the different sectors of the hippocampus, therefore calling for the involvement of other factors.

A possible mechanism that may clarify the MR-dependent enhancement of LTP should take into consideration the activation of mMR (Fig. 8.1). These receptors act through a faster mechanism (de Kloet et al. 2008) and have lower affinities for corticosterone compared to their intracellular cognates (Joels 2008) and similar to that of the iGR (Joels 2008). In addition, MR activation enhances LTP in the VH within 1 h, too short time window to be accounted for by activation of genomic mechanisms (Joels and Krugers 2007; Joels 2008), but compatible with the faster time course of the nongenomic routes. Thus, mMR could be the preferential target for rising concentrations of corticosterone in the VH if one takes into account the similar affinities for corticosterone between mMR and iGR, and the denser distribution of the former over the latter (Robertson et al. 2005) (Fig. 8.1a, b).

MRs are likely to enhance LTP through activation of VGCC. In our experiments, we could not detect any effect of iGR on VGCC LTP. This could most likely be due to the shorter time window of observation in our experiments compared to those done by others (Krugers et al. 2005). In any case, both MR and GR were reported to increase VGCC LTP (Krugers et al. 2005; Maggio and Segal 2007b). This apparent contrast could probably be explained by considering the different time courses of MR and GR enhancement of VGCC LTP. Specifically, MR has an earlier effect than GR and it could be that in the VH stress mediates a fast enhancement of LTP by MR followed by a second, slow increase in LTP due to GR activation. This proposal is compatible with the proposed role of the VH as a key player in the pathway that conveys stressful information to the hypothalamus and the amygdale so as to organize the stress response (Fig. 8.2) (Moser and Moser 1998; Maggio and Segal 2010; Segal et al. 2010).

8.5 Corticosteroid regulation of hippocampal functions

The regulation of LTP by corticosterone in the hippocampus has profound system implications. Following stress, the quick MR-mediated increase in LTP facilitates the flow of the information related to stress from the VH to the ventral hypothalamus and other lower brain centers, so that the autonomic response to stress can be organized. Later on, the MR-mediated response fades away and the effect of GR dominates. As previously mentioned, GR enhancement of VGCC LTP has been shown to have a role in the formation of fear memories in the amygdale (Blair et al.

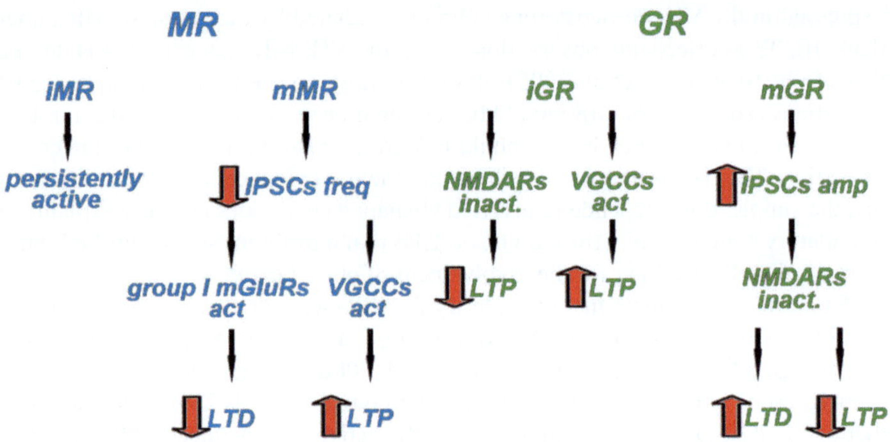

Fig. 8.2 Summary diagram of the main corticosterone effects in the hippocampus, there are intracellular mineralocorticosterone receptors (*iMR*), membrane mineralocorticosterone receptors (*mMR*), and the same for glucocorticosterone receptors (*GR*). Each receptor type has specific effects on the ability of CA1 neurons in the hippocampus to undergo long-term potentiation (*LTP*) in response to afferent stimulation. *MR* mineralocorticoid receptors, *IPSC* inhibitory postsynaptic currents, *LTD* long-term depression, *VGCC* voltage-gated calcium current, *mGluR* metabotropic glutamate receptor, *iGR* intracellular glucocorticosterone receptors, *NMDAR* N-methyl-D-aspartate receptor, *mGR* membrane glucocorticosterone receptor

2001; Bauer et al. 2002). In this respect, GR could play the same function in the VH: the formation of the memory for the stressful event at the VH-amygdala pathway. Indeed, the evidence that MR and GR act on the same mechanism can have different purposes due to the time window of the respective outcomes that take place. Considering this, it could be interesting to study the relationship between the MR and GR responses in the VH.

In the DH, the reduction of LTP is likely to be mediated by GR (Maggio and Segal 2007b). This effect seems to occur in less than 1 h, a relatively quick response that is unlikely to be mediated by a genomic mechanism. GR could reduce NMDA-mediated LTP either by a direct or an indirect mechanism. As far as it concerns the indirect mechanism hypothesis, we have demonstrated that a GR agonist, dexamethasone, increases IPSCs and mIPSCs amplitude in the DH within 10 min (Maggio and Segal 2009a, 2012), consistent with the possible activation of mGR. Therefore, the increase in GABAA conductance could hyperpolarize the membrane, thus preventing the cell from reaching the threshold of depolarization that unlocks NMDA receptors from the Mg2+block (Fig. 8.1b). All in all, our experiments indicate that GR affect LTP through a fast, probably nongenomic mechanism. Even though this hypothesis needs to be explored further, the fast suppression of LTP in the DH can underlie the switch in the weight between the DH and VH; by reducing DH LTP and simultaneously enhancing LTP in the VH, the stressful stimuli could temporarily suppress the cognitive route of the hippocampus to cortical structures and enable the transmission of the emotional information through the VH to the amygdala.

Conversely, LTD induction is facilitated by behavioral stress, through a mechanism that requires GR (Pavlides et al. 1995; Xu et al. 1997; Xu et al. 1998) and their effect on NMDA receptors (Kim et al. 1996; Yang et al. 2005). We replicated previous experiments where both stress and corticosterone facilitate LTD through a GR-dependent mechanism in the DH, but we have also shown that LTD is impaired in the VH through a MR-dependent mechanism (Maggio and Segal 2009b). Specifically in the latter case, LTD is transformed into a slow-onset LTP following the exposure to stressful stimulation (Maggio and Segal 2009b). As is the case for LTP, changes in LTD either in the DH or VH were observed at approximately 1 h after the exposure to the stress, a time window that could be compatible with non-genomic mechanisms. The MR-induced conversion of LTD to LTP in the VH could be due to the activation of VGCC, which will further facilitate the ventral route to the amygdale (Fig. 8.1b). Group I mGluR have been shown to enhance LTD in CA1 (Fitzjohn et al. 2001; Rammes et al. 2003), but, interestingly, they have been reported to induce a slow-onset potentiation in the DG (Manahan-Vaughan and Reymann 1996). In a previous study, we showed that, in the VH, application of DHPG, a group I mGluR agonist, increases the population spike amplitude in response to a baseline stimulation (Maggio and Segal 2007a). Taken together, these observations suggest that in the VH, a decrease in GABAergic inhibition can shift LTD to a slow-onset LTP through a group I mGluR-mediated mechanism (Fig. 8.1b).

Corticosteroid regulation of synaptic plasticity in the hippocampus is affected by several factors. An inverted U-shape effect of corticosterone mainly refers to the activation of intracellular corticosteroid receptors and does not count the contribution of membrane-bound steroid receptors. In fact, mMR, which bears a similar corticosterone affinity to that of iGR, will be activated at similar steroid concentrations. This implies that the effect of mMR appears earlier than that of iGR, thus inducing an enhancement of LTP instead of LTD. This might be the case in the VH. An additional factor to be considered is the distribution of MR and GR in specific brain areas, and the ratio of membrane-bound to intracellular receptors expressed therein. This is because at the same affinity value for corticosterone concentration, the receptor that is highly expressed will lead the effects on synaptic plasticity. Another issue that has to be considered is the clusters of brain areas that are involved in a particular stress situation. Various brain regions have specific properties and are incorporated into unique networks, so that even if corticosterone evokes the same effect at the single cell level, this would not always result in the same effect on network functions such as LTP. For instance, both CA1 pyramidal neurons and granule cells in the DG highly express MR as well as GR (Joels 2007, 2008). In the DH, corticosterone and stress consistently suppress the induction of CA1 LTP in vivo and in vitro, unlike the case for the DG. High concentration of corticosteroid (Pavlides et al. 1993) or tail shocks (Shors and Dryver 1994) can indeed suppress LTP; however, in other situations, either no effect (Bramham et al. 1998; Gerges et al. 2001; Alfarez et al. 2003) or enhancement of LTP has been reported (Kavushansky et al. 2006). This is because LTP in the DG seems to be more dependent on indirect inputs from the amygdale (Akirav and Richter-Levin 2002; Kavushanski et al. 2006, Kavushanski and Richter-Levin 2006). Finally, the response to a stressor is also determined by the history of the organism. For instance, the induction of LTP is impaired in animals that have been exposed to repetitive stress in the weeks prior to the experiment, even if corticosterone levels, at the time of LTP induction, are compatible with the expression of a normal LTP (Alfarez et al. 2003). Studies on the effect of maternal care on synaptic plasticity report that animals that received very little maternal care have poor LTP when they are adult, as opposed to animals that received high maternal care (Champagne et al. 2008). Interestingly, while LTP is suppressed by corticosterone in the latter group, it is enhanced in the former (Champagne et al. 2008).

Long-term effects of stress can produce changes in hippocampal morphology, in addition to an immediate effect on ability to express LTP. For example, Silva-Gomez et al. (2012) found that chronic (5 days) exposure to dexamethasone, a GR agonist, caused a significant reduction in dendritic spine density, primarily in the VH, which was also associated with a shrinkage of dendritic length in these neurons. Thus, the VH appears to be more susceptible to stress than the DH. Another interesting recent difference between DH and VH is in the effects of corticosterone to increase serotonin neurotransmission. Once again, this effect is restricted to the VH (Barr and Forster 2011). Likewise, it has been shown before that physical exercise can counteract the effect of maternal separation. Once again, enhanced locomotion in this experiment has a significant effect to increase synaptic markers only in the VH (Hescham et al. 2009).

Finally, a recent study describes a differential effect of acute stress on glutamate receptors in the DH and VH: while acute stress causes a reduced glutamate synaptic efficacy in the prefrontal cortex and the DH, it causes an augmented glutamate receptor activity in the amygdala and VH (Caudal et al. 2010). This observation complements our proposal for a stress-induced shift in hippocampal control from the DH to the VH (Fig. 8.2). Whether the primary effect of stress is mediated by modulation of the excitatory or the inhibitory synaptic tone in the two sectors of the hippocampus remains to be determined, but evidence for both possibilities has been presented recently. On the other hand, Marrocco et al. (2012) describe a reduction in glutamate release in VH following prenatal stress. Whether these results are congruent with the previous ones remain unclear. These actions may have to do with differences in mode of induction of stress, age of the animals, different receptor distribution, but may also reflect difference in intrinsic properties of the VH neurons compared to the DH counterparts.

8.6 Summary

All in all, corticosteroid modulation of synaptic plasticity in the hippocampus seems to be more complex than previously thought. Additional factors related to the unique spatio-temporal organization of the hippocampus, the different subsets of receptors and intrinsic properties of neurons in the different sectors and their connectivity with the rest of the brain are critical in finalizing the role of stress in neuronal plasticity. Furthermore, the definition of the borders of the VH is not precise, and different studies range from the bottom half of the hippocampus, to the bottom 1/5th of it. Likewise, stress is defined differently in different studies, and it may cause different levels of transient and sustained elevation of corticosterone, which may affect the observed estimation of the role of stress in neuronal plasticity. Thus, a careful evaluation of the regions and specific neurons tested, the behavioral and physical parameters tested and the time course of expected effects should allow a more reliable progress in the understanding of the role of 'stress' in neuronal plasticity. As it may turn out, there may be more than two hippocampi in one structure, and the possibility of three has been proposed recently (Fanselow and Dong 2010). Further studies will elucidate these issues with respect to hippocampal functions.

References

Akirav I, Richter-Levin G. Mechanisms of amygdala modulation of hippocampal plasticity. J Neurosci. 2002;22(22):9912–21.

Alfarez DN, Wiegert O, Joels M, Krugers HJ. Corticosterone and stress reduce synaptic potentiation in mouse hippocampal slices with mild stimulation. Neuroscience. 2002;115:1119–26.

Alfarez DN, Joels M, Krugers HJ. Chronic unpredictable stress impairs long-term potentiation in rat hippocampal CA1 area and dentate gyrus in vitro. Eur J Neurosci. 2003;17(9):1928–34.

Avital A, Segal M, Richter-Levin G. Contrasting roles of corticosteroid receptors in hippocampal plasticity. J Neurosci. 2006;26:9130–4.

Barr JL, Forster GL Serotonergic neurotransmission in the ventral hippocampus is enhanced by corticosterone and altered by chronic amphetamine treatment. Neuroscience. 2011;182:105–14.

Bauer EP, Schafe GE, Ledoux JE. NMDA receptors and L-type voltage-gated calcium channels contribute to long-term potentiation and different components of fear memory formation in the lateral amygdala. J Neurosci. 2002;22:5239–49.

Blair HT, Schafe GE, Bauer EP, Rodrigues SM, Ledoux JE. Synaptic plasticity in the lateral amygdala: a cellular hypothesis of fear conditioning. Learn Mem. 2001;8:229–42.

Bliss TV, Collingridge GL. A synaptic model of memory: long-term potentiation in the hippocampus. Nature. 1993;361:31–9.

Borroni AM, Fichtenholtz H, Woodside BL, Teyler TJ. Role of voltage-dependent calcium channel long-term potentiation (LTP) and NMDA LTP in spatial memory. J Neurosci. 2000;20:9272–6.

Bramham CR, Southard T, Ahlers ST, Sarvey JM. Acute cold stress leading to elevated corticosterone neither enhances synaptic efficacy nor impairs LTP in the dentate gyrus of freely moving rats. Brain Res. 1998;789(2):245–55.

Calabrese F, Guidotti G, Molteni R, Racagni G, Mancini M, Riva MA. Stress-induced changes of hippocampal NMDA receptors: modulation by duloxetine treatment. Neurosci Lett. 2012;521(1):20–5.

Caudal D, Godsil BP, Mailliet F, Bergerot D, Jay TM. Acute stress induces contrasting changes in AMPA receptor subunit phosphorylation within the prefrontal cortex, amygdala and hippocampus. PLoS ONE. 2010;5(12):e15282.

Chameau P, Qin Y, Spijker S, Smit AB, Joels M. Glucocorticoids specifically enhance L-type calcium current amplitude and affect calcium channel subunit expression in the mouse hippocampus. J Neurophysiol. 2007;97:5–14.

Champagne DL, Bagot RC, van Hasselt F, Ramakers G, Meaney MJ, de Kloet ER, Joels M, Krugers H. Maternal care and hippocampal plasticity: evidence for experience-dependent structural plasticity, altered synaptic functioning, and differential responsiveness to glucocorticoids and stress. J Neurosci. 2008;28:6037–45.

Cleren C, Tallarida I, Guiniec EL, Janin F, Nachon O, Canini F, Spennato G, Moreau JL, Garcia R. Low-frequency stimulation of the ventral hippocampus facilitates extinction of contextual fear. Neurobiol Learn Mem. 2013;101:39–45.

Daniels WM, Marais L, Stein DJ, Russell VA. Exercise normalizes altered expression of proteins in the ventral hippocampus of rats subjected to maternal separation. Exp Physiol. 2012;97(2):239–47.

De Kloet ER, Oitzl MS, Joels M. Functional implications of brain corticosteroid receptor diversity. Cell Mol Neurobiol. 1993;13:433–55.

De Kloet ER, Joels M, Holsboer F. Stress and the brain: from adaptation to disease. Nat Rev Neurosci. 2005;6:463–75.

de Kloet ER, Karst H, Joels M. Corticosteroid hormones in the central stress response: quick-and-slow. Front Neuroendocrinol. 2008;29:268–72.

Diamond DM, Bennett MC, Fleshner M, Rose GM. Inverted-U relationship between the level of peripheral corticosterone and the magnitude of hippocampal primed burst potentiation. Hippocampus. 1992;2:421–30.

Dorey R, Piérard C, Chauveau F, David V, Béracochéa D. Stress-induced memory retrieval impairments: different time-course involvement of corticosterone and glucocorticoid receptors in dorsal and ventral hippocampus. Neuropharmacology. 2012;63(8):1380–8.

Dougherty KA, Islam T, Johnston D. Intrinsic excitability of CA1 pyramidal neurones from the rat dorsal and ventral hippocampus. J Physiol. 2012;590(Pt 22):5707–22.

Dougherty KA, Nicholson DA, Diaz LM, Buss EW, Neuman KM, Chetkovich DM, Johnston D. Differential expression of HCN subunits alters voltage-dependent gating of h-channels in CA1 pyramidal neurons from the dorsal and ventral hippocampus. J Neurophysiol. 2013;109:1940–53.

Fanselow MS, Dong HW. Are the dorsal and ventral hippocampus functionally distinct structures? Neuron. 2010;65(1):7–19.

Felice D, O'Leary OF, Pizzo RC, Cryan JF. Blockade of the GABA(B) receptor increases neurogenesis in the ventral but not dorsal adult hippocampus: relevance to antidepressant action. Neuropharmacol. 2012 63:1380–8

Fitzjohn SM, Palmer MJ, May JE, Neeson A, Morris SA, Collingridge GL. A characterisation of long-term depression induced by metabotropic glutamate receptor activation in the rat hippocampus in vitro. J Physiol. 2001;537:421–30.

Foy MR, Stanton ME, Levine S, Thompson RF. Behavioral stress impairs long-term potentiation in rodent hippocampus. Behav Neural Biol. 1987;48:138–49.

Gerges NZ, Stringer JL, Alkadhi KA. Combination of hypothyroidism and stress abolishes early LTP in the CA1 but not dentate gyrus of hippocampus of adult rats. Brain Res. 2001;922(2):250–60.

Grigoryan G, Korkotian E, Segal M. Selective facilitation of LTP in the ventral hippocampus by calcium stores. Hippocampus. 2012;22(7):1635–44.

Hawley DF, Leasure JL. Region-specific response of the hippocampus to chronic unpredictable stress. Hippocampus. 2012;22(6):1338–49.

Hawley DF, Morch K, Christie BR, Leasure JL. Differential response of hippocampal subregions to stress and learning. PLoS ONE. 2012;7(12):e53126.

Hescham S, Grace L, Kellaway LA, Bugarith K, Russell VA. Effect of exercise on synaptophysin and calcium/calmodulin-dependent protein kinase levels in prefrontal cortex and hippocampus of a rat model of developmental stress. Metab Brain Dis. 2009;24(4):701–9.

Joels M. Effects of corticosteriod hormones in the hippocampus. Acta Physiol Scand. 1999;167:A3.

Joels M. Corticosteroid effects in the brain: U-shape it. Trends Pharmacol Sci. 2006;27:244–50.

Joels M. Functional actions of corticosteroids in the hippocampus. Eur J Pharmacol. 2008;583:312–21.

Joels M, Krugers HJ. LTP after stress: up or down? Neural Plast. 2007;2007:93202.

Joels M, Karst H, Derijk R, De Kloet ER. The coming out of the brain mineralocorticoid receptor. Trends Neurosci. 2008;31:1–7.

Joëls M, Baram TZ. The neuro-symphony of stress. Nat Rev Neurosci. 2009;10(6):459–66.

Karst H, Joels M. Corticosterone slowly enhances miniature excitatory postsynaptic current amplitude in mice CA1 hippocampal cells. J Neurophysiol. 2005;94:3479–86.

Karst H, Karten YJ, Reichardt HM, De Kloet ER, Schutz G, Joels M. Corticosteroid actions in hippocampus require DNA binding of glucocorticoid receptor homodimers. Nat Neurosci. 2000;3:977–8.

Karst H, Berger S, Turiault M, Tronche F, Schutz G, Joels M. Mineralocorticoid receptors are indispensable for nongenomic modulation of hippocampal glutamate transmission by corticosterone. Proc Natl Acad Sci U S A. 2005;102:19204–7.

Kavushansky A, Richter-Levin G. Effects of stress and corticosterone on activity and plasticity in the amygdala. J Neurosci Res. 2006;84(7):1580–7.

Kavushansky A, Vouimba RM, Cohen H, Richter-Levin, G, et al. Activity and plasticity in the CA1, the dentate gyrus, and the amygdala following controllable vs. uncontrollable water stress. Hippocampus. 2006;16(1):35–42.

Kim JJ, Foy MR, Thompson RF. Behavioral stress modifies hippocampal plasticity through N-methyl-D-aspartate receptor activation. Proc Natl Acad Sci U S A. 1996;93:4750–3.

Krugers HJ, Alfarez DN, Karst H, Parashkouhi K, Van Gemert N, Joels M. Corticosterone shifts different forms of synaptic potentiation in opposite directions. Hippocampus. 2005;15:697–703.

Maggio N, Segal M. Unique regulation of long term potentiation in the rat ventral hippocampus. Hippocampus. 2007a;17(1):10–25.

Maggio N, Segal M. Striking variations in corticosteroid modulation of long-term potentiation along the septotemporal axis of the hippocampus. J Neurosci. 2007b;27:5757–65.

Maggio N, Segal M. Differential corticosteroid modulation of inhibitory synaptic currents in the dorsal and ventral hippocampus. J Neurosci. 2009a;29:2857–66.

Maggio N, Segal M. Differential modulation of long-term depression by acute stress in the rat dorsal and ventral hippocampus. J Neurosci. 2009b;29:8633–8.

Maggio N, Segal M. Corticosteroid regulation of synaptic plasticity in the hippocampus. ScientificWorldJournal. 2010;10:462–9.

Maggio N, Segal M. Stress and corticosteroid modulation of seizures and synaptic inhibition in the hippocampus. Exp Neurol. 2012;234:200–7.

Manahan-Vaughan D, Reymann KG. Metabotropic glutamate receptor subtype agonists facilitate long-term potentiation within a distinct time window in the dentate gyrus in vivo. Neuroscience. 1996;74(3):723–31.

Marcelin B, Lugo JN, Brewster AL, Liu Z, Lewis AS, McClelland S, Chetkovich DM, Baram TZ, Anderson AE, Becker A, Esclapez M, Bernard C. Differential dorso-ventral distributions of Kv4.2 and HCN proteins confer distinct integrative properties to hippocampal CA1 pyramidal cell distal dendrites. J Biol Chem. 2012;287(21):17656–61.

Marrocco J, Mairesse J, Ngomba RT, Silletti V, Van Camp G, Bouwalerh H, Summa M, Pittaluga A, Nicoletti F, Maccari S, Morley-Fletcher S. Anxiety-like behavior of prenatally stressed rats is associated with a selective reduction of glutamate release in the ventral hippocampus. J Neurosci. 2012;32(48):17143–54.

Moser MB, Moser EI. Functional differentiation in the hippocampus. Hippocampus. 1998;8:608–19.

O'Leary OF, O'Connor RM, Cryan JF. Lithium-induced effects on adult hippocampal neurogenesis are topographically segregated along the dorso-ventral axis of stressed mice. Neuropharmacology. 2012;62(1):247–55.

Papatheodoropoulos C, Kostopoulos G. Decreased ability of rat temporal hippocampal CA1 region to produce long-term potentiation. Neurosci Lett. 2000;279(3):177–80.

Patel J, Fujisawa S, Berényi A, Royer S, Buzsáki G. Traveling theta waves along the entire septotemporal axis of the hippocampus. Neuron. 2012;75(3):410–7.

Pavlides C, Watanabe Y, McEwen BS. Effects of glucocorticoids on hippocampal long-term potentiation. Hippocampus. 1993;3(2):183–92.
Pavlides C, Kimura A, Magarinos AM, Mcewen BS. Hippocampal homosynaptic long-term depression/depotentiation induced by adrenal steroids. Neuroscience. 1995;68:379–85.
Pavlides C, Ogawa S, Kimura A, Mcewen BS. Role of adrenal steroid mineralocorticoid and glucocorticoid receptors in long-term potentiation in the CA1 field of hippocampal slices. Brain Res. 1996;738:229–35.
Rammes G, Palmer M, Eder M, Dodt HU, Zieglgansberger W, Collingridge GL. Activation of mGlu receptors induces LTD without affecting postsynaptic sensitivity of CA1 neurons in rat hippocampal slices. J Physiol. 2003;546(Pt 2):455–60.
Robertson DA, Beattie JE, Reid IC, Balfour DJ. Regulation of corticosteroid receptors in the rat brain: the role of serotonin and stress. Eur J Neurosci. 2005;21:1511–20.
Roth TL, Zoladz PR, Sweatt JD, Diamond DM. Epigenetic modification of hippocampal Bdnf DNA in adult rats in an animal model of post-traumatic stress disorder. J Psychiatr Res. 2011;45(7):919–26.
Segal M, Richter-Levin G, Maggio N. Stress-induced dynamic routing of hippocampal connectivity: a hypothesis. Hippocampus. 2010;20:1332–8.
Shors TJ, Dryver E. Effect of stress and long-term potentiation (LTP) on subsequent LTP and the theta burst response in the dentate gyrus. Brain Res. 1994;666(2):232–8.
Shors TJ, Seib TB, Levine S, Thompson RF. Inescapable versus escapable shock modulates long-term potentiation in the rat hippocampus. Science. 1989;244:224–22.
Silva-Gómez AB, Aguilar-Salgado Y, Reyes-Hernández DO, Flores G. Dexamethasone induces different morphological changes in the dorsal and ventral hippocampus of rats. J Chem Neuroanat. 2012. pii: S0891-0618(12)00080-4.
Steullet P, Cabungcal JH, Kulak A, Kraftsik R, Chen Y, Dalton TP, Cuenod M, Do KQ. Redox dysregulation affects the ventral but not dorsal hippocampus: impairment of parvalbumin neurons, gamma oscillations, and related behaviors. J Neurosci. 2010;30(7):2547–58.
Tanti A, Rainer Q, Minier F, Surget A, Belzung C. Differential environmental regulation of neurogenesis along the septo-temporal axis of the hippocampus. Emotion. 2012;12(1):58–68.
Xia L, Deloménie C, David I, Rainer Q, Marouard M, Delacroix H, David DJ, Gardier AM, Guilloux JP. Ventral hippocampal molecular pathways and impaired neurogenesis associated with 5-HT_1A and 5-HT_1B receptors disruption in mice. Neuropharmacology. 2012;63(3):374–84.
Xu L, Anwyl R, Rowan MJ. Behavioural stress facilitates the induction of long-term depression in the hippocampus. Nature. 1997;387:497–500.
Xu L, Holscher C, Anwyl R, Rowan MJ. Glucocorticoid receptor and protein/RNA synthesis-dependent mechanisms underlie the control of synaptic plasticity by stress. Proc Natl Acad Sci U S A. 1998;95:3204–8.
Yang CH, Huang CC, Hsu KS. Behavioral stress enhances hippocampal CA1 long-term depression through the blockade of the glutamate uptake. J Neurosci. 2005;25:4288–93.

Chapter 9
Neural-Cognitive Effects of Stress in the Hippocampus

Jeansok J. Kim, Blake A. Pellman and Eun Joo Kim

Abstract It is now well-accepted that uncontrollable (i.e., acute traumatic, prolonged) stress can have lingering effects on the hippocampus. At the behavioral level, evidence from human and animal studies indicates that stress generally impedes performance in a variety of hippocampal-dependent memory tasks. At the neural level, animal studies have shown that stress impairs induction of long-term potentiation (LTP), a form of synaptic plasticity, in the hippocampus. Because the hippocampus is important for certain forms of long-term memory and because LTP has properties desirable of an information storage mechanism, it has been hypothesized that stress-induced alterations in hippocampal plasticity contribute to decreased memory functioning following stress exposure. This chapter reviews the effects of stress on three vertically related levels of hippocampal functions—synaptic plasticity, neural activity and memory—and the recent evidence implicating the amygdala as a crucial component of the central stress mechanism.

Abbreviations

ACTH	Adrenocorticotropic hormone
AMYG	Amygdala
APV	DL-2-amino-5-phosphonovaleric acid
CORT	Cortisol/corticosterone
CRF	Corticotropin-releasing factor
GABA	Gamma-aminobutyric acid
GR	Glucocorticoid receptor
HPA-axis	Hypothalamic-pituitary-adrenal axis
I/O	Input/output
LTD	Long-term depression
LTP	Long-term potentiation
mPFC	Medial prefrontal cortex
MR	Mineralocorticoid receptor
NMDA	*N*-methyl-D-aspartate

J. J. Kim (✉) · B. A. Pellman · E. J. Kim
Department of Psychology, Program in Neurobiology and Behavior, University of Washington, Seattle, WA 98195-1525, USA
e-mail: jeansokk@u.washington.edu

M. Popoli et al. (eds.), *Synaptic Stress and Pathogenesis of Neuropsychiatric Disorders*,
DOI 10.1007/978-1-4939-1056-4_9,

PTSD	Post-traumatic stress disorder
S	Stimulus
R	Response

9.1 Introduction

Stress is a biologically significant factor that plays pervasive roles in our lives, from influencing daily behaviors to precipitating symptoms of mental health disorders. Hence, stress presents a natural means to investigate the socio-environmental contributions to various psychopathologies, such as anxiety, panic and posttraumatic stress disorders (PTSD), depression, schizophrenia, and relapse in drug use (Kim and Diamond 2002; Lupien et al. 2009; Sinha et al. 2011).

Semantically, stress describes any significant socio-environmental conditions that require appropriate physiological and/or behavioral readjustment (or adaptation) that serves to preserve the well-being of the organism (Selye 1956, 1973; McEwen and Sapolsky 1995). At present, stress phenomena are conceptually and procedurally dichotomized as physical (real) versus psychological (perceived), early life versus adulthood, and acute versus chronic (e.g., Foy et al. 2005; Kosten et al. 2012). While *stress* refers to an unpleasant state (distress) in colloquial speech, a related concept, *eustress,* has been proposed to represent positive valence of stress (e.g., voluntary exercise), highlighting the conceptual distinction between the emotional perception of stress and the fundamental process underlying physiological and behavioral adaptation (Selye 1974).

A number of putative stress paradigms are utilized in different laboratories, making it sometimes difficult to evaluate experimental findings across studies. To standardize the framework of stress that can be applied across different animal and human models, one proposal (Kim and Diamond 2002; Kim and Haller 2007) suggested that stress must satisfy three conditions: (1) heighten the excitability or arousal of the organism, (2) induce perceived aversiveness, and (3) decrease perceived controllability of the situation. This operational definition makes a clear distinction between stress and other aversive states such as fear. For instance, traffic congestions can elicit arousal, be aversive (but not fearful), and evoke a loss of controllability (if there is no alternative route) in most people, and in such case satisfy the three stipulations of stress. While the *stress response* is an adaptive mechanism, the prolonged stress response can have deleterious physiological and psychological outcomes, such as hypertension, diabetes, gastric-intestinal ulceration, depression, and anxiety disorders (Sapolsky 1992; Rosen and Schulkin 1998).

In recent decades, researchers have focused on the adverse effects of stress on brain-memory systems (Kim and Diamond 2002; Shors 2004). Because the effects of stress on memory are similar between humans and a number of animals, animal models provide a valuable means to investigate the neurocognitive effects of stress. At present, neurobiological studies have found that uncontrollable stress alters syn-

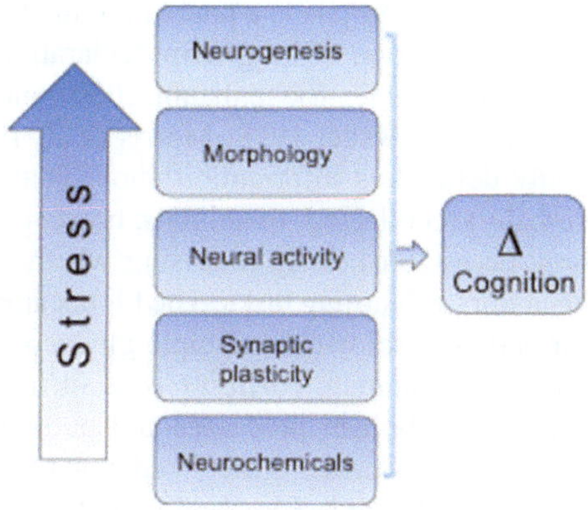

Fig. 9.1 Neurobiological effects of stress in the hippocampus. As the intensity and duration of stress increases, alterations in neurochemicals, synaptic plasticity, neural activity, morphology, and neurogenesis occur in the hippocampus. These changes can contribute to stress-induced dysfunctions in memory

aptic plasticity and neuronal morphology (soma size, dendritic arborization), exacerbates neurotoxicity and suppresses neurogenesis in the hippocampus (Fig. 9.1) (Kim and Yoon 1998). These stress-induced physiological changes, presumably, can influence ensuing learning and memory functions. Accordingly, stress presents a natural means to study the contribution of learning and memory dysfunction to various psychopathologies. While diverse stress paradigms have been shown to influence a number of brain-memory systems, this chapter will highlight the effects of acute, uncontrollable stress on hippocampal plasticity, neural activity and memory, and the role that the amygdala plays in the emergence of stress effects.

9.2 Stress Effects on Hippocampal Memory

Almost a half century ago, Seligman, Maier, and Overmier made the significant discovery that animals that had previously experienced uncontrollable stress (i.e., random, inescapable electric shocks) were impaired in learning to escape from footshocks in the shuttle box task, a phenomenon known as *learned helplessness* (Seligman and Maier 1967; Overmier and Seligman 1967). According to the learned helplessness hypothesis, when an organism learns that its behavior (response, R) and aversive outcomes (stimulus, S) are independent, this learning produces cognitive, emotional, and motivational transformations that later hinder learning of other tasks. In laboratory settings, humans, dogs, cats, rats, and even fish have been shown to demonstrate learned helplessness following exposure to uncontrollable stress (loud noise, electric shock). Importantly, when the cessation of an aversive S is made contingent upon the animals R (e.g., a rat emitting a wheel turn R to terminate a tailshock S), the learning of this S-R association (namely, controllability) protects the animal from developing learned helplessness (Maier and Seligman 1976). Subsequent studies have revealed that stress particularly interferes with behavioral tasks that depend on the hippocampus (Kim and Yoon 1998).

The hippocampus is a part of the medial temporal lobe system, which is crucial for the formation of long-term declarative (explicit) memory in humans (Scoville and Milner 1957; Eichenbaum 2000) and spatial (relational) memory in rodents (OKeefe and Nadel 1978; Morris et al. 1982, 1998). Declarative memory is generally defined as information about facts and events that can be consciously (or verbally) recollected. In animals, however, the human declarative-like memory can only be established by assessing whether hippocampal lesions abolish particular behaviors in learning tasks. The hippocampus is highly concentrated with receptors for corticosteroids-the principle glucocorticoids synthesized by the adrenal cortex (*cortisol* in human, *corticosterone* in rodent; CORT) to regulate general cellular energy metabolism processes-and participates in terminating the stress response through glucocorticoid-mediated negative feedback of the hypothalamic-pituitary-adrenal (HPA) axis (Axelrod and Reisine 1984). Because its secretion is highly responsive to stress, CORT is commonly referred to as the "stress hormone" (or even tacitly believed as a stress-producing hormone). In the rodent hippocampus, CORT has been found to alter the metabolic, physiologic, and genomic functions of neurons (Sapolsky 1992). As a result, the mnemonic functions of the hippocampus appear to be sensitive to stress.

Consistent with this view, a large body of evidence indicates that exposures to stress and/or stress hormones negatively impact hippocampal-dependent memory tasks in humans and animals (see Lupien and McEwen 1997). For example, PTSD patients exhibit deficits in verbal recall tasks when compared to control subjects (Bremner et al. 1993; Utto et al. 1993). Injections of CORT in healthy human subjects have been reported to selectively impair verbal declarative memory, sans affecting nonverbal (nonhippocampal) memory (Newcomer et al. 1994; Kirschbaum et al. 1996; de Quervain et al. 2000; Kuhlmann et al. 2005). Moreover, hypercortisolemia conditions in certain depressive patients and those afflicted with Cushings disease have been implicated in declarative memory impairments (Starkman et al. 1992; Sapolsky 2000). However, administration of CORT has also been reported to selectively enhance the long-term recall of emotionally arousing (but not neutral) pictures (e.g., Buchanan and Lovallo 2001), suggesting that stress hormone effects may be more subtle and complex than previously reported.

Similar to human studies, rats subjected to uncontrollable stress (or administered high doses of CORT) show memory deficits in various hippocampal-dependent behavioral tasks (e.g., Luine et al. 1993; de Quervain et al. 1998). The test par excellence of hippocampal memory in rodents is the spatial memory task, typically utilizing variations of Oltons 8-arm radial maze (Olton and Samuelson 1976) and Morris water maze (Morris 1981). In a series of elegant experiments, Diamond and colleagues have shown that stress impairs hippocampal-dependent spatial working memory while hippocampal-independent spatial reference memory is unaffected (Diamond and Rose 1994; Diamond et al. 1999; Woodson et al. 2003).

Spatial memory deficits have also been reported in transgenic mice with elevated CORT levels caused by the central over-expression of corticotropin-releasing factor (CRF) (Heinrichs et al. 1996). CRF, a neuropeptide secreted by the paraventricular nucleus of the hypothalamus, triggers the release of adrenocorticotropic hormone

(ACTH) from the pituitary gland, and ACTH in turn stimulates the production and secretion of glucocorticoids by the adrenal gland (Sapolsky 1992). Paralleling the spatial memory deficits are recent findings that stress impairs the stability of place cell firing rates (Kim et al. 2007; Passecker et al. 2011). Hippocampal place cells are thought to support spatial learning and navigation by encoding memories of familiar spatial locations (O'Keefe and Distrovsky 1971; OKeefe and Nadel 1978).

The stress effects on hippocampal memory do not seem to be limited to spatial information in rodents. Other studies found that stress also impairs nonspatial (hippocampal-dependent) object recognition memory (Beck and Luine 1999; Baker and Kim 2002). Stress also disrupts medial prefrontal cortex (mPFC)-based spatial working memory on a T-maze task (Arnsten and Goldman-Rakic 1998; Qin et al. 2009) as well as decision-making in a foraging task in rats (Graham et al. 2010).

Interestingly, the same stress that impairs hippocampal memory has been found to enhance the relative use of competing hippocampal-independent memory (e.g., the caudate-dependent response memory) in rats and humans (Kim et al. 2001; Pruessner et al 2008; Wingard and Packard 2008; Quirarte et al. 2009; Lovallo 2010; Schwabe et al. 2007; Schwabe and Wolf 2012). Stress has also been shown to enhance aversive memory, such as fear and eyeblink conditioning (Beylin and Shors 2003; Conrad et al. 1999a; Jackson et al. 2006; Rau et al. 2005). It remains to be determined, however, whether the learning enhancements in other behavioral tasks are due to direct effects of stress on those brain-memory systems or due to indirect effects of stress reducing the hippocampus ability to compete with other brain-memory systems. Thus, although the study of individual memory systems affected by stress has proved to be useful, particularly in the hippocampus, recent data increasingly point towards complex interactions between stress and multiple brain-memory systems (Kim and Baxter 2001).

9.3 Stress Effects on Hippocampal Synaptic Plasticity

Long-term potentiation (LTP) is characterized by an enduring increase in synaptic transmission resulting from high frequency stimulation (or tetanus) of afferent fibers (Bliss and Lomo 1973; Bliss and Gardner-Medwin 1973). Because LTP occurs rapidly, is stable over time, requires cooperativity (i.e., adequate afferents to reach threshold), is strengthened by repetition, and demonstrates input specificity and associativity, LTP has long been proposed as a synaptic model of information storage in the mammalian brain (Bliss and Collingridge 1993; Martin et al. 2000). In 1987, Thompson and colleagues found that hippocampal slices prepared from rats that received 30 min of intermittent tailshocks while being restrained exhibited striking deficits in the Schaffer collateral/commissural-cornu Ammonis 1 (CA1) LTP (Foy et al. 1987). Importantly, hippocampal slices taken from rats that were able to terminate the shock showed relatively normal LTP, while slices from "yoked" animals that received the identical shock schedule without control exhibited severely impaired LTP (Shors et al. 1989). Hence, similar to learned helplessness, the LTP impairment appears to be largely due to the psychological, rather than physical, qualities of stress. Other forms of psychological stress, such as forced exposures to a novel chamber or to a predator, have also been found to impede LTP and/or primed-burst potentiation (a low threshold form of LTP) in behaving rats (Diamond et al. 1990; Xu et al. 1997; Diamond and Park 2000).

Stress-induced LTP impairments have also been observed in other regions of the hippocampus (Shors and Dryver 1994), and following 30-min restraint + shock stress, LTP deficits continue up to 48 h in rats (Shors et al. 1997) and 24 h in mice (Garcia et al. 1997). There seems to be a critical stress threshold for LTP impairment as 10-min restraint + shock stress, while producing robust fear conditioning and elevating corticosterone levels, does not impair LTP (Shors et al. 1989). Other studies indicate a time-dependent, biphasic effect on hippocampal LTP (an enhancing effect on LTP followed by a longer-lasting suppressing effect on LTP) (Akirav and Richter-Levin 1999), and stress has been reported to enhance theta-burst stimulation-induced LTP but impair high-frequency stimulation-induced LTP in the mouse hippocampus (Blank et al. 2002). These findings suggest that differences in stress paradigms, in vitro versus in vivo recordings, tetanus patterns, and species must be considered when evaluating stress effects on hippocampal synaptic plasticity.

The discovery that stress impairs hippocampal LTP is significant because it offers a testable synaptic mechanism to investigate stress-induced memory deficits, and because the LTP impairment can serve as a "neurophysiological marker" to compare behavioral consequences associated with different stress paradigms. For example, not all putative stress procedures would be expected to impair LTP and/or memory. Regardless, the relationship between stress effects on LTP and memory in the hippocampus is consistent with the hypothesis, namely Hebbs (1949) postulate, that memories are stored via changes in the pattern of synaptic connections.

In theory, LTP alone cannot provide a dynamic synaptic model for information storage; decreases in synaptic efficacy are essential to normalize synaptic strength and prevent LTP saturation (Sejnowski 1977). This is accounted for by long-term depression (LTD) characterized by a decrease in synaptic efficacy following low-frequency stimulation of afferent fiber which, like LTP, has several properties desirable for an information storage mechanism (e.g., longevity and input specificity) (Bear and Malenka 1994; Dudek and Bear 1992). When stress effects were examined in the Schaffer collateral/commissural-CA1 pathway, the same stress that impaired LTP was found to enhance LTD (Kim et al. 1996; Xu et al. 1997). Moreover, administration of a competitive *N*-methyl-D-aspartate (NMDA) receptor antagonist prior to stress blocked stress effects on both LTP and LTD (Table 9.1). These findings indicate that stress effects on LTP and LTD are related (see also Coussens et al. 1997; Diamond et al. 2004).

Two possibilities can explain the opposing effects of stress on LTP and LTD (Fig. 9.2). Since LTP is known to be "saturable" (i.e., has an upper limit of potentiation), if LTP or LTP-like changes occur in the hippocampus during stress, then any following LTP will be occluded due to a ceiling-like effect, whereas LTD can now be enhanced because the range for synaptic depression has increased (e.g., Kim et al. 1996; Diamond et al. 2004). This possibility is analogous to learned helpless-

Table 9.1 A summary of stress effects on in vitro LTP and LTD

Hippocampus (CA1)[a]	LTP	LTD
Control (unstressed)	+	−
Stressed	−	+
Control + APV	−	NA
Stressed + APV	NA	−
Control (from LTP state)	NA	−
Control (from LTP state) + APV	NA	+
Stressed (from LTD state)	+	NA
Stressed (from LTD state) + APV	−	NA
Stressed with NMDA antagonist	+	−

+ present or enabled, − absent or attenuated, *NA* not applicable, *LTP* long-term potentiation, *LTD* long-term depression, *APV* DL-2-amino-5-phosphonovaleric acid, *NMDA* N-methyl-D-aspartate
[a] Slices prepared from adult male rats. Modified from Kim et al. 1996

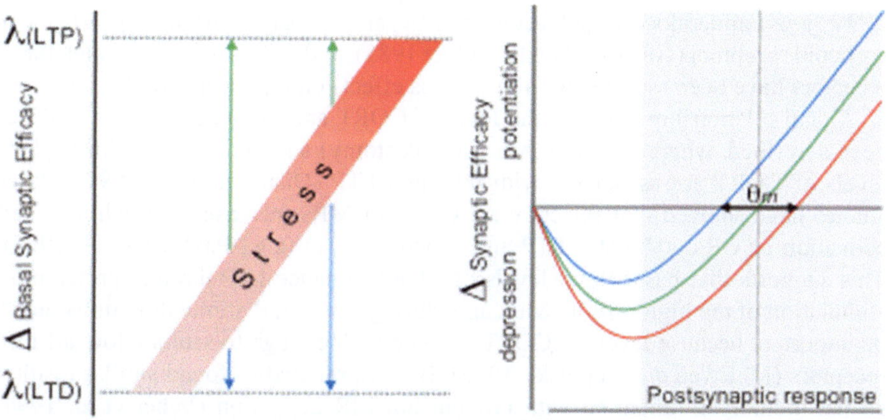

Fig. 9.2 Hypothetical models to account for stress effects on hippocampal synaptic plasticity. *Left*: The saturation hypothesis posits that stress produces long-term potentiation (*LTP*)-like changes in hippocampal synapses which then occlude subsequent LTP but enhance long-term depression (*LTD*) (λ, limit of plasticity). *Right*: The metaplasticity hypothesis proposes that stress shifts the modification threshold, θm, to the right (*represented by the red line*) so that ensuing synaptic changes favor LTD over LTP. (Adapted from Kim and Yoon 1998)

ness, wherein the animals learning of the independence between its behavior and the aversive situation interferes with subsequent memory functioning. A different possibility is that stress produces a "metaplastic" effect (i.e., higher-order plasticity that influences ensuing plasticity) in the hippocampus such that the threshold for LTP and LTD is biased towards LTD over LTP induction (see Abraham and Tate 1997; Kim and Yoon 1998). In order to reveal whether saturation or metaplasticity underlies stress effects on hippocampal plasticity, future studies will need to methodically monitor the input/output (I/O) functions in the hippocampus (e.g., the Schaffer collateral/commissural-CA1 pathway) while the animal transitions from the baseline to during stress to post-stress. If uncontrollable stress produces LTP-like changes, then there should be differences in the baseline synaptic transmission when I/O functions are compared between baseline versus during and after stress. Specifically, the I/O functions should increase during the stress and such change should remain stable after stress. If stress produces metaplastic changes instead, then there should be no differences in I/O functions between baseline versus during and after stress.

9.4 Glucocorticoids and Hippocampal Plasticity

Contemporary stress research has consistently implicated corticosteroids (and other neurochemicals of the HPA-axis) as the main causes of stress effects on the hippocampus (McEwen and Gianaros 2011; Popoli et al. 2012; Ulrich-Lai and Herman 2009; Joels and Baram 2009). The hippocampus is enriched with both the high-affinity *Type-I* mineralocorticoid receptors (MR) and the lower-affinity *Type-II* glucocorticoid receptors (GR) (Reul and de Kloet 1985), and CORT actions through these receptors have been reported to mimic stress effects on hippocampal plasticity.

A dual relationship between the level of CORT and the magnitude of LTP has been described, where both low (via adrenalectomy) and high (via administration) levels of CORT are associated with impaired LTP (Diamond et al. 1992). Other studies have showed that selective activation of MRs increases LTP while added activation of GRs attenuates LTP and enhances LTD (e.g., Pavlides et al. 1995). This suggests that basal (low) levels of CORT enhance LTP through preferential stimulation of the high-affinity MRs and, during stress, GR stimulation turns out to be important because levels of CORT become high enough to saturate low-affinity receptors (McEwen and Sapolsky 1995). Behavioral studies found similar results-spatial memory is impaired with GR but not MR activation (Vaher et al. 1994; Conrad et al. 1999b; Oitzl et al. 2001). Bath application of CORT also prolongs calcium-dependent afterhyperpolarization of CA1 neurons (Kerr et al. 1989; Nair et al. 1998), which would decrease cell excitability and in so doing affect synaptic plasticity.

If corticosteroids are the main contributing factors in the mediation of stress effects, then removing them during stress and directly applying them in absence of behavioral stress, should preclude and produce stress effects, respectively. However, there are behavioral, synaptic plasticity and neural activity data from animal studies inconsistent with this simple linear neurochemical-level stress effect notion (Shors et al. 1989, 1990; Foy et al. 1990; Woodson et al. 2003; Stranahan et al. 2006). Very recent studies have reported that both stress and environmental enrichment significantly and comparably elevate CORT levels but have opposite effects on hippocampal neurogenesis (e.g., Schoenfeld and Gould 2012); findings that are incompatible to those in vivo and in vitro studies where CORT administration mimics behavioral stress effects. It is important to recognize that, like CORT, other hormones, peptides, and neurotransmitters implicated in stress (such as CRF, serotonin, dopamine, enkephalins) also have multifold functions and none are known to respond uniquely to stress, and thus none of them is likely to be a sufficient mediator of stress effects.

9.5 Amygdala and Stress Effects on Hippocampus

Emerging evidence indicates that the amygdala is crucial in mediating stress-related behaviors and modulating hippocampal function. The amygdala is one of the principal structures of the limbic system that has access to sensory inputs from various brain regions (such as the thalamus, the neocortex) and sends projections to autonomic and somatomotor structures involved in defensive responses (such as the bed nucleus of stria terminalis for activating stress hormones, the periaqueductal gray for defensive behavior, the lateral hypothalamus for sympathetic activation) (see LeDoux 1996). Such rich sensory-amygdala-defensive (autonomic and motor) connections can explain how amygdalar lesions can prevent stress-induced gastric erosions (Henke 1981), analgesia (Helmstetter 1992), and anxiety-like behaviors (Adamec et al. 1999).

McGaugh and colleagues (Packard et al. 1994; McGaugh 2000; Roozendaal et al. 2003) have shown that pharmacological manipulations that alter synaptic transmissions in the amygdala (such as GABA, opioid, norepinephrine, and acetylcholine) can modulate memory strength in the hippocampus. Other studies have reported that lesions, stimulations, and drug infusions in the amygdala can also regulate LTP magnitude in the dentate gyrus (Abe 2001; Akirav and Richter-Levin 1999, 2002). Hence, the amygdala, via its (largely ipsilateral) projections to the hippocampus (Krettek and Price 1977; Pikkarainen et al. 1999), might also regulate stress effects on the hippocampus.

Consistent with this notion, amygdalar lesions have been found to block stress effects on hippocampal LTP and spatial memory in rats (Kim et al. 2001). Similarly, temporary inactivation of the amygdala via the $GABA_A$ receptor agonist muscimolprior to stress effectively blocked stress-induced physiological and behavioral effects (Kim et al. 2005). Intra-amygdalar muscimol also blocked spatial memory impairment following predator stress experience (Park et al. 2008). Because immediate post-stress muscimol infusions into the amygdala failed to prevent stress effects on LTP and memory, the critical time window of amygdalar activity is during (and not after) stress (Kim et al. 2005). It should be mentioned that amygdalar lesions/inactivation blocked stress effects on hippocampal LTP and memory despite the increase in corticosterone secretion to stress (Kim et al. 2001, 2005). An earlier study implicated the NMDA receptors in the amygdala in mediating stress-induced facilitation of classical eyeblink conditioning (Shors and Mathew 1998). Thus, it is likely that NMDA receptor-dependent plasticity in the amygdala is somehow involved in mediating stress effects on hippocampal plasticity and memory (Kim et al. 1996). Recently, electrical stimulation of the amygdala was found to selectively suppress CA1 LTP in the hippocampus (Vouimba and Richter-Levin 2005)

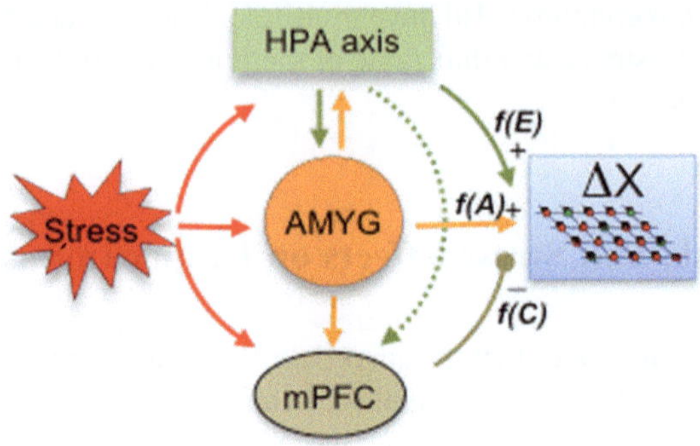

Fig. 9.3 A connectionist model of stress. The hypothalamic-pituitary-adrenal axis (*HPA*) axis (signifying the function of excitability, *f*(*E*)), amygdala (*AMYG*; aversiveness, *f*(*0*)), and medial prefrontal cortex (*mPFC*) (controllability, *f*(*C*)) interact to produce alterations (ΔX) in stress-vulnerable structures (e.g., the hippocampus). The model posits that HPA and AMYG exert excitatory (+) stress influences while mPFC exerts inhibitory (−) stress influence. (Adapted from Kim and Diamond 2002)

and produce stress-like impairment effects on hippocampal place cells (Kim et al. 2012). These findings suggest that the amygdala is a critical component of the central stress mechanism that alters hippocampal functioning (Fig. 9.3).

Stress has also been found to induce LTP and morphological changes in the amygdala. Unlike the hippocampus, which inhibits stress-induced HPA activation, the amygdala enhances glucocorticoid secretion in response to stress (Herman et al. 2005). Moreover, in contrast to hippocampal effects, stress (i.e., chronic immobilization stress) enhances LTP and increases growth of dendrites and spines in amygdalar neurons (Vyas et al. 2002, 2003; Mitra et al. 2005; Radley and Morrison 2005). These changes in the amygdala have been proposed to underlie stress-induced symptoms of chronic anxiety disorders (McEwen 2004). However, because different stress paradigms were used in hippocampal and amygdalar studies, it remains to be investigated whether neurophysiological changes in the amygdala precede and/or are prerequisite to stress-induced changes in the hippocampus. Thus, additional work is necessary to understand the nature of amygdala–hippocampal interaction during stress.

9.6 Summary

Contemporary stress research has focused on the effects of particular hormones (e.g., glucocorticoids), peptides (e.g., CRF, enkephalins), or neurotransmitters (e.g., serotonin, dopamine) on intracellular signaling cascades, synaptic plasticity, structural changes, cell death, and neurogenesis, which has generated a wealth of information. However, given that these chemical messengers are also engaged in nonstress functions, it is likely that focusing on specific chemical messengers cannot provide an adequate representation of how uncontrollable stress impacts brain and behavior. Recent data from stress-amygdala-mPFC studies increasingly point towards complex neural-endocrine interactions in mediating stress effects on the hippocampus. Thus, consideration of multiple stress factors and their dynamics will advance our current understanding of the neural-cognitive effects of stress that may lead to stress-related psychopathology.

References

Abe K. Modulation of hippocampal long-term potentiation by the amygdala: a synaptic mechanism linking emotion and memory. Jpn J Pharmacol. 2001;86:18–22.

Abraham WC, Tate WP. Metaplasticity: a new vista across the field of synaptic plasticity. Prog Neurobiol. 1997;52(4):303–23.

Adamec RE, Burton P, Shallow T, Budgell J. Unilateral block of NMDA receptors in the amygdala prevents predator stress-induced lasting increases in anxiety-like behavior and unconditioned startle-effective hemisphere depends on the Behavior. Physiol Behav. 1999;65:739–51.

Akirav I, Richter-Levin G. Biphasic modulation of hippocampal plasticity by behavioral stress and basolateral amygdalar stimulation in the rat. J Neurosci. 1999;19:10530–5.

Akirav I, Richter-Levin G. Mechanisms of amygdala modulation of hippocampal plasticity. J Neurosci. 2002;22:9912–21.

Arnsten AF, Goldman-Rakic PS. Noise stress impairs prefrontal cortical cognitive function in monkeys: evidence for a hyperdopaminergic mechanism. Arch Gen Psychiatry. 1998;55:362–8.

Axelrod J, Reisine TD. Stress hormones: their interaction and regulation. Science. 1984;224(4648):452–9.

Baker KB, Kim JJ. Effects of stress and hippocampal NMDA receptor antagonism on recognition memory in rats. Learn Mem. 2002;9(2):58–65.

Bear MF, Malenka RC. Synaptic plasticity: LTP and LTD. Curr Opin Neurobio. 1994;4:389–99.

Beck KD, Luine VN. Food deprivation modulates chronic stress effects on object recognition in male rats: role of monoamines and amino acids. Brain Res. 1999;830(1):56–71.

Beylin AV, Shors TJ. Glucocorticoids are necessary for enhancing the acquisition of associative memories after acute stressful experience. Hormones Behav. 2003;43:124–31.

Blank T, Nijholt I, Eckart K, Spiess J. Priming of long-term potentiation in mouse hippocampus by corticotrophin-releasing factor and acute stress: implications for hippocampus-dependent learning. J Neurosci. 2002;22:3788–94.

Bliss TVP, Collingridge GL. A synaptic model of memory: long-term potentiation in the hippocampus. Nature. 1993;361:31–9.

Bliss TVP, Gardner-Medwin AR. Long-lasting potentiation of synaptic transmission in the dentate area of the unanaesthetized rabbit following stimulation of the perforant path. J Physiol. 1973;232:357–74.

Bliss TVP, Lomo T. Long-lasting potentiation of synaptic transmission in the dentate area of the anaesthetized rabbit following stimulation of the perforant path. J Physiol. 1973;232:331–56.

Bremner JD, Scott TM, Delaney RC, Southwick SM, Mason JW, Johnson DR, et al. Deficits in short-term memory in posttraumatic-stress-disorder. Am J Psychiatry. 1993;150:1015–9.

Buchanan TW, Lovallo WR. Enhanced memory for emotional material following stress-level cortisol treatment in humans. Psychoneuroendocrinology. 2001;26(3):307–17.

Conrad CD, LeDoux JE, Magarinos AM, McEwen BS. Repeated restraint stress facilitates fear conditioning independently of causing hippocampal CA3 dendritic atrophy. Behav Neurosci. 1999a;113:902–13.

Conrad CD, Lupien SJ, McEwen BS. Support for a bimodal role for type II adrenal steroid receptors in spatial memory. Neurobiol Learn Mem. 1999b;72(1):39–46.

Coussens CM, Kerr DS, Abraham WC. Glucocorticoid receptor activation lowers the threshold for NMDA-receptor-dependent homosynaptic long-term depression in the hippocampus through activation of voltage-dependent calcium channels. J Neurophysiol. 1997;78:1–9.

de Quervain DJ-F, Roozendaal B, McGaugh JL. Stress and glucocorticoids impair retrieval of long-term spatial memory. Nature. 1998;394:787–90.

de Quervain DJ-F, Roozendaal B, Nitsch RM, McGaugh JL, Hock C. Acute cortisone administration impairs retrieval of long-term declarative memory in humans. Nat Neurosci. 2000;3(4):313–4.

Diamond DM, Park CR. Predator exposure produces retrograde amnesia and blocks synaptic plasticity. Progress toward understanding how the hippocampus is affected by stress. Ann N Y Acad Sci. 2000;911:453–5.

Diamond DM, Rose GM. Stress impairs LTP and hippocampal-dependent memory. Hippocampus. 1994;746:411–4.

Diamond DM, Bennett MC, Stevens KE, Wilson RL, Rose GM. Exposure to a novel environment interferes with the induction of hippocampal primed burst potentiation in the behaving rats. Psychobiology. 1990;18:273–81.

Diamond DM, Bennett MC, Fleshner M, Rose GM. Inverted-U relationship between the level of peripheral corticosterone and the magnitude of hippocampal primed burst potentiation. Hippocampus. 1992;2:421–30.

Diamond DM, Park CR, Heman KL, Rose GM. Exposing rats to a predator impairs spatial working memory in the radial arm water maze. Hippocampus. 1999;9:542–52.

Diamond DM, Park CR, Woodson JC. Stress generates emotional memories and retrograde amnesia by inducing an endogenous form of hippocampal LTP. Hippocampus. 2004;14:281–91.

Dudek SM, Bear MF. Homosynaptic long-term depression in area CA1 of hippocampus and the effects of NMDA receptor blockade. Proc Natl Acad Sci U S A. 1992;89:4363–7.

Eichenbaum H. Hippocampus: mapping or memory? Curr Biol. 2000;21:R785–R787.

Foy MR, Stanton ME, Levine S, Thompson RF. Behavioral stress impairs long-term potentiation in rodent hippocampus. Behav Neural Biol. 1987;48(1):138–49.

Foy MR, Foy JG, Levine S, Thompson RF. Manipulation of pituitary-adrenal activity affects neural plasticity in rodent hippocampus. Psychol Sci. 1990;3:201–4.

Foy MR, Kim JJ, Shors TJ, Thompson RF. Neurobiological foundations of stress. In: Yehuda S, Mostofsky DI, editors. Nutrients, stress, and medical disorders. Totowa: Humana; 2005. p 37–65.

Garcia R, Musleh W, Tocco G, Thompson RF, Baudry M. Time-dependent blockade of STD and LTP in hippocampal slices following acute stress in mice. Neurosci Lett. 1997;233:41–4.

Graham LK, Yoon T, Kim JJ. Stress impairs optimal behavior in a water foraging choice task in rats. Learn Mem. 2010;17:790–3.

Hebb DO. The organization of behavior: a neuropsychological theory. New York: Wiley; 1949.

Heinrichs SC, Lapsansky J, Behan DP, Chan RK, Sawchenko PE, Lorang M, et al. Corticotropin-releasing factor-binding protein ligand inhibitor blunts excessive weight gain in genetically obese Zucker rats and rats during nicotine withdrawal. Proc Natl Acad Sci U S A. 1996;93(26):15475–80.

Helmstetter FJ. Contribution of the amygdala to learning and performance of conditional fear. Physiol Behav. 1992;51(6):1271–6.

Henke PG. Attenuation of shock-induced ulcers after lesions in the medial amygdala. Physiol Behav. 1981;27:143–6.

Herman JP, Ostrander MM, Mueller NK, Figueiredo H. Limbic system mechanisms of stress regulation: hypothalamo-pituitary-adrenocortical axis. Prog Neuropsychopharmacol Biol Psychiatry. 2005;29:1201–13.

Jackson ED, Payne JD, Nadel L, Jacobs WJ. Stress differentially modulates fear conditioning in healthy men and women. Biol Psychiatry. 2006;59:516–22.

Joels M, Baram TZ. The neuro-symphony of stress. Nat Rev Neurosci. 2009;10:459–66.

Kerr DS, Campbell LW, Hao SY, Landfield PW. Corticosteroid modulation of hippocampal potentials: increased effects with aging. Science. 1989;245:1505–9.
Kim JJ, Baxter MG. Multiple brain-memory systems: the whole does not equal the sum of its parts. Trends Neurosci. 2001;24:324–30.
Kim JJ, Diamond DM. The stressed hippocampus, synaptic plasticity and lost memories. Nat Rev Neurosci. 2002;3(6):453–62.
Kim JJ, Haller J. Glucocorticoid hyper- and hypofunction: stress effects on cognition and aggression. Ann N Y Acad Sci. 2007;1113:291–303.
Kim JJ, Yoon KS. Stress: metaplastic effects in the hippocampus. Trends Neurosci. 1998;21(12):505–9.
Kim JJ, Foy MR, Thompson RF. Behavioral stress modifies hippocampal plasticity through N-methyl-D-aspartate receptor activation. Proc Natl Acad Sci U S A. 1996;93:4750–3.
Kim JJ, Lee HJ, Han J-S, Packard MG. Amygdala is critical for stress-induced modulation of hippocampal long-term potentiation (LTP) and learning. J Neurosci. 2001;21(14):5222–8.
Kim JJ, Koo JW, Lee HJ, Han J-S. Amygdalar inactivation blocks stress-induced impairments in hippocampal long-term potentiation and spatial memory. J Neurosci. 2005;25:1532–9.
Kim JJ, Lee HJ, Welday AC, Song EY, Cho J, Sharp PE, et al. Stress-induced alterations in hippocampal plasticity, place cells and spatial memory. Proc Nat Acad Sci U S A. 2007;104:18297–302.
Kim EJ, Kim E, Park M, Cho J, Kim JJ. Amygdalar stimulation produces alterations on firing properties of hippocampal place cells. J Neurosci. 2012;32:11424–34.
Kirschbaum C, Wolf OT, May M, Wippich W, Hellhammer DH. Stress- and treatment-induced elevations of cortisol levels associated with impaired declarative memory in healthy adults. Life Sci. 1996;58:1475–83.
Kosten TA, Kim JJ, Lee HJ. Early life manipulations alter learning and memory in rats. Neurosci Biobehav Rev. 2012;36:1985–2006.
Krettek JE, Price JL. Projections from the amygdaloid complex to the cerebral cortex and thalamus in the rat and cat. J Comp Neurol. 1977;172:687–722.
Kuhlmann S, Piel M, Wolf OT. Impaired memory retrieval after psychosocial stress in healthy young men. J Neurosci. 2005;25:2977–82.
LeDoux JE. The emotional brain. New York: Simon & Schuster; 1996.
Lovallo WR, Robinson JL, Glahn DC, Fox PT. Acute effects of hydrocortisoneon the human brain: an fMRI study. Psychoneuroendocrinology. 2010;35:15–20.
Luine VN, Spencer RL, McEwen BS. Effects of chronic corticosterone ingestion on spatial memory performance and hippocampal serotonergic function. Brain Res. 1993;616:65–70.
Lupien SJ, McEwen BS. The acute effects of corticosteroids on cognition: integration of animal and human models studies. Brain Res. 1997;24:1–27.
Lupien SJ, McEwen BS, Gunnar MR, Heim C. Effects of stress throughout the lifespan on the brain, behaviour and cognition. Nat Rev Neurosci. 2009;10:434–45.
Maier SF, Seligman MEP. Learned helplessness: Theory and evidence. J Exp Psychol Gen. 1976;105:3–45.
Martin SJ, Grimwood PD, Morris RG. Synaptic plasticity and memory: an evaluation of the hypothesis. Annu Rev Neurosci. 2000;23:649–711.
McEwen BS. Protection and damage from acute and chronic stress: allostasis and allostatic overload and relevance to the pathophysiology of psychiatric disorders. Ann N Y Acad Sci. 2004;1032:1–7.
McEwen BS, Gianaros PJ. Stress- and allostasis-induced brain plasticity. Ann Rev Med. 2011;62:431–45.
McEwen BS, Sapolsky RM. Stress and cognitive function. Curr Opin Neurobiol. 1995;5:205–16.
McGaugh JL. Memory-a century of consolidation. Science. 2000;287:248–51.
Mitra R, Jadhav S, McEwen BS, Vyas A, Chattarji S. Stress duration modulates the spatiotemporal patterns of spine formation in the basolateral amygdala. Proc Natl Acad Sci U S A. 2005;102:9371–6.
Morris RGM. Spatial localization does not require the presence of local cues. Learn Motiv. 1981;12:239–60.

Morris RGM, Garrud P, Rawlins JNP, O'Keefe J. Place navigation impaired in rats with hippocampal lesions. Nature 1982;297:681–683.
Moser EI, Krobert KA, Moser M-B, Morris RGM. Impaired spatial learning after saturation of long-term potentiation. Science. 1998;281:2038–42.
Nair VD, Savelli JE, Mishra RK. Modulation of dopamine D2 receptor expression by an NMDA receptor antagonist in rat brain. J Mol Neurosci. 1998;11(2):121–6.
Newcomer JW, Craft S, Hershey T, Askins K, Bardgett ME. Glucocorticoid-induced impairment in declarative memory performance in adult humans. J Neurosci. 1994;14:2047–53.
Oitzl MS, Reichardt HM, Joels M, de Kloet ER. Point mutation in the mouse glucocorticoid receptor preventing DNA binding impairs spatial memory. Proc Natl Acad Sci U S A. 2001;98(22):12790–5.
O'Keefe J, Dostrovsky J. The hippocampus as a spatial map. Preliminary evidence from unit activity in the freely-moving rat. Brain Res. 1971;34:171–5.
O'Keefe J, Nadel L. The hippocampus as a cognitive map. New York: Oxford University Press; 1978.
Olton DS, Samuelson RJ. Rememberance of places passed: spatial memory in rats. J Exp Psychol Animal Behav Process. 1976;2:97–116.
Overmaier JB, Seligman MEP. Effects of inescapable shock upon subsequent escape and avoidance responding. J Comp Physiol Psychol. 1967;63:28–33.
Packard MG, Cahill L, McGaugh JL. Amygdala modulation of hippocampal-dependent and caudate nucleus-dependent memory processes. Proc Nat Acad Sci U S A. 1994;91:8477–81.
Park CR, Zoladz PR, Conrad CD, Fleshner M, Diamond DM. Acute predator stress impairs the consolidation and retrieval of hippocampus-dependent memory in male and female rats. Learn Mem. 2008;15:271–80.
Passecker J, Hok V, Della-Chiesa A, Chah E, O'Mara SM. Dissociation of dorsal hippocampal regional activation under the influence of stress in freely behaving rats. Front Behav Neurosci. 2011;5:1–7.
Pavlides C, Watanabe Y, Margarinos AM, McEwen BS. Opposing roles of Type I and Type II adrenal steroid receptors in hippocampal long-term potentiation. Neuroscience. 1995;68:387–94.
Pikkarainen M, Ronkko S, Savander V, Insausti R, Pitkanen A. Projections from the lateral, basal, and accessory basal nuclei of the amygdala to the hippocampal formation in rat. J Comp Neurol. 1999;403:229–60.
Popoli M, Yan Z, McEwen BS, Sanacora G. The stressed synapse: the impact of stress and glucocorticoids on glutamate transmission. Nat Rev Neurosci. 2012;13:22–37.
Pruessner JC, Dedovic K, Khalili-Mahani N, Engert V, Pruessner M, Buss C, Renwick R, Dagher A, Meaney MJ, Lupien S. Deactivation of the limbic system during acute psychosocial stress: evidence from positron emission tomography and functional magnetic resonance imaging studies. Biol Psychiatry. 2008;63:234–40.
Qin S, Hermans EJ, van Marle HJ, Luo J, Fernandez G. Acute psychological stress reduces working memory-related activity in the dorsolateral prefrontal cortex. Biol Psychiatry. 2009;66:25–32.
Quirarte GL, de la TIS, Casillas M, Serafín N, Prado-Alcala´ RA, Roozendaal B. Corticosterone infused into the dorsal striatum selectively enhances memory consolidation of cued water-maze training. Learn Mem. 2009;16:586–9.
Rau V, DeCola JP, Fanselow MS. Stress-induced enhancement of fear learning: an animal model of posttraumatic stress disorder. Neurosci Biobehav Rev. 2005;29:1207–23.
Radley JJ, Morrison JH. Repeated stress and structural plasticity in the brain. Ageing Res Rev. 2005;4:271–87.
Reul JM, de Kloet ER. Two receptor systems for corticosterone in rat brain: microdistribution and differential occupation. Endocrinology. 1985;117(6):2505–11.
Roozendaal B, Griffith QK, Buranday J, de Quervain J-F, McGaugh JL. The hippocampus mediates glucocorticoid-induced impairment of spatial memory retrieval: dependence on the basolateral amygdala. Proc Natl Acad Sci U S A. 2003;100:1328–33.
Rosen JB, Schulkin J. From normal fear to pathalogical anxiety. Psychol Rev. 1998;105:325–50.
Sapolsky RM. Stress: the aging brain and the mechanisms of neuron death. Cambridge: MIT Press; 1992.

Sapolsky RM. The possibility of neurotoxicity in the hippocampus in major depression: a primer on neuron death. Biol Psychiatry. 2000;48(8):755–65.

Schoenfeld TJ, Gould E. Stress, stress hormones, and adult neurogenesis. Exp Neurol. 2012;233:12–21.

Schwabe L, Oitzl MS, Philippsen C, Richter S, Bohringer A, Wippich W, et al. Stress modulates the use of spatial and stimulus-response learning strategies in humans. Learn Mem. 2007;14:109–16.

Schwabe L, Wolf OT. Stress modulates the engagement of multiple memory systems in classification learning. J Neurosci. 2012;32:11042–9.

Scoville WB, Milner B. Loss of recent memory after bilateral hippocampal lesions. J Neurol Neurosurg Psychiatry. 1957;20:11–21.

Sejnowski TJ. Statistical constraints on synaptic plasticity. J Theoretical Biol. 1977;69:385–938.

Seligman MEP, Maier SF. Failure to escape traumatic shock. J Exp Psychol. 1967;74:1–9.

Selye H. The stress of life. New York: McGraw-Hill; 1956.

Selye H. The evolution of the stress concept. Am Psychol. 1973;61:692–9.

Selye H. Stress without distress. New York: New American Library; 1974.

Shors TJ. Learning during stressful times. Learn Mem. 2004;11:137–44.

Shors TJ, Dryver E. Effects of stress and long-term potentiation (LTP) on subsequent LTP and the theta burst response in the dentate gyrus. Brain Res. 1994;666(2):232–8.

Shors TJ, Mathew PR. NMDA receptor antagonism in the lateral/basolateral but not central nucleus of the amygdala prevents the induction of facilitated learning in response to stress. Learn Mem. 1998;5(220):230.

Shors TJ, Levine S, Thompson RF. Effect of adrenalectomy and demedullation on the stress-induced impairment of long-term potentiation. Neuroendocrinology. 1990;51:70–5.

Shors TJ, Gallegos RA, Breindl A. Transient and persistent consequences of acute stress on long-term potentiation (LTP), synaptic efficacy, theta rhythms and bursts in area CA1 of the hippocampus. Synapse. 1997;26(3):209–17.

Shors TJ, Seib TB, Levine S, Thompson RF. Inescapable versus escapable shock modulates long-term potentiation in the rat hippocampus. Science. 1989;244:224–6.

Sinha R, Shaham Y, Heilig M. Translational and reverse translational research on the role of stress in drug craving and relapse. Psychopharm. 2011;218:69–82.

Starkman MN, Gebarski SS, Berent S, Schteingart DE. Hippocampal formation volume, memory dysfunction, and cortisol levels in patients with Cushing's syndrome. Biol Psychiatry. 1992;32(9):756–65.

Stranahan AM, Khalil D, Gould E. Social isolation delays the positive effects of running on adult neurogenesis. Nat Neurosci. 2006;9:526–33.

Utto M, Vasterling JJ, Brailey K, Sutker PB. Memory and attention in combat-related posttraumatic-stress-disorder (PTSD). J Psychopathol Behav Assess. 1993;15:43–52.

Ulrich-Lai YM, Herman JP. Neural regulation of endocrine and autonomic stress responses. Nat Rev Neurosci. 2009;10:397–409.

Vaher PR, Luine VN, Gould E, McEwen BS. Effects of adrenalectomy on spatial memory performance and dentate gyrus morphology. Brain Res. 1994;656:71–8.

Vouimba RM, Richter-Levin G. Physiological dissociation in hippocampal subregions in response to amygdala stimulation. Cereb Cortex. 2005;15:1815–21.

Vyas A, Mitra R, Shankaranarayana R, Chattarji S. Chronic stress induces contrasting patterns of dendritic remodeling in hippocampal and amygdaloid neurons. J Neurosci. 2002;22:6810–8.

Vyas A, Bernal S, Chattarji S. Effects of chronic stress on dendritic arborization in the central and extended amygdala. Brain Res. 2003;965:290–4.

Wingard JC, Packard MG. The amygdala and emotional modulation of competition between cognitive and habit memory. Behav Brain Res. 2008;193:126–31.

Woodson JC, Macintosh D, Fleshner M, Diamond DM. Emotion-induced amnesia in rats: working memory-specific impairment, corticosterone-memory correlation, and fear versus arousal effects on memory. Learn Mem. 2003;10(5):326–36.

Xu L, Anwyl R, Rowan MJ. Behavioural stress facilitates the induction of long-term depression in the hippocampus. Nature. 1997;387:497–500.

Chapter 10
Evolutionary, Historical and Mechanistic Perspectives on How Stress Affects Memory and Hippocampal Synaptic Plasticity

George E. Farmer, Collin R. Park, Laura A. Bullard and David M. Diamond

Abstract We have reviewed research on stress effects on brain and memory processing from evolutionary, historic, and mechanistic perspectives. Our view is that the stress response has been refined through the process of natural selection to provide a rapid activation of attention and memory-related neural systems in response to a threat to survival. Specifically, stress enhances synaptic plasticity in the hippocampus (in conjunction with amygdala activation) to generate a rapid, but time-restricted, enhancement of memory. The activation period, lasting only seconds to minutes, is followed by a period in which the hippocampus is relatively resistant to developing excitatory plasticity. One consequence of this rapid, but brief, activation of the hippocampus in response to intense stress is that life-threatening experiences can produce abnormal memories which represent only small fragments of the original experience. These fragmented memories of trauma are highly resistant to extinction, and underlie the intrusive memories commonly reported in people suffering from posttraumatic stress disorder (PTSD). This evolutionary-based perspective may provide insight into the neurobiological basis of traumatic memories and aid in the development of more effective treatments for individuals diagnosed with PTSD.

D. M. Diamond (✉) · G. E. Farmer · C. R. Park · L. A. Bullard
Medical Research Service, VA Hospital, Tampa, FL, USA
e-mail: ddiamond@mail.usf.edu

G. E. Farmer · C. R. Park · L. A. Bullard · D. M. Diamond
Department of Psychology, University of South Florida, Tampa, FL 33620, USA

D. M. Diamond
Department of Molecular Pharmacology and Physiology, University of South Florida, Tampa, FL 33620, USA

G. E. Farmer · C. R. Park · L. A. Bullard · D. M. Diamond
Center for Preclinical and Clinical Research on PTSD, University of South Florida, Tampa, FL 33620, USA

M. Popoli et al. (eds.), *Synaptic Stress and Pathogenesis of Neuropsychiatric Disorders*,
DOI 10.1007/978-1-4939-1056-4_10, © Springer Science+Business Media New York 2014

Abbreviations

AMPA	α-amino-3-hydroxy-5-methyl-4-isoxazolepropionic acid
CaMKII	Ca^{2+}/calmodulin-dependent protein kinases II
CRH	Corticotropin-releasing hormone
GR	Glucocorticoid receptor
LTP	Long-term potentiation
MR	Mineralocorticoid receptor
NE	Norepinephrine
NMDA	*N*-methyl D-aspartate
PB	Primed burst
pMAPK2	Phosphorylated mitogen-activated protein kinase 2
PTSD	Posttraumatic stress disorder
VTA	Ventral tegmental area

10.1 Introduction: Evolutionary Perspective on Stress-Memory Dynamics

From an evolutionary perspective, the behavioral and physiological responses to stress have all developed to accomplish one goal: to maximize the likelihood an individual will survive a life-threatening experience. In particular, the stress response appears highly efficient at enhancing survival in response to an attack which has a high likelihood of producing structural damage. This stress adaptation is illustrated, for example, by the rapid stress-induced increase in blood glucose which mobilizes energy reserves to maximize an effective escape or attack. Moreover, stress promotes activation of the immune system and blood coagulation factors, processes that prepare an individual for wounds which may be inflicted during an attack (Sapolsky 1994). From a neuroethological perspective, a critical component of the stress response is activation of brain attention systems to maximize the processing of sensory components, which enable an individual to respond effectively to a threat. It is therefore of heuristic value to consider all components of the stress response to have been refined by the forces of natural selection to maximize survival in response to current and future life-threatening experiences.

How the brain forms memories of a stressful experience, however, is a challenge to understand from an evolutionary perspective. One may hypothesize that when a life-threatening experience occurs stress should provoke brain memory systems to generate highly accurate and durable memories, which can be of value if the individual survives the assault and then is faced with a similar threat in the future. This hypothesis is supported by the observation that intense stress can produce such powerful memories of the experience that they achieve a pathological status, as in the intrusive memories commonly reported in traumatized people diagnosed with posttraumatic stress disorder (PTSD) (Bryant et al. 2011; Ehlers et al. 2004). This perspective on emotional memory suggests that the cognitive component of PTSD symptoms reflects an evolutionarily adaptive process, albeit, a process that has the capacity to go horribly awry.

A milder version of emotion-induced modulation of memory is described in the extensive literature on "flashbulb memories," which describes the phenomenon of enhanced memory processing for events and circumstances coincident with periods of high arousal (Brown and Kulik 1977). At a later time, the reappearance of cues which had been present at the time of the arousing experience is interpreted by the brain as a potential reemergence of the same threat to the individual's life. The memory of the original experience is activated, thereby enabling the individual to respond more effectively to the same situation, for example, by avoiding a place which was associated with predators. Although the precision with which flashbulb memories represent an accurate representation of the original experience has been debated (Laney and Loftus 2005; Loftus 2005; Schmidt 2004; Tekcan et al. 2003), their general accuracy and durability, which can span decades, is remarkable (Berntsen and Thomsen 2005; Tekcan and Peynircioglu 2002; Van der Kolk 1997). Therefore, findings from human and animal research indicate that an experience that evokes strong arousal, particularly in life threatening conditions, generates enduring memories of the event.

Although the findings of the veracity and durability of emotional memories are consistent with the evolutionary value of enhanced memory processing in response to life-threatening experiences, a thorough review of the literature reveals a more complex story on the modulation of memory by stress. Over a century of research has provided a vast and seemingly conflicting literature providing evidence that stress not only enhances memory, it can also impair memory in rodents and people (Buchanan et al. 2006; Diamond et al. 2007; Elzinga et al. 2005; Kim et al. 2006; Kirschbaum et al. 1996; Payne et al. 2002, 2006; Roozendaal et al. 2009; Schwabe et al. 2010, 2012; Wolf 2009). This more comprehensive assessment of the complexity of the stress-memory literature is not consistent with the hypothesis that stress nonspecifically enhances memory storage.

Despite the complexity of the stress-memory literature, we remain guided by the principal that stress-memory interactions, in step with all other physiological processes, have been refined by natural selection to maximize survival in response to a life-threatening stimulus. In this chapter, we discuss a refined hypothesis which takes into account the adaptive value of the complexity of stress-memory interactions. Specifically, we consider the initiation of an attack to be the moment when an individual's survival is at greatest jeopardy, which thereby makes this relatively brief period of time crucial for optimizing brain attention and memory processing. Our hypothesis is that memory storage is optimal for events occurring during the brief period of time (seconds to minutes) around the onset of an experience that generates a sudden increase in attention and arousal. In contrast to this brief memory enhancing period at stress onset, events that occur well before or long after the initiation of the stress experience would not be remembered as well. This time-dependent dynamic shift in memory processing provides an ethologically relevant approach toward understanding the complexity of memory processing in response to stress.

Our hypothesizing on the time-dependency of memory processing during intense stress provides a foundation for enhancing our understanding of stress-related psychiatric disorders. For example, a core feature of PTSD includes pathologically intense, intrusive, and extinction-resistant memories of the traumatic experience (Debiec et al. 2011; Milad et al. 2009; Rougemont-Bucking et al. 2011). To improve our understanding of PTSD and to provide a background on memory, stress and psychopathology, in the next section we review research which has examined how stress affects the hippocampus, a structure which is central to emotional and non-emotional memory processing (Eichenbaum 2004). We conclude this chapter with a discussion of physiological mechanisms which appear to underlie the dynamic time-dependent shifts in brain-memory processing that determine whether events occurring during heightened emotion will be remembered or forgotten.

10.2 Historical Perspective on How Acute Stress Affects Hippocampal Functioning

Pioneering studies on stress and the brain were performed by Bruce McEwen and his colleagues who determined that the hippocampus has the greatest density of corticosteroid receptors in the brain (McEwen et al. 1969, 1968). These findings indicated that the hippocampus, in addition to its crucial role in memory formation, was also highly sensitive to stress. In related work, McEwen's group suggested that prolonged stress, via glucocorticoid receptor (GR) activation, impairs hippocampal function (Micco Jr. et al. 1979). The view of stress interfering with hippocampal functioning was incorporated into theorizing on hippocampal functioning by Jacobs and Nadel (Jacobs and Nadel 1985) who suggested that the stress-induced disruption of hippocampal functioning contributed to the expression of psychiatric disorders. Hence, early studies implicated acute stress as having a detrimental influence on hippocampal functioning.

In the decades since McEwen's pioneering research, studies on stress and synaptic plasticity have further supported the view that stress impairs hippocampal functioning. The first such evidence from electrophysiological studies on synaptic plasticity was provided by Thompson and coworkers, who demonstrated in 1987 that acute stress blocked the induction of hippocampal long-term potentiation (LTP) in vitro (Foy et al. 1987), a physiological model of memory formation (Miller and Mayford 1999; Muller et al. 2002). At that time our group was investigating how acute stress or corticosterone affected a low threshold form of LTP, referred to as primed burst (PB) potentiation, in vivo (Diamond et al. 1988; Rose and Dunwiddie 1986). We reported that adrenalectomized, and therefore corticosterone-depleted, rats exhibited a greater magnitude of PB potentiation than adrenal intact rats (Diamond et al. 1989), which suggested that corticosterone exerted an inhibitory influence on hippocampal plasticity. We then extended this work with the finding of an overall inverted U-shaped function between corticosterone levels and PB potentiation (Diamond et al. 1992), thereby providing strong support for the hypothesis that stress levels of corticosterone exerted a profound inhibitory effect on hippocampal functioning.

In behavioral work, we reported that the induction of PB potentiation was blocked in rats that were exposed to an unfamiliar, and therefore stress-provoking, environment (Diamond et al. 1990, 1994). We also showed that when rats were explicitly acclimated to the environment, as indicated by a significant reduction in their levels of serum corticosterone, the blockade of PB potentiation was no longer present (Diamond et al. 1994). Importantly, when these same rats were then exposed to a second, stress provoking (corticosterone-elevating) environment, once again, PB potentiation was suppressed. These findings demonstrated that the capacity for the hippocampus to generate plasticity, and presumably its memory storage functioning, was continuously influenced by an animal's emotional state; under stress conditions hippocampal functioning was impaired and when the stress abated hippocampal functioning resumed its normal capacity to process and store memories.

Subsequent work conducted over the past two decades by our laboratory, as well as work from numerous other groups have replicated the finding of a stress- or corticosterone-induced suppression of hippocampal synaptic plasticity. For example, we demonstrated that stress blocked the induction of PB potentiation in vivo (Diamond et al. 1999a; Vouimba et al. 2006) and in vitro (Mesches et al. 1999). Complementary findings from other groups have shown that acute stress or corticosterone administration can block hippocampal LTP (Cazakoff and Howland 2010; Diamond et al. 2007; Huang et al. 2005; Joels and Krugers 2007; Schmidt et al. 2013; Schwabe et al. 2012; Segal et al. 2010); (see Segal et al. (2010) for discussion of differences in stress and corticosterone effects on hippocampal plasticity in the dorsal versus ventral hippocampus).

In addition to work on synaptic plasticity, studies on learning and memory in rodents and people have provided strong evidence that stress impairs cognitive aspects of hippocampal functioning. For almost two decades our group has shown that stress, involving exposure of rats to either an unfamiliar environment or to a live cat, impairs hippocampus-dependent spatial memory (Campbell et al. 2008; Conboy et al. 2009; Diamond et al. 1996, 1999b, 2006; Sandi et al. 2005; Woodson et al. 2003). Our findings are consistent with work from other laboratories indicating that acute stress or corticosterone administration can impair hippocampus-specific memory processing in rats and people (Joels et al. 2008, 2011; Schwabe et al. 2012; Yehuda et al. 2010).

This brief overview of studies on stress and synaptic plasticity summarizes the prevailing view that strong stress inhibits hippocampal functioning (Acheson et al. 2012; Brewin 2001; Diamond et al. 2005; Jacobs and Nadel 1985; Joseph 1999; Kim and Yoon 1998; Kim and Diamond 2002; Kim et al. 2006; Layton and Krikorian 2002; LeDoux 1996; Metcalfe and Jacobs 1998; Nadel and Jacobs 1998; Van der Kolk 1996). It can therefore be stated with certainty that stress can impair the capacity for the hippocampus to generate excitatory synaptic plasticity, and that stress interferes with the involvement of the hippocampus in the storage of information.

10.3 Temporal Dynamics of Stress-Plasticity Interactions: Resolving the Paradox of How the Hippocampus is Involved in the Formation of Stressful Memories

The attentive reader may be forgiven for being perplexed by the historical perspective we just provided as to how stress affects hippocampal synaptic plasticity and memory. In the first section of this chapter we emphasized the evolutionary value of enhancing memory under stressful conditions, which was reinforced by our brief review of the durability and accuracy of emotional (flashbulb) memories. We also referred to the vast research literature confirming that the integrity of the hippocampus has long been demonstrated to be essential for the formation of declarative (fact-based, episodic) memories. The paradox is that stress produces intense and durable episodic memories, as exemplified by flashbulb and intrusive memory phenomena, and yet, the literature provides strong evidence that stress impairs the functioning of the hippocampus, a structure at the center of brain memory circuitry. To resolve this paradox, we will revisit the hypothesis we presented in the introductory section regarding dynamic changes in memory processing in response to stress. We speculated that it is the onset of an intense emotional experience, as in the immediate response to an attack by a predator, which is the critical time to optimize memory storage. Hence, focusing on memory, and specifically hippocampal processing, for events occurring around the onset of a stressor may resolve the inconsistencies in the literature as to how stress affects the brain and memory.

Empirical research relevant to our hypothesis has been provided in the work by Ehlers et al. (2002) in their analysis of intrusive memories reported by traumatized people. These investigators examined the relation between intrusive memories for trauma and the timing of events occurring during traumatic experience. People who had experienced severe trauma identified features of their intrusive memories (a core symptom of PTSD). Most subjects reported visual intrusive memories of stimuli or events that occurred immediately before or at the onset of the traumatic event. For example, one patient who had experienced a head-on car crash at night saw headlights coming towards her as a prominent component of her intrusive memories of the experience. Ehlers and coworkers suggested that because these stimuli occurred in close temporal proximity to the traumatic event, they became "warning signals," or stimuli that, if encountered in the future, would indicate something dangerous is about to happen. These authors noted that events occurring more distant from the initiation or peak period of trauma were less likely to be incorporated into intrusive memories.

At extreme levels of emotionality the memory storage process underlying the "warning signal" phenomenon can become pathological, as in the intrusive memories which interfere with the traumatized person's sleep quality, and more globally, with the person's quality of life. Nevertheless, from a neuroethological perspective, the intrusive memories suffered by a traumatized person represent an adaptive process since the repeated rehearsals of the traumatic experience (via intrusive memory reactivation) primes the individual to be more sensitive to the warning signal in the future. Even impaired sleep quality, which is a central feature of the PTSD diagnosis, is adaptive from an evolutionary perspective; suppressing sleep is a strategy with which the brain can ensure that the individual is always on-guard to respond more effectively to warning signals which were associated with a threat.

Although the "warning signal" hypothesis of Ehlers and coworkers was not presented in a neurobiological framework, its primary emphasis, of maximal memory storage for events occurring at the onset of a stress experience, has been addressed in experimental and theoretical work in behavioral neuroscience research. Specifically, there is a small, and perhaps overlooked, subset of electrophysiological research that has demonstrated that manipulations, which produce strong emotionality in rats, can *enhance* hippocampal LTP. This finding was first described by Seidenbecher et al. (1995), who showed that water-deprived rats given access to water around the time of tetanizing stimulation exhibited an *increase* in the duration of hippocampal LTP. Other studies have replicated and extended this finding to show that a variety of arousing experiences, such as water immersion, exposure to novel places and objects, and spatial learning occurring around the time of the delivery of tetanizing stimulation, all increased the duration of LTP (Ahmed et al. 2006; Almaguer-Melian et al. 2005; Davis et al. 2004; Frey 2001; Li et al. 2003; e.g., Seidenbecher et al. 1997; Straube et al. 2003; Uzakov et al. 2005); but see (Tabassum and Frey 2013).

The rapid effects of stress on enhancing hippocampal plasticity appear to be mediated, in part, by amygdala–hippocampus interactions (Kim and Diamond 2002). Studies demonstrating the enhancing effect of amygdala activation effects on hippocampal LTP were originally provided by Akirav and Richter-Levin (Akirav and Richter-Levin 2006; Bergado et al. 2011; Richter-Levin and Akirav 2003; Richter-Levin 2004). These investigators showed that stimulation of the amygdala 30 s, but not 1 h, prior to perforant path stimulation of the hippocampus enhanced LTP in the DG. These findings of a time-dependent modulation of hippocampal plasticity by amygdala stimulation or stress are consistent with our work in which stress blocked the induction of PB potentiation in vivo and in vitro (discussed above); in our research tetanizing stimulation has always been delivered at least 1 h, and as many as 4 h, after the stress manipulation began. Work from other laboratories, as well, that have shown inhibitory effects of stress on LTP involve necessary amygdala activation (Kim et al. 2001, 2005), in conjunction with prolonged stress (at least 30 min) prior to the delivery of tetanizing stimulation (Alfarez et al. 2002; Foy et al. 1987; Garcia et al. 1997; Shors et al. 1997). Overall, these findings indicate that for a relatively brief period of time, stress via amygdala activation enhances the hippocampal synaptic plasticity, followed by a later developing phase when the induction of LTP is suppressed.

Therefore, the dominant theme of stress uniformly impairing hippocampal LTP has not incorporated conflicting findings, which have demonstrated that stress can enhance, as well as impair, the induction of hippocampal synaptic plasticity. The enhancement of LTP by stress appears to be confined to conditions in which the stress and tetanizing stimulation occur in close temporal proximity; in contrast, the suppression of LTP occurs when there is a prolonged delay between the time of stress onset and the delivery of tetanizing stimulation.

This view of dynamic temporal shifts in processing by the hippocampus has been a topic of extensive theorizing in the past decade. For example, Joels et al. (2006) theorized regarding the role of corticosterone in the time-dependent effects of stress on memory and LTP. In related work, Richter-Levin and coworkers (Bergado et al. 2011; Richter-Levin and Akirav 2003; Richter-Levin 2004) proposed the "emotional tagging" hypothesis, which states that there is a selective activation of synapses in the hippocampus and amygdala in response to arousing experiences. In related theorizing, we proposed the temporal dynamics model (Diamond et al. 2007), which addressed the implications of strong emotionality briefly activating hippocampal mechanisms of synaptic plasticity, thereby increasing the duration of LTP, followed by a prolonged period of inhibition. We speculated that the relatively brief stress-induced enhancement of hippocampal functioning underlies the declarative component of flashbulb and traumatic memories in people, and contextual fear conditioning in rodents. In theory, following the brief period in which hippocampal plasticity is activated is a refractory period, in which there is an increase in the threshold for the induction of new plasticity and new learning. We provided support for our hypothesis with the finding that brief (2 min) stress coincident with the time of spatial learning strengthened spatial memory, but more prolonged stress impaired spatial memory, as well as contextual (hippocampal-dependent), but not cued (hippocampal-independent), fear memory (Diamond et al. 2007). Recently, Schwabe et al. (2012) elaborated on these issues with a comprehensive review of the temporal dynamics of stress–memory–brain interactions.

The mechanisms underlying the enhancement of hippocampal plasticity by stress act, in part, by modulating NMDA receptor-based synaptic plasticity. Rapid stress-induced increases in hippocampal glutamate levels (Bagley and Moghaddam 1997; Musazzi et al. 2011; Piroli et al. 2013) increase AMPA receptor-mediated postsynaptic depolarization, followed by the transient removal of the magnesium block on the NMDA channel. Continued glutamate-mediated activation of the AMPA and NMDA receptors enables calcium ions to enter the NMDA channel, thereby increasing postsynaptic calcium concentration, triggering a cascade of events (including CaMKII activation and autophosphorylation) involved in the strengthening of synaptic activity (Nicoll and Malenka 1999).

The extensive series of studies conducted by Joels and coworkers is relevant to the rapid stress-induced modulation of NMDA- and non-NMDA-dependent synaptic plasticity. These investigators have shown that brief application of corticosterone around the time of tetanizing stimulation enhanced LTP in CA1 in vitro via nongenomic activation of mineralocorticoid receptors (MRs) (Karst et al. 2005; Wiegert et al. 2006), which rapidly enhance mEPSP frequency and glutamatergic neurotransmission. In addition, activation of membrane MRs facilitates lateral diffusion of GluA1 and GluA2 subunits and enhances activity dependent insertion of AMPA receptors (Groc et al. 2008).

Complementary work by Ahmed et al. (2006) demonstrated that brief stress transforms protein synthesis-independent LTP into a long-lasting protein synthesis-dependent form of LTP, via activation of MRs. This group also showed that stress rapidly initiated dynamic changes in gene expression (Morsink et al. 2006), and levels of cellular signaling molecules in the hippocampus, including phosphorylated mitogen-activated protein kinase 2 (pMAPK2) and calcium/calmodulin-dependent protein kinase II (pCaMKII). Conversely, stress levels of corticosterone applied for a longer period of time (>20 min) increased the magnitude of inhibitory components of electrophysiological activity, such as the afterhyperpolarization (Joels and Kloet 1989, 1991; Karst et al. 1991) and reduced NMDA receptor-mediated plasticity (Krugers et al. 2005), thereby suppressing the induction of LTP (Alfarez et al. 2002; Kerr et al. 1994; Krugers et al. 2005; Pavlides et al. 1993, 1995a, 1995b, 1996; Rey et al. 1994; Zhou et al. 2000).

In addition to corticosterone, other neuromodulators contribute to the rapid, but brief, stress-induced enhancement of synaptic plasticity. For example, the dopaminergic innervation of the hippocampus from the ventral tegmental area (VTA) produces a rapid enhancement of hippocampal synaptic plasticity (Li et al. 2003; Lisman and Grace 2005). Moreover, brief exposure of rats to a novel environment (something considered to be a mild stressor) produced a dopamine-dependent enhancement in CA1 LTP (Li et al. 2003). In addition, projections from the locus coeruleus, in response to an arousing experience, produce a rapid release of norepinephrine (NE) into the hippocampus and amygdala, which interact with elevated levels of glucocorticoids, to enhance hippocampal excitability, plasticity, and overall function (Kitchigina et al. 1997; McGaugh et al. 1996; McIntyre et al. 2003; Roozendaal et al. 2006; Sara et al. 1994; Valentino and Van Bockstaele 2008). Specifically, the stress induced activation of the locus coeruleus has been shown to enhance excitability in the dentate gyrus of the hippocampus (Harley and Sara 1992; Kitchigina et al. 1997), which is dependent on adrenergic β-receptor activation (Hopkins and Johnston 1988; Sarvey et al. 1989). In addition to the NMDA-mediated calcium influx discussed previously, activation of β-receptors enhances calcium influx through voltage dependent L-type calcium channels via upregulation of cAMP (Gray and Johnston 1987). This NE-mediated calcium influx contributes to enhanced LTP, in part, through β-receptor dependent increases in cAMP levels and enhanced activity of PKA and CaMKII which have been shown to enhance phosphorylation of GluR1 subunits and facilitate synaptic insertion of AMPA receptors (Hu et al. 2007). Together with the MR-mediated insertion of AMPA receptors (discussed above), NE release in response to a stressful event further enhances excitability in the hippocampus.

Finally, corticotropin-releasing hormone (CRH) is a critical factor in neuroendocrine modulation of brain activity. CRH is released from hippocampal interneurons in response to stress (Chen et al. 2004) and has been shown to rapidly influence hippocampal electrophysiological activity (Aldenhoff et al. 1983). CRH has also been shown to enhance synaptic efficacy in the dentate gyrus of the hippocampus in (Wang et al. 1998). Though brief application of CRH has been shown to enhance excitability and LTP in the hippocampus (Kratzer et al. 2013), prolonged application of CRH, perhaps mimicking delayed effects of stress, has been shown to impair hippocampal LTP (Rebaudo et al. 2001). Thus, CRH, as well as corticosterone, exhibit rapid and delayed effects on hippocampal synaptic activity, which reflect their participation in the dynamic time-dependent modulation of hippocampal functioning by stress.

Ultimately, rapid stress-induced elevations in glutamate levels in the hippocampus followed by increased influx of intracellular calcium are necessary for memory formation, but continued influx of postsynaptic calcium can lead to excitotoxicity (Foster and Kumar 2002). Therefore, following the rapid enhancement of plasticity, NMDA receptors desensitize to reduce calcium influx and prevent glutamate-induced neurotoxicity (Zorumski and Thio 1992). The desensitization of NMDA receptors would serve the dual purpose to protect the neurons from excitotoxicity, as well as to minimize the corruption of the memory from events occurring long after the onset of the stress initiation (Laney and Loftus 2005).

10.4 Summary

We have provided our perspective on how stress affects memory, in general, and specifically, how the hippocampus is affected by acute stress. We have critiqued the global hypothesis that a stress response involves a global enhancement of attention and memory processing. Instead, we have suggested that there is a relatively brief period of time around the initiation of a stress experience in which maximal memory processing occurs. Our discussion of dynamic shifts in the processing of synaptic plasticity, and therefore optimal memory processing, addresses the complexity and heterogeneity of the literature on how stress affects memory and synaptic plasticity. The apparent paradox that stress produces flashbulb and traumatic memories that can last a lifetime, and yet, stress blocks hippocampal synaptic plasticity, is resolved by taking into account the temporal dynamics of changes in hippocampal functioning following stress onset. That is, is a rapid stress-induced enhancement of hippocampal plasticity, followed soon after by a prolonged period of inhibition of plasticity. This time-based shift in hippocampal functioning creates an isolated (temporally fragmented) memory of events that were coincident with the onset of the stress. This perspective on the neural basis of emotional memories is relevant to the finding that traumatic intrusive memories reported by people with PTSD are described as representing only temporally disjointed fragments of the trauma, rather than as a continuous representation of the entire experience (Rubin et al. 2004).

Acknowledgments The researchers were supported by Career Scientist and Merit Review Awards from the Veterans Affairs Department during the production of this chapter. The opinions expressed in this chapter are those of the authors and not of the Department of Veterans Affairs or the US government.

Reference

Acheson DT, Gresack JE, Risbrough VB. Hippocampal dysfunction effects on context memory: possible etiology for posttraumatic stress disorder. Neuropharmacology. 2012;62(2):674–85.
Ahmed T, Frey JU, Korz V. Long-term effects of brief acute stress on cellular signaling and hippocampal LTP. J Neurosci. 2006;26(15):3951–8.
Akirav I, Richter-Levin G. Factors that determine the non-linear amygdala influence on hippocampus-dependent memory. Dose Response. 2006;4(1):22–37.
Aldenhoff JB, Gruol DL, Rivier J, Vale W, Siggins GR. Corticotropin releasing factor decreases postburst hyperpolarizations and excites hippocampal neurons. Science. 1983;221(4613):875–7.
Alfarez DN, Wiegert O, Joels M, Krugers HJ. Corticosterone and stress reduce synaptic potentiation in mouse hippocampal slices with mild stimulation. Neuroscience. 2002;115(4):1119–26.
Almaguer-Melian W, Cruz-Aguado R, Riva CL, Kendrick KM, Frey JU, Bergado J. Effect of LTP-reinforcing paradigms on neurotransmitter release in the dentate gyrus of young and aged rats. Biochem Biophys Res Commun. 2005;327(3):877–83.
Bagley J, Moghaddam B. Temporal dynamics of glutamate efflux in the prefrontal cortex and in the hippocampus following repeated stress: effects of pretreatment with saline or diazepam. Neuroscience. 1997;77(1):65–73.
Bergado JA, Lucas M, Richter-Levin G. Emotional tagging-a simple hypothesis in a complex reality. Prog Neurobiol. 2011;94(1):64–76.
Berntsen D, Thomsen DK. Personal memories for remote historical events: Accuracy and clarity of flashbulb memories related to World War II. J Exp Psychol Gen 2005;134(2):242–57.
Brewin CR. A cognitive neuroscience account of posttraumatic stress disorder and its treatment. Behav Res Ther. 2001;39(4):373–93.
Brown R, Kulik J. Flashbulb memories. Cognition. 1977;5(1):73–99.
Bryant RA, O'Donnell ML, Creamer M, McFarlane AC, Silove D. Posttraumatic intrusive symptoms across psychiatric disorders. J Psychiatr Res. 2011;45(6):842–7.
Buchanan TW, Tranel D, Adolphs R. Impaired memory retrieval correlates with individual differences in cortisol response but not autonomic response. Learn Mem. 2006;13(3):382–7.
Campbell AM, Park CR, Zoladz PR, Munoz C, Fleshner M, Diamond DM. Pre-training administration of tianeptine, but not propranolol, protects hippocampus-dependent memory from being impaired by predator stress. Eur Neuropsychopharmacol. 2008;18(2):87–98.
Cazakoff BN, Howland JG. Acute stress disrupts paired pulse facilitation and long-term potentiation in rat dorsal hippocampus through activation of glucocorticoid receptors. Hippocampus. 2010;20(12):1327–31.
Chen Y, Brunson KL, Adelmann G, Bender RA, Frotscher M, Baram TZ. Hippocampal corticotropin releasing hormone: pre- and postsynaptic location and release by stress. Neuroscience. 2004;126(3):533–40.
Conboy L, Tanrikut C, Zoladz PR, Campbell AM, Park CR, Gabriel C, et al. The antidepressant agomelatine blocks the adverse effects of stress on memory and enables spatial learning to rapidly increase neural cell adhesion molecule (NCAM) expression in the hippocampus of rats. Int J Neuropsychopharmacol. 2009;12(3):329–41.
Davis CD, Jones FL, Derrick BE. Novel environments enhance the induction and maintenance of long-term potentiation in the dentate gyrus. J Neurosci. 2004;24(29):6497–506.
Debiec J, Bush DE, LeDoux JE. Noradrenergic enhancement of reconsolidation in the amygdala impairs extinction of conditioned fear in rats-a possible mechanism for the persistence of traumatic memories in PTSD. Depress Anxiety. 2011;28(3):186–93.
Diamond DM, Dunwiddie TV, Rose GM. Characteristics of hippocampal primed burst potentiation in vitro and in the awake rat. J Neurosci. 1988;8(11):4079–88.
Diamond DM, Bennett MC, Engstrom DA, Fleshner M, Rose GM. Adrenalectomy reduces the threshold for hippocampal primed burst potentiation in the anesthetized rat. Brain Res. 1989;492(1–2):356–60.

Diamond DM, Bennett MC, Stevens KE, Wilson RL, Rose GM. Exposure to a novel environment interferes with the induction of hippocampal primed burst potentiation in the behaving rat. Psychobiology. 1990;18(3):273–81.

Diamond DM, Bennett MC, Fleshner M, Rose GM. Inverted-U relationship between the level of peripheral corticosterone and the magnitude of hippocampal primed burst potentiation. Hippocampus. 1992;2(4):421–30.

Diamond DM, Fleshner M, Rose GM. Psychological stress repeatedly blocks hippocampal primed burst potentiation in behaving rats. Behav Brain Res. 1994;62(1):1–9.

Diamond DM, Fleshner M, Ingersoll N, Rose GM. Psychological stress impairs spatial working memory: relevance to electrophysiological studies of hippocampal function. Behav Neurosci. 1996;110(4):661–72.

Diamond DM, Fleshner M, Rose GM. The enhancement of hippocampal primed burst potentiation by dehydroepiandrosterone sulfate (DHEAS) is blocked by psychological stress. Stress. 1999a;3(2):107–21.

Diamond DM, Park CR, Heman KL, Rose GM. Exposing rats to a predator impairs spatial working memory in the radial arm water maze. Hippocampus. 1999b;9(5):542–52.

Diamond DM, Park CR, Campbell AM, Woodson JC. Competitive interactions between endogenous LTD and LTP in the hippocampus underlie the storage of emotional memories and stress-induced amnesia. Hippocampus. 2005;15(8):1006–25.

Diamond DM, Campbell AM, Park CR, Woodson JC, Conrad CD, Bachstetter AD, et al. Influence of predator stress on the consolidation versus retrieval of long-term spatial memory and hippocampal spinogenesis. Hippocampus. 2006;16(7):571–6.

Diamond DM, Campbell AM, Park CR, Halonen J, Zoladz PR. The temporal dynamics model of emotional memory processing: a synthesis on the neurobiological basis of stress-induced amnesia, flashbulb and traumatic memories, and the Yerkes-Dodson Law. Neural Plast 2007;60803.

Ehlers A, Hackmann A, Steil R, Clohessy S, Wenninger K, Winter H. The nature of intrusive memories after trauma: the warning signal hypothesis. Behav Res Ther. 2002;40(9):995–1002.

Ehlers A, Hackmann A, Michael T. Intrusive re-experiencing in post-traumatic stress disorder: phenomenology, theory, and therapy. Memory. 2004;12(4):403–15.

Eichenbaum H. Hippocampus: cognitive processes and neural representations that underlie declarative memory. Neuron. 2004;44(1):109–20.

Elzinga BM, Bakker A, Bremner JD. Stress-induced cortisol elevations are associated with impaired delayed, but not immediate recall. Psychiatry Res. 2005;134(3):211–23.

Foster TC, Kumar A. Calcium dysregulation in the aging brain. Neuroscientist. 2002;8(4):297–301.

Foy MR, Stanton ME, Levine S, Thompson RF. Behavioral stress impairs long-term potentiation in rodent hippocampus. Behav Neural Biol. 1987;48(1):138–149.

Frey JU. Long-lasting hippocampal plasticity: cellular model for memory consolidation? Results Probl Cell Differ. 2001;34:27–40.

Garcia R, Musleh W, Tocco G, Thompson RF, Baudry M. Time-dependent blockade of STP and LTP in hippocampal slices following acute stress in mice. Neurosci Lett. 1997;233(1):41–4.

Gray R, Johnston D. Noradrenaline and beta-adrenoceptor agonists increase activity of voltage-dependent calcium channels in hippocampal neurons. Nature. 1987;327(6123):620–2.

Groc L, Choquet D, Chaouloff F. The stress hormone corticosterone conditions AMPAR surface trafficking and synaptic potentiation. Nat Neurosci. 2008;11(8):868–70.

Harley CW, Sara SJ. Locus coeruleus bursts induced by glutamate trigger delayed perforant path spike amplitude potentiation in the dentate gyrus. Exp Brain Res. 1992;89(3):581–7.

Hopkins WF, Johnston D. Noradrenergic enhancement of long-term potentiation at mossy fiber synapses in the hippocampus. J Neurophysiol. 1988;59(2):667–87.

Hu H, Real E, Takamiya K, Kang MG, LeDoux J, Huganir RL, et al. Emotion enhances learning via norepinephrine regulation of AMPA-receptor trafficking. Cell. 2007;131(1):160–73.

Huang CC, Yang CH, Hsu KS. Do stress and long-term potentiation share the same molecular mechanisms? Mol Neurobiol. 2005;32(3):223–35.

Jacobs WJ, Nadel L. Stress-induced recovery of fears and phobias. Psychol Rev. 1985;92(4):512–31.
Joels M, de Kloet ER. Effects of glucocorticoids and norepinephrine on the excitability in the hippocampus. Science. 1989;245(4925):1502–5.
Joels M, de Kloet ER. Effect of corticosteroid hormones on electrical activity in rat hippocampus. J Steroid Biochem Mol Biol. 1991;40(1–3):83–6.
Joels M, Krugers HJ. LTP after stress: up or down? Neural Plast 2007;93202.
Joels M, Pu Z, Wiegert O, Oitzl MS, Krugers HJ. Learning under stress: how does it work? Trends Cogn Sci. 2006;10(4):152–8.
Joels M, Krugers H, Karst H. Stress-induced changes in hippocampal function. Prog Brain Res. 2008;167:3–15.
Joels M, Fernandez G, Roozendaal B. Stress and emotional memory: a matter of timing. Trends Cogn Sci. 2011;15(6):280–8.
Joseph R. The neurology of traumatic "dissociative" amnesia: commentary and literature review. Child Abuse Negl. 1999;23(8):715–27.
Karst H, Joels M. The induction of corticosteroid actions on membrane properties of hippocampal CA1 neurons requires protein synthesis. Neurosci Lett. 1991;130(1):27–31.
Karst H, Berger S, Turiault M, Tronche F, Schutz G, Joels M. Mineralocorticoid receptors are indispensable for nongenomic modulation of hippocampal glutamate transmission by corticosterone. Proc Natl Acad Sci U S A. 2005;102(52):19204–7.
Kerr DS, Huggett AM, Abraham WC. Modulation of hippocampal long-term potentiation and long-term depression by corticosteroid receptor activation. Psychobiology. 1994;22(2):123–33.
Kim JJ, Diamond DM. The stressed hippocampus, synaptic plasticity and lost memories. Nat Rev Neurosci. 2002;3(6):453–62.
Kim JJ, Yoon KS. Stress: metaplastic effects in the hippocampus. Trends Neurosci. 1998;21(12):505–9.
Kim JJ, Lee HJ, Han JS, Packard MG. Amygdala is critical for stress-induced modulation of hippocampal long-term potentiation and learning. J Neurosci. 2001;21(14):5222–8.
Kim JJ, Koo JW, Lee HJ, Han JS. Amygdalar inactivation blocks stress-induced impairments in hippocampal long-term potentiation and spatial memory. J Neurosci. 2005;25(6):1532–9.
Kim JJ, Song EY, Kosten TA. Stress effects in the hippocampus: synaptic plasticity and memory. Stress. 2006;9(1):1–11.
Kirschbaum C, Wolf OT, May M, Wippich W, Hellhammer DH. Stress- and treatment-induced elevations of cortisol levels associated with impaired declarative memory in healthy adults. Life Sci. 1996;58(17):1475–83.
Kitchigina V, Vankov A, Harley C, Sara SJ. Novelty-elicited, noradrenaline-dependent enhancement of excitability in the dentate gyrus. Eur J Neurosci. 1997;9(1):41–7.
Kratzer S, Mattusch C, Metzger MW, Dedic N, Noll-Hussong M, Kafitz KW, et al. Activation of CRH receptor type 1 expressed on glutamatergic neurons increases excitability of CA1 pyramidal neurons by the modulation of voltage-gated ion channels. Front Cell Neurosci. 2013;7:91.
Krugers HJ, Alfarez DN, Karst H, Parashkouhi K, van Gemert N, Joels M. Corticosterone shifts different forms of synaptic potentiation in opposite directions. Hippocampus. 2005;15(6):697–703.
Laney C, Loftus EF. Traumatic memories are not necessarily accurate memories. Canadian J Psychiatry. 2005;50(13):823–8.
Layton B, Krikorian R. Memory mechanisms in posttraumatic stress disorder. J Neuropsychiatry Clin Neurosci. 2002;14(3):254–61.
LeDoux JE. The emotional brain: the mysterious underpinnings of emotional life. New York:Simon and Schuster; 1996.
Li S, Cullen WK, Anwyl R, Rowan MJ. Dopamine-dependent facilitation of LTP induction in hippocampal CA1 by exposure to spatial novelty. Nat Neurosci. 2003;6(5):526–31.
Lisman JE, Grace AA. The hippocampal-VTA loop: controlling the entry of information into long-term memory. Neuron. 2005;46(5):703–13.
Loftus EF. Searching for the neurobiology of the misinformation effect. Learn Mem. 2005;12(1):1–2.

McEwen BS, Weiss JM, Schwartz LS. Selective retention of corticosterone by limbic structures in rat brain. Nature. 1968;220(5170):911–2.
McEwen BS, Weiss JM, Schwartz LS. Uptake of corticosterone by rat brain and its concentration by certain limbic structures. Brain Res. 1969;16(1):227–41.
McGaugh JL, Cahill L, Roozendaal B. Involvement of the amygdala in memory storage: interaction with other brain systems. Proc Natl Acad Sci U S A. 1996;93(24):13508–14.
McIntyre CK, Power AE, Roozendaal B, McGaugh JL. Role of the basolateral amygdala in memory consolidation. Ann N Y Acad Sci. 2003;985:273–93.
Mesches MH, Fleshner M, Heman KL, Rose GM, Diamond DM. Exposing rats to a predator blocks primed burst potentiation in the hippocampus in vitro. J Neurosci. 1999;19(14):RC18.
Metcalfe J, Jacobs WJ. Emotional memory: the effects of stress on "cool" and "hot" memory systems. Psychology Learn Motiv. 1998;38:187–222.
Micco DJ Jr, McEwen BS, Shein W. Modulation of behavioral inhibition in appetitive extinction following manipulation of adrenal steroids in rats: implications for involvement of the hippocampus. J Comp Physiol Psychol. 1979;93(2):323–9.
Milad MR, Pitman RK, Ellis CB, Gold AL, Shin LM, Lasko NB, et al. Neurobiological basis of failure to recall extinction memory in posttraumatic stress disorder. Biol Psychiatry. 2009;66(12):1075–82.
Miller S, Mayford M. Cellular and molecular mechanisms of memory: the LTP connection. Curr Opin Genet Dev. 1999;9(3):333–7.
Morsink MC, Steenbergen PJ, Vos JB, Karst H, Joels M, de Kloet ER, et al. Acute activation of hippocampal glucocorticoid receptors results in different waves of gene expression throughout time. J Neuroendocrinol. 2006;18(4):239–52.
Muller D, Nikonenko I, Jourdain P, Alberi S. LTP, memory and structural plasticity. Curr Mol Med. 2002;2(7):605–11.
Musazzi L, Racagni G, Popoli M. Stress, glucocorticoids and glutamate release: effects of antidepressant drugs. Neurochem Int. 2011;59(2):138–49.
Nadel L, Jacobs WJ. Traumatic memory is special. Current directions. Psychol Sci. 1998;7(5):154–7.
Nicoll RA, Malenka RC. Expression mechanisms underlying NMDA receptor-dependent long-term potentiation. Ann N Y Acad Sci. 1999;868:515–25.
Pavlides C, Watanabe Y, McEwen BS. Effects of glucocorticoids on hippocampal long-term potentiation. Hippocampus. 1993;3(2):183–92.
Pavlides C, Kimura A, Magarinos AM, McEwen BS. Hippocampal homosynaptic long-term depression/depotentiation induced by adrenal steroids. Neuroscience. 1995a;68(2):379–85.
Pavlides C, Watanabe Y, Magarinos AM, McEwen BS. Opposing roles of type I and type II adrenal steroid receptors in hippocampal long-term potentiation. Neuroscience. 1995b;68(2):387–94.
Pavlides C, Ogawa S, Kimura A, McEwen BS. Role of adrenal steroid mineralocorticoid and glucocorticoid receptors in long-term potentiation in the CA1 field of hippocampal slices. Brain Res. 1996;738(2):229–35.
Payne JD, Nadel L, Allen JJ, Thomas KG, Jacobs WJ. The effects of experimentally induced stress on false recognition. Memory. 2002;10(1):1–6.
Payne JD, Jackson ED, Ryan L, Hoscheidt S, Jacobs JW, Nadel L. The impact of stress on neutral and emotional aspects of episodic memory. Memory. 2006;14(1):1–16.
Piroli GG, Reznikov LR, Grillo CA, Hagar JM, Fadel JR, Reagan LP. Tianeptine modulates amygdalar glutamate neurochemistry and synaptic proteins in rats subjected to repeated stress. Exp Neurol. 2013;241:184–93.
Rebaudo R, Melani R, Balestrino M, Izvarina N. Electrophysiological effects of sustained delivery of CRF and its receptor agonists in hippocampal slices. Brain Res. 2001;922(1):112–7.
Rey M, Carlier E, Talmi M, Soumireu-Mourat B. Corticosterone effects on long-term potentiation in mouse hippocampal slices. Neuroendocrinology. 1994;60(1):36–41.
Richter-Levin G. The amygdala, the hippocampus, and emotional modulation of memory. Neuroscientist. 2004;10(1):31–9.
Richter-Levin G, Akirav I. Emotional tagging of memory formation-in the search for neural mechanisms. Brain Res Brain Res Rev. 2003;43(3):247–56.

Roozendaal B, Okuda S, de Quervain DJ, McGaugh JL. Glucocorticoids interact with emotion-induced noradrenergic activation in influencing different memory functions. Neuroscience. 2006;138(3):901–10.
Roozendaal B, McEwen BS, Chattarji S. Stress, memory and the amygdala. Nat Rev Neurosci. 2009;10(6):423–33.
Rose GM, Dunwiddie TV. Induction of hippocampal long-term potentiation using physiologically patterned stimulation. Neurosci Lett. 1986;69(3):244–8.
Rougemont-Bucking A, Linnman C, Zeffiro TA, Zeidan MA, Lebron-Milad K, Rodriguez-Romaguera J, et al. Altered processing of contextual information during fear extinction in PTSD: an fMRI study. CNS Neurosci Ther. 2011;17(4):227–36.
Rubin DC, Feldman ME, Beckham JC. Reliving, emotions, and fragmentation in the autobiographical memories of veterans diagnosed with PTSD. Appl Cogn Psychol. 2004;18(1):17–35.
Sandi C, Woodson JC, Haynes VF, Park CR, Touyarot K, Lopez-Fernandez MA, et al. Acute stress-induced impairment of spatial memory is associated with decreased expression of neural cell adhesion molecule in the hippocampus and prefrontal cortex. Biol Psychiatry. 2005;57(8):856–64.
Sapolsky RM. Why zebras don't get ulcers. New York:St. Martin's; 1994.
Sara SJ, Vankov A, Herve A. Locus coeruleus-evoked responses in behaving rats: a clue to the role of noradrenaline in memory. Brain Res Bull. 1994;35(5-6):457–65.
Sarvey JM, Burgard EC, Decker G. Long-term potentiation: studies in the hippocampal slice. J Neurosci Methods. 1989;28(1–2):109–24.
Schmidt SR. Autobiographical memories for the September 11th attacks: reconstructive errors and emotional impairment of memory. Mem Cognit. 2004;32(3):443–54.
Schmidt MV, Abraham WC, Maroun M, Stork O, Richter-Levin G. Stress-induced metaplasticity: from synapses to behavior. Neuroscience. 2013;250C:112–20.
Schwabe L, Wolf OT, Oitzl MS. Memory formation under stress: quantity and quality. Neurosci Biobehav Rev. 2010;34(4):584–91.
Schwabe L, Joels M, Roozendaal B, Wolf OT, Oitzl MS. Stress effects on memory: an update and integration. Neurosci Biobehav Rev. 2012;36(7):1740–9.
Segal M, Richter-Levin G, Maggio N. Stress-induced dynamic routing of hippocampal connectivity: a hypothesis. Hippocampus. 2010;20(12):1332–8.
Seidenbecher T, Balschun D, Reymann KG. Drinking after water deprivation prolongs "unsaturated" LTP in the dentate gyrus of rats. Physiol Behav. 1995;57:1001–4.
Seidenbecher T, Reymann KG, Balschun D. A post-tetanic time window for the reinforcement of long-term potentiation by appetitive and aversive stimuli. Proc Natl Acad Sci U S A. 1997;94(4):1494–9.
Shors TJ, Gallegos RA, Breindl A. Transient and persistent consequences of acute stress on long-term potentiation (LTP), synaptic efficacy, theta rhythms and bursts in area CA1 of the hippocampus. Synapse. 1997;26(3):209–17.
Straube T, Korz V, Frey JU. Bidirectional modulation of long-term potentiation by novelty-exploration in rat dentate gyrus. Neurosci Lett. 2003;344(1):5–8.
Tabassum H, Frey JU. The effect of acute swim stress and training in the water maze on hippocampal synaptic activity as well as plasticity in the dentate gyrus of freely moving rats: revisiting swim-induced LTP reinforcement. Hippocampus. 2013;23(12):1291–8.
Tekcan AI, Peynircioglu ZF. Effects of age on flashbulb memories. Psychol Aging. 2002;17(3):416–22.
Tekcan AI, Ece B, Gulgoz S, Er N. Autobiographical and event memory for 9/11: changes across one year. Appl Cogn Psychol. 2003;17(9):1057–66.
Uzakov S, Frey JU, Korz V. Reinforcement of rat hippocampal LTP by holeboard training. Learn Mem. 2005;12(2):165–71.
Valentino RJ, Van Bockstaele E. Convergent regulation of locus coeruleus activity as an adaptive response to stress. Eur J Pharmacol. 2008;583(2–3):194–203.
Van der Kolk BA. Trauma and memory. In: Van der Kolk BA, McFarlane AC, Weisaeth L, editors. Traumatic stress. New York: Guilford; 1996. 279–302.

Van der Kolk BA. The psychobiology of posttraumatic stress disorder. [Review] [75 refs]. J Clin Psychiatry. 1997;58:Suppl-24.

Vouimba RM, Munoz C, Diamond DM. Differential effects of predator stress and the antidepressant tianeptine on physiological plasticity in the hippocampus and basolateral amygdala. Stress. 2006;9(1):29–40.

Wang HL, Wayner MJ, Chai CY, Lee EH. Corticotrophin-releasing factor produces a long-lasting enhancement of synaptic efficacy in the hippocampus. Eur J Neurosci. 1998;10(11):3428–37.

Wiegert O, Joels M, Krugers H. Timing is essential for rapid effects of corticosterone on synaptic potentiation in the mouse hippocampus. Learn Mem. 2006;13(2):110–3.

Wolf OT. Stress and memory in humans: twelve years of progress? Brain Res. 2009;13(1293):142–54.

Woodson JC, Macintosh D, Fleshner M, Diamond DM. Emotion-induced amnesia in rats: working memory-specific impairment, corticosterone-memory correlation, and fear versus arousal effects on memory. Learn Mem. 2003;10(5):326–36.

Yehuda R, Joels M, Morris RG. The memory paradox. Nat Rev Neurosci. 2010;11(12):837–9.

Zhou J, Zhang F, Zhang Y. Corticosterone inhibits generation of long-term potentiation in rat hippocampal slice: involvement of brain-derived neurotrophic factor. Brain Res. 2000;885(2):182–91.

Zorumski CF, Thio LL. Properties of vertebrate glutamate receptors: calcium mobilization and desensitization. Prog Neurobiol. 1992;39(3):295–336.

Chapter 11
Acute Stress Disrupts Short- and Long-Term Patterns of Synaptic Plasticity in Dorsal Hippocampus and Subiculum: Implications for Hippocampal Output and Behaviour

John G. Howland and Don A. Davies

Abstract A period of acute stress has complex effects on hippocampal-dependent cognition in the minutes and hours following its occurrence. The neural mechanisms mediating these effects have been the focus of intense investigation for the past several decades. Much of this research has examined the role of acute stress-induced changes in long-term synaptic plasticity in the CA1 region of the dorsal hippocampus. However, numerous experiments demonstrate that acute stress also impairs short-term plasticity in the hippocampus. In addition, the effects of acute stress on short- and long-term plasticity in the dorsal subiculum, the main output area of the hippocampus, has recently been explored. The goals of this chapter are to thoroughly review these data and integrate them with theories regarding the mechanisms underlying the effects of acute stress on hippocampal-dependent cognition. We conclude that acute stress-induced alterations in synaptic plasticity at both CA1 and subiculum synapses likely contribute to the effects of acute stress on declarative-like learning and memory.

Abbreviations

AMPA	α-amino-3-hydroxy-5-methylisoxazole-4-propionic acid
BDNF	Brain derived neurotrophic factor
CA	Cornu Ammonis
Cort	Corticosterone
GR	Glucocorticoid receptor
HPC	Hippocampus
HPA	Hypothalamic-pituitary-adrenal
LDP	Late developing potentiation
LE	Long-Evans

J. G. Howland (✉) · D. A. Davies
Department of Physiology, University of Saskatchewan, GB33, Health Sciences Building, 107 Wiggins Road, Saskatoon, SK S7N 5E5, Canada
Tel.: 306.966.2032
e-mail: john.howland@usask.ca

M. Popoli et al. (eds.), *Synaptic Stress and Pathogenesis of Neuropsychiatric Disorders*,
DOI 10.1007/978-1-4939-1056-4_11,

LTP	Long-term potentiation
LTD	Long-term depression
MR	Mineralocorticoid receptor
N/C	No change
NMDA	*N*-methyl-D-aspartate
PPF	Paired pulse facilitation
SD	Sprague Dawley

11.1 Introduction

In 2008, Howland co-wrote a review paper relating the effects of acute stress on hippocampal synaptic plasticity to learning and memory (Howland and Wang 2008). It focused on the effects of acute stress on *long-term* synaptic plasticity, principally in the Cornu Ammonis (CA)1 subregion. As reviewed in that paper and numerous others (Kim and Diamond 2002; Shors 2004; Joels et al. 2006; Kim et al. 2006; Diamond et al. 2007; Collingridge et al. 2010; Cazakoff et al. 2010; Schwabe et al. 2012), there is strong evidence to support the role of altered long-term potentiation (LTP) and long-term depression (LTD) in the effects of acute stress on cognition, particularly hippocampal-dependent learning and memory. However, the effects of acute stress on patterns of *short-term* hippocampal synaptic plasticity have also been demonstrated in a number of different laboratories (Zhou et al. 2000; Commins et al. 2001; Karst et al. 2005; Gao et al. 2008; Cazakoff and Howland 2010; MacDougall and Howland 2013a;b). These observations raise questions regarding: (1) the exclusive role of altered long-term synaptic plasticity in the effects of acute stress on cognition and (2) whether distinct forms of cognition are disturbed by the effects of acute stress on short-term hippocampal synaptic plasticity. The present review will integrate findings related to short-term synaptic plasticity into existing theories regarding the effects of acute stress on hippocampal-dependent learning and memory. In addition, the effects of acute stress on synaptic plasticity in the subiculum, arguably the major output of the hippocampus (Naber et al. 2000; Behr et al. 2009; O'Mara et al. 2009), have been largely neglected in previous reviews. Thus, the acute stress effects on synaptic plasticity in the CA1 and subiculum regions will be compared.

11.2 Acute Stress

The term stress has been used historically to describe the rather vague range of perceived stimuli or conditions that disturb an organism's homeostasis (Kim and Diamond 2002). While physical threats are commonly considered stressful, psychological aspects of an organism's experience of given stimuli or conditions, such as level of aversiveness or controllability, are also critical in determining whether a given experience is perceived as "stressful" (Kim and Diamond 2002). Stress causes rapid physiological changes in the body and brain that enable organisms to overcome short periods of challenge; however, chronic stress exposure has negative effects on a number of physiological systems (McEwen and Sapolsky 1995; Sapolsky 2000). Exposure to stress results in activation of the hypothalamic-pituitary-adrenal (HPA) axis leading to the release of glucocorticoid hormones (cortisol in humans; corticosterone in most rodents) from the adrenal glands as well as the release of other mediators such as catecholamine neurotransmitters and cytokines (Herman et al. 2005; Joels and Baram 2009). In the brain, these signalling molecules activate their respective receptors, which produce an array of functional changes such as alterations in synaptic activity, dendritic organization, and neurogenesis (de Kloet et al. 2005; Kim et al. 2006; Howland and Wang 2008; Holmes and Wellman 2009). One brain region that is particularly responsive to acute stress and critically involved in regulating the responsiveness of the HPA axis to acute stress is the hippocampus (Herman et al. 2005).

This review will focus on findings concerning the effects of acute stress on hippocampal synaptic plasticity and related cognitive processes within minutes to hours of the acute stressor. Such effects are the result of short-term changes in the functionality of existing neural circuits prior to the structural remodelling of circuits that occurs in the hours-days following stress. We will also focus on the role of corticosterone in mediating these effects via actions on its two known receptors subtypes: the high-affinity mineralocorticoid receptors (MRs) and lower affinity (approximately tenfold) glucocorticoid receptors (GRs; de Kloet et al. 2005; Joels and Baram 2009; Joels et al. 2012). Both receptor subtypes are expressed in the dorsal hippocampus and subiculum, with expression of MRs particularly high and GR expression more moderate (Reul and de Kloet 1985). Evidence suggests that signalling by MRs and GRs occurs through classical genomic mechanisms and more recently appreciated non-genomic mechanisms to regulate the brain's responsiveness to activation of the HPA axis (Tasker et al. 2006; Joels et al. 2012). As will be discussed below, both of these modes of action are likely involved in regulating the effects of acute stress on synaptic plasticity and learning and memory.

11.3 Hippocampal Synaptic Plasticity

The mammalian hippocampal formation consists of several anatomically distinct subregions including the entorhinal cortex, dentate gyrus, hippocampus proper (CA3 and CA1 subfields), and subiculum (O'Mara et al. 2001; Andersen et al. 2006; van Strien et al. 2009). Standard anatomical views hold that a number of major glutamatergic pathways direct information flow through the hippocampal formation (Andersen et al. 2006; van Strien et al. 2009). Accordingly, highly integrated sensory information from entorhinal cortex (layer II) arrives at dentate gyrus via the perforant path or the CA3 and CA1 regions via the temporoammonic pathway

(Behr et al. 2009; van Strien et al. 2009). Dentate gyrus granular cells direct information to CA3 neurons via the mossy fibers which in turn project to the CA1 region through the Schaffer collaterals. Lastly, CA1 pyramidal cells project either directly back to the entorhinal cortex or to a topographically organized projection to subiculum (Amaral et al. 1991; O'Mara et al. 2001; Andersen et al. 2006). The majority of subicular cells conserve their topographic input along the transverse axis from CA1 and transmit information to the deep layers (layers V and VI) of entorhinal cortex (van Strien et al. 2009), although notable reciprocal projections to other cortical areas also exist (Naber et al. 2001; Behr et al. 2009; O'Mara et al. 2009). Thus, both CA1 and subiculum function as major output structures for the hippocampal formation and are therefore integral for hippocampal-cortical information processing (Naber et al. 2000; Behr et al. 2009; O'Mara et al. 2009). Given availability of experimental data, the effects of acute stress on synaptic plasticity in the monosynaptic Schaffer collateral-CA1 and CA1-subiculum pathways will be the focus of the following discussion.

The characteristics and molecular mechanisms of synaptic plasticity in the hippocampal formation have been intensely investigated given the hypothesized role of synaptic plasticity in normal cognition and brain disorders (Citri and Malenka 2008; Howland and Wang 2008; Collingridge et al. 2010). In this review, a distinction will be drawn between *short-term synaptic plasticity*, plasticity lasting for milliseconds to minutes (Zucker and Regehr 2002), and *long-term synaptic plasticity*, plasticity lasting for hours to days or longer (Martin et al. 2000; Collingridge et al. 2010). A number of models of short- and long-term synaptic plasticity are routinely studied in the rodent hippocampus using in vitro and in vivo electrophysiological recording techniques (Citri and Malenka 2008). Paired pulse facilitation (PPF) is one of the most commonly studied models of short-term plasticity; furthermore, several reports suggest that mechanisms consistent with PPF have an integral role in cognitive processing and memory (Cao and Leung 1991; Silva et al. 1996; Matilla et al. 1998; Dobrunz and Stevens 1999; Ferguson et al. 2004; Kushner et al. 2005). Paired pulse facilitation refers to an increase in the evoked amplitude of the second field potential following the application of two stimuli in close succession (~10–200 ms apart) (Zucker and Regehr 2002; Citri and Malenka 2008). Synapses in both the Schaffer collateral-CA1 and CA1-subiculum pathways exhibit PPF under normal recording conditions (Cazakoff and Howland 2010; MacDougall and Howland 2013a;b). The mechanisms underlying PPF are complex and difficult to specify directly, although residual presynaptic calcium from the first stimulus increasing the probability of neurotransmitter (glutamate) release to the second stimulus is likely involved (Zucker and Regehr 2002; Citri and Malenka 2008).

The most well-characterized models of long-term synaptic plasticity are LTP, a persistent increase in synaptic potential, and LTD, a persistent decrease in synaptic potential, following application of a tetanus. Long-term potentiation and LTD have received a great deal of attention as cellular models for learning and memory (Martin et al. 2000; Malenka and Bear 2004; Citri and Malenka 2008; Collingridge et al. 2010). In the CA1 and subiculum, LTP and LTD are induced by the activation of postsynaptic N-methyl-D-aspartate (NMDA) receptors (Bliss and Collingridge 1993; Malenka and Bear 2004; Citri and Malenka 2008; Howland and Wang 2008; Behr et al. 2009; Collingridge et al. 2010), although other pre- and postsynaptic mechanisms also contribute (Malenka and Bear 2004; Lisman and Raghavachari 2006; Behr et al. 2009; Kullmann 2012). One important mechanism for the expression of LTP and LTD involves trafficking of postsynaptic α-amino-3-hydroxy-5-methylisoxazole-4-propionic acid (AMPA) receptors (Collingridge et al. 2004, 2010; Derkach et al. 2007; Kessels and Malinow 2009). Other forms of long-term hippocampal synaptic plasticity include primed burst potentiation, a low threshold form of synaptic potentiation (Diamond et al. 1988), and late developing or low-frequency induced potentiation (Habib and Dringenberg 2010). Differences have been noted in the effects of low frequency stimulation on synaptic responses in the CA1 and subiculum, particularly in vivo. In the adult rodent CA1 region, low frequency stimulation (1–3 Hz) often fails to induce LTD, as is commonly reported in slices from younger rodents (Xu et al. 1997; Fox et al. 2007; Wong et al. 2007). In contrast, low frequency stimulation of the CA1-subiculum pathway induces a late developing potentiation in the subiculum (Anderson et al. 2000; Huang and Kandel 2005; MacDougall and Howland 2013a;b) and, if paired with postsynaptic depolarization, a muscarinic-dependent form of LTD (Li et al. 2005). In the next two sections, the reported effects of acute stress on these forms of short- and long-term synaptic plasticity will be reviewed. Data related to long-term synaptic plasticity will be reviewed first as the effects of acute stress on these forms of plasticity have been studied more comprehensively.

11.4 Effects of Acute Stress on Long-Term Synaptic Plasticity in the CA1 and Subiculum

11.4.1 CA1 Region

The majority of the research regarding the effects of acute stress on long-term synaptic plasticity in the hippocampal CA1 region has been reviewed (Kim and Yoon 1998; Kim and Diamond 2002; Diamond et al. 2005, 2007; Howland and Wang 2008; Collingridge et al. 2010). The initial report showing that exposure to acute stress impaired LTP in the CA1 region of hippocampal slices was published in 1987 (Foy et al. 1987), a finding that has been consistently replicated using both in vitro and in vivo preparations (Shors and Thompson 1992; Kim et al. 1996; Xu et al. 1997; Kim et al. 2001; Yang et al. 2004; Li et al. 2008; Cazakoff and Howland 2010; MacDougall and Howland 2013a; for reviews see Kim and Diamond 2002; Howland and Wang 2008). Importantly, the regulation of LTP by acute stress differs along the septo-temporal axis of the CA1 region, with disruptions in LTP occurring in the dorsal CA1 region and a surprising facilitation of a voltage-gated calcium channel-dependent form of LTP in the ventral hippocampus following acute stress (Maggio and Segal 2007) that coincides with an increase in PPF ratios (Maggio and Segal 2012). Primed burst potentiation, a low threshold form of synaptic potentiation, is also impaired in the dorsal CA1 region of rats following exposure to acute stress (Diamond et al. 1990), even under conditions when LTP is not impaired (Mesches et al. 1999). Acute stress has also been widely reported to facilitate the induction of LTD in the CA1 region (Xu et al. 1997; Xu et al. 1998; Wong et al. 2007; Li et al. 2008; Dong et al. 2013; for reviews see Diamond et al. 2005; Howland and Wang 2008; Collingridge et al. 2010). The alterations in dorsal hippocampal LTP and LTD depend on activation of GRs (Xu et al. 1998; Yang et al. 2004, 2005; Cazakoff and Howland 2010), NMDA receptors (Kim et al. 1996; Wang et al. 2006; Wong et al. 2007), and intracellular signalling cascades including the extracellular signal-regulated kinase/mitogen-activated protein kinase (Yang et al. 2004). Whether these changes reflect a form of meta-plasticity or occur independently has been the subject of debate (Kim and Yoon 1998; Kim and Diamond 2002; Howland and Wang 2008), although increased glutamate release may contribute to the changes in hippocampal LTP and LTD following acute stress (Yang et al. 2005; Wong et al. 2007; Howland and Wang 2008; Reagan et al. 2012).

11.4.2 Subiculum

In contrast to the extensive characterization of the changes in long-term patterns of synaptic plasticity in the CA1 region in response to stress, scarce research has been conducted regarding the subiculum. Three in vivo studies in anesthetized rats have shown that acute stress disrupts LTP in the dorsal subiculum of rats, two using an acute restraint procedure (MacDougall and Howland 2013a;b) and the other systemic administration of the bacterial endotoxin lipopolysaccharide (Commins et al. 2001). While the changes in synaptic plasticity were shown to depend on GRs, acute injection of corticosterone alone failed to significantly alter plasticity in the subiculum even though the levels of circulating corticosterone were similar in acutely stressed and corticosterone-injected rats (MacDougall and Howland 2013b). In the same manner, late developing potentiation induced by low-frequency stimulation of the CA1-subiculum pathway was impaired by acute stress, but not corticosterone, due to activation of GRs (MacDougall and Howland 2013b). Interestingly, as LTP in the CA1-dorsal subiculum pathway appears to involve a presynaptic component (Commins et al. 1998a; MacDougall and Howland 2013b), the mechanisms by which this disruption occurs may be distinct from those in the CA3-CA1 synapse where postsynaptic modifications involving postsynaptic AMPA receptor trafficking may be more important (Fox et al. 2007; Wong et al. 2007; Dong et al. 2013).

11.5 Effects of Acute Stress on Short-Term Synaptic Plasticity in the CA1 and Subiculum

11.5.1 CA1 Region

Table 11.1 summarizes the published findings regarding the effects of acute stress on PPF and includes details related to the exact methodological parameters used in the experiments. In the CA1 region, studies have used both in vitro and in vivo preparations. Two studies have tested the effects of acute stress on PPF in the CA1 region of hippocampal slices. One study that used a severe stressor combining restraint with inescapable tail shocks found a disruption in CA1 LTP with no effect on PPF ratios in hippocampal slices from Long Evans rats (Shors and Thompson 1992). A second study exposed Wistar rats to ten shocks in a novel chamber and reported decreased PPF ratios and facilitated LTD in the CA1 region of hippocampal slices (Gao et al. 2008). Using an in vivo preparation in anesthetized rats, Cazakoff and Howland observed that 30 min of exposure to an elevated platform disrupted both PPF and LTP in the CA1 region that could be blocked with a GR antagonist (RU38486) administered before the acute stress (Cazakoff and Howland 2010 ; see also MacDougall and Howland 2013a). In contrast to the results observed in the subiculum (see below), reduced PPF was observed both before and after the high frequency tetanus was administered to induce LTP (Cazakoff and Howland 2010).

Three additional studies have tested the effects of bath application of corticosterone on PPF in the dorsal CA1 region of hippocampal slices. Karst and colleagues observed a rapid disruption in PPF and enhanced frequency of miniature excitatory postsynaptic currents following 10 min of corticosterone (100 nM) perfusion that depended on MR activation (Karst et al. 2005) and likely presynaptic activation of the extracellular signal-regulated kinase 1/2 pathway (Olijslagers et al. 2008). No change in PPF is observed 1–4 h following corticosterone perfusion (100 nM for 20 min; Karst and Joels 2005). Perfusion of a higher dose of corticosterone (1 or 10 μM) for longer (3 h) impaired PPF and LTP in another study, an effect related to decreases in brain-derived neurotrophic factor (Zhou et al. 2000).

11.5.2 Subiculum

To our knowledge, three in vivo studies have examined the effects of acute stress on PPF in the CA1-subiculum pathway while no data exist from in vitro experiments (Table 11.1). In one study, exploration of a novel box failed to alter PPF in the CA1-subiculum pathway (Commins and O'Mara 2000) while a second study showed that administration of the bacterial endotoxin LPS 4 h prior to in vivo recordings impaired PPF prior to delivery of a tetanus (Commins et al. 2001). The third study demonstrated that acute restraint stress (30 min), but not corticosterone injections

Table 11.1 The effects of acute stress or corticosterone administration on prepulse facilitation (PPF) and long-term synaptic plasticity in the CA1 and subiculum of the dorsal hippocampus. The mechanism involved in the reduction of PPF is noted where data exist. See the text for further details.

Strain/species	Stressor	E-phys protocol	PPF effect	Long-term effect	Reference
CA3/Schaffer collateral-CA1 pathway, in vitro					
LE/rat	Restraint and tail shocks (60 shocks in 60 min)	HPC slices; stratum radiatum/CA1 pathway; PPF @ 50, 75, 100, 200 ms	N/C PPF	↓ LTP	Shors and Thompson (1992)
Wistar/rat	Shocks in novel chamber (10 shocks in 10 min)	Coronal HPC slices; stratum radiatum/CA1 pathway; PPF @ 60 ms	↓ PPF	↑ LTD	Gao et al. (2008)
C57BL6/mouse	Cort (100 nM, 10 min)	Transverse HPC slices; CA3/Schaffer collateral-CA1 pathway; PPF @ 100 ms	↓ PPF (MR)	no data	Karst et al. (2005)
C57BL6/mouse	Cort (100 nM, 20 min)	Transverse HPC slices; CA3/Schaffer collateral-CA1 pathway; PPF @ 100 ms 1–4 h following Cort	N/C PPF	no data	Karst and Joels (2005)
SD/rat	Cort (1 or 10 μM, 3 h)	Transverse HPC slices; CA3/Schaffer collateral-CA1 pathway; PPF @ 100 ms immediately following	↓ PPF (BDNF)	↓ LTP	Zhou et al. (2000)
CA3/Schaffer collateral-CA1 pathway, in vivo					
SD/rat	Elevated platform (30 min)	Urethane anesthetized; CA3/Schaffer collateral-CA1 pathway; PPF @ 25, 50, 100, 200 ms	↓ PPF (GR)	↓ LTP	Cazakoff and Howland (2010)
CA1-subiculum pathway, in vivo					
SD/rat	Restraint (30 min)	Urethane anesthetized; CA1-SUB pathway; PPF @ 25, 50, 100, 200 ms	↓ PPF (GR)	↓ LTP/LDP	MacDougall and Howland(2013b)
Wistar/rat	Exposure to a novel environment	Sodium pentobarbitone/urethane anesthetized; CA1-SUB pathway; PPF @ 50, 100 ms	N/C PPF	↑ LTD	Commins and O'Mara (2000)
Wistar/rat	LPS (4 h prior to recordings)	Sodium pentobarbitone/urethane anesthetized; CA1-SUB pathway; PPF @ 50, 100 ms	↓ PPF	↓ LTP	Commins et al. (2001)
SD/rat	Cort (3 mg/kg)	Urethane anesthetized; CA1-SUB pathway; PPF @ 25, 50, 100, 200 ms	N/C PPF	N/C LTP/LDP	MacDougall and Howland(2013b)

BDNF brain derived neurotrophic factor, *Cort* corticosterone, *GR* glucocorticoid receptor, *HPC* hippocampus, *LDP* late developing potentiation, *LE* Long-Evans, *LTP* long-term potentiation, *LTD* long-term depression, *MR* mineralocorticoid receptor, *N/C* no change, *SD* Sprague Dawley

(3 mg/kg), disrupted PPF prior to delivery of a tetanus (MacDougall and Howland 2013b; see also MacDougall and Howland 2013a. In both studies that showed PPF disruptions following acute stress, LTP was also disrupted by the stressor (Commins et al. 2001; MacDougall and Howland 2013b). As previously mentioned, the induction of LTP in the CA1-subiculum pathway has been shown to reduce PPF ratios (Commins et al. 1998; MacDougall and Howland 2013b), which may be indicative of a presynaptic locus for the mechanism(s) underlying LTP in this pathway (Commins et al. 1998; Behr et al. 2009). Importantly, acute stress was also shown to disrupt this reduction in PPF observed following administration of a tetanus, suggesting that acute stress may have effects on distinct forms of LTP observed in the CA1 and subiculum (MacDougall and Howland 2013b). Injections of the GR antagonist RU38486 prior to the stressor blocked the effects of acute stress both before and after administration of the tetanus (MacDougall and Howland 2013b).

11.6 Integration of the Effect of Acute Stress on Short- and Long-Term Forms of Synaptic Plasticity

Inspection of Table 11.1 reveals a complex set of findings related to short- and long-term synaptic plasticity in the CA1 and subiculum following acute stress or corticosterone treatment. Alterations in PPF are observed in six of the nine studies; however, the role of MRs and GRs in mediating the changes in PPF differed among the studies. One factor that likely contributed to these differences is the delay between the stressor/corticosterone treatment and electrophysiological measurements as the effects of acute stress on cognition and related brain circuits are well-known to be time dependent (de Quervain et al. 1998; Joels et al. 2006, 2012). Differences related to the timing of the stressor relative to the recordings in the studies can be illustrated by considering the demonstrated role of MRs in causing the reduced PPF following acute stress/corticosterone administration in some studies (Karst et al. 2005) versus GRs in others (Cazakoff and Howland 2010; MacDougall and Howland 2013b). Karst and colleagues used hippocampal slices and bath applied corticosterone for 10 min before measuring PPF (Karst et al. 2005). Under these conditions, the disrupted CA1 PPF depended on MR activation. Given the short time period for the MR-dependent reductions in PPF to be observed, these researchers proposed that a non-genomic effect of MR activation must be involved (Karst et al. 2005). In contrast, evidence that GR activation is necessary for the PPF disruptions in the CA1 and subiculum by acute stress was gained using in vivo recordings (Cazakoff and Howland 2010; MacDougall and Howland 2013b). In these experiments, the animals were exposed to acute stress for 30 min before being anesthetized. Once anesthetized, 60–90 min were needed to prepare the animal for recordings and lower the electrodes. Thus, the PPF measurements would have been taken 90–120 min after the HPA axis was activated and corticosterone was initially released in the response to the stressor. Previous studies suggest that GR activation significantly affects gene expression within a time frame of 1–3 h (Zhou et al. 2000;

Morsink et al. 2006, 2007). Thus, PPF may be altered over a broad timescale after acute stress: initially by the rapid non-genomic actions of MR activation and subsequently by the slower genomic changes following GR activation.

Glucocorticoid receptor-dependent disruptions of PPF following acute stress have also been reported for the perforant path to dentate gyrus pathway in vivo (Avital et al. 2006; although see also Bramham et al. 1998; Spyrka et al. 2011) and the medial prefrontal cortex in vitro (Musazzi et al. 2010; Popoli et al. 2012) in rats. Similarly to the studies described above that also noted a GR-dependent reduction in PPF (Cazakoff and Howland 2010; MacDougall and Howland 2013b), the electrophysiological recordings would have been performed hours after the stressor. Taken together, these findings indicate that while corticosterone has extremely rapid effects on PPF in the CA1 region (i.e., in minutes) that are caused by non-genomic actions of MRs (Karst et al. 2005), periods of acute stress recruit a GR-dependent change in PPF in a number of areas, including the CA1 and subiculum (Avital et al. 2006; Cazakoff and Howland 2010; Musazzi et al. 2010; MacDougall and Howland 2013b).

Similar timeframes for MR and GR-dependent effects of corticosterone have been noted in a study testing the effect of corticosterone on AMPA receptor trafficking using quantum-dot imaging, a technique which allows the diffusion of receptors to be quantified (Groc et al. 2008; Krugers et al. 2010). A rapid (< 10 min), MR-dependent increase in membrane surface diffusion of GluA2 subunit-containing AMPA receptors was observed following application of corticosterone. Importantly, this effect likely depended on membrane bound MRs as a membrane impermeable BSA-corticosterone conjugate also produced the effect. In additional experiments, a slower (150 min) GR-dependent increase in GluA2-subunit containing surface expression was observed following corticosterone exposure (Groc et al. 2008; see also Martin et al. 2009).

Other differences among the studies summarized in Table 11.1 may explain why altered PPF following acute stress/corticosterone was reported in some (Zhou et al. 2000; Commins et al. 2001; Karst et al. 2005; Gao et al. 2008; Cazakoff and Howland 2010; MacDougall and Howland 2013b; Maggio and Segal 2012) but not others (Shors and Thompson 1992; Commins and O'Mara 2000; Karst and Joels 2005). While it is tempting to speculate that differences in the species/strain of rodents or in vitro/in vivo preparation used may contribute, the effects of acute stress on long-term synaptic plasticity are generally resistant to these factors. Secondly, the effects of corticosterone generally follow an inverted U-shaped relationship (Lupien and McEwen 1997; Park et al. 2006; Diamond et al. 2007) so the differences in doses of corticosterone must be taken into account. For example, application of high doses of corticosterone (1–10 μM) for multiple hours reduced CA1 PPF and LTP in hippocampal slices (Zhou et al. 2000) whereas application of 100 nM of corticosterone for 20 min had no effect on PPF assessed 1–4 h later (Karst and Joels 2005). Different effects of "acute stress" versus "elevations in corticosterone" have been noted in both electrophysiological and behavioural experiments related to the hippocampus (Kim et al. 2001, 2005; Kim and Diamond 2002; Woodson et al. 2003; MacDougall and Howland 2013b). Thus, elevations in corticosterone may be necessary, but not sufficient, to alter synaptic plasticity. The transmission of emotional information regarding the stressor by the amygdala may be an additional critical factor necessary for acute stress to affect synaptic plasticity and cognition (Kim et al. 2001, 2005; Kim and Diamond 2002; Schwabe et al. 2012).

It is not surprising that acute stress has effects on short- and long-term patterns of synaptic plasticity given the established effects of acute stress on presynaptic and postsynaptic aspects of the glutamate signalling in the hippocampus and other areas including the prefrontal cortex (Popoli et al. 2012; Sanacora et al. 2012). One remaining issue relates to whether the effects of acute stress on short-term plasticity are due to the same or distinct mechanisms from those that cause the effects of acute stress on long-term synaptic plasticity. If the mechanisms are distinct, the possibility exists that alterations in short- and long-term synaptic plasticity following acute stress may underlie different effects of acute stress on cognition. Table 11.1 summarizes the findings related to long-term plasticity from the studies that also observed changes in PPF following acute stress in an effort to address this issue. In every study where both short and long-term synaptic plasticity were measured and PPF was impaired, long-term plasticity was also altered. Reduced PPF correlated with reduced LTP in four of the studies (Zhou et al. 2000; Commins et al. 2001; Cazakoff and Howland 2010; MacDougall and Howland 2013b) and increased LTD in one of the studies (Gao et al. 2008). In two of the studies, long-term plasticity was altered by acute stress while PPF was unaffected (Shors and Thompson 1992; Commins and O'Mara 2000). In two of the studies, a GR antagonist blocked the effects of acute stress on both PPF and long-term synaptic plasticity in the CA1 (Cazakoff and Howland 2010) and subiculum (MacDougall and Howland 2013b). Thus, these data suggest that the alteration in short- and long-term forms of synaptic plasticity is initiated by activation of GRs. Whether the signalling pathways downstream of GRs mediating these effects on short- and long-term plasticity are the same or different remains an open question.

11.7 Linking the Effects of Acute Stress on Synaptic Plasticity in CA1 and Subiculum to Hippocampal-Dependent Behaviour

The effects of acute stress on cognition are complex and influenced by a variety of factors including the type of cognition examined, specifics of the stressor, timing of the stressor, level of intrinsic arousal associated with the task, and characteristics of the subject examined (Kim and Diamond 2002; Joels et al. 2006; Shors 2006; Diamond et al. 2007; Sandi and Pinelo-Nava 2007; Holmes and Wellman 2009; Cazakoff et al. 2010; Schwabe et al. 2012). The focus of the following discussion will be effects of acute, extrinsic stress (i.e., stress not directly associated with the task) on spatial and recognition memory in rodents. In most cases, extrinsic stress disrupts hippocampal-dependent spatial learning and memory (for review, see Cazakoff et al. 2010), effects that are hypothesized to be caused by alterations in long-term synaptic plasticity caused by acute stress (Kim and Diamond 2002; Diamond et al. 2005, 2007; Wong et al. 2007; Howland and Wang 2008; Cazakoff et al. 2010). Importantly, both the dorsal CA1 and subiculum are both involved in processing spatial information and memory (Morris et al. 1990; McNaughton et al. 1996; O'Mara et al. 2009); however, their anatomical positions and behavioural data (Deadwyler and Hampson 2004) suggest that their roles are likely distinct (Behr et al. 2009). While the dorsal CA1 receives strong input via the glutamatergic Schaffer collaterals from CA3 and inputs from the cortex via the temporoammonic pathway (Behr et al. 2009), the subiculum receives strong projections from the CA1 (Amaral et al. 1991) and cortical areas including the entorhinal, perirhinal, and postrhinal areas (Naber et al. 2001; Behr et al. 2009; O'Mara et al. 2009). Thus, the subiculum is in a privileged position to receive both highly processed information that has made its way through the hippocampus and "raw" sensory information directly from the cortex (Behr et al. 2009). As reviewed above, acute stress disrupts short- and long-term patterns of synaptic plasticity in both the CA1 and subiculum. These studies have examined the traditional pathways of information flow through the hippocampal system, the CA3-CA1 pathway and the CA1-subiculum pathway. Given the role of both regions in spatial memory formation, it is reasonable to conclude that the impairments in synaptic plasticity in both regions of the circuit contribute to the deficits in spatial memory retrieval observed following acute stress (O'Mara 2006; Cazakoff et al. 2010; MacDougall and Howland 2013b). One interesting test of this hypothesis would be to assess whether the pharmacological agents reported to block the effects of acute stress on CA1 synaptic plasticity and spatial memory retrieval (Howland and Wang 2008; Cazakoff et al. 2010) also block the effects of acute stress on synaptic plasticity in the subiculum. Two examples of such agents are the GluN2B subunit-selective NMDA receptor antagonist Ro25–6981 (Wang et al. 2006; Wong et al. 2007; Howland and Cazakoff 2010) and transient receptor potential vanilloid 1 agonist capsaicin (Li et al. 2008).

The role of corticosteroid receptors in the effects of acute stress on hippocampal-mediated behaviour is also of interest given their roles in the acute stress effects on synaptic plasticity. Interestingly, convergence between the time-dependent involvement of MRs and GRs in the alterations of synaptic plasticity and spatial learning and memory by acute stress has been gained from recent studies (Dorey et al. 2011; Dorey et al. 2012). The studies used a delayed alternation procedure on a T maze that involved forcing mice to enter one arm of the maze twice during a training period. In a test session 24 h later, mice were allowed to enter either the arm they had visited during training or the opposite "novel" arm. Control mice displayed robust preference for entering the arm they had not entered during training. Exposure to acute stress 15 min before the test trial disrupted alternation behaviour, an effect that was mimicked by injecting the mice with membrane impermeable corticosterone injections suggesting that a membrane bound corticosteroid receptor was involved in the effect (Dorey et al. 2011). Intra-hippocampal microinfusions of an MR, but not a GR, antagonist before acute stress or corticosterone injections block their effects on delayed alternation. In a subsequent study, the same researchers showed that blockade of MRs in the dorsal hippocampus prevented the stress induced disruptions in delayed alternation at short (15 min), but not long (60–105 min), delays. Blocking GRs prevented the memory deficit at 60 min (dorsal hippocampus) and 105 min (ventral hippocampus), but not the short (15 min) delay (Dorey et al. 2012). In another study, the disruptive effects of corticosterone administration on spatial memory retrieval in a water maze task were also reversed by an MR antagonist, but not a GR antagonist or protein synthesis inhibitor, suggesting a non-genomic action of MRs in mediating the effect of corticosterone or acute stress on spatial memory retrieval (Khaksari et al. 2007). These behavioural data may appear to conflict with the studies reviewed showing that the effects of acute stress on synaptic plasticity in the CA1 and subiculum depend on GR activation (Xu et al. 1998; Cazakoff and Howland 2010; MacDougall and Howland 2013b); however, two points are worth emphasizing in this regard: (1) To our knowledge, no published data are available assessing the effects of MR antagonists on the alterations in hippocampal synaptic plasticity caused by acute stress in a time frame of minutes and (2) the time frame after stress assessed in the studies on synaptic plasticity is consistent with the effects of GR antagonists on stress-induced memory disruptions (i.e., 60 min or longer; Dorey et al. 2012). Thus, one critical experiment will be to assess the potential time-dependent effects of MR and GR antagonists on the alterations in synaptic plasticity caused by acute stress. Because the time required for preparing the animals for recordings in brain slices or under anaesthesia is too long to assess the potential effects of MR antagonists on the alterations of synaptic plasticity caused by acute stress, field potential recordings in freely moving rodents will be necessary.

Recognition memory is routinely assessed for a variety of stimuli including objects and spatial locations in different paradigms (Dere et al. 2007; Winters et al. 2008). While the neural substrates mediating recognition memory remain controversial, roles for the perirhinal cortex in object recognition and hippocampus in spatial recognition tasks are supported by the literature (Dere et al. 2007; Howland et al. 2008; Winters et al. 2008). Recordings of local field potentials from the CA1 and subiculum during an object recognition task showed increased theta power in the subiculum, but not CA1 region, during object recognition (Chang and Huerta 2012), which is interesting in light of the direct input the subiculum receives from perirhinal cortex (Behr et al. 2009; O'Mara et al. 2009). Object recognition and object-place recognition are both susceptible to disruption by acute stress (Baker and Kim 2002; Cazakoff et al. 2010; Howland and Cazakoff 2010; Li et al. 2012); however, the potential role of alterations in synaptic plasticity by acute stress in mediating these effects has received scant attention. The mechanisms in perirhinal cortex that support object recognition memory are distinct from those typically ascribed to spatial memory in the hippocampus. Long-term depression caused by AMPA receptor endocytosis in perirhinal cortex is implicated in object recognition memory under normal conditions (Griffiths et al. 2008; Cazakoff and Howland 2011) whereas AMPA receptor endocytosis in the CA1 region has been reported to mediate the effects of acute stress on memory retrieval (Wong et al. 2007). The disruptive effects of acute stress on both spatial memory retrieval and object recognition can be blocked by systemic injections of the GluN2B subunit-selective NMDA receptor antagonist Ro25–6981 (Howland and Cazakoff 2010). Future studies examining the effects of acute stress on synaptic plasticity in the reciprocal pathway connecting the subiculum to perirhinal cortex will be critical for fully appreciating the potential role of alterations in synaptic plasticity in mediating the effects of acute stress on recognition memory.

11.8 Conclusion

Periods of acute stress have significant effects on different types of synaptic plasticity in the dorsal hippocampus. This chapter reviewed evidence that acute stress alters short-term synaptic plasticity by impairing PPF ratios in both the CA1 and subiculum. The mechanisms mediating these effects appear to involve release of the hormone corticosterone acting at its two main receptors in a time-dependent manner. Rapid disruptions in PPF in the minutes following corticosterone application are caused by activation of MRs, likely signalling through a non-genomic pathway. Disruption of PPF later in time (in hours after the stressor) appears to involve GR activation. The effects of acute stress on long-term synaptic plasticity in both the CA1 and subiculum should be taken into account when developing theories regarding the neural circuitry underlying the effects of acute stress on hippocampal-dependent tasks.

References

Amaral DG, Dolorfo C, Alvarez-Royo P. Organization of CA1 projections to the subiculum: a PHA-L analysis in the rat. Hippocampus. 1991;1:415–35.

Anderson M, Commins S, O'Mara SM. The effects of low frequency and two-pulse stimulation protocols on synaptic transmission in the CA1-subiculum pathway in the anaesthetized rat. Neurosci Lett. 2000;279:181–4.

Andersen P, Morris R, Bliss T, Amaral D, O'Keefe J. The hippocampus book. New York: Oxford University Press; 2006.

Avital A, Segal M, Richter-Levin G. Contrasting roles of corticosteroid receptors in hippocampal plasticity. J Neurosci. 2006;26:9130–4.

Baker KB, Kim JJ. Effects of stress and hippocampal NMDA receptor antagonism on recognition memory in rats. Learn Mem. 2002;9:58–65.

Behr J, Wozny C, Fidzinski P, Schmitz D. Synaptic plasticity in the subiculum. Prog Neurobiol. 2009;89:334–42.

Bliss TV, Collingridge GL. A synaptic model of memory: long-term potentiation in the hippocampus. Nature. 1993;361:31–9.

Bramham CR, Southard T, Ahlers ST, Sarvey JM. Acute cold stress leading to elevated corticosterone neither enhances synaptic efficacy nor impairs LTP in the dentate gyrus of freely moving rats. Brain Res. 1998;789:245–55.

Cao F, Leung LS. Behavior-dependent paired-pulse responses in the hippocampal CA1 region. Exp Brain Res. 1991;87:553–61.

Cazakoff BN, Howland JG. Acute stress disrupts paired pulse facilitation and long-term potentiation in rat dorsal hippocampus through activation of glucocorticoid receptors. Hippocampus. 2010;20:1327–31.

Cazakoff BN, Howland JG. AMPA receptor endocytosis in rat perirhinal cortex underlies retrieval of object memory. Learn Mem. 2011;18:688–92.

Cazakoff BN, Johnson KJ, Howland JG. Converging effects of acute stress on spatial and recognition memory in rodents: a review of recent behavioural and pharmacological findings. Prog Neuropsychopharmacol Biol Psychiatry. 2010;34:733–41.

Chang EH, Huerta PT. Neurophysiological correlates of object recognition in the dorsal subiculum. Front Behav Neurosci. 2012;6:46.

Citri A, Malenka RC. Synaptic plasticity: multiple forms, functions, and mechanisms. Neuropsychopharmacology. 2008;33:18–41.

Collingridge GL, Isaac JT, Wang YT. Receptor trafficking and synaptic plasticity. Nat Rev Neurosci. 2004;5:952–62.

Collingridge GL, Peineau S, Howland JG, Wang YT. Long-term depression in the CNS. Nat Rev Neurosci. 2010;11:459–73.

Commins S, O'Mara SM. Interaction between paired-pulse facilitation, low-frequencey stimulation, and behavioral stress in the pathway from hippocampal area CA1 to the subiculum: dissocation of baseline synaptic transmission from paired-pulse faciliation and depression of the same pathway. Psychobiology. 2000;28:1–11.

Commins S, Gigg J, Anderson M, O'Mara SM. Interaction between paired-pulse facilitation and long-term potentiation in the projection from hippocampal area CA1 to the subiculum. Neuroreport. 1998a;9:4109–13.

Commins S, Gigg J, Anderson M, O'Mara SM. The projection from hippocampal area CA1 to the subiculum sustains long-term potentiation. Neuroreport. 1998b;9:847–50.

Commins S, O'Neill LA, O'Mara SM. The effects of the bacterial endotoxin lipopolysaccharide on synaptic transmission and plasticity in the CA1-subiculum pathway in vivo. Neuroscience. 2001;102:273–80.

de Kloet ER, Joels M, Holsboer F. Stress and the brain: from adaptation to disease. Nat Rev Neurosci. 2005;6:463–75.

de Quervain DJ, Roozendaal B, McGaugh JL. Stress and glucocorticoids impair retrieval of long-term spatial memory. Nature. 1998;394:787–90.

Deadwyler SA, Hampson RE. Differential but complementary mnemonic functions of the hippocampus and subiculum. Neuron. 2004;42:465–76.

Dere E, Huston JP, De Souza Silva MA. The pharmacology, neuroanatomy and neurogenetics of one-trial object recognition in rodents. Neurosci Biobehav Rev. 2007;31:673–704.

Derkach VA, Oh MC, Guire ES, Soderling TR. Regulatory mechanisms of AMPA receptors in synaptic plasticity. Nat Rev Neurosci. 2007;8:101–13.

Diamond DM, Dunwiddie TV, Rose GM. Characteristics of hippocampal primed burst potentiation in vitro and in the awake rat. J Neurosci. 1988;8:4079–88.

Diamond DM, Bennett MC, Stevens KE, Wilson RL, Rose GM. Exposure to a novel environment interferes with the induction of hippocampal primed burst potentiation in the behaving rat. Psychobiology. 1990;18:273–81.

Diamond DM, Park CR, Campbell AM, Woodson JC. Competitive interactions between endogenous LTD and LTP in the hippocampus underlie the storage of emotional memories and stress-induced amnesia. Hippocampus. 2005;15:1006–25.

Diamond DM, Campbell AM, Park CR, Halonen J, Zoladz PR. The temporal dynamics model of emotional memory processing: a synthesis on the neurobiological basis of stress-induced amnesia, flashbulb and traumatic memories, and the Yerkes-Dodson law. Neural Plast. 2007;2007:60803.

Dobrunz LE, Stevens CF. Response of hippocampal synapses to natural stimulation patterns. Neuron. 1999;22:157–66.

Dong Z, Bai Y, Wu X, Li H, Gong B, Howland JG, et al. Hippocampal long-term depression mediates spatial reversal learning in the Morris water maze. Neuropharmacology. 2013;13(164):73.

Dorey R, Pierard C, Shinkaruk S, Tronche C, Chauveau F, Baudonnat M, et al. Membrane mineralocorticoid but not glucocorticoid receptors of the dorsal hippocampus mediate the rapid effects of corticosterone on memory retrieval. Neuropsychopharmacology. 2011;36:2639–49.

Dorey R, Pierard C, Chauveau F, David V, Beracochea D. Stress-induced memory retrieval impairments: different time-course involvement of corticosterone and glucocorticoid receptors in dorsal and ventral hippocampus. Neuropsychopharmacology. 2012;37:2870–80.

Ferguson GD, Wang H, Herschman HR, Storm DR. Altered hippocampal short-term plasticity and associative memory in synaptotagmin IV (-/-) mice. Hippocampus. 2004;14:964–74.

Fox CJ, Russell K, Titterness AK, Wang YT, Christie BR. Tyrosine phosphorylation of the GluR2 subunit is required for long-term depression of synaptic efficacy in young animals in vivo. Hippocampus. 2007;17:600–5.

Foy MR, Stanton ME, Levine S, Thompson RF. Behavioral stress impairs long-term potentiation in rodent hippocampus. Behav Neural Biol. 1987;48:138–49.

Gao Y, Han H, Xu R, Cao J, Luo J, Xu L. Effects of prolonged exposure to context following contextual fear conditioning on synaptic properties in rat hippocampal slices. Neurosci Res. 2008;61:385–9.

Griffiths S, Scott H, Glover C, Bienemann A, Ghorbel MT, Uney J, et al. Expression of long-term depression underlies visual recognition memory. Neuron. 2008;58:186–94.

Groc L, Choquet D, Chaouloff F. The stress hormone corticosterone conditions AMPAR surface trafficking and synaptic potentiation. Nat Neurosci. 2008;11:868–70.

Habib D, Dringenberg HC. Low-frequency-induced synaptic potentiation: a paradigm shift in the field of memory-related plasticity mechanisms? Hippocampus. 2010;20:29–35.

Herman JP, Ostrander MM, Mueller NK, Figueiredo H. Limbic system mechanisms of stress regulation: hypothalamo-pituitary-adrenocortical axis. Prog Neuropsychopharmacol Biol Psychiatry. 2005;29:1201–13.

Holmes A, Wellman CL. Stress-induced prefrontal reorganization and executive dysfunction in rodents. Neurosci Biobehav Rev. 2009;33:773–83.

Howland JG, Cazakoff BN. Effects of acute stress and GluN2B-containing NMDA receptor antagonism on object and object-place recognition memory. Neurobiol Learn Mem. 2010;93:261–7.

Howland JG, Wang YT. Synaptic plasticity in learning and memory: stress effects in the hippocampus. Prog Brain Res. 2008;169:145–58.

Howland JG, Harrison RA, Hannesson DK, Phillips AG. Ventral hippocampal involvement in temporal order, but not recognition, memory for spatial information. Hippocampus. 2008;18:251–7.

Huang YY, Kandel ER. Theta frequency stimulation up-regulates the synaptic strength of the pathway from CA1 to subiculum region of hippocampus. Proc Natl Acad Sci U S A. 2005;102:232–7.

Joels M, Baram TZ. The neuro-symphony of stress. Nat Rev Neurosci. 2009;10(6):459–66.

Joels M, Pu Z, Wiegert O, Oitzl MS, Krugers HJ. Learning under stress: how does it work? Trends Cogn Sci. 2006;10:152–8.

Joels M, Sarabdjitsingh RA, Karst H. Unraveling the time domains of corticosteroid hormone influences on brain activity: rapid, slow, and chronic modes. Pharmacol Rev. 2012;64:901–38.

Karst H, Joels M. Corticosterone slowly enhances miniature excitatory postsynaptic current amplitude in mice CA1 hippocampal cells. J Neurophysiol. 2005;94:3479–86.

Karst H, Berger S, Turiault M, Tronche F, Schutz G, Joels M. Mineralocorticoid receptors are indispensable for nongenomic modulation of hippocampal glutamate transmission by corticosterone. Proc Natl Acad Sci U S A. 2005;102:19204–7.

Kessels HW, Malinow R. Synaptic AMPA receptor plasticity and behavior. Neuron. 2009;61:340–50.

Khaksari M, Rashidy-Pour A, Vafaei AA. Central mineralocorticoid receptors are indispensable for corticosterone-induced impairment of memory retrieval in rats. Neuroscience. 2007;149:729–38.

Kim JJ, Diamond DM. The stressed hippocampus, synaptic plasticity and lost memories. Nat Rev Neurosci. 2002;3:453–62.

Kim JJ, Yoon KS. Stress: metaplastic effects in the hippocampus. Trends Neurosci. 1998;21:505–9.

Kim JJ, Foy MR, Thompson RF. Behavioral stress modifies hippocampal plasticity through N-methyl-D-aspartate receptor activation. Proc Natl Acad Sci U S A. 1996;93:4750–3.

Kim JJ, Lee HJ, Han JS, Packard MG. Amygdala is critical for stress-induced modulation of hippocampal long-term potentiation and learning. J Neurosci. 2001;21:5222–8.

Kim JJ, Koo JW, Lee HJ, Han JS. Amygdalar inactivation blocks stress-induced impairments in hippocampal long-term potentiation and spatial memory. J Neurosci. 2005;25:1532–9.
Kim JJ, Song EY, Kosten TA. Stress effects in the hippocampus: synaptic plasticity and memory. Stress. 2006;9:1–11.
Krugers HJ, Hoogenraad CC, Groc L. Stress hormones and AMPA receptor trafficking in synaptic plasticity and memory. Nat Rev Neurosci. 2010;11:675–81.
Kullmann DM. The Mother of All Battles 20 years on: is LTP expressed pre- or postsynaptically? J Physiol. 2012;590:2213–6.
Kushner SA, Elgersma Y, Murphy GG, Jaarsma D, van Woerden GM, Hojjati MR, et al. Modulation of presynaptic plasticity and learning by the H-ras/extracellular signal-regulated kinase/synapsin I signaling pathway. J Neurosci. 2005;25:9721–34.
Li H, Zhang J, Xiong W, Xu T, Cao J, Xu L. Long-term depression in rat CA1-subicular synapses depends on the G-protein coupled mACh receptors. Neurosci Res. 2005;52:287–94.
Li HB, Mao RR, Zhang JC, Yang Y, Cao J, Xu L. Antistress effect of TRPV1 channel on synaptic plasticity and spatial memory. Biol Psychiatry. 2008;64:286–92.
Li S, Fan YX, Wang W, Tang YY. Effects of acute restraint stress on different components of memory as assessed by object-recognition and object-location tasks in mice. Behav Brain Res. 2012;227:199–207.
Lisman J, Raghavachari S. A unified model of the presynaptic and postsynaptic changes during LTP at CA1 synapses. Sci STKE. 2006;2006:re11.
Lupien SJ, McEwen BS. The acute effects of corticosteroids on cognition: integration of animal and human model studies. Brain Res Brain Res Rev. 1997;24:1–27.
MacDougall MJ, Howland JG. Acute stress and hippocampal output: exploring dorsal CA1 and subicular synaptic plasticity simultaneously in anesthetized rats. Physiol Rep. 2013a;1:e00035.
MacDougall MJ, Howland JG. Acute stress, but not corticosterone, disrupts short- and long-term synaptic plasticity in rat dorsal subiculum via glucocorticoid receptor activation. Cereb Cortex. 2013b;23:2611–9.
Maggio N, Segal M. Striking variations in corticosteroid modulation of long-term potentiation along the septotemporal axis of the hippocampus. J Neurosci. 2007;27:5757–65.
Maggio N, Segal M. Cellular basis of a rapid effect of mineralocorticosteroid receptors activation on LTP in ventral hippocampal slices. Hippocampus. 2012;22:267–75.
Malenka RC, Bear MF. LTP and LTD: an embarrassment of riches. Neuron. 2004;44:5–21.
Martin SJ, Grimwood PD, Morris RG. Synaptic plasticity and memory: an evaluation of the hypothesis. Annu Rev Neurosci. 2000;23:649–711.
Martin S, Henley JM, Holman D, Zhou M, Wiegert O, van SM, et al. Corticosterone alters AMPAR mobility and facilitates bidirectional synaptic plasticity. PLoS ONE. 2009;4:e4714.
Matilla A, Roberson ED, Banfi S, Morales J, Armstrong DL, Burright EN, et al. Mice lacking ataxin-1 display learning deficits and decreased hippocampal paired-pulse facilitation. J Neurosci. 1998;18:5508–16.
McEwen BS, Sapolsky RM. Stress and cognitive function. Curr Opin Neurobiol. 1995;5:205–16.
McNaughton BL, Barnes CA, Gerrard JL, Gothard K, Jung MW, Knierim JJ, et al. Deciphering the hippocampal polyglot: the hippocampus as a path integration system. J Exp Biol. 1996;199:173–85.
Mesches MH, Fleshner M, Heman KL, Rose GM, Diamond DM. Exposing rats to a predator blocks primed burst potentiation in the hippocampus in vitro. J Neurosci. 1999;19:RC18.
Morris RG, Schenk F, Tweedie F, Jarrard LE. Ibotenate lesions of hippocampus and/or subiculum: dissociating components of allocentric spatial learning. Eur J Neurosci. 1990;2:1016–28.
Morsink MC, Steenbergen PJ, Vos JB, Karst H, Joels M, de Kloet ER, et al. Acute activation of hippocampal glucocorticoid receptors results in different waves of gene expression throughout time. J Neuroendocrinol. 2006;18:239–52.
Morsink MC, Van Gemert NG, Steenbergen PJ, Joels M, de Kloet ER, Datson NA. Rapid glucocorticoid effects on the expression of hippocampal neurotransmission-related genes. Brain Res. 2007;1150:14–20.

Musazzi L, Milanese M, Farisello P, Zappettini S, Tardito D, Barbiero VS, et al. Acute stress increases depolarization-evoked glutamate release in the rat prefrontal/frontal cortex: the dampening action of antidepressants. PLoS ONE. 2010;5:e8566.
Naber PA, Witter MP, Lopes da Silva FH. Evidence for a direct projection from the postrhinal cortex to the subiculum in the rat. Hippocampus. 2001;11:105–17.
Naber PA, Witter MP, Lopes Silva FH. Networks of the hippocampal memory system of the rat. The pivotal role of the subiculum. Ann N Y Acad Sci. 2000;911:392–403.
Olijslagers JE, de Kloet ER, Elgersma Y, van Woerden GM, Joels M, Karst H. Rapid changes in hippocampal CA1 pyramidal cell function via pre- as well as postsynaptic membrane mineralocorticoid receptors. Eur J Neurosci. 2008;27:2542–50.
O'Mara S. Controlling hippocampal output: the central role of subiculum in hippocampal information processing. Behav Brain Res. 2006;174:304–12.
O'Mara SM, Commins S, Anderson M, Gigg J. The subiculum: a review of form, physiology and function. Prog Neurobiol. 2001;64:129–55.
O'Mara SM, Sanchez-Vives MV, Brotons-Mas JR, O'Hare E. Roles for the subiculum in spatial information processing, memory, motivation and the temporal control of behaviour. Prog Neuropsychopharmacol Biol Psychiatry. 2009;33:782–90.
Park CR, Campbell AM, Woodson JC, Smith TP, Fleshner M, Diamond DM. Permissive influence of stress in the expression of a u-shaped relationship between serum corticosterone levels and spatial memory errors in rats. Dose Response. 2006;4:55–74.
Popoli M, Yan Z, McEwen BS, Sanacora G. The stressed synapse: the impact of stress and glucocorticoids on glutamate transmission. Nat Rev Neurosci. 2012;13:22–37.
Reagan LP, Reznikov LR, Evans AN, Gabriel C, Mocaer E, Fadel JR. The antidepressant agomelatine inhibits stress-mediated changes in amino acid efflux in the rat hippocampus and amygdala. Brain Res. 2012;1466:91–8.
Reul JM, de Kloet ER. Two receptor systems for corticosterone in rat brain: microdistribution and differential occupation. Endocrinology. 1985;117:2505–11.
Sanacora G, Treccani G, Popoli M. Towards a glutamate hypothesis of depression: an emerging frontier of neuropsychopharmacology for mood disorders. Neuropharmacology. 2012;62:63–77.
Sandi C, Pinelo-Nava MT. Stress and memory: behavioral effects and neurobiological mechanisms. Neural Plast. 2007;2007:78970.
Sapolsky RM. Glucocorticoids and hippocampal atrophy in neuropsychiatric disorders. Arch Gen Psychiatry. 2000;57:925–35.
Schwabe L, Joels M, Roozendaal B, Wolf OT, Oitzl MS. Stress effects on memory: an update and integration. Neurosci Biobehav Rev. 2012;36:1740–9.
Shors TJ. Learning during stressful times. Learn Mem. 2004;11:137–44.
Shors TJ. Stressful experience and learning across the lifespan. Annu Rev Psychol. 2006;57:55–85.
Shors TJ, Thompson RF. Acute stress impairs (or induces) synaptic long-term potentiation (LTP) but does not affect paired-pulse facilitation in the stratum radiatum of rat hippocampus. Synapse. 1992;11:262–5.
Silva AJ, Rosahl TW, Chapman PF, Marowitz Z, Friedman E, Frankland PW, et al. Impaired learning in mice with abnormal short-lived plasticity. Curr Biol. 1996;6:1509–18.
Spyrka J, Danielewicz J, Hess G. Brief neck restraint stress enhances long-term potentiation and suppresses long-term depression in the dentate gyrus of the mouse. Brain Res Bull. 2011;85:363–7.
Tasker JG, Di S, Malcher-Lopes R. Minireview: rapid glucocorticoid signaling via membrane-associated receptors. Endocrinology. 2006;147:5549–56.
van Strien NM, Cappaert NL, Witter MP. The anatomy of memory: an interactive overview of the parahippocampal-hippocampal network. Nat Rev Neurosci. 2009;10:272–82.
Wang M, Yang Y, Dong Z, Cao J, Xu L. NR2B-containing N-methyl-D-aspartate subtype glutamate receptors regulate the acute stress effect on hippocampal long-term potentiation/long-term depression in vivo. Neuroreport. 2006;17:1343–6.

Winters BD, Saksida LM, Bussey TJ. Object recognition memory: neurobiological mechanisms of encoding, consolidation and retrieval. Neurosci Biobehav Rev. 2008;32:1055–70.

Wong TP, Howland JG, Robillard JM, Ge Y, Yu W, Titterness AK, et al. Hippocampal long-term depression mediates acute stress-induced spatial memory retrieval impairment. Proc Natl Acad Sci U S A. 2007;104:11471–6.

Woodson JC, Macintosh D, Fleshner M, Diamond DM. Emotion-induced amnesia in rats: working memory-specific impairment, corticosterone-memory correlation, and fear versus arousal effects on memory. Learn Mem. 2003;10:326–36.

Xu L, Anwyl R, Rowan MJ. Behavioural stress facilitates the induction of long-term depression in the hippocampus. Nature. 1997;387:497–500.

Xu L, Holscher C, Anwyl R, Rowan MJ. Glucocorticoid receptor and protein/RNA synthesis-dependent mechanisms underlie the control of synaptic plasticity by stress. Proc Natl Acad Sci U S A. 1998;95:3204–8.

Yang CH, Huang CC, Hsu KS. Behavioral stress modifies hippocampal synaptic plasticity through corticosterone-induced sustained extracellular signal-regulated kinase/mitogen-activated protein kinase activation. J Neurosci. 2004;24:11029–34.

Yang CH, Huang CC, Hsu KS. Behavioral stress enhances hippocampal CA1 long-term depression through the blockade of the glutamate uptake. J Neurosci. 2005;25:4288–93.

Zhou J, Zhang F, Zhang Y. Corticosterone inhibits generation of long-term potentiation in rat hippocampal slice: involvement of brain-derived neurotrophic factor. Brain Res. 2000;885:182–91.

Zucker RS, Regehr WG. Short-term synaptic plasticity. Annu Rev Physiol. 2002;64:355–405.

Chapter 12
Synaptic Mechanisms and Cognitive Computations Underlying Stress Effects on Cognitive Function

Gediminas Luksys and Carmen Sandi

Abstract The cognitive effects of stress vary depending on a number of factors related to the characteristics of the stressor, the cognitive function under study and individual differences. Identifying the unifying principles that can explain this diversity is one of the main challenges in the field. Here, we attempt to define how variations in stressor intensity affect cognitive function. At the phenomenological level, we confirm the existence of an inverted-U-shaped function to account for varying stress intensities and cognitive performance under certain conditions. At the mechanistic level, we revise potential synaptic mechanisms and computations underlying these diverging effects of stress. Among the synaptic mechanisms, we discuss strong evidence implicating glutamatergic pathways and neural cell adhesion molecules as key mediators of the varying cognitive effects of stress on memory. As computational modeling is emerging as a useful approach to integrate and to reveal neural and cognitive computations underlying complex behaviors, we introduce its basic concepts and explain its recent applications to the field of stress and cognition.

Abbreviations

AMPAR	α-Amino-3-hydroxy-5-methylisoxazole-4-propionic acid receptor
BCM	Bienenstock–Cooper–Munro
Ca^{2+}	Calcium
NMDAR	GluN2B subunit-containing N-Methyl-D-aspartate receptor

C. Sandi (✉)
Laboratory of Behavioral Genetics, Brain Mind Institute, Ecole Polytechnique Federale de Lausanne (EPFL), Station 19, CH-1015 Lausanne, Switzerland
e-mail: carmen.sandi@epfl.ch

G. Luksys
Division of Cognitive Neuroscience, University of Basel, Birmannsgasse 8, CH-4055 Basel, Switzerland

Division of Molecular Psychology, University of Basel, Missionsstrasse 60/62A, CH-4055 Basel, Switzerland

M. Popoli et al. (eds.), *Synaptic Stress and Pathogenesis of Neuropsychiatric Disorders*,
DOI 10.1007/978-1-4939-1056-4_12, © Springer Science+Business Media New York 2014

LTD	Long-term depression
LTP	Long-term potentiation
NCAM	Neural cell adhesion molecule
NE	Norepinephrine
PSA-NCAM	Polysialylated neural cell adhesion molecule
TDRL	Temporal difference reinforcement learning
HPA axis	Hypothalamus-pituitary-adrenocortical axis
DA	Dopamine

12.1 Introduction

In the past decades, the field of stress and cognition has exploded, confirming the enormous power of stress to affect cognitive function and synaptic plasticity. However, the emerging picture is that the results are not uniform (Sandi 2013). Instead, the large numbers of accumulated findings describe a differential impact of a number of stress conditions on specific plasticity and cognitive processes. Although stress effects are frequently deleterious, in many occasions cognitive and synaptic functions are not compromised by stress and in many others they are even improved.

Systematic reviews of the literature have shown that the specific effect of stress on cognitive function depends on a number of factors related to both the stress characteristics and to specific aspects of the cognitive function under consideration (Sandi and Pinelo-Nava 2007; Sandi 2013). In addition, there are important individual-related factors that modulate, as well, the way individuals are affected in their cognitive capabilities when exposed to particular stress conditions.

Regarding stress-related factors, the key ones identified so far as critical to define stress effects in cognitive function are stress "intensity," its contingency with regards to actual performance in a cognitive task (Sandi 1998; de Kloet et al. 1999; Joels et al. 2006), and its "duration" (e.g., whether acutely or chronically experienced) (Sandi 2013). Here, we will focus on the modulatory role of the factor stress intensity, as its importance has been acknowledged in the literature for a long time. Stress-induced changes in synaptic plasticity (Kim and Diamond 2002; Joëls et al. 2008) as well as in the connectivity and dynamic interactions between brain regions (Schwabe and Wolf 2012) have been identified as crucial mechanisms translating stress into behavioral changes. One of the current challenges in the field is to develop an integrated approach that allows explaining these varying effects of stress. The recent attempts to apply computational modeling appear as promising developments to reveal the fundamental computations affected by different degrees of stress in different individuals (Luksys and Sandi 2011). In the last part of the chapter, we will introduce recent modeling studies attempting to explain the computations underlying stress effects in plasticity and learning.

12.2 The Varying Effects of Stress in Cognitive Function

12.2.1 Stress Intensity

Although stress is a vague concept and there is no absolute consensus in the literature as to its ultimate meaning, a classical view considers that stress implies any challenge to the homeostasis of an individual that requires an adaptive response from that individual (Steckler 2005). Despite notable recent attempts to reconceptualize the term "to be restricted to conditions where an environmental demand exceeds the natural regulatory capacity of an organism, in particular situations that include unpredictability and uncontrollability" (Koolhaas et al. 2011), the term stress is typically and widely used to refer to conditions ranging from mild challenges to extremely aversive conditions.

Recent reviews of the literature have identified consistent findings in the studies relating changes in stress intensity with either Pavlovian conditioning or with cognitive "effortful" tasks (Sandi and Pinelo-Nava 2007; Sandi 2013). Regarding Pavlovian conditioning, they highlight the existence of a "linear relationship" between stressor intensity and the strength of the conditioned memory (i.e., fear conditioning, eye-blink conditioning) formed; that is, the higher the stressor intensity, the stronger the memory formed. Importantly, these facilitating effects of stress in conditioning processes were reported both when the variation in the physiological stress responses is triggered by the task (Cordero et al. 1998; Merino et al. 2000; Laxmi et al. 2003; Rau et al. 2005) and when it is elicited by other stressful conditions experienced before task exposure (Cordero et al. 2003; Shors 2004). These observations suggest that in Pavlovian conditioning it is the stress level experienced by the individual around the training experience that counts, and that the origin or "source of stress" (i.e., whether triggered by the task or outside the task) might not be so relevant to define cognitive effects of stress.

Contrary to the linear effects observed for conditioning processes, an inverted-U-shaped function seems to account for the relationship between stress intensity and performance in cognitive "effortful" tasks. In other words, low and high stress levels typically impair cognitive performance, whereas intermediate levels tend to facilitate it (Yerkes and Dodson 1908; Mendl 1999). This function was originally proposed from experiments published by Yerkes and Dodson in 1908. On the one side, the highly intuitive appealing of the Yerkes and Dodson law had a tremendous impact in the field. Despite important methodological flaws in Yerkes and Dodson's original experiments that led several authors throughout the twentieth century to question the validity of the law (Diamond 2005), the high intuitive power behind the idea that stress affects cognition following an inverted-U-shaped function favored its pervasiveness despite the lack of solid experimental evidence to support it. Until very recently, claims on the existence of an inverted-U-shaped function in hippocampus-dependent learning processes in the animal literature were typically made from the integration for the argument of partial findings obtained in separate studies and combining disparate approaches regarding stress timing with regards to the cognitive task (Morris 2006; Park et al. 2008; Sandi and Pinelo-Nava 2007). However, a recent study (Salehi et al. 2010) has confirmed the existence of an inverted-U-shaped function for performance in an effortful task under the same experimental conditions. Using a radial-arm water maze validated as a hippocampus-dependent spatial learning task (Diamond et al. 1999), stress levels were applied by changing the temperature of the water maze. Rats trained under moderate stress conditions were found to learn better and to show better memory for the task than those trained under either high or low stress conditions. Importantly, the study found as well an interaction between certain personality-like traits (i.e., anxiety and exploration traits) and the way individuals were affected by stress in their learning abilities. In human studies looking at decision making in the Iowa Gambling Task, an inverted-U-shaped relationship was found between the level of cortisol and performance in participants (van den Bos et al. 2009).

In fact, several studies tackling glucocorticoids (Diamond et al. 1992; Lupien and McEwen 1997; Conrad 2005; Joëls 2006) and the noradrenergic system (Introini-Collison et al. 1995) have provided convincing evidence for a key role of these hormonal and neurochemical stress systems in the inverted-U-shaped function for stress effects in cognition and synaptic plasticity. However, the story seems to be more complicated than expected, as recent studies have shown that stress and corticosterone can have opposite effects on LTP expression in the dorsal and ventral hippocampus (Maggio and Segal 2010). In addition, the amygdala seems to be critically implicated in the biphasic effects of stress on hippocampal synaptic plasticity (Akirav and Richter-Levin 2002). Understanding how differential effects in different brain regions lead to specific cognitive effects of stress as a function of varying levels of stress intensity is currently one of the key challenges of the field.

12.2.2 Individual Differences

Evidence from both animal and human literature highlights the existence of significant differences in the way individuals are affected in their cognitive capabilities when exposed to particular stress conditions. Several factors have been identified as critical to define differences in vulnerability to the cognitive impact of stress. One of them, sex, appears to be extremely influential. In a recent review, Andreano and Cahill (2009) have concluded that, generally, stress effects in conditioning tasks are more facilitating in males than in females; however, stress effects in relational and working memory tasks are varying: while, in rodents, males tend to be more impaired by stress than females (Park et al. 2008), working memory in humans was shown to follow a positive relationship with cortisol in men while a negative one in women (McCormick et al. 2007). Further studies are warranted to understand the complex effects of stress according to sex differences. Another important source of differential vulnerability to stress is the genetic and epigenetic endowment (Palumbo et al. 2010; Booij et al. 2013). The study of the interactions between genome and epigenome in the context of stress is an exploding and highly promising new field to mechanistically understand the molecular basis for individual differences in stress effects (Klengel et al. 2013).

Another important factor influencing differential vulnerability to stress is the individual's personality and, more specifically, the personality anxiety trait or the neuroticism personality factor (Holmes 2008; Sandi et al. 2008; Sandi and Richter-Levin 2009). As anxiety trait reflects how dispositionally anxious an individual is across time and situations, it seems logical to assume that it will play a key modulatory role on the behavioral effects of stress (Herrero et al. 2006; Sandi et al. 2008; Salehi et al. 2010; Castro et al. 2010, 2012; for a neurocognitive model of the mediating role of anxiety on stress effects, see Sandi and Richter-Levin 2009). In relational learning tasks, highly anxious rats typically show poorer performance than rats with low anxiety (Herrero et al. 2006; Salehi et al. 2010). However, different levels of trait anxiety interact with differences in stressor intensity to define the actual cognitive effect of anxiety. Thus, whereas analyses are focused in low-exploratory rats, performance of highly anxious individuals is at its best under low-stress conditions, individuals with low anxiety show superior performance under high-stress conditions (Salehi et al. 2010). The mechanisms underlying these differences have not been to date revealed. As the activation of the hypothalamus-pituitary-adrenocortical (HPA) axis has been found to not consistently reflect differences in anxiety in rodents (Armario et al. 2012), other mediating pathways are supposed to contribute. Given its critical role in the modulation of emotional states and particularly in relation to stress and anxiety, as well as the growing literature indicating a key role for the noradrenergic system in memory modulation (Sara 2009; Roozendaal and McGaugh 2011), the noradrenergic system appears as a plausible key system to regulate trait anxiety-related individual differences in the interactions between stress and cognition. In fact, glucocorticoids and the noradrenergic system have been found to interact in modulating cognitive function in a number of tasks (Roozendaal et al. 2009; McGaugh and Roozendaal 2002; Quirarte et al. 2009). Changes in the dynamic pattern of brain activity (e.g., such a deactivation of prefrontal cortical areas) are believed to mediate some of the concerted actions exerted by glucocorticoids and the noradrenergic system in cognitive function (van Stegeren et al. 2010).

12.3 Synaptic Mechanisms Mediating Stress Effects on Cognition

Different brain systems (de Quervain et al. 2009; Brown and Morey 2012; Schwabe and Wolf 2012) as well as diverse synaptic mechanisms (Kim et al. 2006; Sandi 2004 2011; Roozendaal et al. 2009; Chen et al. 2012) have been implicated in the cognitive effects of stress. Here, we will refer to two pathways, glutamatergic mechanisms and neural cell adhesion molecules, whose regulation at the interface between stress and learning has been reported to follow a coherent pattern within the framework of the inverted-U-shaped effects described above.

12.3.1 Glutamatergic Systems

Increasing evidence highlights glutamatergic mechanisms as crucial mediators of the cognitive actions of acute stress (Sandi 2011). Following in vitro evidence indicating that glucocorticoids can facilitate glutamate transmission (Joëls et al. 2008), the potentiation of α-amino-3-hydroxy-5-methylisoxazole-4-propionic acid receptor (AMPAR) trafficking leading to increased synaptic surface GluA2 content was implicated in the effect (Groc et al. 2008). Importantly, this mechanism was implicated in vivo as a key underlying mechanism of the left–ascending-side of the inverted-U-shaped curve, linking differences in stress intensity with cognitive performance. The study (Conboy and Sandi 2010), performed in mice, used a water maze spatial task with varying water temperatures, including a low stress (water at 30 °C) and a more stressful (water at 22 °C) condition; the latter leading to better performance at both learning and memory phases of the task. This facilitation of learning by stress was found along with enhanced synaptic GluA2 content that was not observed in mice trained under lower stress. The causal involvement of stress-released corticosterone was established in experiments in which inhibiting glucocorticoid release at training prevented both the stress-induced facilitation of memory and the enhancement of GluA2-AMPAR trafficking. The causal involvement of GluA2 trafficking in stress-induced facilitation of spatial learning and memory was claimed on the basis of pharmacological experiments addressed to block GluA2 synaptic trafficking and successful in interfering with stress-facilitating effects in learning and memory. Interestingly, individual differences in vulnerability to develop depression symptoms following stress in an outbred strain of mice was also shown to be related to genetic variations in the GluA1-AMPAR subunit (Schmidt et al. 2010), highlighting again an important role of AMPARs in differential cognitive vulnerability to stress.

In a recent review, Sandi (2011) has hypothesized that glucocorticoid effects at various levels within glutamatergic pathways may represent the principle underlying the variety of cellular mechanisms by which glucocorticoids affect cognition. The review proposes a two-component model implying that "positive effects of glucocorticoids will occur when there is a coupling between neural activity related to information processing in relevant circuits and moderate-to-high glucocorticoid-induced enhanced glutamate levels and/or AMPAR synaptic delivery." Conversely, the model states that "negative effects will take place when high-to-very-high corticosterone-induced high extracellular glutamate levels are uncoupled, but closely linked in time to neural activity" (Sandi 2011, p. 173). Regarding the latter, mechanisms underlying the induction of long-term depression (LTD) have been identified among those that mediate the impairing effects of stress and glucocorticoids in the retrieval of information, including the activation of extrasynaptic GluN2B subunit-containing N-methyl-D-aspartate receptors (NMDARs) and the endocytosis of the GluA2 AMPAR subunit (Wong et al. 2007).

12.3.2 Neural Cell Adhesion Molecules

Among the myriad synaptic proteins potentially involved in synaptic plasticity and memory (Leslie and Nedivi 2011), the key role played by the neural cell adhesion molecule (NCAM) not only during early neural development but also in synaptic plasticity and cognitive function in adulthood (Conboy et al. 2010) placed it as a good candidate to mediate stress effects. Indeed, over the past decade strong evidence accumulated for the involvement of NCAM in both facilitating (Lopez-Fernandez et al. 2007) and impairing (Bisaz et al. 2011) effects of stress in memory function (Sandi 2004; Bisaz et al. 2009).

The first observation linking NCAM with the cognitive effects of stress was obtained in the passive avoidance learning model in 1-day-old chicks (Sandi et al. 1995). Corticosterone injections given after training chicks in a task leading to a weak memory resulted in a facilitation of long-term memory that was blocked by administration of an NCAM antibody. As strong training in this model leads to higher plasma corticosterone levels than weak training (Sandi and Rose 1994), these results strongly implicated NCAM as a critical molecular mechanism underlying the memory facilitating effects of stress and glucocorticoids. Further evidence was obtained using biochemical approaches in rats, with NCAM expression levels in the hippocampus varying as a function of stressor intensity during training (Merino et al. 2000).

The posttranslational modification of NCAM involving its polysialylation (PSA-NCAM) has also been described as critically involved in synaptic remodeling and synaptogenesis, synaptic plasticity, memory formation, and in the link between stress and memory (Sandi 2004; Bisaz et al. 2009). Training-related regulation of hippocampal PSA-NCAM has been linked to differences in stressor intensity (Merino et al. 2000; Sandi et al. 2003) and to individual differences in cognitive performance; for example, the highest increase on PSA-NCAM hippocampal expression following water maze training in rats was found in the animals that showed the slowest acquisition rate (Sandi et al. 2004), which are the ones that show highest anxiety and stress responses while performing the task (Sandi et al. 2004; Venero et al. 2004).

The causal implication of NCAM on memory strength has been established with a variety of approaches, including the administration of antibodies (Doyle et al. 1992; Scholey et al. 1993) or peptides (Foley et al. 2000; Cambon et al. 2003;

Venero et al. 2006) that were found to interfere with NCAM function and memory formation, as well as the administration of NCAM mimetic peptides that were found to effectively facilitate memories established through weak training protocols (Cambon et al. 2004). Interestingly, the NCAM mimetic peptide FGL was identified to facilitate memory and synaptic plasticity by facilitating synaptic delivery of AMPARs (Knafo et al. 2012). This finding opens the possibility that the stress-triggered actions on glutamatergic pathways (described in the previous section) and on NCAM remodeling are, in fact, convergent mechanisms translating stress effects in cognition.

12.4 Computations Underlying Stress Effects on Cognition

In this section, we address the emerging computational models attempting to explain how stress affects plasticity and cognition. As described above, neurophysiological and behavioral studies provide important insights into stress' effects on synaptic plasticity mechanisms, yet a more complete picture of their functional implications can be achieved only if many different parts of this complex biological system are taken into account at different levels. As this would be too challenging, expensive, and time consuming to accomplish using merely experimental methods, employing in-silico simulation approaches is indispensable. First, we describe the general computational approaches to model synaptic plasticity. Then, we focus on computational approaches to reinforcement learning and, finally, we review the recently developed methodology of model-based analysis and its applications to studying how stress affects cognition.

12.4.1 Computational Modeling of Synaptic Plasticity

It has been shown that essential bits of knowledge about synaptic plasticity could be described by a few, relatively simple equations, which could in turn be used to simulate how circuits of neurons and connections between them are shaped by various patterns of stimulation. Ideas of Hebbian learning (Hebb 1949)—that connection strengths between neurons with correlated firing should increase and those between neurons with uncorrelated firing should decrease—provided the basis for computational models of LTP and LTD. For example, if x and y are the firing frequencies of two connected neurons and $<x>$ and $<y>$ the respective averages, then the synaptic weight between these neurons should change as follows (Sejnowski and Tesauro 1989):

$$\Delta w_{xy} = \alpha \cdot (x - <x>) \cdot (y - <y>).$$

Here, parameter α controls the learning rate, and its different settings can lead to substantially different learning and activity dynamics of the modeled neuronal network. For example, with high α values new information leads to rapid network update that depending on input statistics may lead to instability, whereas with low α values the neural network is more robust but may not be capable to respond to sudden environmental changes. For this reason, it has been suggested that neuromodulators could play a role in changing the settings of such model parameters (Doya 2002).

As stress acts through neuromodulators such as norepinephrine (NE), its effects on synaptic plasticity could be modeled using the learning rate as a dependent variable. Another way of modeling effects of stress is using a biologically realistic Bienenstock–Cooper–Munro (BCM) rule (Bienenstock et al. 1982), according to which the synaptic weight is updated based on presynaptic firing frequency x and a nonlinear function φ of the postsynaptic firing frequency y:

$$\Delta w_{xy} = \alpha \cdot x \cdot \varphi(y, \theta_{\mathrm{m}}).$$

Here, θ_{m} is a threshold that separates potentiation ($y > \theta_{\mathrm{m}}$) and depression ($y < \theta_{\mathrm{m}}$). This threshold can be affected by several factors at different levels, such as average postsynaptic activity or Ca^{2+} concentration. It has been suggested that high-intensity stress, through glucocorticoid action, increases intracellular Ca^{2+} levels in the hippocampus (Joëls 2006), thereby shifting the threshold θ_{m} to the right and the relative balance of synaptic plasticity from LTP to LTD (Kim and Yoon 1998).

12.5 Computational Models of Reinforcement Learning

In order to relate processes at the cellular level that can be described by models of synaptic plasticity to the behavioral level, additional computational techniques are necessary. As learning, memory and decision making in both animals are humans are influenced by rewards and punishments, computational models ought to include the reinforcement factor as well. After the discovery that dopamine neurons code for the difference between actual and predicted reinforcement (Schultz et al. 1997), it has been suggested that including a third, reinforcement-related factor to synaptic plasticity learning rules is biologically realistic, as it corresponds to dopaminergic modulation of plasticity in the striatum and other brain areas (Reynolds et al. 2001; Wörgötter and Porr 2005). For instance, the BCM rule can be modified as follows:

$$\Delta w_{xy} = \alpha \cdot (r - \langle r \rangle) \cdot x \cdot \varphi(y, \theta_{\mathrm{m}}).$$

Here, $\langle r \rangle$ is the average reinforcement (rewards being positive and punishments negative), and r the actually received reinforcement. Three-factor learning rules combining Hebbian learning with a reinforcement-related factor were recently used to model spatial learning in rodents (Foster et al. 2000; Strösslin et al. 2005; Vasilaki et al. 2009) that is based on synaptic plasticity between hippocampal-like place cells coding animal's location and putative action cells coding direction of its movement.

Although computational modeling of synaptic plasticity provides an important tool for neurophysiological studies, its level of detail may not be necessary in some behavioral, pharmacological, and neuroimaging studies. Too-detailed models that aim to simulate all relevant neural systems and relate the resulting model's performance to actual behavior may contain too many parameters, which may lead either to arbitrary and unjustified choices of parameter values, or overfitting and lack of generalization, if many parameters are estimated from the data. For this reason, behavior in many reinforcement learning tasks is modeled at a higher level of abstraction. The theory of temporal difference reinforcement learning (TDRL; Sutton and Barto 1998), originating from the artificial intelligence field, has received increasing empirical support from electrophysiological and neuroimaging studies (Schultz et al. 1997; O'Doherty et al. 2003; Samejima et al. 2005), and has since become widely used in modeling reinforcement-based behavior and studying its neural correlates.

TDRL relies on the assumption that intelligent agents learn the consequences of actions performed at different *states* of their environment, and using this knowledge they select *actions* that lead to the optimal outcome. More specifically, a key quantity to be learned in TDRL is the so-called Q-value that for action a performed from state s at time t describes the expected cumulative future reinforcement r. If future reinforcements are discounted exponentially at a rate of γ per time unit, the Q-values can be written as follows:

$$Q(s_t, a_t) = \mathrm{E}[r_t + \gamma \cdot r_{t+1} + \gamma^2 \cdot r_{t+2} + \ldots] = E[r_t] + \gamma \cdot Q(s_{t+1}, a_{t+1}).$$

The setting of discounting rate γ is crucial because before any consideration of whether and how fast optimal actions can be learned, it defines the value function itself, either by prioritizing immediate outcomes (in the case of small γ values) or considering outcomes occurring over a longer time span similarly important.

The temporal difference learning rule can be derived from the definition of Q-values by following the principle that their update should be proportional to the difference between actual and predicted rewards r_t and $E(r_t)$ respectively:

$$\Delta Q(s_t, a_t) = \alpha\left(r_t - E[r_t]\right) = \alpha\left(r_t - Q(s_t, a_t) + \gamma \cdot Q(s_{t+1}, a_{t+1})\right).$$

Here α is the learning rate. Note that the TDRL rule is similar to three-factor learning, except for the absence of explicit presynaptic and postsynaptic terms. Instead, their function is accomplished by updating only the values of currently visited state and performed action (their "pre" and "post" terms equal to 1) and not changing values of all other state and action pairs (their "pre" and "post" terms equal to 0). In more sophisticated TDRL implementations, where states and actions are encoded by a neural population, explicit "pre" and "post" terms become necessary as they indicate the extent to which each neuron is encoding a particular state or action

(Strösslin et al. 2005; Vasilaki et al. 2009). The resulting rule becomes essentially a three-factor learning rule.

One of the key problems in reinforcement learning is addressing the exploration–exploitation dilemma: should the actions that currently have the highest value be selected expecting the most positive reinforcement or should other actions be explored? The main benefit of exploration is gathering more accurate information about action outcomes, as the Q-values, particularly during early stages of learning, may be inaccurate. The most common method of action selection that takes into account exploration is the "softmax":

$$p(a) = \exp(\beta \cdot Q(s,a)) / \Sigma_i \left(\exp(\beta \cdot Q(s,a_i))\right).$$

Here, actions are chosen probabilistically with probability p of action a dependent on its Q-value, exploration–exploitation factor β (also called inverse temperature), and Q-values of all other actions a_i available from state s (Σ_i is the sum over the exponential terms for all these actions). If the parameter β is set high, the action with the highest Q-value is selected nearly always, whereas low β values allow more exploration (with $\beta = 0$ the action choice is totally random).

The discussed TDRL parameters such as learning rate α, exploration–exploitation factor β and discounting rate γ can have various impacts on modeled learning behavior, ranging from acceleration or slowing down of the learning process to qualitative changes in adopted behavioral strategies (see Fig. 12.1). From a theoretical perspective of achieving optimal learning in a stationary environment, α should be gradually decreased and β increased (Doya 2002), because such strategy ensures sufficient flexibility at early stages of learning and preservation as well as use of the acquired knowledge at later stages. It is important to note that in dynamic environments or when learning complex tasks this simple rule may not always lead to optimal outcomes.

a. Under relatively steep discounting, e.g., $\gamma = 0.5$, the modeled animal will remain at the outer wall, as walking through an open space for three steps to get the large reward has a lower value ($\gamma^3 \cdot R_{t+3} = 0.5^3 \cdot 10 = 1.25$) than staying at or walking around the wall ($r_t + \gamma \cdot r_{t+1} + \gamma^2 \cdot r_{t+2} + \gamma^3 \cdot r_{t+3} = 1 + 0.5 + 0.5^2 + 0.5^3 = 1.875$). If discounting is more shallow, e.g., $\gamma = 0.9$, walking towards the target ($0.9^3 \cdot 10 = 7.29$) becomes preferable to remaining at the wall ($1 + 0.9 + 0.9^2 + 0.9^3 = 3.439$). Note that the same preferences are achieved if instead of rewarding locations next to the wall, visiting open locations is modeled by a small punishment ($r = -1$).
b. If the exploration–exploitation factor is high (e.g., $\beta = 10$) from the beginning, the modeled animal will keep choosing the actions where it experienced even small rewards first, i.e., if starting at location (S) it first chose to walk along or remain at the wall, it will never explore the route towards the target because it does not know that the reward is there. In this case, modeling open locations with a small punishment instead of a small reward for outer wall locations may lead to a slightly different outcome, as the punishment would be learned only after visiting the open locations.

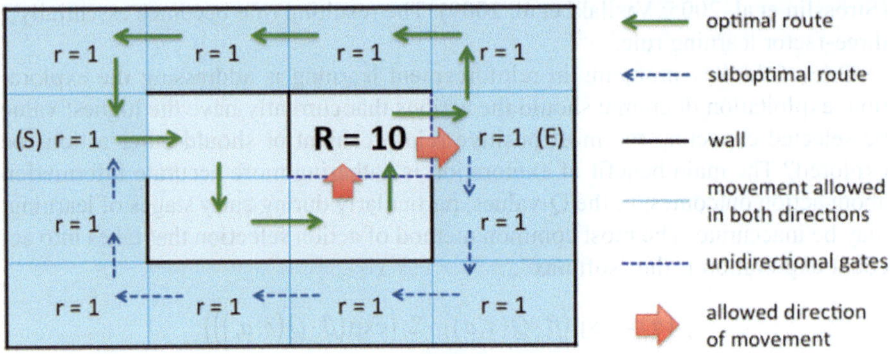

Fig. 12.1 Effects of TDRL parameters on modeled behavioral strategies, as illustrated by a virtual maze (adapted with permission from Luksys and Sandi 2011). The maze environment consists of 16 states (4 × 4), with a target location in its center containing a large reward (R=10). The maze also contains walls and two unidirectional gates, because of which the target can only be reached via a single route from the left (and exited to the right). Actions are performed by moving between the 16 states whenever the movement is allowed (i.e., it does not cross a wall or enter a gate in the wrong direction), starting from the location marked with (S). As animals tend to avoid open spaces, presence in locations around the outer wall is modeled by a small reward (r=1). The optimal strategy, which allows maximizing the reward per time period, is visiting the platform via a route from the left and returning along the upper wall (which is a shorter route than along the bottom). However, as the modeled animal initially does not know the target location or that it contains a large reward, trial-and-error learning, modeled by TDRL, can lead to different behaviors depending on parameter settings (see accompanying text with a,b,c options)

c. If exploration (e.g., β=10) is maintained throughout the learning process, the modeled animal may learn going to the platform but not taking the optimal route because at the exit location (E), where this decision is made, the difference between reward R=10 discounted for 9 steps (route along the upper wall) and for 11 steps (route along the lower wall) is likely to be small. Such small differences in value can only be distinguished under high values of β and γ

12.6 Model-Based Analyses of Stress, Learning, and Memory

Although the main rationale behind computational models of synaptic plasticity and learning has been the study of computational mechanisms underlying these processes, during the last decade a new approach emerged: using computational models (mostly well studied ones like TDRL) to analyze neurobehavioral data (Corrado and Doya 2007). For instance, a TDRL model in a given environment produces a sequence of actions and reinforcements that depend on model parameters (α, β, γ). If such actions and reinforcements can be recorded from an individual performing the modeled task, one can estimate the parameter settings under which the experimentally observed sequences are the most likely. If actual sequences of actions or reinforcements are not available or when fitting them is impractical, model parameters can also be estimated from secondary statistics, such as reward rates and counts or frequencies of certain behaviorally relevant actions. This approach allows making inferences about internal variables (such as Q-values) or parameters (such as acquisition and forgetting rates, preferences regarding exploration versus exploitation and immediate versus delayed reinforcement) that can be more easily related to cognitive processes of interest than classical behavioral variables (such as escape latencies, response times, and numbers of recalled items).

Using the classical approach, many cognitive processes can be studied only thanks to specialized experimental setups designed to extract the cleanest possible signal for the process of interest. In contrast, model-based analyses enable studying many processes of interest that are elements of the model simultaneously, taking into account their interaction and possibly even using complex behavioral phenotypes outside the traditional laboratory setting (such as games), whose analysis using the classical approach would be way too superficial and remote from neurobiological mechanisms.

In effect, model-based analysis performs a transformation of observed behavioral variables, but unlike in the applications of principal component analysis (Clément et al. 2007), the transformation is usually nonlinear and biased towards variables of interest, which facilitates interpretability at a cost of missing aspects of the data not addressed by the model. For this reason, it is important to ensure that the chosen model is biologically plausible, and in cases where several candidate models or model settings might be similarly applicable, determine empirically which of them produces the best fit to experimental data (Mars et al. 2012). For example, a number of model-based studies estimate only one or two parameters of the model that are of interest for the study and related to experimental manipulations in the modeled task, while keeping the rest fixed, which is only justified if it can be shown that these latter parameters would not improve the model fit to experimental data. In general, provided that sufficiently rich experimental data are available to avoid overfitting, it is best to keep all essential model parameters flexible, as in case of subjective selection the parameter(s) chosen to be flexible may not be the best one(s) to account for experimental results. For instance, many learning and memory studies tend to ignore the exploration–exploitation aspect, which is a part of almost any active learning experience and thus it, not the learning per se (or its rate), may be responsible for differences between certain experimental manipulations.

Model-based analyses became popular as a result of rapid developments in the neuroscience of reward learning, sparked by the landmark discovery (Schultz et al. 1997) of midbrain dopamine (DA) neurons coding the reward prediction error (received minus expected reward), a key quantity in TDRL. Soon after, neural codes for many different variables and parameters of reward learning and decision-making were discovered using a variety of approaches, both experimental and computational (for a review see Doya 2008). Model-based analysis studies related steepness of discounting to the serotonin (5HT) levels (Schweighofer et al. 2008), explorative choices to the frontopolar cortex activity (Daw et al. 2006), and learning rates to the

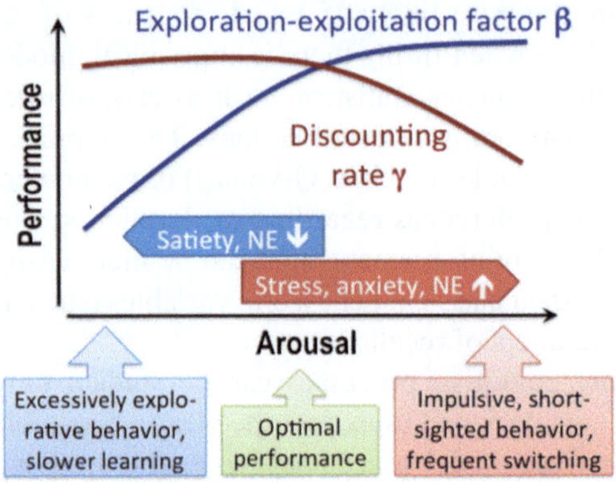

Fig. 12.2 Computational interpretation of the inverted-U-shape (adapted with permission from Luksys and Sandi 2011)

anterior cingulate activity (Behrens et al. 2007). Polymorphisms in genes regulating different aspects of dopaminergic activity were linked to differences in learning rates and uncertainty-based exploration (Frank et al. 2007, 2009).

The role of stress in learning was addressed by a model-based analysis study (Luksys et al. 2009) that investigated behavior of two genetic strains of mice (the "calm" C57BL/6 and the more anxious DBA/2) learning to make nose pokes in response to light onset for the delivery of food. Individually estimated model parameters were compared between two genetic strains exposed to different stress conditions, and correlated with anxiety and motivation of each mouse. The results indicated that for more anxious animals stress led to steeper discounting, which impaired learning of delayed rewards, whereas for less anxious mice stress increased exploitation, improving their performance. Their analysis suggests that in order to achieve optimal performance at the middle of the inverted U-shape both sufficient exploitation and shallow discounting are needed (Fig. 12.2). Results from model-based analysis in Luksys et al. 2009 suggest that the inverted-U-shape relationship between arousal and performance in many tasks (aside from Pavlovian conditioning) may arise because of changes in two TDRL parameters (β and γ) as a result of stress, anxiety or increased norepinephrine (*NE*) levels: a shift from exploration to exploitation at intermediate arousal levels and increasingly steep discounting at high arousal levels. Conversely, satiety or decreased NE levels may lead to a shallower discounting and a shift to exploration. It is important to note that frequent behavioral switches, observed at high levels of arousal and NE (Aston-Jones and Cohen 2005), may be caused by increasingly steep discounting and high exploitation in the following way. Q-values, defined as the expected cumulative future reward, depend not only on the reward history but also on the discounting rate γ; therefore even without changes in the received rewards, Q-values become smaller if γ is decreased. As Q-values are updated only for the currently exploited actions, these actions become less favorable compared to the alternative ones, leading to a switch. Then these alternative actions are exploited and their Q-values decrease as well, leading to further switches. Consistent with this interpretation, stress has linear relationship to performance in simple tasks (which typically do not require learning over delays) and an inverted-U-shaped relationship in more complex tasks (Diamond et al. 2007). Furthermore, high exploitation combined with increasingly steep discounting that occurs under high stress/NE levels, can lead to behavior that might be interpreted as exploration (Aston-Jones and Cohen 2005) but in TDRL terms is rather an exploitative switching due to relative devaluation of the current best action compared to its alternatives (for more explanations see Fig. 12.2).

12.7 Concluding remarks

The exploding field of stress and cognition has accumulated numerous studies showing varying effects of stress across different experimental conditions. There is however a critical need to develop a unifying model that allows understanding a large variety of the effects and, ideally, in the future to incorporate as well the growing information in terms of the mechanisms involved at the different (molecular, cellular, network, systems) levels of analyses. Here, we have organized the existing literature on acute effects of stress on cognitive function by examining the impact of variations in stressor intensity and in individual characteristics. We have revisited the evidence in support of the existence of an inverted-U-shaped function to account for varying effects of different stress intensities on cognitive performance and discussed potential synaptic mechanisms and computations underlying the diversity of effects. At the synaptic level, strong evidence implicates glutamatergic pathways and neural cell adhesion molecules among the key mechanisms mediating the diversity of effects induced by varying levels of stress at learning. As to the computational approaches, recent modeling attempts offer illuminating explanations to the fundamental cognitive computations affected by different degrees of stress in different individuals that should be tested experimentally to determine their generalization outside the model and the task used. We propose that using model-based analyses can help identifying neural mechanisms underlying specific cognitive operations, and that their application to the field of stress and cognition can improve our understanding and predictability of the diverse effects that stress exerts not only in the healthy but also in the dysfunctional brain.

References

Akirav I, Richter-Levin G. Mechanisms of amygdala modulation of hippocampal plasticity. J Neurosci. 2002;22:9912–21.

Andreano JM, Cahill L. Sex influences on the neurobiology of learning and memory. Learn Mem. 2009;16:248–66.

Armario A, Daviu N, Muñoz-Abellán C, Rabasa C, Fuentes S, Belda X, Gagliano H, Nadal R. What can we know from pituitary-adrenal hormones about the nature and consequences of exposure to emotional stressors? Cell Mol Neurobiol. 2012;32:749–58.

Aston-Jones G, Cohen JD. An integrative theory of locus coeruleus-norepinephrine function: adaptive gain and optimal performance. Annu Rev Neurosci. 2005;28:403–50.
Behrens TE, Woolrich MW, Walton ME, Rushworth MF. Learning the value of information in an uncertain world. Nat Neurosci. 2007;10:1214–21.
Bienenstock EL, Cooper LN, Munroe PW. Theory of the development of neuron selectivity: orientation specificity and binocular interaction in visual cortex. J Neurosci. 1982;2:32–48.
Bisaz R, Conboy L, Sandi C. Learning under stress: a role for the neural cell adhesion molecule NCAM. Neurobiol Learn Mem. 2009;91:333–42.
Bisaz R, Schachner M, Sandi C. Causal evidence for the involvement of the neural cell adhesion molecule, NCAM, in chronic stress-induced cognitive impairments. Hippocampus. 2011;21:56–71.
Booij L, Wang D, Lévesque ML, Tremblay RE, Szyf M. Looking beyond the DNA sequence: the relevance of DNA methylation processes for the stress-diathesis model of depression. Philos Trans R Soc Lond B Biol Sci. 2013;368:20120–251.
Brown VM, Morey RA. Neural systems for cognitive and emotional processing in posttraumatic stress disorder. Front Psychol. 2012;3:449.
Cambon K, Hansen SM, Venero C, Herrero AI, Skibo G, Berezin V, Bock E, Sandi C. A synthetic neural cell adhesion molecule mimetic peptide promotes synaptogenesis, enhances presynaptic function, and facilitates memory consolidation. J Neurosci. 2004;24:4197–204.
Cambon K, Venero C, Berezin V, Bock E, Sandi C. Post-training administration of a synthetic peptide ligand of the neural cell adhesion molecule, C3d, attenuates long-term expression of contextual fear conditioning. Neuroscience. 2003;122:183–91.
Castro JE, Varea E, Márquez C, Cordero MI, Poirier G, Sandi C. Role of the amygdala in antidepressant effects on hippocampal cell proliferation and survival and on depression-like behavior in the rat. PLoS ONE. 2010;5:e8618.
Castro JE, Diessler S, Varea E, Márquez C, Larsen MH, Cordero MI, Sandi C. Personality traits in rats predict vulnerability and resilience to developing stress-induced depression-like behaviors, HPA axis hyper-reactivity and brain changes in pERK1/2 activity. Psychoneuroendocrinology. 2012;37:1209–23.
Chen DY, Bambah-Mukku D, Pollonini G, Alberini CM. Glucocorticoid receptors recruit the CaMKIIα-BDNF-CREB pathways to mediate memory consolidation. Nat Neurosci. 2012;15:1707–14.
Clément Y, Joubert C, Kopp C, Lepicard EM, Venault P, Misslin R, Cadot M, Chapouthier G. Anxiety in mice: a principal component analysis study. Neural Plast. 2007;2007:35457.
Conboy L, Bisaz R, Markram K, Sandi C. Role of NCAM in emotion and learning. Adv Exp Med Biol. 2010;663:271–96.
Conboy L, Sandi C. Stress at learning facilitates memory formation by regulating AMPA receptor trafficking through a glucocorticoid action. Neuropsychopharmacology. 2010;35:674–85.
Conrad CD. The relationship between acute glucocorticoid levels and hippocampal function depends upon task aversiveness and memory processing stage. Nonlinearity Biol Toxicol Med. 2005;3:57–78.
Cordero MI, Merino JJ, Sandi C. Correlational relationship between shock intensity and corticosterone secretion on the establishment and subsequent expression of contextual fear conditioning. Behav Neurosci. 1998;112:885–91.
Cordero MI, Venero C, Kruyt ND, Sandi C. Prior exposure to a single stress session facilitates subsequent contextual fear conditioning in rats. Evidence for a role of corticosterone. Horm Behav. 2003;44:338–45.
Corrado G, Doya K. Understanding neural coding through the model-based analysis of decision making. J Neurosci. 2007;27:8178–80.
Daw ND, O'Doherty JP, Dayan P, Seymour B, Dolan RJ. Cortical substrates for exploratory decisions in humans. Nature. 2006;441:876–9.
de Kloet ER, Oitzl MS, Joëls M. Stress and cognition: are corticosteroids good or bad guys? Trends Neurosci. 1999;22:422–6.

de Quervain DJ, Aerni A, Schelling G, Roozendaal B. Glucocorticoids and the regulation of memory in health and disease. Front Neuroendocrinol. 2009;30:358–70.
Diamond DM, Bennett MC, Fleshner M, Rose GM. Inverted-U relationship between the level of peripheral corticosterone and the magnitude of hippocampal primed burst potentiation. Hippocampus. 1992;2:421–30.
Diamond DM, Campbell AM, Park CR, Halonen J, Zoladz PR. The temporal dynamics model of emotional memory processing: a synthesis on the neurobiological basis of stress-induced amnesia, flashbulb and traumatic memories, and the Yerkes-Dodson law. Neural Plast. 2007;2007:60803.
Diamond DM, Park CR, Heman KL, Rose GM. Exposing rats to a predator impairs spatial working memory in the radial arm water maze. Hippocampus. 1999;9:542–52.
Diamond DM. Cognitive, endocrine and mechanistic perspectives on non-linear relationships between arousal and brain function. Nonlinearity Biol Toxicol Med. 2005;3:1–7.
Doya K. Metalearning and neuromodulation. Neural Netw. 2002;15:495–506.
Doya K. Modulators of decision making. Nat Neurosci. 2008;11:410–6.
Doyle E, Nolan PM, Bell R, Regan CM. Intraventricular infusions of anti-neural cell adhesion molecules in a discrete posttraining period impair consolidation of a passive avoidance response in the rat. J Neurochem. 1992;59:1570–3.
Foley AG, Hartz BP, Gallagher HC, Rønn LC, Berezin V, Bock E, Regan CM. A synthetic peptide ligand of neural cell adhesion molecule (NCAM) IgI domain prevents NCAM internalization and disrupts passive avoidance learning. J Neurochem. 2000;74:2607–13.
Foster DJ, Morris RG, Dayan P. A model of hippocampally dependent navigation, using the temporal difference learning rule. Hippocampus. 2000;10:1–16.
Frank MJ, Doll BB, Oas-Terpstra J, Moreno F. Prefrontal and striatal dopaminergic genes predict individual differences in exploration and exploitation. Nat Neurosci. 2009;12:1062–8.
Frank MJ, Moustafa AA, Haughey HM, Curran T, Hutchison KE. Genetic triple dissociation reveals multiple roles for dopamine in reinforcement learning. Proc Natl Acad Sci U S A. 2007;104:16311–6.
Groc L, Choquet D, Chaouloff F. The stress hormone corticosterone conditions AMPAR surface trafficking and synaptic potentiation. Nat Neurosci. 2008;11:868–70.
Hebb DO The organization of behavior. New York: Wiley; 1949.
Herrero AI, Sandi C, Venero C. Individual differences in anxiety trait are related to spatial learning abilities and hippocampal expression of mineralocorticoid receptors. Neurobiol Learn Mem. 2006;86:150–9.
Holmes A. Genetic variation in cortico-amygdala serotonin function and risk for stress-related disease. Neurosci Biobehav Rev. 2008;32:1293–314.
Introini-Collison IB, Ford L, McGaugh JL. Memory impairment induced by intraamygdala beta-endorphin is mediated by noradrenergic influences. Neurobiol Learn Mem. 1995;63:200–5.
Joëls M, Krugers H, Karst H. Stress-induced changes in hippocampal function. Prog Brain Res. 2008;167:3–15.
Joëls M. Corticosteroid effects in the brain: U-shape it. Trends Pharmacol Sci. 2006;27:244–50.
Kim JJ, Diamond DM. The stressed hippocampus, synaptic plasticity and lost memories. Nat Rev Neurosci. 2002;3:453–62.
Kim JJ, Song EY, Kosten TA. Stress effects in the hippocampus: synaptic plasticity and memory. Stress. 2006;9:1–11.
Kim JJ, Yoon KS. Stress: metaplastic effects in the hippocampus. Trends Neurosci. 1998;21:505–9.
Klengel T, Mehta D, Anacker C, Rex-Haffner M, Pruessner JC, Pariante CM, Pace TW, Mercer KB, Mayberg HS, Bradley B, Nemeroff CB, Holsboer F, Heim CM, Ressler KJ, Rein T, Binder EB. Allele-specific FKBP5 DNA demethylation mediates gene-childhood trauma interactions. Nat Neurosci. 2013;16:33–41.
Knafo S, Venero C, Sánchez-Puelles C, Pereda-Peréz I, Franco A, Sandi C, Suárez LM, Solís JM, Alonso-Nanclares L, Martín ED, Merino-Serrais P, Borcel E, Li S, Chen Y, Gonzalez-Soriano J, Berezin V, Bock E, Defelipe J, Esteban JA. Facilitation of AMPA receptor synaptic delivery as a molecular mechanism for cognitive enhancement. PLoS Biol. 2012;10:e1001262.

Koolhaas JM, Bartolomucci A, Buwalda B, de Boer SF, Flügge G, Korte SM, Meerlo P, Murison R, Olivier B, Palanza P, Richter-Levin G, Sgoifo A, Steimer T, Stiedl O, van Dijk G, Wöhr M, Fuchs E. Stress revisited: a critical evaluation of the stress concept. Neurosci Biobehav Rev. 2011;35:1291–301.

Laxmi TR, Stork O, Pape HC. Generalisation of conditioned fear and its behavioural expression in mice. Behav Brain Res. 2003;145:89–98.

Leslie JH, Nedivi E. Activity-regulated genes as mediators of neural circuit plasticity. Prog Neurobiol. 2011;94:223–37.

Lopez-Fernandez MA, Montaron MF, Varea E, Rougon G, Venero C, Abrous DN, Sandi C. Upregulation of polysialylated neural cell adhesion molecule in the dorsal hippocampus after contextual fear conditioning is involved in long-term memory formation. J Neurosci. 2007;27:4552–61.

Luksys G, Gerstner W, Sandi C. Stress, genotype and norepinephrine in the prediction of mouse behavior using reinforcement learning. Nat Neurosci. 2009;12:1180–6.

Luksys G, Sandi C. Neural mechanisms and computations underlying stress effects on learning and memory. Curr Opin Neurobiol. 2011;21:502–8.

Lupien SJ, McEwen BS. The acute effects of corticosteroids on cognition: integration of animal and human model studies. Brain Res Rev. 1997;24:1–27.

Maggio N, Segal M. Corticosteroid regulation of synaptic plasticity in the hippocampus. Sci World J. 2010 Mar 16;10:462–9. doi: 10.1100/tsw.2010.48.

Mars RB, Shea NJ, Kolling N, Rushworth MFS. Model-based analyses: promises, pitfalls, and example applications to the study of cognitive control. Q J Exp Psychol. 2012;65:252–67.

McCormick CM, Lewis E, Somley B, Kahan TA. Individual differences in cortisol levels and performance on a test of executive function in men and women. Physiol Behav. 2007;91:87–94.

McGaugh JL, Roozendaal B. Role of adrenal stress hormones in forming lasting memories in the brain. Curr Opin Neurobiol. 2002;12:205–10.

Mendl M. Performing under pressure: stress and cognitive function. Appl Anim Behav Sci. 1999;65:221–44.

Merino JJ, Cordero MI, Sandi C. Regulation of hippocampal cell adhesion molecules NCAM and L1 by contextual fear conditioning is dependent upon time and stressor intensity. Eur J Neurosci. 2000;12:3283–90.

Morris R. Stress and the hippocampus. In Bliss T, editor. The hippocampus book. New York: Oxford University Press; 2006. p. 751–67.

O'Doherty JP, Dayan P, Friston K, Critchley H, Dolan RJ. Temporal difference models and reward-related learning in the human brain. Neuron. 2003;28:329–37.

Palumbo ML, Canzobre MC, Pascuan CG, Ríos H, Wald M, Genaro AM. Stress induced cognitive deficit is differentially modulated in BALB/c and C57Bl/6 mice: correlation with Th1/Th2 balance after stress exposure. J Neuroimmunol. 2010;218:12–20.

Park CR, Zoladz PR, Conrad CD, Fleshner M, Diamond DM. Acute predator stress impairs the consolidation and retrieval of hippocampus-dependent memory in male and female rats. Learn Mem. 2008;15:271–80.

Quirarte GL, de la TIS, Casillas M, Serafín N, Prado-Alcalá RA, Roozendaal B. Corticosterone infused into the dorsal striatum selectively enhances memory consolidation of cued water-maze training. Learn Mem. 2009;16:586–9.

Rau V, DeCola JP, Fanselow MS. Stress-induced enhancement of fear learning: an animal model of posttraumatic stress disorder. Neurosci Biobehav Rev. 2005;29:1207–23.

Reynolds JN, Hyland BI, Wickens JR. A cellular mechanism of reward-related learning. Nature. 2001;413:67–70.

Roozendaal B, McEwen BS, Chattarji S. Stress, memory and the amygdala. Nat Rev Neurosci. 2009;10:423–33.

Roozendaal B, McGaugh JL. Memory modulation. Behav Neurosci. 2011;125:797–824.

Salehi B, Cordero MI, Sandi C. Learning under stress: the inverted-U-shape function revisited. Learn Mem. 2010;17:522–30.

Samejima K, Ueda Y, Doya K, Kimura M. Representation of action-specific reward values in the striatum. Science. 2005;310:1337–40.
Sandi C, Merino JJ, Cordero MI, Kruyt ND, Murphy KJ, Regan CM. Modulation of hippocampal NCAM polysialylation and spatial memory consolidation by fear conditioning. Biol Psychiatry. 2003 Sep 15;54(6):599–607.
Sandi C, Cordero MI, Merino JJ, Kruyt ND, Regan CM, Murphy KJ. Neurobiological and endocrine correlates of individual differences in spatial learning ability. Learn Mem. 2004;11:244–52.
Sandi C, Cordero MI, Ugolini A, Varea E, Caberlotto L, Large CH. Chronic stress-induced alterations in amygdala responsiveness and behavior-modulation by trait anxiety and corticotropin-releasing factor systems. Eur J Neurosci. 2008;28:1836–48.
Sandi C, Pinelo-Nava MT. Stress and memory: behavioral effects and neurobiological mechanisms. Neural Plast. 2007;2007:78970.
Sandi C, Richter-Levin G. From high anxiety trait to depression: a neurocognitive hypothesis. Trends Neurosci. 2009;32:312–20.
Sandi C, Rose SP, Mileusnic R, Lancashire C. Corticosterone facilitates long-term memory formation via enhanced glycoprotein synthesis. Neuroscience. 1995;69:1087–93.
Sandi C, Rose SP. Corticosterone enhances long-term retention in one-day-old chicks trained in a weak passive avoidance learning paradigm. Brain Res. 1994;647:106–12.
Sandi C. The role and mechanisms of action of glucocorticoid involvement in memory storage. Neural Plast. 1998;6:41–52.
Sandi C. Stress, cognitive impairment and cell adhesion molecules. Nat Rev Neurosci. 2004;5:917–30.
Sandi C. Glucocorticoids act through glutamatergic pathways to affect memory processes. Trends Neurosci. 2011;34:165–76.
Sandi C. Stress and Cognition. WIRES Cognitive Sci. 2013;4:245–61.
Sara SJ. The locus coeruleus and noradrenergic modulation of cognition. Nat Rev Neurosci. 2009;10:211–23.
Schmidt MV, Trümbach D, Weber P, Wagner K, Scharf SH, Liebl C, Datson N, Namendorf C, Gerlach T, Kühne C, Uhr M, Deussing JM, Wurst W, Binder EB, Holsboer F, Müller MB. Individual stress vulnerability is predicted by short-term memory and AMPA receptor subunit ratio in the hippocampus. J Neurosci. 2010;30:16949–58.
Scholey AB, Rose SP, Zamani MR, Bock E, Schachner M. A role for the neural cell adhesion molecule in a late, consolidating phase of glycoprotein synthesis six hours following passive avoidance training of the young chick. Neuroscience. 1993;55:499–509.
Schultz W, Dayan P, Montague PR. A neural substrate of prediction and reward. Science. 1997;275:1593–9.
Schwabe L, Wolf OT. Stress modulates the engagement of multiple memory systems in classification learning. J Neurosci. 2012;32:11042–9.
Schweighofer N, Bertin M, Shishida K, Okamoto Y, Tanaka SC, Yamawaki S, Doya K. Low-serotonin levels increase delayed reward discounting in humans. J Neurosci. 2008;28:4528–32.
Sejnowski TJ, Tesauro G. The Hebb rule for synaptic plasticity: algorithms and implementations. In Byrne JH, Berry WO, editors. Neural models of plasticity. New York: Academic; 1989. pp. 94–103.
Shors TJ. Learning during stressful times. Learn Mem. 2004;11:137–44.
Steckler T. The neuropsychology of stress. In: Steckler T, Kalin NH, Reul JMHM, editors. Handbook of stress and the brain part 1: the neurobiology of stress. Amsterdam: Elsevier; 2005. p. 25.
Strösslin T, Sheynikhovich D, Chavarriaga R, Gerstner W. Robust self-localisation and navigation based on hippocampal place cells. Neural Netw. 2005;18:1125–40.
Sutton RS, Barto AG. Reinforcement Learning: An Introduction. Cambridge: MIT Press; 1998.
van den Bos R, Harteveld M, Stoop H. Stress and decision-making in humans: performance is related to cortisol reactivity, albeit differently in men and women. Psychoneuroendocrinology. 2009;34:1449–58.

van Stegeren AH, Roozendaal B, Kindt M, Wolf OT, Joëls M. Interacting noradrenergic and corticosteroid systems shift human brain activation patterns during encoding. Neurobiol Learn Mem. 2010;93:56–65.

Vasilaki E, Frémaux N, Urbanczik R, Senn W, Gerstner W. Spike-based reinforcement learning in continuous state and action space: when policy gradient methods fail. PLoS Comput Biol. 2009;5:e1000586.

Venero C, Herrero AI, Touyarot K, Cambon K, López-Fernández MA, Berezin V, Bock E, Sandi C. Hippocampal up-regulation of NCAM expression and polysialylation plays a key role on spatial memory. Eur J Neurosci. 2006;23:1585–95.

Venero C, Tilling T, Hermans-Borgmeyer I, Herrero AI, Schachner M, Sandi C. Water maze learning and forebrain mRNA expression of the neural cell adhesion molecule L1. J Neurosci Res. 2004;75:172–81.

Wong TP, Howland JG, Robillard JM, Ge Y, Yu W, Titterness AK, Brebner K, Liu L, Weinberg J, Christie BR, Phillips AG, Wang YT. Hippocampal long-term depression mediates acute stress-induced spatial memory retrieval impairment. Proc Natl Acad Sci U S A. 2007;104:11471–6.

Wörgötter F, Porr B. Temporal sequence learning, prediction, and control: a review of different models and their relation to biological mechanisms. Neural Comput. 2005;17:245–319.

Yerkes RM, Dodson JD. The relation of strength of stimulus to rapidity of habit-formation. J Comp Neurol Psychol. 1908;18:459–82.

Chapter 13
Altered GABA function in Major Depression

Beverly French, Marianne L. Seney and Etienne Sibille

Abstract Disrupted information transfer and processing at gamma-aminobutyric acid (GABA) and glutamate synapses, especially in corticolimbic circuits, has been proposed as a critical component of the pathophysiology of mood disorders. Here we review evidence of the primary pathology from human postmortem brains, supported by imaging studies in living subjects, for alterations in pyramidal excitatory neurons, GABA inhibitory neurons, and supporting glia, including oligodendrocytes and astrocytes. The data suggest combinatorial changes in most investigated components, converge on putative functional changes at glutamate and GABA synapses, and indicate that a subset of GABA neurons, which express specific cellular markers (calbindin, somatostatin, neuropeptide Y) and target distal dendrites of pyramidal neurons, may be more selectively and robustly affected in major depression. Pathologies in this subset of GABA neurons display a continuum of changes across brain disorders, may significantly contribute to deregulated GABA-containing tripartite synapses providing dendritic inhibition, and have implications for corticolimbic information processing in major depression and other brain disorders sharing similar pathologies.

Abbreviations

ABAT	4-Aminobutyrate aminotransferase
ACC	Anterior cingulate cortex
ALDH1L1	Aldehyde hydrogenase 1 family, member L1
BA	Brodmann area
BDNF	Brain-derived neurotrophic factor
CB	Calbindin
CCK	Cholecystokinin
CORT	Cortistatin

E. Sibille (✉)
Center for Neuroscience, University of Pittsburgh, Bridgeside Point II, Suite 231, 450 Technology Drive, Pittsburgh, PA 15219, USA
e-mail: sibilleel@upmc.edu

B. French · M. L. Seney · E. Sibille
Department of Psychiatry, University of Pittsburgh, Bridgeside Point II, Suite 231, 450 Technology Drive, Pittsburgh, PA 15219, USA

M. Popoli et al. (eds.), *Synaptic Stress and Pathogenesis of Neuropsychiatric Disorders*, DOI 10.1007/978-1-4939-1056-4_13,

CR	Calretinin
dlPFC	Dorsal lateral prefrontal cortex
EAAT1	Excitatory amino acid transporter1 glutamate clearance transporter 1
EAAT2	Excitatory amino acid transporter1glutamate clearance transporter 2
GABA	Gamma-aminobutyric acid
$GABA_AR$	GABA A receptor
$GABA_BR$	GABA B receptor
GABA-T	GABA transaminase
GABBR1	GABA B receptor 1
GABBR2	GABA B receptor 2
GAD	Glutamic acid decarboxylase
GAT	GABA transporter
GFAP	Glial fibrillary acidic protein
GLS	Glutaminase
GLUL	Glutamate ammonia ligase
GS	Glutamine synthetase
^1H MRS	Proton magnetic resonance spectroscopy
NPY	Neuropeptide Y
OFC	Orbital frontal cortex
PFC	Prefrontal cortex
PV	Parvalbumin
qPCR	Quantitative polymerase chain reaction
sgACC	Subgenual anterior cingulate cortex
SNAT3	Astrocytic system N transporter
SNAT1/SNAT2	System A transporters
SST	Somatostatin
TRKB	Tropomyosin related kinase B
VGAT	Vesicular GABA transporter

13.1 Introduction

Major depressive disorder, or major depression, is a common, devastating mood disorder characterized by low mood and a reduced ability to experience pleasure from previously enjoyable activities (anhedonia), which occurs in the presence of additional cognitive and physiological symptoms, such as loss of attention and concentration, recurrent thoughts of suicide, changes in weight, sleep patterns, and psychomotor retardation (American Psychiatric Association 2000). The costs at the individual and societal levels of this disorder are profound: the lifetime prevalence of major depression in the USA is approximately 17% (Kessler et al. 2005), and depression is currently considered the leading cause of years of healthy life lost due to disability, or "time spent in less than full health" among both men and women worldwide, as defined by the World Health Organization (WHO 2008). Women are disproportionately at risk, with twice as many women affected as men (Kessler et al. 2005). Despite this dire public health concern, current treatments have low efficacy, and only one out of two patients who meet criteria for major depression is expected to achieve remission (Huynh and McIntyre 2008; Kennedy and Giacobbe 2007).

Recent neural models of emotion perception have implicated the amygdala, anterior cingulate cortex (ACC), and dorsal lateral prefrontal cortex (dlPFC) as core regions of a neural network of identification and regulation of emotion (Phillips et al. 2008). Functional and morphological alterations have been reported in all three of these regions in mood disorders, and increased activation in the subgenual ACC (sgACC), an anatomical subdivision of the ACC, and amygdala are frequently reported in major depression (Mayberg et al. 1999; Siegle et al. 2007; Suslow et al. 2010). Within this neural network, several lines of evidence, from human postmortem brains to molecular imaging studies in live subjects, suggest that the pathophysiology of major depression may involve altered gamma-aminobutyric acid (GABA) and glutamate signaling. Specifically, disrupted information transfer and processing at GABAergic and glutamatergic synapses in major depression may occur at several points throughout the signaling process, from the movement of information (i.e., an excitatory signal) down a glutamatergic axon, to neurotransmitter release and recycling at the synapse, and to postsynaptic modulation of transferred signal by GABAergic inhibition. Here we review postmortem studies for specific cell types (pyramidal and GABA neurons, astrocytes, and oligodendrocytes) and genes, which together provide evidence for putative changes in glutamate and/or GABA structural tripartite synapses, involving presynaptic neurons, postsynaptic targets, and astrocytic support. The nature of affected genes and cellular markers further suggests that GABAergic signaling targeting distal dendrites of pyramidal neurons may be more selectively and robustly affected in major depression. Implications of altered dendritic GABAergic inhibition for corticolimbic information processing in major depression and other brain disorders sharing similar pathologies are discussed. Aspects of this chapter were presented in Sibille and French (2013b).

13.2 Altered Components of the GABA Tripartite Synapse in Major Depression

13.2.1 The GABA Structural Tripartite Synapse

Before discussing cellular findings from postmortem investigations in major depression, we briefly review the major cell components and biochemical pathways engaged in GABA homeostasis (Fig. 13.1). GABA is the principal neurotransmitter responsible for neural inhibition and is present in approximately 20 % of all neurons in the adult neocortex (Hendry et al. 1987; Sahara et al. 2012). In inhibitory neurons

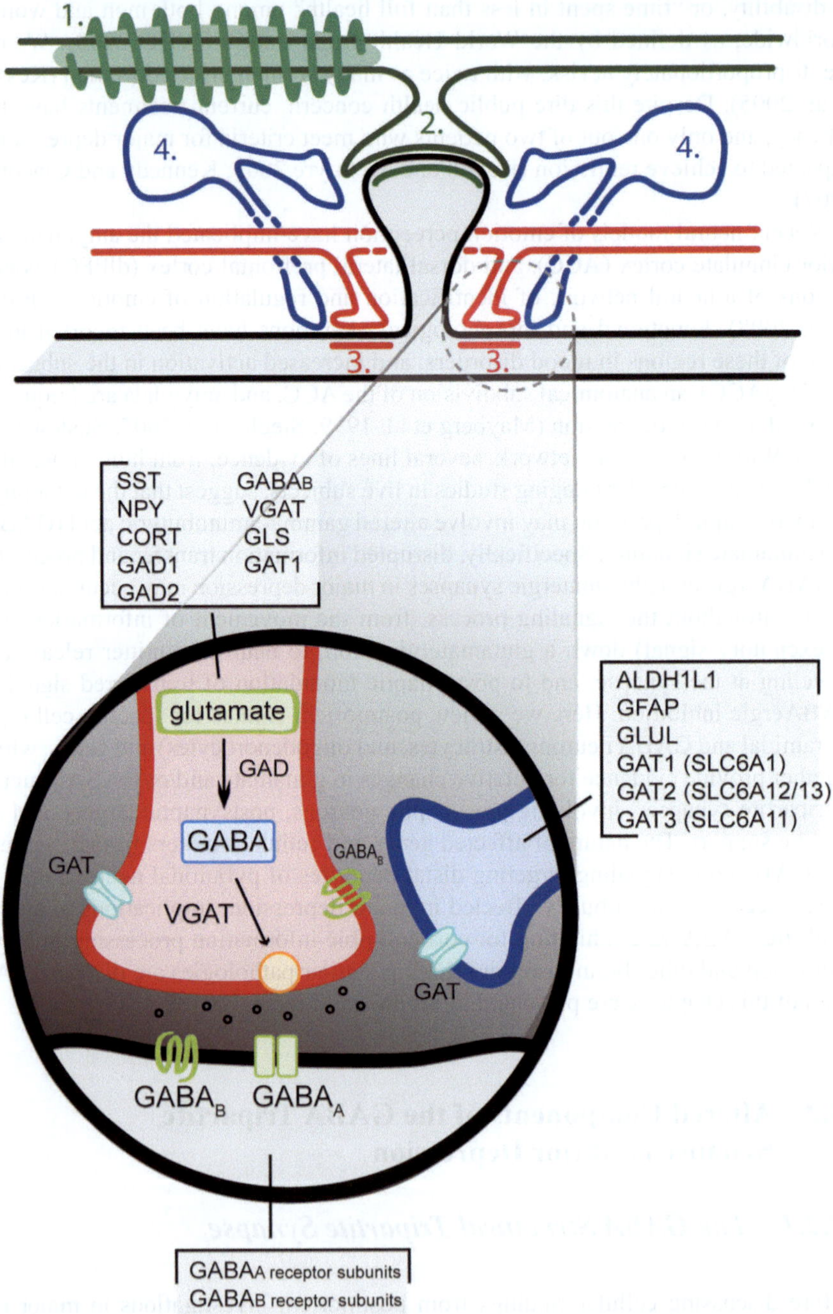

Fig. 13.1 *Gamma-aminobutyric acid (GABA) and glutamate tripartite synapses*. The *top panel* represents synapses between an excitatory axonal terminal (*top*) and a GABAergic inhibitory terminal (*middle*; *red*) onto a pyramidal dendritic spine (*bottom*; *black*). Astrocytes (*right* and

at a GABA synapse, the enzyme glutamic acid decarboxylase (GAD), found in two forms, GAD65 and GAD67 (Erlander et al. 1991), synthesizes GABA through decarboxylation of glutamate, which is then packaged into synaptic vesicles by the vesicular GABA transporter (VGAT) (Fon and Edwards 2001). Upon release of GABA from the synapse, the neurotransmitter acts at two main classes of receptors: (i) the ionotropic $GABA_A$Rs, heteropentameric ligand-gated chloride channels which mediate fast inhibitory actions of GABA, and (ii) the metabotropic $GABA_B$Rs, G-protein coupled receptors that, on a slower time scale, modulate synaptic transmission through second messenger systems. Termination of GABA transmission occurs when GABA transporters (GAT) on both GABAergic presynaptic terminals and neighboring glia remove GABA from the synaptic cleft. While neurons can recycle GABA by direct reuptake from the synaptic cleft, GABA is also metabolized in glial cells by GABA transaminase (GABA-T) to succinate, which enters the TCA cycle and is converted to glutamate. This glutamate is further converted by glutamine synthetase (GS) into glutamine, which is shuttled back to GABA neurons. In GABA neurons, glutamine released by astrocytic system N transporters (SNAT3) is taken by system A transporters (SNAT1/SNAT2) (Broer and Brookes 2001; Chaudhry et al. 2002), converted to glutamate by glutaminase (GLS), and to GABA by GAD, as reviewed in Bak et al. (2006) and Owens and Kriegstein (2002).

---

left middle; *blue*) are present at both excitatory and inhibitory synapses, and myelination of the excitatory axon by oligodendrocytes is shown (*green spiral*). In major depression, the integrity of information transfer and processing could be compromised at several compartments (numbers correspond to figure components): (1) Decreased oligodendrocyte support of axonal function leading to suboptimal conduction of action potentials along the axon; (2) Disruption of synaptic transfer of information, due to changes in the structure of pyramidal neurons and the availability of glutamate; (3) Suboptimal modulation or "fine-tuning" of excitatory postsynaptic signals onto dendritic spines due to reduced somatostatin (*SST*)-positive GABAergic dendritic targeted inhibition; and (4) Impaired astrocyte function resulting in altered extracellular neurotransmitter clearance, affecting the GABA/glutamate balance and cycling.
Molecular correlates of the GABA tripartite synapse. The *bottom panel* is a close-up on the GABA tripartite synapse. Genes whose products are associated with the presynaptic GABAergic neuron (*top; red*) are listed. SST, neuropeptide Y (*NPY*), and cortistatin (*CORT*) encode for neuropeptide markers for a specific subset of dendritic-targeting inhibitory interneurons (see Fig. 13.2). Glutamate decarboxylase 1 (*GAD1*) and *GAD2* encode the 67- and 65-kDa forms of GAD, which are responsible for synthesizing GABA from L-glutamic acid. SLC32A1 (vesicular GABA transporter, *VGAT*) encodes for an integral membrane protein that packages GABA into synaptic vesicles for release into the synaptic cleft. SLC6A1 (GABA transporter 1; *GAT1*) encodes a GABA transporter that removes GABA from the synaptic cleft. Glutaminase (*GLS*) encodes glutaminase, which generates glutamate from glutamine. $GABA_B$ receptors are also found on the presynaptic neuron. At the postsynaptic neuron (*bottom*), both $GABA_A$ and $GABA_B$ receptors are present, exhibiting various subunit combinations. For astrocytes (*blue*), both aldehyde hydrogenase 1 family, member L1 (*ALDH1L1*) and glial fibrillary acidic protein (*GFAP*) can be used as markers. Glutamate ammonia ligase (*GLUL*) encodes for the GS protein that is important for glutamate-glutamine-GABA cycling; GS catalyzes synthesis of glutamine from glutamate. ABAT encodes for the enzyme 4-aminobutyrate aminotransferase, which catabolizes GABA. SLC6A1 (*GAT1*), SLC6A13 (*GAT2*), and SLC6A11 (*GAT3*) all encode for various GABA transporters present in astrocytes. Although GAT1 is primarily considered a neuronal GABA transporter, it is present on some astrocytic processes. (Figure adapted from Sibille and French 2013b)

13.2.2 Glial Pathology in Major Depression

Glial cells are nonneuronal cells of the adult brain, which provide support and protection for neurons and are traditionally classified into three main groups: astrocytes, oligodendrocytes, and microglia. Observations of a 24% decrease in mean number of subgenual PFC glial cells in patients with familial major depression provided early evidence for glial cell changes in depression (Ongur et al. 1998). Reduced glial cell density was also observed in the dlPFC and ACC in depression (Rajkowska et al. 1999; Cotter et al. 2001), alongside reports of low glial numbers in the amygdala (Bowley et al. 2002). Although there are negative reports citing no changes in glial cells in orbital frontal cortex (OFC) and dlPFC in late-life depression (Khundakar et al. 2009, 2011), the majority of studies point towards reductions in glial density and number in subjects with major depression. Technological challenges associated with isolating and identifying homogenous cell types partially account for the absence of cell-specific findings in the earliest reports, but more recent studies do distinguish between the glial classes. Evidence suggesting astrocyte- and oligodendrocyte-specific pathologies in major depression is discussed here.

13.2.3 Reduced Oligodendrocyte Structure and Numbers in Major Depression

Oligodendrocytes ensheathe neurons with myelin and provide critical axonal insulation to facilitate the conduction of electrical impulses and enable saltatory conduction, together ensuring integrity of information transfer along axons. Using morphological cell-type determination, it has been suggested that decreased glial cell numbers in the amygdala and PFC may be attributed to a specific reduction in oligodendrocytes (Hamidi et al. 2004; Uranova et al. 2004). Decreased oligodendrocytes have also been reported by flow cytometry, using fluorescently labeled suspended nuclei from the frontopolar cortex in major depression (Hayashi et al. 2011). Although negative findings were reported in the ACC (Sibille et al. 2009) and amygdala (Guilloux et al. 2012) in female subjects, a gene array study in the amygdala of male subjects with major depression showed reduced expression of numerous genes related to oligodendrocyte structure and function (Sibille et al. 2009), consistent with reduced transcripts in the parietal and prefrontal cortices in other studies (Klempan et al. 2009; Kim and Webster 2010). Pathological oligodendrocyte function may result in impaired communication and altered integrity of neuronal information transfer in major depression (Edgar and Sibille 2012). This hypothesis is grounded in the basis of the role of oligodendrocytes in facilitating axonal conduction (Brodal 2010), and of specific dysregulation of genes coding for proteins located at the nodes of Ranvier (Sibille et al. 2009), the site of functional interaction between oligodendrocytes and neurons. Hence, altered conduction and/or maintenance of axonal signaling integrity through altered oligodendrocyte structure and function may represent in some cases an early deregulated component (i.e., incoming information) contributing to altered information flow in major depression (*green* spindles in Fig. 13.1).

13.2.4 Astrocyte-Related Findings in Major Depression

Divided into two primary categories, protoplasmic (found in gray matter) and fibrous (found in white matter), astrocytes are principally responsible for regulating the neuronal chemical environment, and play an important role in extracellular clearance and recycling of neurotransmitters, including glutamate and GABA, as discussed above (Brodal 2010). Controversial evidence also suggests a more active role for astrocytes in neurotransmission than previously considered, as they may release gliotransmitters, including glutamate, when depolarized (Araque et al. 1999). However, recent findings that G_q-coupled metabotropic receptors mediating calcium influx (i.e., mGluR5) may not be expressed by adult human cortical astrocytes (Sun et al. 2013), indicate that the role of depolarization and calcium-triggered neurotransmission of astrocytes may need to be revisited. Astrocyte-specific investigations of glial pathology in major depression suggest cellular hypertrophy in ACC white matter (Torres-Platas et al. 2011). In addition, decreases in GFAP, a commonly used astrocytic marker, and in glutamate clearance transporters (EAAT1, EAAT2) expressed in astrocytes have been observed in the PFC of subjects with major depression (Miguel-Hidalgo et al. 2000; Si et al. 2004; Choudary et al. 2005; Miguel-Hidalgo et al. 2010). In one study, GFAP levels between subjects with major depression and controls under 60 years of age showed a tenfold reduction; in contrast, no difference was seen between depressed subjects and controls over 60 years of age (Si et al. 2004). Astrocyte connexins 43 and 30, which mediate gap junction-based direct communication between astrocytes, and also participate in astrocyte-oligodendrocyte junctions (Orthmann-Murphy et al. 2008), were observed to have decreased expression in suicide victims with a range of psychiatric diagnoses, including bipolar disorder, schizophrenia, and major depression (Ernst et al. 2011). Together, the current model for altered glutamate and GABA neurotransmission in major depression includes the contribution of dysregulated astrocytic neurotransmitter recycling and homeostasis (Valentine and Sanacora 2009; Oh et al. 2012) (*blue* cells in Fig. 13.1).

13.2.5 Altered Cortical Neuronal Structure and Function in Major Depression

Several reviews have been published on the topic of morphological and cellular changes in the context of depression (Hercher et al. 2009; Rajkowska 2003; Rajkowska and Miguel-Hidalgo 2007). Using morphological techniques to investigate changes in cell size and/or number, Rajkowska et al. have reported reduced density of neuronal cell bodies with large cell body size in cortical layers 2 through 5 of the OFC and in layers 2, 3, and 6 of the dlPFC in subjects with major depression (Rajkowska et al. 1999; Rajkowska 2000). These findings were concurrent with increased density of neurons with small body size in layer 3 (OFC) and layers 3 and 6 (dlPFC). Decreased mean neuronal cell body size was also reported in layers 3 and 6 (dlPFC), layers 2 and 3 (OFC), and layer 6 (ACC) (Rajkowska et al. 1999; Cotter et al. 2001). Reduced neuronal size in layer 6 of the dlPFC (Cotter et al. 2002b) and lower combined neuron density in both dorsal and frontal PFC (Underwood et al. 2012) were confirmed in later studies. Although post hoc comparisons between controls and depressed subjects were not significant, one study found the lowest density of pyramidal neurons in subjects with major depression compared to two other psychiatric disorders (bipolar disorder, schizophrenia) (Law and Harrison 2003). Interestingly, in elderly depressed subjects, reduced pyramidal density in cortical layers 3 and 5 of the OFC (Rajkowska et al. 2005), but not the dlPFC (Van Otterloo et al. 2009) was reported, potentially reflecting a separate etiology for late-life depression in elderly patients. More recently, a reduction in pyramidal neuron density was identified in layer 5b of the ACC in a cohort of primarily older mood disorder subjects, of which five subjects were diagnosed with major depression and two of whom suffered from bipolar disorder (Gittins and Harrison 2011).

Despite earlier speculation that the reductions in neuronal density observed in dlPFC involved glutamatergic pyramidal neurons (due to location of the reductions in pyramidal layers), no difference in packing density of pyramidal neurons labeled using NF200, a neurofilament protein marker, was observed between depressed and control subjects (Miguel-Hidalgo et al. 2005). While the authors concluded that this presented lack of evidence for neurofilament-related cytoskeleton deficiencies in NF200 immunoreactive neurons in PFC circuitry in depression, it should be noted that an earlier study using a variation of the same antibody (N200), found that even with the three antibodies used to label subpopulations of pyramidal cells (FNPY, SMI32, N200), at least half of all pyramidal neurons remained unlabeled (Law and Harrison 2003).

Although it is unlikely that neuronal loss underlies these changes, it remains to be resolved whether the decreased neuronal density reflects changes in neuropil or dendritic complexity. Supporting the hypothesis of reduced dendritic complexity in depression, a decrease in total dendritic length and somal areas was observed in deep and superficial layer 3 in a cohort enriched in patients with major depression (Yung et al. 2000). Reduced numbers of third-order basilar dendritic branches were also reported in ACC layer 6 of depressed suicide victims using Golgi staining of neuronal processes (Hercher et al. 2010). All together, despite evidence for reduced density and dendritic length of pyramidal neurons, findings of potential pathological changes in pyramidal cells in major depression are often subtle, depend on the age of the subjects, are regionally specific, mixed, and are overall in need of further confirmation in large cohorts.

13.2.6 Glutamate Levels in Major Depression

In support of functional changes related to glutamatergic signaling in major depression, investigators have reported altered in vivo levels of glutamate and glutamate-related metabolites in subjects with major depression using proton magnetic resonance spectroscopy (^{1}H MRS). These studies frequently report their findings using the term Glx, a measurement which primarily reflects glutamate and glutamine, but does contain small contributions from other metabolites, including GABA (Valentine and Sanacora 2009; Maddock and Buonocore 2012). Moreover, it is important to keep in perspective that MRS findings do not reflect what is occurring at the level of neurotransmission, as only a small part of these measurements reflect synaptic levels and the majority of metabolites measured with ^{1}H MRS are intracellular (Sanacora et al. 2012). Nevertheless, diagnostic differences reported in these studies suggest changes in neurotransmitter cycling and clearance, which may affect GABA/glutamate homeostasis. As reviewed in Yuksel and Ongur (2010), MRS studies have shown reduced Glx and/or glutamate concentrations across multiple cortical and subcortical brain regions including ACC (Auer et al. 2000; Pfleiderer et al. 2003; Zhang et al. 2012), PFC (Michael et al. 2003a; Hasler et al. 2007), amygdala (Michael et al. 2003b), and hippocampus (Block et al. 2009). However, no differences in hippocampus (Milne et al. 2009) and occipital cortex (Price et al. 2009) and even increased glutamate levels have also been reported in the occipital cortex (Sanacora et al. 2004), suggesting region-specific pathological effects, and opposing findings in frontal cortex and cingulate regions (reduced glutamate) versus occipital and parietal/occipital regions (increased glutamate) (Sanacora et al. 2012). Notably, both glutamate (Zhang et al. 2012) and Glx (Pfleiderer et al. 2003; Michael et al. 2003a, b) have been shown to increase with electroconvulsive treatment.

13.2.7 Low GABA Levels and Reduced Markers of Dendritic-Targeting GABA Neurons in Major Depression

GABA neurons can be divided into subpopulations not only based on diverse morphology, but also on the calcium-binding proteins or neuropeptides that they express (Fig. 13.2). GABA neurons expressing SST, NPY, and CORT are known to target and inhibit the distal dendrites of pyramidal neurons; whereas interneurons expressing parvalbumin (PV) and cholecystokinin (CCK) target the cell body and axon initial segment, and calretinin (CR)-expressing neurons target other GABA neurons. Reduced density of GABA neurons immunoreactive for specific calcium-binding proteins has been reported in the dlPFC in major depression (Rajkowska et al. 2007), but see also Beasley et al. (2002) and Cotter et al. (2002a) for negative findings in the dlPFC and ACC. In Rajkowska et al. (2007), the density of calbindin (CB)-positive neurons was reduced by 50 % in dlPFC, and no differences in PV-positive neurons were observed. Reductions in the density of CB-positive neurons were also reported in the occipital cortex (Maciag et al. 2010).

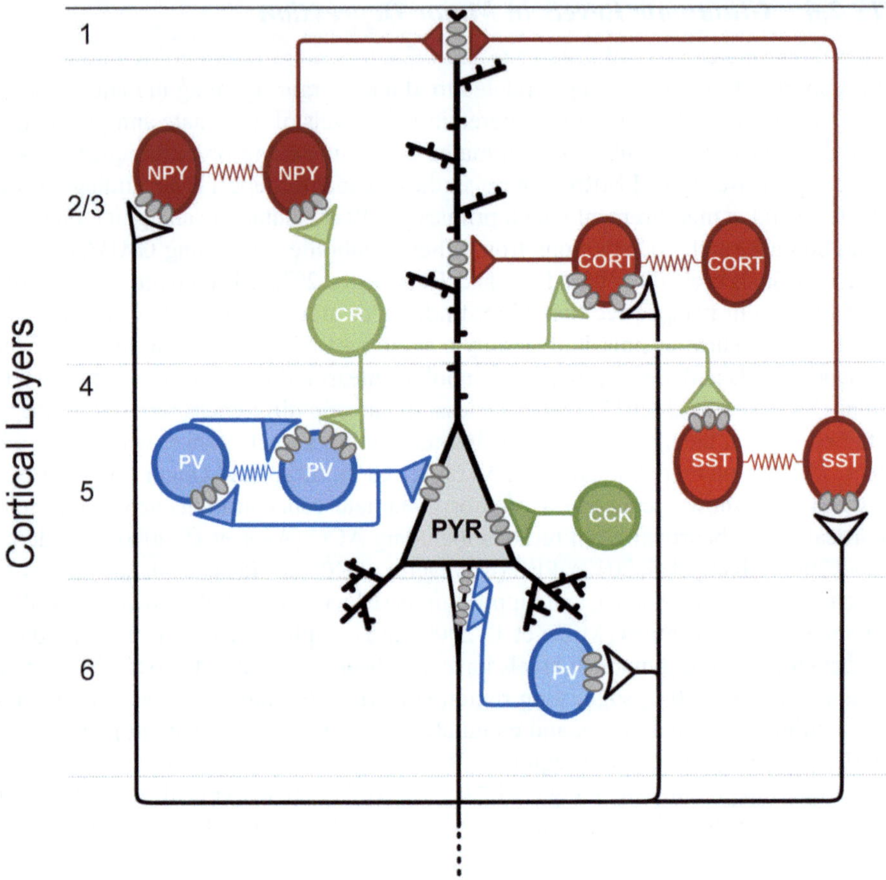

Fig. 13.2 GABA microcircuitry. GABA neurons expressing somatostatin (*SST*), neuropeptide Y (*NPY*), and cortistatin (*CORT*) target and inhibit the distal dendrites of pyramidal neurons (PYR); whereas interneurons expressing parvalbumin (*PV*) and cholecystokinin (*CCK*) target the cell body and axon initial segment, and calretinin (*CR*)-expressing neurons target other GABA neurons. Evidence from human postmortem studies suggest that the subsets of GABA neurons that selectively target the dendrites of pyramidal neurons (i.e., SST-, NPY-, and CORT-positive) are affected in major depression, while evidence for changes in other GABA neuron subtypes are fewer (i.e., PV) or mostly (i.e., CCK, CR) negative. (Figure adapted from Sibille 2013a)

Recently, our group has reported reduced SST, a modulatory neuropeptide, in the dlPFC (Sibille et al. 2011), subgenual ACC (Tripp et al. 2011; Tripp et al. 2012), and amygdala (Guilloux et al. 2012) of subjects with major depression. These findings are consistent with previous studies, as SST is mostly expressed in CB-positive cells in the cortex (Martel et al. 2012). Intriguingly, based on their common target of distal dendrites of excitatory pyramidal neurons, *NPY* and *CORT* expression was also lower in the sgACC and amygdala in subjects with major depression (Guilloux et al. 2012; Tripp et al. 2012). In contrast, *CCK* and *CR* were unaffected in the

ACC and amygdala, and *PV* expression was lower in the ACC, but not the dlPFC (Sibille et al. 2011; Tripp et al. 2012). Consistent with an elevated female vulnerability for major depression, analysis of these GABA interneuron markers stratified by sex revealed more robust downregulation of SST in female subjects with major depression in the subgenual ACC compared to males patients (Tripp et al. 2011, 2012). Additionally, *SST* was reduced in the amygdala of females (Guilloux et al. 2012), but not males with major depression (Sibille et al. 2009). Interestingly, control female subjects (i.e., not depressed) already displayed lower expression of *SST*, *CORT*, and *NPY*, compared to male control subjects, suggesting that baseline expression of these genes is already close to the low levels observed in depressed subjects in female subjects. Reduced levels of GAD67, an enzyme responsible for synthesis of GABA from glutamate, have not been consistently reported, but were observed in some studies, including at the protein level in the dlPFC (Karolewicz et al. 2010), and at the mRNA levels coding for both *GAD65* and *GAD67* in the sgACC (Tripp et al. 2012). Adding another layer of complexity, expression profiles of *SST*, *NPY*, and *CORT* also decrease with age in multiple brain regions (Erraji-Benchekroun et al. 2005). For instance, *SST* levels are approximately 40–50 % lower at age 70 compared to age 20 (Erraji-Benchekroun et al. 2005; Tripp et al. 2011); the age-related reduction in *SST* is present in both controls and subjects with major depression, with depressed subjects consistently exhibiting lower levels across all ages (Sibille et al. 2011; Tripp et al. 2011).

Notably, these cellular findings are consistent with reports of decreased GABA concentration in major depression, as observed by ^{1}H MRS or by transcranial magnetic stimulation occipital and frontal cortices (Hasler et al. 2007; Levinson et al. 2010; Gabbay et al. 2012; Hasler and Northoff 2011). Selective serotonin reuptake inhibitors or electroconvulsive therapy reverse these changes (Sanacora et al. 2002, 2003). It has been suggested that the concentration of GABA in the ACC mediates default-mode network negative responses during emotion processing by studies that combine functional imaging and resting-state MRS (Northoff et al. 2007), and interestingly, reduced GABA levels in the ACC correlate with measures of anhedonia across depressed and control adolescents (Gabbay et al. 2012). These data provide brain-based evidence in human subjects with major depression for the GABA hypothesis of emotion dysregulation in depression (Brambilla et al. 2003; Luscher et al. 2011), originally proposed in 1980 as a broader GABAergic dysfunction hypothesis of affective disorders, based on the efficacy of sodium valproate, a GABAergic anticonvulsant, in treatment of mania (Emrich et al. 1980). This hypothesis was supported by reports of low GABA levels in the plasma and cerebrospinal fluid of depressed subjects (Gold et al. 1980; Petty and Schlesser 1981; Petty and Sherman 1984; Gerner and Hare 1981), and later by the association between GABAergic transmission and control of stress, reviewed in Luscher et al. (2011), the effect of monoaminergic antidepressants on GABAergic transmission (Sanacora et al. 2002), and genetic manipulation studies in rodents (Earnheart et al. 2007).

The combined results from these studies provide supporting evidence for reduced GABA levels and for selective cellular changes potentially affecting neuropeptide- and/or GABA-related functions of the CB/SST-positive interneuron subtype (red cells in Fig. 13.1), which in some studies paralleled the heightened female vulnerability to suffer from depressive episodes. Together, these converging results suggest that GABA neurons regulating dendritic inhibition may represent a "weak link" within the biological module formed by the GABA tripartite synapse, which is frequently affected in major depression and potentially moderated by age and sex.

13.2.8 GABA Receptors in Major Depression and Other Mood Disorders

Deficits in $GABA_A$R-binding sites have been implicated by studies of benzodiazepine receptor binding sites in various anxiety disorders, such as generalized anxiety disorder (Tiihonen et al. 1997) and posttraumatic stress disorder (Bremner et al. 2000). An absence of alterations in benzodiazepine receptor binding was found in depressed subjects (Kugaya et al. 2003); although microarray expression profiling and analysis of gene expression by quantitative polymerase chain reaction (qPCR) have reported altered expression and subunit composition of specific $GABA_A$R subunits in depressed suicides and in major depression in various brain regions, including among others, multiple areas of the frontal and motor cortices and inferior temporal gyrus (Merali et al. 2004; Sequeira et al. 2009, 2007; Aston et al. 2005; Choudary et al. 2005; Klempan et al. 2009). In major depression, decreased expression of the β2 and δ subunits of the $GABA_A$ receptor has been reported by microarray analysis in the middle temporal area (Brodmann area (BA) 21) (Aston et al. 2005) and dlPFC (BA9/46) (Choudary et al. 2005), respectively, along with increased expression of the β3 and γ2 subunits in the dlPFC (BA9/46) (Choudary et al. 2005). Reports of dysregulation in a number of $GABA_A$ receptors by microarray analysis of the ACC and dlPFC in subjects with major depression were made in conjunction with changes in glutamate receptor subunits, and lowered expression of *GS* and glial glutamate transporters (*EAAT1*, *EAAT2*) (Choudary et al. 2005). In depressed suicides, decreased expression of the α1, α3, and α4 subunits were found by either qPCR or microarray in BA8, BA9, BA10, and BA24 (Merali et al. 2004; Klempan et al. 2009; Sequeira et al. 2007); β1 was reported up in BA24 (Sequeira et al. 2007, 2009), but down in BA46; β3 was reported increased in BA 6, 10, and 38 (Sequeira et al. 2009); δ was up in BA6, BA44, and BA46 (Sequeira et al. 2009; Klempan et al. 2009); decreased expression of γ1 was found for BA 10, 21, and 46 (Merali et al. 2004; Sequeira et al. 2009; Klempan et al. 2009); γ2 was found increased in BA 20 (Sequeira et al. 2009; Klempan et al. 2009); and finally, decreases in ρ1 expression were found for BA21 and BA44 (Sequeira et al. 2009; Klempan et al. 2009); as reviewed in (Luscher et al. 2011). Upregulation in expression of the α5 subunit of the $GABA_A$ receptor, which is selective to dendritic branches, was reported in bipolar disorder. Although one study reported an upregulation of the α5 subunit in BA46 in major depression (Sequeira et al. 2009), these changes were not evident in other studies, or elevated levels were restricted to depressed suicides compared to suicide victims with no lifetime history of major depression (Klempan et al. 2009; Choudary et al. 2005). Alterations in $GABA_B$ receptor subunits have also been implicated in psychiatric disorders. For instance, upregulation in expression of *GABBR1* ($GABA_BR1$) has been reported for bipolar disorder (Choudary et al. 2005), and up-regulation in expression of *GABBR2 (*$GABA_BR2$) has been reported for depressed suicides (Klempan et al. 2009).

Altogether, reports of changes in the levels of GABA receptors in mood disorders are complex, differ depending on the brain region investigated, and are not consistently replicated across studies. This may reflect variable attempts to adapt and/or compensate to deregulations in GABA signaling and local circuits, and the limitations of postmortem studies to capture events that occur on time frame of hours, as noted for adaptive changes in GABA receptors (Jacob et al. 2008). Further studies are needed to characterize the role of changes in GABA receptors, including of dynamic changes over time and more systematic investigation in cohorts with equal representation of male and female subjects with major depression.

13.3 Implications of Altered GABA Function in Major Depression

13.3.1 Summary of Postmortem Findings

Molecular and cellular evidence from postmortem studies, combined with imaging data, suggest alterations in several components of the local cell circuitry in major depression, which may affect GABA and glutamate homeostasis, including changes to the structure and function of glutamatergic neurons, dendritic-targeting GABAergic neurons, astrocytes, and oligodendrocytes. Based on the findings summarized here, a speculative set of events contributing to dysregulated information processing and transfer in depression may occur in corticolimbic circuits as follows (Fig. 13.1): first, suboptimal conduction of action potentials along the axon could be caused by decreased oligodendrocyte support, leading to decreased integrity of information input or output. Second, changes in pyramidal neuron structure and in the availability of glutamate could disrupt the synaptic transfer of information. Third, reduced inhibition by dendritic targeting SST-positive GABA interneurons may lead to suboptimal modulation of excitatory postsynaptic signals onto dendritic spines. Finally, impaired astrocyte function may cause altered extracellular neurotransmitter clearance and recycling, which in turn may lead to an imbalance in GABA and glutamate homeostasis within their respective tripartite synapses. Although the situation in the diseased condition is more complex than proposed here, disrupted information transfer may result from pathologies occurring in any of these components. The glutamate component of this model is discussed elsewhere in this book, and this chapter has focused primarily on evidence suggesting a robust deregulated GABA component, specifically affecting dendritic inhibition.

13.3.2 *GABA-Related Dendritic Inhibition, as a Vulnerable Biological Component of the Local Cell Circuitry in Major Depression; Continuum with Other Brain Disorders and Implications for Information Processing*

The fact that the above-mentioned GABA-related findings were identified in corticolimbic brain regions suggests that these local circuit changes may affect the function of several nodes within a critical neural network of mood regulation, potentially resulting in altered sensing and processing of emotionally salient information. Current models of emotion regulation identify the amygdala as critical for sensing and assessing emotionally-salient stimuli, the dlPFC as one of the regions responsible for top-down cognitive assessment of emotional salience, and the ACC as the site of integration of information between bottom-up amygdala information and top-down dlPFC control, together providing cognitive control over emotional and motivational states (Phillips et al. 2008). Reduced GABA-mediated inhibition may contribute to increased activity in respective brain areas, including increased amygdala and/or sgACC activities, as frequently reported in major depression (Surguladze et al. 2005; Fu et al. 2004). So, restoring GABA-mediated dendritic inhibitory function may reduce pyramidal cell activation and excitatory tone, contributing to reduced ACC activation with positive response to therapeutic intervention such as deep brain stimulation, for instance (Mayberg et al. 2005).

At the local circuit level, the converging evidence points towards selective deficits in the subtype of GABA neurons that targets the dendrites of excitatory glutamatergic neurons, whereas evidence for reduced markers of other GABA neuron subtypes are sparse or negative. On the other hand, cortical inhibitory deficits are frequent in other neuropsychiatric disorders, and alterations in SST levels have also been identified in schizophrenia (Morris et al. 2008), bipolar disorder (Konradi et al. 2004, 2010), and in Huntington's (Timmers et al. 1996), Alzheimer's (Davies et al. 1980; Epelbaum et al. 2009), and Parkinson's diseases (Epelbaum et al. 1989). This could suggest the presence of intrinsic vulnerability factors within SST and related GABA neurons, and that common biological insults may similarly affect this cell population across several brain disorders. This dimensional perspective on disease pathology is more consistent with biological principles than with the categorical definition of psychiatric syndromes, although the functional output of similar pathologies may vary based on the biological context. For instance, the downstream effects of reduced dendritic inhibition may depend on the presence of additional pathological entities, such as robust downregulation of markers of soma-targeting GABA neurons, i.e., PV-positive, in schizophrenia (Lewis and Sweet 2009), or the presence of neurodegenerative processes in other neurological disorders. These more complex inhibitory deficits, compared to evidence of more focused reductions in markers of GABA neurons mediating dendritic inhibition in major depression, may complicate the interpretation across disorders. Consequently, while the GABA tripartite synapse may represent a vulnerable biological module in the brain and accordingly across brain disorders, it may not lead to a unique behavioral endophenotype across these disorders. In major depression, the observation that reduced SST levels were more robust in female patients provides an interesting parallel with the increased female vulnerability to depression, although this putative causal link will need to be tested directly, potentially through the use of rodent models.

Etiological pathways leading to reduced markers of dendritic inhibition may also vary across disorders. Brain-derived neurotrophic factor (BDNF), a necessary trophic factor for SST and NPY expression (Glorioso et al. 2006), shows reduced expression in neuropsychiatric disorders, such as depression, schizophrenia, and bipolar disorder (Lu and Martinowich 2008; Rakofsky et al. 2012). Altered BDNF signaling in major depression, as evidenced by reduced expression in amygdala (Guilloux et al. 2012) and reduced receptor (*TRKB*) expression in the sgACC (Tripp et al. 2012), is expected to impact *SST* expression. SST cells may also be particularly vulnerable to oxidative stress due to their expression of neuronal nitric oxide synthase (Jaglin et al. 2012). In schizophrenia, reduced SST levels occur more systematically in the context of reduced PV, suggestive of a role for etiological pathways involving early developmental disturbances, such as deficits in transcription factors known to regulate the ontogeny of both neuron subtypes (Volk et al. 2012). It is also interesting to note that a recent report in the rodent cortex revealed that a small subset of SST-positive GABA neurons differs from the traditional dendritic targeting model, and in fact disinhibits layer 4 neurons, through the targeted inhibition of PV-positive GABA neurons in that layer (Xu et al. 2013). If confirmed in other cortical areas and across species, the implication of this finding could be of opposing effects of low SST-positive GABA neuron function on excitatory neurons (i.e., less inhibition in layers 2 and 5, less disinhibition in layer 4). This would also indicate that an even downregulation across cortical layers in major depression (Tripp and Sibille, unpublished report) may reflect the presence of common upstream causal factors, rather than compensatory mechanisms to maintain local circuit homeostasis across cortical layers which could take complex patterns, together consistent with a model of SST-positive GABA neuron subtype intrinsic vulnerabilities.

13.4 Conclusion, Future Directions

How does the framework of the tripartite synapse, informed by a robust identification of reduced GABA-mediated inhibition in major depression, enable us to move forward in uncovering the pathophysiology of major depression? The evidence reviewed above suggests that a dimensional approach considering the biological modules of the glutamate and GABA tripartite synapses (Fig. 13.1) and their subcellular targeted components (Fig. 13.2; e.g., dendritic versus perisomatic) may contribute to the identification of biological vulnerabilities (i.e., weak links) in major depression that will also have implications across several brain-related disorders. Further molecular characterization of the primary cellular pathology in the human postmortem brain of patients and control subjects is needed; for example, by using a combination of laser capture microscopy with gene expression panels and targeted proteomic approaches for groups of genes and gene products related to the glutamate, GABA, and astrocyte components of local circuit modules (Fig. 13.1). Imaging genetics for panels of genetic variants corresponding to these local circuit components may serve to bridge those basic cellular and gene studies with functional outcomes to systematically assess specificities and/or continuum between putative cellular pathologies and symptom dimensions in major depression and across other categorically defined neuropsychiatric disorders.

Acknowledgements The authors have no disclosure. This work was supported by National Institute of Mental Health (NIMH) MH084060 (ES) and MH077159 (ES) grants. The funding agency had no role in the study design, data collection and analysis, decision to publish, and preparation of the manuscript. The content is solely the responsibility of the authors and does not necessarily represent the official views of the NIMH or the National Institutes of Health.

References

American Psychiatric Association. Diagnostic and statistical manual of mental disorders: DSM-IV-TR. 4th ed. Washington, DC:American Psychiatric Association; 2000.

Araque A, Parpura V, Sanzgiri RP, Haydon PG. Tripartite synapses: glia, the unacknowledged partner. Trends Neurosci. 1999;22(5):208–15.

Aston C, Jiang L, Sokolov BP. Transcriptional profiling reveals evidence for signaling and oligodendroglial abnormalities in the temporal cortex from patients with major depressive disorder. Mol Psychiatry. 2005;10(3):309–22.

Auer DP, Putz B, Kraft E, Lipinski B, Schill J, Holsboer F. Reduced glutamate in the anterior cingulate cortex in depression: an in vivo proton magnetic resonance spectroscopy study. Biol Psychiatry. 2000;47(4):305–13.

Bak LK, Schousboe A, Waagepetersen HS. The glutamate/GABA-glutamine cycle: aspects of transport, neurotransmitter homeostasis and ammonia transfer. J Neurochem. 2006;98(3):641–53.

Beasley CL, Zhang ZJ, Patten I, Reynolds GP. Selective deficits in prefrontal cortical GABAergic neurons in schizophrenia defined by the presence of calcium-binding proteins. Biol Psychiatry. 2002;52(7):708–15.

Block W, Traber F, von Widdern O, Metten M, Schild H, Maier W, Zobel A, Jessen F. Proton MR spectroscopy of the hippocampus at 3 T in patients with unipolar major depressive disorder: correlates and predictors of treatment response. Int J Neuropsychopharmacol. 2009;12(3):415–22.

Bowley MP, Drevets WC, Ongur D, Price JL. Low glial numbers in the amygdala in major depressive disorder. Biol Psychiatry. 2002;52(5):404–12.

Brambilla P, Perez J, Barale F, Schettini G, Soares JC. GABAergic dysfunction in mood disorders. Mol Psychiatry. 2003;8(8):721–37, 715.

Bremner JD, Innis RB, Southwick SM, Staib L, Zoghbi S, Charney DS. Decreased benzodiazepine receptor binding in prefrontal cortex in combat-related posttraumatic stress disorder. Am J Psychiatry. 2000;157(7):1120–6.

Brodal P. The central nervous system: structure and function. 4th ed. New York: Oxford University Press; 2010.

Broer S, Brookes N. Transfer of glutamine between astrocytes and neurons. J Neurochem. 2001;77(3):705–19.

Chaudhry FA, Reimer RJ, Edwards RH. The glutamine commute: take the N line and transfer to the A. J Cell Biol. 2002;157(3):349–55.

Choudary PV, Molnar M, Evans SJ, Tomita H, Li JZ, Vawter MP, Myers RM, Bunney WE Jr, Akil H, Watson SJ, Jones EG. Altered cortical glutamatergic and GABAergic signal transmission with glial involvement in depression. Proc Natl Acad Sci U S A. 2005;102(43):15653–8.

Cotter D, Mackay D, Landau S, Kerwin R, Everall I. Reduced glial cell density and neuronal size in the anterior cingulate cortex in major depressive disorder. Arch Gen Psychiatry. 2001;58(6):545–53.

Cotter D, Landau S, Beasley C, Stevenson R, Chana G, MacMillan L, Everall I. The density and spatial distribution of GABAergic neurons, labelled using calcium binding proteins, in the anterior cingulate cortex in major depressive disorder, bipolar disorder, and schizophrenia. Biol Psychiatry. 2002a;51(5):377–86.

Cotter D, Mackay D, Chana G, Beasley C, Landau S, Everall IP. Reduced neuronal size and glial cell density in area 9 of the dorsolateral prefrontal cortex in subjects with major depressive disorder. Cereb Cortex. 2002b;12(4):386–94.

Davies P, Katzman R, Terry RD. Reduced somatostatin-like immunoreactivity in cerebral cortex from cases of Alzheimer disease and Alzheimer senile dementa. Nature. 1980;288(5788):279–80.

Earnheart JC, Schweizer C, Crestani F, Iwasato T, Itohara S, Mohler H, Luscher B. GABAergic control of adult hippocampal neurogenesis in relation to behavior indicative of trait anxiety and depression states. J Neurosci. 2007;27(14):3845–54.

Edgar N, Sibille E. A putative functional role for oligodendrocytes in mood regulation. Transl Psychiatry. 2012;2:e109.

Emrich HM, von Zerssen D, Kissling W, Moller HJ, Windorfer A. Effect of sodium valproate on mania. The GABA-hypothesis of affective disorders. Arch Psychiatr Nervenkr. 1980;229(1):1–16.

Epelbaum J, Agid F, Agid Y, Beaudet A, Bertrand P, Enjalbert A, Heidet V, Kordon C, Krantic S, Leonard JF, et al. Somatostatin receptors in brain and pituitary. Horm Res. 1989;31(1-2):45–50.

Epelbaum J, Guillou JL, Gastambide F, Hoyer D, Duron E, Viollet C. Somatostatin, Alzheimer's disease and cognition: an old story coming of age? Prog Neurobiol. 2009;89(2):153–61.

Erlander MG, Tillakaratne NJ, Feldblum S, Patel N, Tobin AJ. Two genes encode distinct glutamate decarboxylases. Neuron. 1991;7(1):91–100.

Ernst C, Nagy C, Kim S, Yang JP, Deng X, Hellstrom IC, Choi KH, Gershenfeld H, Meaney MJ, Turecki G. Dysfunction of astrocyte connexins 30 and 43 in dorsal lateral prefrontal cortex of suicide completers. Biol Psychiatry. 2011;70(4):312–9.

Erraji-Benchekroun L, Underwood MD, Arango V, Galfalvy H, Pavlidis P, Smyrniotopoulos P, Mann JJ, Sibille E. Molecular aging in human prefrontal cortex is selective and continuous throughout adult life. Biol Psychiatry. 2005;57(5):549–58.

Fon EA, Edwards RH. Molecular mechanisms of neurotransmitter release. Muscle Nerve. 2001;24(5):581–601.

Fu CH, Williams SC, Cleare AJ, Brammer MJ, Walsh ND, Kim J, Andrew CM, Pich EM, Williams PM, Reed LJ, Mitterschiffthaler MT, Suckling J, Bullmore ET. Attenuation of the neural response to sad faces in major depression by antidepressant treatment: a prospective, event-related functional magnetic resonance imaging study. Arch Gen Psychiatry. 2004;61(9):877–89.

Gabbay V, Mao X, Klein RG, Ely BA, Babb JS, Panzer AM, Alonso CM, Shungu DC. Anterior cingulate cortex gamma-aminobutyric acid in depressed adolescents: relationship to anhedonia. Arch Gen Psychiatry. 2012;69(2):139–49.

Gerner RH, Hare TA. CSF GABA in normal subjects and patients with depression, schizophrenia, mania, and anorexia nervosa. Am J Psychiatry. 1981;138(8):1098–101.

Gittins RA, Harrison PJ. A morphometric study of glia and neurons in the anterior cingulate cortex in mood disorder. J Affect Disord. 2011;133(1-2):328–32.

Glorioso C, Sabatini M, Unger T, Hashimoto T, Monteggia LM, Lewis DA, Mirnics K. Specificity and timing of neocortical transcriptome changes in response to BDNF gene ablation during embryogenesis or adulthood. Mol Psychiatry. 2006;11(7):633–48.

Gold BI, Bowers MB, Roth RH, Sweeney DW. Gaba levels in CSF of patients with psychiatric-disorders. Am J Psychiat. 1980;137(3):362–4.

Guilloux JP, Douillard-Guilloux G, Kota R, Wang X, Gardier AM, Martinowich K, Tseng GC, Lewis DA, Sibille E. Molecular evidence for BDNF- and GABA-related dysfunctions in the amygdala of female subjects with major depression. Mol Psychiatry. 2012;17(11):1130–42.

Hamidi M, Drevets WC, Price JL. Glial reduction in amygdala in major depressive disorder is due to oligodendrocytes. Biol Psychiatry. 2004;55(6):563–9.

Hasler G, Northoff G. Discovering imaging endophenotypes for major depression. Mol Psychiatry. 2011;16(6):604–19.

Hasler G, van der Veen JW, Tumonis T, Meyers N, Shen J, Drevets WC. Reduced prefrontal glutamate/glutamine and gamma-aminobutyric acid levels in major depression determined using proton magnetic resonance spectroscopy. Arch Gen Psychiatry. 2007;64(2):193–200.

Hayashi Y, Nihonmatsu-Kikuchi N, Yu X, Ishimoto K, Hisanaga SI, Tatebayashi Y. A novel, rapid, quantitative cell-counting method reveals oligodendroglial reduction in the frontopolar cortex in major depressive disorder. Mol Psychiatry. 2011;16(12):1155–8.

Hendry SH, Schwark HD, Jones EG, Yan J. Numbers and proportions of GABA-immunoreactive neurons in different areas of monkey cerebral cortex. J Neurosci. 1987;7(5):1503–19.

Hercher C, Turecki G, Mechawar N. Through the looking glass: examining neuroanatomical evidence for cellular alterations in major depression. J Psychiatr Res. 2009;43(11):947–61.

Hercher C, Canetti L, Turecki G, Mechawar N. Anterior cingulate pyramidal neurons display altered dendritic branching in depressed suicides. J Psychiatr Res. 2010;44(5):286–93.

Huynh NN, McIntyre RS. What Are the Implications of the STAR*D Trial for Primary Care? A Review and Synthesis. Prim Care Companion J Clin Psychiatry. 2008;10(2):91–6.

Jacob TC, Moss SJ, Jurd R. GABA(A) receptor trafficking and its role in the dynamic modulation of neuronal inhibition. Nat Rev Neurosci. 2008;9(5):331–43.

Jaglin XH, Hjerling-Leffler J, Fishell G, Batista-Brito R. The origin of neocortical nitric oxide synthase-expressing inhibitory neurons. Front Neural Circuits. 2012;6:44.

Karolewicz B, Maciag D, O'Dwyer G, Stockmeier CA, Feyissa AM, Rajkowska G. Reduced level of glutamic acid decarboxylase-67 kDa in the prefrontal cortex in major depression. Int J Neuropsychopharmacol. 2010;13(4):411–20.

Kennedy SH, Giacobbe P. Treatment resistant depression-advances in somatic therapies. Ann Clin Psychiatry. 2007;19(4):279–87.

Kessler RC, Berglund P, Demler O, Jin R, Merikangas KR, Walters EE. Lifetime prevalence and age-of-onset distributions of DSM-IV disorders in the National Comorbidity Survey Replication. Arch Gen Psychiatry. 2005;62(6):593–602.

Khundakar A, Morris C, Oakley A, McMeekin W, Thomas AJ. Morphometric analysis of neuronal and glial cell pathology in the dorsolateral prefrontal cortex in late-life depression. Br J Psychiatry. 2009;195(2):163–9.

Khundakar A, Morris C, Oakley A, Thomas AJ. A morphometric examination of neuronal and glial cell pathology in the orbitofrontal cortex in late-life depression. Int Psychogeriatr. 2011;23(1):132–40.

Kim S, Webster MJ. Correlation analysis between genome-wide expression profiles and cytoarchitectural abnormalities in the prefrontal cortex of psychiatric disorders. Mol Psychiatry. 2010;15(3):326–36.

Klempan TA, Sequeira A, Canetti L, Lalovic A, Ernst C, Ffrench-Mullen J, Turecki G. Altered expression of genes involved in ATP biosynthesis and GABAergic neurotransmission in the ventral prefrontal cortex of suicides with and without major depression. Mol Psychiatry. 2009;14(2):175–89.

Konradi C, Eaton M, MacDonald ML, Walsh J, Benes FM, Heckers S. Molecular evidence for mitochondrial dysfunction in bipolar disorder. Arch Gen Psychiatry. 2004;61(3):300–8.

Konradi C, Zimmerman EI, Yang CK, Lohmann KM, Gresch P, Pantazopoulos H, Berretta S, Heckers S. Hippocampal interneurons in bipolar disorder. Arch Gen Psychiatry. 2011;68(4):340–50.

Kugaya A, Sanacora G, Verhoeff NP, Fujita M, Mason GF, Seneca NM, Bozkurt A, Khan SA, Anand A, Degen K, Charney DS, Zoghbi SS, Baldwin RM, Seibyl JP, Innis RB. Cerebral benzodiazepine receptors in depressed patients measured with [123I]iomazenil SPECT. Biol Psychiatry. 2003;54(8):792–9.

Law AJ, Harrison PJ. The distribution and morphology of prefrontal cortex pyramidal neurons identified using anti-neurofilament antibodies SMI32, N200 and FNP7. Normative data and a comparison in subjects with schizophrenia, bipolar disorder or major depression. J Psychiatr Res. 2003;37(6):487–99.

Levinson AJ, Fitzgerald PB, Favalli G, Blumberger DM, Daigle M, Daskalakis ZJ. Evidence of cortical inhibitory deficits in major depressive disorder. Biol Psychiatry. 2010;67(5):458–64.

Lewis DA, Sweet RA. Schizophrenia from a neural circuitry perspective: advancing toward rational pharmacological therapies. J Clin Invest. 2009;119(4):706–16.

Lu B, Martinowich K. Cell biology of BDNF and its relevance to schizophrenia. Novartis Found Symp. 2008;289:119–29; discussion 29–35, 93–5.

Luscher B, Shen Q, Sahir N. The GABAergic deficit hypothesis of major depressive disorder. Mol Psychiatry. 2011;16(4):383–406.

Maciag D, Hughes J, O'Dwyer G, Pride Y, Stockmeier CA, Sanacora G, Rajkowska G. Reduced density of calbindin immunoreactive GABAergic neurons in the occipital cortex in major depression: relevance to neuroimaging studies. Biol Psychiatry. 2010;67(5):465–70.

Maddock RJ, Buonocore MH. MR spectroscopic studies of the brain in psychiatric disorders. Curr Top Behav Neurosci. 2012; 11:199–251.

Martel G, Dutar P, Epelbaum J, Viollet C. Somatostatinergic systems: an update on brain functions in normal and pathological aging. Front Endocrinol (Lausanne). 2012;3:154.

Mayberg HS, Liotti M, Brannan SK, McGinnis S, Mahurin RK, Jerabek PA, Silva JA, Tekell JL, Martin CC, Lancaster JL, Fox PT. Reciprocal limbic-cortical function and negative mood: converging PET findings in depression and normal sadness. Am J Psychiatry. 1999;156(5):675–82.

Mayberg HS, Lozano AM, Voon V, McNeely HE, Seminowicz D, Hamani C, Schwalb JM, Kennedy SH. Deep brain stimulation for treatment-resistant depression. Neuron. 2005;45(5):651–60.

Merali Z, Du L, Hrdina P, Palkovits M, Faludi G, Poulter MO, Anisman H. Dysregulation in the suicide brain: mRNA expression of corticotropin-releasing hormone receptors and GABA(A) receptor subunits in frontal cortical brain region. J Neurosci. 2004;24(6):1478–85.

Michael N, Erfurth A, Ohrmann P, Arolt V, Heindel W, Pfleiderer B Metabolic changes within the left dorsolateral prefrontal cortex occurring with electroconvulsive therapy in patients with treatment resistant unipolar depression. Psychol Med. 2003a;33(7):1277–84.

Michael N, Erfurth A, Ohrmann P, Arolt V, Heindel W, Pfleiderer B Neurotrophic effects of electroconvulsive therapy: A proton magnetic resonance study of the left amygdalar region in patients with treatment-resistant depression. Neuropsychopharmacology. 2003b;28(4):720–5.

Miguel-Hidalgo JJ, Baucom C, Dilley G, Overholser JC, Meltzer HY, Stockmeier CA, Rajkowska G. Glial fibrillary acidic protein immunoreactivity in the prefrontal cortex distinguishes younger from older adults in major depressive disorder. Biol Psychiatry. 2000;48(8):861–73.

Miguel-Hidalgo JJ, Dubey P, Shao Q, Stockmeier C, Rajkowska G. Unchanged packing density but altered size of neurofilament immunoreactive neurons in the prefrontal cortex in schizophrenia and major depression. Schizophr Res. 2005;76(2–3):159–71.

Miguel-Hidalgo JJ, Waltzer R, Whittom AA, Austin MC, Rajkowska G, Stockmeier CA. Glial and glutamatergic markers in depression, alcoholism, and their comorbidity. J Affect Disord. 2010;127(1-3):230–40.

Milne A, MacQueen GM, Yucel K, Soreni N, Hall GB. Hippocampal metabolic abnormalities at first onset and with recurrent episodes of a major depressive disorder: a proton magnetic resonance spectroscopy study. Neuroimage. 2009;47(1):36–41.

Morris HM, Hashimoto T, Lewis DA. Alterations in somatostatin mRNA expression in the dorsolateral prefrontal cortex of subjects with schizophrenia or schizoaffective disorder. Cereb Cortex. 2008;18(7):1575–87.

Northoff G, Walter M, Schulte RF, Beck J, Dydak U, Henning A, Boeker H, Grimm S, Boesiger P. GABA concentrations in the human anterior cingulate cortex predict negative BOLD responses in fMRI. Nat Neurosci. 2007;10(12):1515–7.

Oh DH, Son H, Hwang S, Kim SH. Neuropathological abnormalities of astrocytes, GABAergic neurons, and pyramidal neurons in the dorsolateral prefrontal cortices of patients with major depressive disorder. Eur Neuropsychopharmacol. 2012;22(5):330–8.

Ongur D, Drevets WC, Price JL. Glial reduction in the subgenual prefrontal cortex in mood disorders. Proc Natl Acad Sci U S A. 1998;95(22):13290–5.

Orthmann-Murphy JL, Abrams CK, Scherer SS. Gap junctions couple astrocytes and oligodendrocytes. J Mol Neurosci. 2008;35(1):101–16.

Owens DF, Kriegstein AR. Is there more to GABA than synaptic inhibition? Nat Rev Neurosci. 2002;3(9):715–27.

Petty F, Schlesser MA. Plasma GABA in affective illness. A preliminary investigation. J Affect Disord. 1981;3(4):339–43.

Petty F, Sherman AD. Plasma GABA levels in psychiatric illness. J Affect Disord. 1984;6(2):131–8.

Pfleiderer B, Michael N, Erfurth A, Ohrmann P, Hohmann U, Wolgast M, Fiebich M, Arolt V, Heindel W. Effective electroconvulsive therapy reverses glutamate/glutamine deficit in the left anterior cingulum of unipolar depressed patients. Psychiat Res-Neuroim. 2003;122(3):185–92.

Phillips ML, Ladouceur CD, Drevets WC. A neural model of voluntary and automatic emotion regulation: implications for understanding the pathophysiology and neurodevelopment of bipolar disorder. Mol Psychiatry. 2008;13(9):829, 833–57.

Price RB, Shungu DC, Mao X, Nestadt P, Kelly C, Collins KA, Murrough JW, Charney DS, Mathew SJ. Amino acid neurotransmitters assessed by proton magnetic resonance spectroscopy: relationship to treatment resistance in major depressive disorder. Biol Psychiatry. 2009;65(9):792–800.

Rajkowska G. Postmortem studies in mood disorders indicate altered numbers of neurons and glial cells. Biol Psychiatry. 2000;48(8):766–77.

Rajkowska G. Depression: what we can learn from postmortem studies. Neuroscientist. 2003;9(4):273–84.

Rajkowska G, Miguel-Hidalgo JJ. Gliogenesis and glial pathology in depression. CNS Neurol Disord Drug Targets. 2007;6(3):219–33.

Rajkowska G, Miguel-Hidalgo JJ, Wei J, Dilley G, Pittman SD, Meltzer HY, Overholser JC, Roth BL, Stockmeier CA. Morphometric evidence for neuronal and glial prefrontal cell pathology in major depression. Biol Psychiatry. 1999;45(9):1085–98.

Rajkowska G, Miguel-Hidalgo JJ, Dubey P, Stockmeier CA, Krishnan KR. Prominent reduction in pyramidal neurons density in the orbitofrontal cortex of elderly depressed patients. Biol Psychiatry. 2005;58(4):297–306.

Rajkowska G, O'Dwyer G, Teleki Z, Stockmeier CA, Miguel-Hidalgo JJ. GABAergic neurons immunoreactive for calcium binding proteins are reduced in the prefrontal cortex in major depression. Neuropsychopharmacology. 2007;32(2):471–82.

Rakofsky JJ, Ressler KJ, Dunlop BW. BDNF function as a potential mediator of bipolar disorder and post-traumatic stress disorder comorbidity. Mol Psychiatry. 2012;17(1):22–35.

Sahara S, Yanagawa Y, O'Leary DD, Stevens CF. The fraction of cortical GABAergic neurons is constant from near the start of cortical neurogenesis to adulthood. J Neurosci. 2012;32(14):4755–61.

Sanacora G, Mason GF, Rothman DL, Krystal JH. Increased occipital cortex GABA concentrations in depressed patients after therapy with selective serotonin reuptake inhibitors. Am J Psychiatry. 2002;159(4):663–5.

Sanacora G, Mason GF, Rothman DL, Hyder F, Ciarcia JJ, Ostroff RB, Berman RM, Krystal JH. Increased cortical GABA concentrations in depressed patients receiving ECT. Am J Psychiatry. 2003;160(3):577–9.

Sanacora G, Gueorguieva R, Epperson CN, Wu YT, Appel M, Rothman DL, Krystal JH, Mason GF. Subtype-specific alterations of gamma-aminobutyric acid and glutamate in patients with major depression. Arch Gen Psychiatry. 2004;61(7):705–13.

Sanacora G, Treccani G, Popoli M. Towards a glutamate hypothesis of depression: an emerging frontier of neuropsychopharmacology for mood disorders. Neuropharmacology. 2012;62(1):63–77.

Sequeira A, Klempan T, Canetti L, ffrench-Mullen J, Benkelfat C, Rouleau GA, Turecki G. Patterns of gene expression in the limbic system of suicides with and without major depression. Mol Psychiatry. 2007;12(7):640–55.

Sequeira A, Mamdani F, Ernst C, Vawter MP, Bunney WE, Lebel V, Rehal S, Klempan T, Gratton A, Benkelfat C, Rouleau GA, Mechawar N, Turecki G. Global brain gene expression analysis links glutamatergic and GABAergic alterations to suicide and major depression. PLoS ONE. 2009;4(8):e6585.

Si X, Miguel-Hidalgo JJ, O'Dwyer G, Stockmeier CA, Rajkowska G. Age-dependent reductions in the level of glial fibrillary acidic protein in the prefrontal cortex in major depression. Neuropsychopharmacology. 2004;29(11):2088–96.

Sibille E. Molecular aging of the brain, neuroplasticity, and vulnerability to depression and other brain-related disorders. Dialogues Clin Neurosci. 2013a; 15:53–65.

Sibille E, French B. Biological substrates underpinning diagnosis of major depression. Int J Neuropsychopharmacol. 2013b; 16:1893–1909.

Sibille E, Wang Y, Joeyen-Waldorf J, Gaiteri C, Surget A, Oh S, Belzung C, Tseng GC, Lewis DA. A molecular signature of depression in the amygdala. Am J Psychiatry. 2009;166(9):1011–24.

Sibille E, Morris HM, Kota RS, Lewis DA. GABA-related transcripts in the dorsolateral prefrontal cortex in mood disorders. Int J Neuropsychopharmacol. 2011:1–14.

Siegle GJ, Thompson W, Carter CS, Steinhauer SR, Thase ME. Increased amygdala and decreased dorsolateral prefrontal BOLD responses in unipolar depression: related and independent features. Biol Psychiatry. 2007;61(2):198–209.

Sun W, McConnell E, Pare JF, Xu Q, Chen M, Peng W, Lovatt D, Han X, Smith Y, Nedergaard M. Glutamate-dependent neuroglial calcium signaling differs between young and adult brain. Science. 2013;339(6116):197–200.

Surguladze S, Brammer MJ, Keedwell P, Giampietro V, Young AW, Travis MJ, Williams SC, Phillips ML. A differential pattern of neural response toward sad versus happy facial expressions in major depressive disorder. Biol Psychiatry. 2005;57(3):201–9.

Suslow T, Konrad C, Kugel H, Rumstadt D, Zwitserlood P, Schoning S, Ohrmann P, Bauer J, Pyka M, Kersting A, Arolt V, Heindel W, Dannlowski U. Automatic mood-congruent amygdala responses to masked facial expressions in major depression. Biol Psychiatry. 2010;67(2):155–60.

Tiihonen J, Kuikka J, Rasanen P, Lepola U, Koponen H, Liuska A, Lehmusvaara A, Vainio P, Kononen M, Bergstrom K, Yu M, Kinnunen I, Akerman K, Karhu J. Cerebral benzodiazepine receptor binding and distribution in generalized anxiety disorder: a fractal analysis. Mol Psychiatry. 1997;2(6):463–71.

Timmers HJ, Swaab DF, van de Nes JA, Kremer HP. Somatostatin 1-12 immunoreactivity is decreased in the hypothalamic lateral tuberal nucleus of Huntington's disease patients. Brain Res. 1996;728(2):141–8.

Torres-Platas SG, Hercher C, Davoli MA, Maussion G, Labonte B, Turecki G, Mechawar N. Astrocytic hypertrophy in anterior cingulate white matter of depressed suicides. Neuropsychopharmacology. 2011;36(13):2650–8.

Tripp A, Kota RS, Lewis DA, Sibille E. Reduced somatostatin in subgenual anterior cingulate cortex in major depression. Neurobiol Dis. 2011;42(1):116–24.

Tripp A, Oh H, Guilloux JP, Martinowich K, Lewis DA, Sibille E. Brain-derived neurotrophic factor signaling and subgenual anterior cingulate cortex dysfunction in major depressive disorder. Am J Psychiatry. 2012;169(11):1194–202.

Underwood MD, Kassir SA, Bakalian MJ, Galfalvy H, Mann JJ, Arango V. Neuron density and serotonin receptor binding in prefrontal cortex in suicide. Int J Neuropsychopharmacol. 2012;15(4):435–47.

Uranova NA, Vostrikov VM, Orlovskaya DD, Rachmanova VI. Oligodendroglial density in the prefrontal cortex in schizophrenia and mood disorders: a study from the Stanley Neuropathology Consortium. Schizophr Res. 2004;67(2-3):269–75.

Valentine GW, Sanacora G. Targeting glial physiology and glutamate cycling in the treatment of depression. Biochem Pharmacol. 2009;78(5):431–9.

Van Otterloo E, O'Dwyer G, Stockmeier CA, Steffens DC, Krishnan RR, Rajkowska G. Reductions in neuronal density in elderly depressed are region specific. Int J Geriatr Psychiatry. 2009;24(8):856–64.

Volk DW, Matsubara T, Li S, Sengupta EJ, Georgiev D, Minabe Y, Sampson A, Hashimoto T, Lewis DA. Deficits in transcriptional regulators of cortical parvalbumin neurons in schizophrenia. Am J Psychiatry. 2012;169(10):1082–91.

WHO. World Health Organization—The Global Burden of Disease—2004 update. Geneva: WHO Library; 2008.

Xu H, Jeong HY, Tremblay R, Rudy B. Neocortical somatostatin-expressing GABAergic interneurons disinhibit the thalamorecipient layer 4. Neuron. 2013;77(1):155–67.
Yuksel C, Ongur D. Magnetic resonance spectroscopy studies of glutamate-related abnormalities in mood disorders. Biol Psychiatry. 2010;68(9):785–94.
Yung WK, Albright RE, Olson J, Fredericks R, Fink K, Prados MD, Brada M, Spence A, Hohl RJ, Shapiro W, Glantz M, Greenberg H, Selker RG, Vick NA, Rampling R, Friedman H, Phillips P, Bruner J, Yue N, Osoba D, Zaknoen S, Levin VA. A phase II study of temozolomide vs. procarbazine in patients with glioblastoma multiforme at first relapse. Br J Cancer. 2000;83(5):588–93.
Zhang J, Narr KL, Woods RP, Phillips OR, Alger JR, Espinoza RT. Glutamate normalization with ECT treatment response in major depression. Mol Psychiatry. 2013;18(3):268–70.

Chapter 14
Pathology in Astroglia, Glutamate, and GABA in Major Depressive Disorder: Evidence from Studies of Human Postmortem Tissue

Grazyna Rajkowska

Abstract Evidence will be reviewed for pathology in astroglial cells, and for glutamate and γ-aminobutyric acid (GABA) neurons, their receptors and transporters in human postmortem brain tissue from subjects diagnosed with major depressive disorder (MDD). These observations will be compared with similar endpoints in preclinical animal models of chronic stress. Repeated stressful experiences or stressful life events can be risk factors for the onset or relapse of depressive episodes. Thus, animal studies on the behavioral and biological responses to exposure to chronic stress may shed light on underlying pathological mechanisms relevant to findings in postmortem brain tissue from subjects that experienced depression. Moreover, dysfunction of astrocytes, glutamate, and GABA—vital components of the tripartite synapse—will be proposed as a major source of fundamental pathology in depression and related animal behavioral models. Finally, the role of glutamate-based drugs in treating depressive symptoms will be discussed. In summary, evidence from postmortem brain tissue in MDD and animal models related to depression supports the hypothesis that pathology in astrocytes, glutamate, and GABA systems may be fundamental to the pathophysiology of depression.

Abbreviations

AQP4	Aquaporin 4
AMPA	Alpha-amino-3-hydroxy-5-methyl-4-isoxazole propionic acid
BDNF	Brain derived neurotrophic factor
Ca^{+2}	Calcium
CA1	Ammoni horn region 1
CA3	Ammoni horn region 3
CNS	Central nervous system
EAAT1	Excitatory amino acid transporter-1
EAAT2	Excitatory amino acid transporter-2

G. Rajkowska (✉)
Department of Psychiatry and Human Behavior, University of Mississippi Medical Center, 39216 Jackson, MS, USA
e-mail: grajkowska@umc.edu

M. Popoli et al. (eds.), *Synaptic Stress and Pathogenesis of Neuropsychiatric Disorders*, DOI 10.1007/978-1-4939-1056-4_14,

GABA	γ-aminobutyric acid
GAD	Glutamic acid decarboxylase
GC1	Mitochondrial glutamate carrier
GFAP	Glial fibrillary acidic protein
GLAST	Glutamate–aspartate transporter
GLT1	Glutamate transporter 1
GluR1	AMPA receptor subunit 1
GluR2	AMPA receptor subunit 2
GluR3	AMPA receptor subunit 3
GluR4	AMPA receptor subunit 4
GluR5	Kainate receptor subunit 5
GRINA	Glutamate receptor ionotropic NMDA-associated protein 1
IR	Immunoreactive
MDD	Major depressive disorder
mGluR5	Metabotropic glutamate receptor 5
mRNA	Messenger ribonucleic acid
mTOR	Mammalian target of rapamycin
NeuN	Neuronal nuclei (neuron-specific nuclear protein)
NMDA	*N*-methyl-D-aspartate
NR1	NMDA receptor 1
NR2A	NMDA receptor 2A
NR2B	NMDA receptor 2B
NR2C	NMDA receptor 2C
PSD95	Postsynaptic density protein 95
SAP102	Synapse-associated protein 102
SSRI	Serotonin-selective reuptake inhibitor

14.1 Astrocyte Pathology in MDD

Cell counting studies in postmortem brain tissue revealed prominent glial pathology in MDD. Early studies examined the entire population of glial cells (astrocytes, oligodendrocytes and microglia) by using a routine stain for Nissl substance. The packing density or number of glial cells was decreased in subjects retrospectively diagnosed with MDD, as compared to nonpsychiatric control subjects (Ongür et al. 1998; Rajkowska et al. 1999; Cotter et al. 2001; Bowley et al. 2002; Torres-Platas et al. 2002; Cotter et al. 2002a; Gittins and Harrison 2011). Such changes were observed in fronto-limbic brain regions including the dorsolateral prefrontal cortex (Rajkowska et al. 1999; Torres-Platas et al. 2002; Cotter et al. 2002a), orbitofrontal cortex (Rajkowska et al. 1999), subgenual cortex (Ongür et al. 1998), anterior cingulate cortex (Cotter et al. 2001; Gittins and Harrison 2011) and amygdala (Bowley et al. 2002). However, in examining elderly subjects with MDD, Khundakar et al. (2011a, 2011b) noted no change in glial density in the orbitofrontal cortex or anterior cingulate cortex.

In addition to reductions in glial cell density and number in MDD, the average size of the nuclei of glial cells was also increased in the gray matter of dorsolateral prefrontal cortex (Rajkowska et al. 1999). However, one study in the dorsolateral prefrontal cortex reported no change in the size of glial nuclei in MDD (Cotter et al. 2002a). A detailed analysis of astrocytes stained with the Golgi method reported hypertrophy of astrocytic cell bodies and processes in the white matter of the anterior cingulate cortex in depressed subjects dying by suicide (Torres-Platas et al. 2011). These authors interpret astrocytic hypertrophy as a reflection of local inflammation in support of the neuroinflammatory theory of depression (Maes et al. 2009).

Of the three types of glial cells in the CNS, astrocytes have been implicated most often as a source of glial pathology in MDD (reviewed in Rajkowska and Stockmeier 2013). This astrocytic pathology may be directly responsible for alterations in glutamate noted in MDD as astrocytes are active in the clearance and metabolism of glutamate at the tripartite glutamatergic synapse (discussed in detail below). Astrocytes have been localized in postmortem brain tissue by antibodies to glial fibrillary acidic protein (GFAP), gap junctions proteins such as connexin 30 and 43, the aquaporin-4 (AQP4) water channel and glutamatergic markers including the excitatory amino acid transporters 1 and 2 (EAAT1 and EAAT2), and the enzyme glutamine synthetase. As outlined below, each of these markers related to astrocytes is affected in postmortem tissues from subjects with depression.

GFAP is the principle component of cytoskeletal intermediate filaments and is strongly expressed in the CNS by mature and reactive astrocyte cells (Jacque et al. 1978; Middeldorp and Hol 2011). The expression of GFAP in depression has been quantified in gray matter by measuring the density of GFAP-immunoreactive (IR) astrocytes or so-called area fraction, the area covered by GFAP-IR cell bodies and processes. There was a significant decrease in the density of GFAP-IR astrocytes and GFAP area fraction in gray matter of the dorsolateral prefrontal cortex in younger depressed subjects (< 60 years' old), as compared to age-matched nonpsychiatric control subjects (Miguel-Hidalgo et al. 2000). In addition, GFAP-IR area fraction was significantly decreased in the gray matter of the orbitofrontal cortex in a mixture of younger and older subjects with MDD (Miguel-Hidalgo et al. 2010). In contrast, older subjects with late-onset depression showed increases in GFAP-IR area fraction and cell density in the gray matter of dorsolateral prefrontal cortex (Miguel-Hidalgo et al. 2000; Davis et al. 2002), suggesting a compensatory response to neuronal damage reported in older subjects with MDD (Rajkowska et al. 2005). Thus, there appears to be a unique pattern of astrocyte pathology in cortical gray matter in younger versus older subjects with depression (Rajkowska and Miguel-Hidalgo 2007; Khundakar and Thomas 2009; Paradise et al. 2012).

Expression of GFAP protein and mRNA has also been examined in MDD. As determined by Western blotting, levels of GFAP protein were decreased in gray matter from the dorsolateral prefrontal and orbitofrontal cortex in MDD (Si et al. 2004; Miguel-Hidalgo et al. 2010). GFAP mRNA was also under-expressed in MDD in the anterior cingulate (Webster et al. 2005) and orbitofrontal cortex (Newton and Rajkowska, unpublished observations). There is a consistent under-expression of GFAP markers in MDD, whether measuring immunohistochemical cell density or area fraction, protein levels or mRNA expression.

Astrocytes are also altered in depression in limbic brain regions and related structures. A reduced density of GFAP-IR astrocytes was found in amygdala of subjects with MDD (Altshuler et al. 2010). In a semiquantitative study, Müller et al. (2001) detected a significant decrease in GFAP-IR astrocytes in the CA1 and CA2 subregions of the hippocampus in depression. A similar decrease in GFAP-IR astrocytes was noted in subjects that had been treated with steroids, suggesting that elevated glucocorticoid hormones acting at glucocorticoid receptors on astrocytes may have contributed to the reduction in GFAP expression in astrocytes (Müller et al. 2001; Wang et al. 2013). In a three-dimensional quantitative study, a significant reduction in the density of GFAP-IR astrocytes was recently observed in the hilus of the hippocampus in subjects with MDD not treated with antidepressant medications (Stockmeier et al. 2010). Bernard et al. (2011) noted a significant decrease in the expression of the mRNA for GFAP in the locus coeruleus in MDD while Chandley et al. (2013) isolated astrocytes from sections of this nucleus and noted a decrease in expression of GFAP mRNA and protein in this subpopulation of glial cells in MDD. In summary, reductions in the density and area fraction of GFAP-IR astrocytes and in the levels of GFAP protein and mRNA reveal dysfunctional astrocytes in MDD in fronto-limbic cortical regions.

Other markers of astrocytes located on astrocytic endfeet include connexin 30, connexin 43, and AQP4, and are also involved in the pathology of depression. Connexin 30 and connexin 43 form gap junctions that allow communication between astrocytes (Giaume and Theis 2010). The expression of genes and proteins for connexin 30 and connexin 43 was reduced in dorsolateral prefrontal cortex and orbitofrontal cortex in MDD (Ernst et al. 2011; Miguel-Hidalgo et al. 2012). The consequences of decreased expression of connexin 30 and connexin 43 alter calcium wave propagation and may affect communication between astrocytes (Blomstrand et al. 1999). In another study, reduced coverage of blood vessels by AQP4, which is a water channel expressed in astrocytic endfeet, was observed in the orbitofrontal cortex in MDD (Rajkowska et al. 2013). Finally, a decrease in the expression of mRNA for AQP4 was identified in locus coeruleus in MDD (Bernard et al. 2011). These decreases in AQP4 in depression could affect many brain functions as AQP4, in addition to its role in water redistribution, also regulates cerebral blood flow (Paulson and Newman 1987; Koehler et al. 2009), glucose transport and metabolism (Kimelberg 2004), integrity of the blood–brain barrier (Nico et al. 2001; Meshorer et al. 2005), glutamate turnover (Zeng et al. 2007), and synaptic plasticity (Li et al. 2012).

14.2 Astrocyte Pathology in Animal Models of Stress and Depression

Studies in preclinical animal models provide evidence for the involvement of GFAP and astrocytes in stress and depression-related behaviors. Various types of stress cause reductions in measures of GFAP-IR astrocytes. For example, the stress of separating juveniles from their family diminished the density of GFAP-IR astrocytes in the rodent medial prefrontal cortex (Braun et al. 2009). The stress of chronic social defeat in tree shrews reduced the number and soma volume of GFAP-IR astrocytes in the hippocampus (Czéh et al. 2006) and social defeat stress decreased the level of GFAP protein in rat hippocampus (Araya-Callís et al. 2012). Early life stress also resulted in a reduced density of GFAP-IR astrocytes in adult rats in various prefrontal and frontal cortical regions, hippocampus, and the basolateral amygdala (Leventopoulos et al. 2007). Furthermore, chronic unpredictable stress significantly decreased expression of GFAP mRNA in rat medial prefrontal cortex (Banasr et al. 2010). Interestingly, infusion of L-α aminoadipic acid in rodent prefrontal cortex, thought to selectively lesion glial cells including GFAP-IR astrocytes but not neurons, induced depressive-like behaviors (Banasr and Duman 2008; Lee et al. 2013). Assuming specificity of the toxin for glia, these two lesion studies appear to support the hypothesis that the loss of glia contributes to the pathology of depression (Rajkowska and Miguel-Hidalgo 2007). There is also support for a correlation between GFAP-IR astrocytes and depressive-like behavior in Wistar-Kyoto rats, a strain of rats that is genetically predisposed to anxiety-like and depressive-like behavior (Will et al. 2003). Significant reductions in the density of GFAP-IR astrocytes but not NeuN-IR neurons were observed in the prefrontal cortex, anterior cingulate cortex, amygdala, and hippocampus in Wistar-Kyoto rats as compared to Sprague-Dawley rats serving as controls (Gosselin et al. 2009). Thus, specific astrocytic deficits in the expression of GFAP in cortico-limbic circuits are associated with depressive-like behavior.

Astrocytes have been suggested as a target for therapeutic interventions in depression (Czéh and Di Benedetto 2013; Sanacora and Banasr 2013). Several animal studies reveal an influence of different classes of antidepressant medications on astrocytes. For example, treatment with fluoxetine, a serotonin-selective reuptake inhibitor (SSRI), prevented the psychosocial stress-induced reduction in astrocyte number in the hippocampus (Czéh et al. 2006). Riluzole, a glutamate modulating drug, also prevented the chronic, unpredictable stress-induced reduction in the expression of GFAP mRNA in the rat prefrontal cortex (Banasr et al. 2010). The beneficial effects of the SSRI antidepressants such as citalopram and fluoxetine may involve their ability to induce calcium signals in astrocytes in the prefrontal cortex (Schipke et al. 2011). However, not all studies show reversibility of the number of astrocytes or GFAP levels by an antidepressant drug. For example, a 4-week treatment with citalopram, also an SSRI, did not restore the social defeat-induced reduction in GFAP protein in the rat hippocampus, although the behavior of the animals was normalized within this treatment period (Araya-Callís et al. 2012). Likewise, imipramine, a tricyclic antidepressant drug, could not reverse the effects of learned helplessness on hippocampal astrocytes (Iwata et al. 2011).

In summary, models of chronic stress in experimental animals significantly diminish cortical and hippocampal astrocytes as measured by GFAP while lesions of cortical glia, including astrocytes, yield behavioral deficits comparable to those seen following chronic stress. The effects of chronic stress on GFAP-IR astrocytes can be reversed by chronic treatment with some, but not all, antidepressant medications. Thus, in light of astrocytic deficits noted in MDD and stress being a risk factor for depression, as well as astrocytic deficits in animal models of chronic stress, astrocytes may indeed be potential targets for the action of novel antidepressant medications.

14.3 Astrocyte Pathology and Glutamate Dysfunction in MDD

Astrocyte pathology described above could be related to dysfunction of the glutamate system, as reported in MDD. Astrocytes are a vital component of the tripartite glutamate synapse which consists of the (1) presynaptic neuronal terminal, (2) postsynaptic neuronal membrane, and (3) surrounding astrocyte processes (Araque et al. 1999; Nedergaard and Verkhratsky 2012). Synaptically associated astrocytes respond to neuronal activity by elevating their internal Ca^{2+} concentrations to trigger the release of glial transmitters which, in turn, regulate neuronal activity (Araque et al. 1999; Nedergaard and Verkhratsky 2012). Astrocytes also control the formation, maturation, function, and elimination of synapses through various secreted and contact-mediated signals (Clarke and Barres 2013). Moreover, astrocytes are actively involved in the uptake, metabolism, and recycling of glutamate. Levels of extracellular glutamate are regulated by removal of this neurotransmitter from the synaptic cleft via specialized transporters located on astrocytic processes (Anderson and Swanson 2000). In the human brain, these glutamate transporters include the EAAT1 and EAAT2, which in rodents are known as the glutamate–aspartate transporter (GLAST) and the glutamate transporter 1 (GLT1), respectively (Bezzi et al. 2004; Furuta et al. 2005). Glutamate internalized within astrocytes is subsequently converted to glutamine by the enzyme, glutamine synthetase (Toro et al. 2006). Glutamine then leaves astrocytes to be taken up by neurons where it can be converted into glutamate or GABA. Thus, astrocytes play a critical role in several aspects of glutamate neurotransmission.

Glutamate transporters and glutamine synthetase associated with astrocytes appear to be dysregulated in postmortem brain tissue from subjects with MDD. For example, reduced expression of mRNA for EAAT1, EAAT2, and glutamine synthetase was noted in the anterior cingulate and dorsolateral prefrontal cortex in subjects with MDD (Choudary et al. 2005). Expression of the mRNA for glutamine synthetase was also down-regulated in the dorsolateral prefrontal cortex, premotor cortex, and the amygdala of depressed suicide victims (Sequeira et al. 2009). Moreover, the expression of EAAT1, EAAT2, and glutamine synthetase protein was reduced in the orbitofrontal cortex in immunohistochemical and Western blotting studies of subjects with MDD (Miguel-Hidalgo et al. 2010). Finally, glutamate signaling and astrocyte-associated genes were under-expressed in locus coeruleus in MDD (Bernard et al. 2011; Chandley et al. 2013; Ordway et al. 2012), suggesting more global dysfunction of glutamate signaling and astrocyte pathology in MDD. Support for disease-specific astroglial pathology in MDD comes from Bernard et al. (2011) demonstrating that these changes in glutamate-related gene expression do not occur in neurons. Other evidence supporting a role for dysregulated uptake of glutamate by astrocytes in depression comes from studies in rats where the pharmacological blockade of glutamate uptake into astrocytes in the amygdala (Lee et al. 2007), ventral tegmental area (Herberg and Rose 1990), or in the prefrontal cortex (John et al. 2012) is sufficient to decrease sucrose consumption, a behavioral marker thought to be related to anhedonia, a core symptom of depression. Finally, animal studies reveal that astrocytic GFAP plays a key role in the trafficking of glutamate transporters and protecting the brain against glutamate-mediated excitotoxicity (Hughes et al. 2004; Sullivan et al. 2007).

14.4 Glutamate Neurons and Receptors in Postmortem Tissues in MDD

Other studies of postmortem tissue reveal a link between neuronal pathology and glutamate dysfunction in MDD. Alterations in glutamatergic neurons' density, levels of their receptors, and other proteins involved in glutamate signaling are reported in MDD. Prominent reductions in the density of glutamatergic, pyramidal neurons were observed in the orbitofrontal cortex in elderly depressed subjects (Rajkowska et al. 2005).

Glutamatergic neurons and astrocytes directly control synaptic and extrasynaptic glutamate levels and release through integrative effects that target glutamate transporters, postsynaptic density proteins, ionotropic receptors (N-methyl-D-aspartate, NMDA;, alpha-amino-3-hydroxy-5-methyl-4-isoxazole propionic acid, AMPA; kainate) as well as metabotropic receptors. Recent studies in postmortem tissue implicate the NMDA class of glutamate receptors in the pathophysiology of MDD. Significant reductions in the protein expression of NMDA receptor subunits, NR2A and NR2B, and PSD-95 were observed in the anterior pole of prefrontal cortex from subjects with MDD as compared to psychiatrically normal control subjects (Feyissa et al. 2009). PSD-95 is linked to the NMDA receptor and plays a role in mediating trafficking and clustering of the receptor and related downstream signaling events. Reduced expression of NR2A transcript in the dorsolateral prefrontal cortex and reductions in expression of both NR2A and NR2B transcripts were noted in the perirhinal cortex in subjects with MDD (Beneyto et al. 2007; Beneyto and Meador-Woodruff 2008). In addition, there is a significant upregulation of genes coding for mitochondrial glutamate carrier (GC1) and the glutamate receptor ionotropic NMDA-associated protein 1 (GRINA) in the anterior pole of prefrontal cortex from subjects with MDD (Goswami et al. 2013). There are conflicting reports on whether expression of mRNA and/or proteins related to the NMDA receptor subunits are altered in the hippocampus in depression. A reduction in the expression of mRNA for the NR1 subunit of the NMDA receptor was noted in the dentate gyrus of the hippocampus in depression (Law and Deakin 2001). In contrast, no change was noted in gene expression for several NMDA receptor subunits (including NR1) in either dentate gyrus or CA1 regions of the hippocampus in MDD (Duric et al. 2013). In the superior temporal cortex, while a decrease in radioligand binding to the glycine site of the NR1 subunit was observed in depression, the expression of the NR1 protein was not significantly different from control subjects (Nudmamud-Thanoi and Reynolds 2004). Furthermore, in MDD, protein expression of the NR1 subunit was also unchanged versus control subjects in the anterior pole of prefrontal cortex, amygdala, locus coeruleus, and cerebellum (Feyissa et al. 2009; Karolewicz et al. 2005; Karolewicz et al. 2009). Thus, NR2A and NR2B subunits, but not the NR1 subunit, appear to be consistently under-expressed in MDD.

Alterations in components of glutamate system in MDD are not restricted to limbic cortical regions (i.e., prefrontal cortex, hippocampus, temporal cortex) but are also found in the brainstem, striatum, and amygdala, regions that receive glutamatergic projections from the cerebral cortex. Increases in the expression of NR2C subunit were observed in the locus coeruleus and NR2A subunit in amygdala (Karolewicz et al. 2005; Karolewicz et al. 2009). There were significant changes in the expression of other glutamate signaling genes in the locus coeruleus in MDD (Bernard et al. 2011). Decreased expression of the mRNA transcript encoding the NMDA interacting postsynaptic density protein SAP 102 was detected in the striatum of depressed subjects (Kristiansen and Meador-Woodruff 2005). Thus, glutamate pathology in MDD affects limbic cortical regions and their subcortical projection areas. Taken together, the above studies provide evidence for pathology of the NMDA receptor in specific brain regions and support hypotheses that drugs altering NMDA receptor signaling may be effective in treating depression.

Fewer studies have been undertaken in depression on non-NMDA receptors such as the ionotropic AMPA and kainate receptors. Radioligand binding to the AMPA receptor was increased in the anterior cingulate cortex but not in the dorsolateral prefrontal cortex in MDD (Gibbons et al. 2012). In the same study, there was no significant depression-related change in radioligand binding to the kainate receptor in either prefrontal or cingulate cortex. However, mRNA expression of the GluR5 subunit of the kainate receptor was decreased in the prefrontal cortex in subjects with MDD (Knable et al. 2001). The expression of mRNA for subunits of the AMPA receptors (GluR1 and GluR3) was downregulated in both dentate gyrus and CA1 whereas mRNA for the GluR4 subunit was decreased only in dentate gyrus in MDD (Duric et al. 2013). Levels of GluR3 were significantly decreased in the dorsolateral prefrontal cortex in subjects with MDD (Beneyto and Meador-Woodruff 2006).

Finally, a reduction in radioligand binding to metabotropic glutamate receptor 5 (mGluR5) was reported by neuroimaging study in multiple brain regions including anterior prefrontal cortex in living depressed subjects (Deschwanden et al. 2011). There was a comparable reduction in protein level of this receptor in the same brain region in postmortem tissue from subjects with MDD (Deschwanden et al. 2011). Thus, reduced binding to mGluR5 receptors in MDD suggests reduced density of functional receptors because of decreased levels of mGluR5 protein. Moreover, as the mGluR5 receptor is present on postsynaptic neurons and on glia, it may modulate extrasynaptic NMDA receptors (D'Ascenzo et al. 2007).

Generally, the aforementioned studies suggest pathology of various components of the glutamate system in depression. Alterations in NMDA, AMPA, kainate, and metabotropic glutamate receptors are found in several areas of postmortem brain tissue in MDD as compared to age- and gender-matched psychiatrically normal control subjects. Reduced levels of glial glutamate transporters and glutamine synthetase suggest enhanced synaptic and/or perhaps presynaptic concentrations of glutamate in MDD. A study of postmortem tissue supporting this hypothesis reported increased tissue levels of glutamate in the frontal cortex in subjects with MDD (Hashimoto et al. 2007). However, several neuroimaging studies of prefrontal and anterior cingulate cortex using magnetic resonance spectroscopy report a significant decrease in glutamate or glutamate/glutamine levels in depressed patients (Auer et al. 2000; Michael et al. 2003; Pfleiderer et al. 2003; Mirza et al. 2004; Hasler et al. 2007), while one study notes an increase in glutamate levels in the occipital cortex in depression (Sanacora et al. 2004). In spite of these discrepancies in whether glutamate levels increase or decrease, other clinical studies support the relevance of glutamate in depression.

There is a growing body of preclinical and clinical research implicating riluzole, an inhibitor of glutamate release, and ketamine, an antagonist of the NMDA receptor, as potent antidepressant medications (reviewed in Pilc et al. 2013). There are several reports that a single low dose of ketamine induces a rapid (within hours), long lasting (up to 7 days), and robust antidepressant effect in treatment-resistant patients with MDD (Berman et al. 2000; Zarate et al. 2006; Pilc et al. 2013). Potential mechanisms underlying the rapid action of ketamine are being identified. Li et al. (2010) reported that ketamine rapidly activated the mammalian target of rapamycin (mTOR) pathway, leading to increased signaling proteins at the synapse and increased number and function of new spine synapses in the prefrontal cortex of rat. Moreover, acute administration of ketamine in rats increased brain-derived neurotrophic factor (BDNF) and mTOR levels in the hippocampus during forced swimming (Yang et al. 2013). Interestingly, a recent study in postmortem prefrontal cortex on the expression of mTOR protein and its core downstream signaling targets reported a decrease in the expression of mTOR, p70S6K, eIF4B, and p-eIF4B proteins in subjects with MDD as compared to nonpsychiatric control subjects (Jernigan et al. 2011). Thus, a deficit in the initiation of mTOR-dependent protein expression may occur in depression and suggests an association between deficits in synaptic proteins and dysregulation of mTOR signaling in this disorder. Other components of the glutamate system also appear to be targets for antidepressant medications. For example, enhanced transmission through glutamatergic AMPA receptor may provide a common mechanism of antidepressant actions (reviewed by Sanacora et al. 2008).

14.5 Preclinical Studies on Stress and Glutamate

The pathology noted in the glutamate system in depression may be related to the effects of chronic stress. MDD is often preceded by exposure to chronic stress or stressful life events. There is evidence that both the onset of and relapse into depression are precipitated by severe repeated stressful experiences (Kessler 1997; Mazure et al. 2000; Kendler et al. 2001; Hammen 2005; Monroe et al. 2006; Pittenger and Duman 2008; Venzala et al. 2012).

Preclinical studies show that stress influences glutamate neurotransmission and metabolism and morphology of glutamate neurons. Consistent with studies in MDD, unpredictable chronic mild stress decreased expression of proteins for NR2A and NR2B subunits of NMDA receptor in the frontal cortex and hippocampus in rats (Feyissa et al. 2009; Lou et al. 2010). Repeated stress in young rats also significantly decreased expression of NMDA (NR1) and AMPA (GluR1) receptor subunits in pyramidal neurons of the prefrontal cortex and had a detrimental effect on cognitive processes dependent on this brain region (Yuen et al. 2012). Thus, glutamate receptors appear to be crucial neural substrates related to the effects of stress on synaptic plasticity and memory (Krugers et al. 2010; Yuen et al. 2012). No consensus has emerged on the effects of chronic mild stress on synaptic and vesicular levels of glutamate (reviewed in Hill et al. 2012). Chronic mild stress increased the expression of the glial glutamate transporter-2 and the vesicular glutamate transporter-1 protein and doubled the vesicular levels of glutamate in rat hippocampus (Raudensky and Yamamoto 2007; Garcia-Garcia et al. 2009). In contrast, reduced levels of mRNA for vesicular glutamate transporter-1 were noted in rat hippocampal subfield CA1 but not CA3 or dentate gyrus (Elizalde et al. 2010a). Within the frontal cortex, expression of both glial glutamate transporter-2 and the vesicular glutamate transporter-1 was not significantly changed by chronic stress in two studies (Garcia-Garcia et al. 2009; Banasr et al. 2010); however, a third study reported reduced levels of mRNA for vesicular glutamate transporter-1 (Elizalde et al. 2010a). Protein levels of the mGluR5 receptor were increased in the hippocampal CA1 subregion in rat in response to chronic mild stress but the receptor was decreased in the CA3 subregion and unchanged in the dentate gyrus (Wierońska et al. 2001). The above data reveal that stress influences glutamate receptors and transporters and these changes are region specific.

The pathology of glutamate systems in depression and chronic stress appears to involve several levels of neuronal morphology. Exposure to chronic unpredictable stress results in a reduction in the length and branching of apical dendrites of glutamate pyramidal neurons in layer V and decreases the number of synapses on these neurons in rat medial prefrontal cortex (Li et al. 2011; Duman & Aghajanian 2012). These observations may parallel findings from human postmortem studies in depression showing a reduction in glutamate, pyramidal neurons density in layer V of prefrontal cortex and smaller sizes of neurons in this and other prefrontal layers (Rajkowska et al.1999; Cotter et al. 2001; Rajkowska et al. 2005). The decreased number of synapses observed in prefrontal cortex of stressed rats is consistent with the recent study of postmortem tissue showing significant decreases in the number of synapses and expression of synapse-related genes in the prefrontal cortex from subjects with MDD (Kang et al. 2012). The expression of several synapse- and glutamate-related genes was also decreased in the dentate gyrus and CA1 regions of hippocampus in MDD (Duric et al. 2013). This synaptic pathology may be related in part to the pathology of astrocytes in MDD since astrocytes control the formation, maturation, function, and elimination of synapses in the brain (Clarke and Barres 2013). In sum, the above findings clearly point to the pathology of glutamate synapses in MDD.

14.6 GABA Dysfunction in Postmortem Tissues in MDD

While neuronal pathology in MDD appears to be less prominent than glial pathology, several studies of postmortem tissue show reductions in the packing density and/or size of a general (Nissl-stained) population of cortical neurons (Rajkowska et al.1999; Cotter et al. 2001; Cotter et al. 2002a; Rajkowska et al. 2005). The most prominent neuronal changes in MDD have been observed in superficial layers of the prefrontal cortex (Rajkowska et al. 1999). Interestingly, these cortical layers are highly populated by GABA neurons. GABA dysfunction in MDD has been suggested by neuroimaging studies showing decreased levels of GABA in occipital and dorsolateral prefrontal cortex (Sanacora et al. 2004; Hasler et al. 2007). Also, some studies of postmortem tissue clearly demonstrate 30–50 % reductions in the density of a subpopulation of GABA neurons, calbindin-IR neurons, in MDD. These decreases, noted only for calbindin- and not parvalbumin-IR GABA neurons, were observed in upper cortical layers (II and upper III) in the dorsolateral prefrontal cortex and in occipital cortex (Rajkowska et al. 2007; Maciag et al. 2010). In both of these studies, reductions in the soma size of calbindin-IR neurons were also noted in MDD. Thus, the studies in postmortem tissue support neuroimaging observations of alterations in GABA neurotransmission in the same brain regions. However, one study of postmortem tissue, examining all three populations of GABA neurons IR for calcium binding proteins, noted no changes in these neurons in the anterior cingulate cortex in MDD (Cotter et al. 2002b). The differences between studies showing alterations in GABA neurons (Rajkowska et al. 2007; Maciag et al. 2010) and that of Cotter et al. (2002b) may be explained by differences in hemispheres and brain regions studied as well as clinical features of the patient cohorts.

A reduction in the density and size of GABA neurons in dorsolateral prefrontal cortex in depression suggests that the synthesis of GABA may also be affected in that region. Glutamic acid decarboxylase (GAD), the enzyme that converts glutamate to GABA, exists as two isoforms, GAD-65 kDa and GAD-67 kDa, which are encoded by two distinct genes (Erlander et al. 1991; Kaufman et al. 1991). There was a significant decrease in the expression of GAD-67, but not GAD-65, in the dorsolateral prefrontal cortex of many of the same depressed subjects used for the calbindin studies (Rajkowska et al. 2007; Karolewicz et al. 2010). The decrease in GAD-67 was only noted in depressed subjects in which antidepressant drugs were absent from postmortem blood. In contrast, subjects with an antidepressant drug in postmortem blood showed no change in protein levels of GAD-67 in comparison to psychiatrically normal control subjects. Antidepressant drugs may either promote synthesis of GAD-67 or prevent the depression-related decrease in GAD-67.

14.7 Preclinical Data on Stress and GABA

The pathology described in the GABA system in depression may be related to effect of chronic stress, which is considered a risk factor for depression. Some animal studies suggest that chronic mild stress and chronic unpredictable stress have a significant effect on the GABA system (reviewed in Hill et al. 2012). For example, the content of GABA is consistently decreased in the hippocampus and frontal cortex following chronic mild stress in the rat (Gronli et al. 2007; Garcia-Garcia et al. 2009; Elizalde et al. 2010b). In contrast, chronic mild stress has highly inconsistent effects on mRNA and protein expression of the GAD-65 and GAD-67 isoforms of the GABA synthetic enzyme in various limbic brain regions. Expression of mRNA for GAD-65 was decreased by this stress and chronic unpredictable stress in the bed nucleus of stria terminalis and preoptic area of the hypothalamus whereas expression of GAD-67 mRNA was decreased in rat prefrontal cortex (Herman and Larson 2001; Lepack et al. 2013). Reduced level of GAD-65 protein has been observed in the ventral hippocampus and frontal cortex following chronic mild stress (Garcia-Garcia et al. 2009; Elizalde et al. 2010b). In contrast, there was a report of increased expression of GAD-65 mRNA and GAD-67 mRNA in the hypothalamus, the bed nucleus of the stria terminalis and the hippocampus following chronic stress (Bowers et al. 1998), whereas, others report that expression of these two markers was unchanged in the same brain regions and in the amygdala, septum, and frontal cortex (Herman and Larson 2001; Herman et al. 2003). Additional studies are necessary to clarify the impact of chronic stress on measures of GAD.

Studies on the influence of chronic mild stress and chronic unpredictable stress on the density of GABA neurons reveal a more consistent effect. In these models of chronic stress, the density of calbindin-IR GABA neurons was decreased in the prefrontal cortex and hippocampus in two studies, whereas the density of parvalbumin-IR GABA neurons was unchanged in these brain regions (Herman and Larson 2001; Nowak et al. 2010; Zadrozna et al. 2011; Lepack et al. 2013). Thus, these studies in a rodent model of chronic stress closely correspond to studies in human postmortem tissue showing a decrease in calbindin-IR GABA neurons but not in parvalbumin-IR GABA neurons in the prefrontal cortex in MDD (Rajkowska et al. 2007). Decreased expression of GAD-67 protein but not GAD-65 was also observed in the same prefrontal cortical region in MDD (Karolewicz et al. 2010).

In summary, chronic stress, neuroimaging studies of depressed patients, and studies of postmortem tissue from depressed subjects show consistent decreases in GABA levels and the density of GABA IR neurons. These reports strongly support a hypothesis of GABA pathology in depression.

14.8 Conclusions

Studies of human postmortem tissue reveal prominent astrocyte pathology in fronto-limbic brain regions in MDD. The mechanisms regulating astrocyte pathology in depression are being explored in preclinical studies which show, in many cases, similar pathology of GFAP and astrocytes in animal models of stress and depressive-like behavior. Astrocyte pathology in MDD appears to be linked to the dysfunction of glutamate and GABA systems as astrocytes are vital components of glutamatergic tripartite synapses. Reductions in the expression of glutamate transporters and enzymes, exclusively found in astrocytes, are detected in studies of postmortem brain tissue from subjects with MDD. Other components of the tripartite synapse, such as postsynaptic glutamate receptors, and glutamate and GABA neurons, are also altered in brain tissue from subjects with MDD. These studies in humans are paralleled by studies in animal models related to depression that show dysregulation of similar components of glutamate and GABA systems as well as astrocytes after exposure to chronic mild and/or chronic unpredictable stress. Moreover, reductions in the density of prefrontal cortical synapses and in the expression of synapse-related genes have been reported in MDD and in animals experiencing chronic stress. This synaptic pathology may be related, in part, to the pathology of astrocytes in MDD since astrocytes control the formation, maturation, function, and elimination of synapses in the brain. Finally, numerous studies implicate glutamate-based drugs as antidepressant in the treatment of depression. Taken together these data suggest that the glutamate synapse is an important substrate in the pathology of MDD. The observations that chronic stress and depression exhibit many similar pathologies in astrocytes and glutamate and GABA support mechanistic studies to identify potential novel targets for new avenues in the treatment of depression.

Supported by NIH grant P30 GM103328

References

Altshuler LL, Abulseoud OA, Foland-Ross L, Bartzokis G, Chang S, Mintz J, et al. Amygdala astrocyte reduction in subjects with major depressive disorder but not bipolar disorder. Bipolar Disord. 2010;12:541–9.

Anderson CM, Swanson RA. Astrocyte glutamate transport: review of properties, regulation, and physiological functions. Glia. 2000;32:1–14.

Araque A, Parpura V, Sanzgiri RP, Haydon PG. Tripartite synapses: glia, the unacknowledged partner. Trends Neurosci. 1999;22:208–15.

Araya-Callís C, Hiemke C, Abumaria N, Flugge G. Chronic psychosocial stress and citalopram modulate the expression of the glial proteins GFAP and NDRG2 in the hippocampus. Psychopharmacology (Berl). 2012;224:209–22.
Auer DP, Putz B, Kraft E, Lipinski B, Schill J, Holsboer F. Reduced glutamate in the anterior cingulate cortex in depression: an in vivo proton magnetic resonance spectroscopy study. Biol Psychiatry. 2000;47:305–13.
Banasr M, Chowdhury GM, Terwilliger R, Newton SS, Duman RS, Behar KL, Sanacora G. Glial pathology in an animal model of depression: reversal of stress-induced cellular, metabolic and behavioral deficits by the glutamate-modulating drug riluzole. Mol Psychiatry. 2010;15:501–11.
Banasr M, Duman RS. Glial loss in the prefrontal cortex is sufficient to induce depressive-like behaviors. Biol Psychiatry. 2008;64:863–70.
Beneyto M, Kristiansen LV, Oni-Orisan A, McCullumsmith RE, Meador-Woodruff JH. Abnormal glutamate receptor expression in the medial temporal lobe in schizophrenia and mood disorders. Neuropsychopharmacol. 2007;32:1888–902.
Beneyto M, Meador-Woodruff JH. Lamina-specific abnormalities of AMPA receptor trafficking and signaling molecule transcripts in the prefrontal cortex in schizophrenia. Synapse. 2006;60:585–98.
Beneyto M, Meador-Woodruff JH. Lamina-specific abnormalities of NMDA receptor-associated postsynaptic protein transcripts in the prefrontal cortex in schizophrenia and bipolar disorder. Neuropsychopharmacol. 2008;33:2175–86.
Berman RM, Cappiello A, Anand A, Oren DA, Heninger GR, Charney DS, et al. Antidepressant effects of ketamine in depressed patients. Biol Psychiatry. 2000;47:351–4.
Bernard R, Kerman IA, Thompson RC, Jones EG, Bunney WE, Barchas JD, Schatzberg AF, Myers RM, Akil H, Watson SJ. Altered expression of glutamate signaling, growth factor, and glia genes in the locus coeruleus of patients with major depression. Mol Psychiatry. 2011;16:634–46.
Bezzi P, Gundersen V, Galbete JL, Seifert G, Steinhäuser C, Pilati E, Volterra A. Astrocytes contain a vesicular compartment that is competent for regulated exocytosis of glutamate. Nat Neurosci. 2004;7:613–20.
Blomstrand F, Aberg ND, Eriksson PS, Hansson E, Rönnbäck L. Extent of intercellular calcium wave propagation is related to gap junction permeability and level of connexin-43 expression in astrocytes in primary cultures from four brain regions. Neuroscience. 1999;92:255–65.
Bowers G, Cullinan WE, Herman JP. Region-specific regulation of glutamic acid decarboxylase (GAD) mrna expression in central stress circuits. J Neurosci. 1998;18:5938–47.
Bowley MP, Drevets WC, Ongür D, Price JL. Low glial numbers in the amygdala in major depressive disorder. Biol Psychiatry. 2002;52:404–12.
Braun K, Antemano R, Helmeke C, Büchner M, Poeggel G. Juvenile separation stress induces rapid region- and layer-specific changes in S100ss- and glial fibrillary acidic protein-immunoreactivity in astrocytes of the rodent medial prefrontal cortex. Neuroscience. 2009;160:629–38.
Chandley MJ, Szebeni K, Szebeni A, Crawford J, Stockmeier CA, Turecki G, et al. Gene expression deficits in pontine locus coeruleus astrocytes in men with major depressive disorder. J Psychiatry Neurosci. 2013;38:276–84.
Choudary PV, Molnar M, Evans SJ, Tomita H, Li JZ, Vawter MP, Myers RM, Bunney WE Jr, Akil H, Watson SJ, Jones EG. Altered cortical glutamatergic and GABAergic signal transmission with glial involvement in depression. Proc Natl Acad Sci USA. 2005;102:15653–58.
Clarke LE, Barres BA. Emerging roles of astrocytes in neural circuit development. Nat Rev Neurosci. 2013;14:311–21.
Cotter D, Landau S, Beasley C, Stevenson R, Chana G, MacMillan L, et al. The density and spatial distribution of GABAergic neurons, labelled using calcium binding proteins, in the anterior cingulate cortex in major depressive disorder, bipolar disorder, and schizophrenia. Biol Psychiatry. 2002b;51:377–86.
Cotter D, Mackay D, Chana G, Beasley C, Landau S, Everall IP. Reduced neuronal size and glial cell density in area 9 of the dorsolateral prefrontal cortex in subjects with major depressive disorder. Cereb Cortex. 2002a;12:386–94.

Cotter D, Mackay D, Landau S, Kerwin R, Everall I. Reduced glial cell density and neuronal size in the anterior cingulate cortex in major depressive disorder. Arch Gen Psychiatry. 2001;58:545–53.

Czéh B, Di Benedetto B. Antidepressants act directly on astrocytes: Evidences and functional consequences. Eur Neuropsychopharmacol. 2013;23:171–85.

Czéh B, Simon M, Schmelting B, Hiemke C, Fuchs E. Astroglial plasticity in the hippocampus is affected by chronic psychosocial stress and concomitant fluoxetine treatment. Neuropsychopharmacology. 2006;31:1616–26.

D'Ascenzo M, Fellin T, Terunuma M, Revilla-Sanchez R, Meaney DF, Auberson YP, et al. mGluR5 stimulates gliotransmission in the nucleus accumbens. Proc Natl Acad Sci U S A. 2007;104:1995–2000.

Davis S, Thomas A, Perry R, Oakley A, Kalaria RN, O'Brien JT. Glial fibrillary acidic protein in late life major depressive disorder: an immunocytochemical study. J Neurol Neurosurg Psychiatry. 2002;73:556–60.

Deschwanden A, Karolewicz B, Feyissa AM, Treyer V, Ametamey SM, Johayem A, et al. Reduced metabotropic glutamate receptor 5 density in major depression determined by [(11)C]ABP688 PET and postmortem study. Am J Psychiatry. 2011;168:727–34.

Duman RS, Aghajanian GK. Synaptic dysfunction in depression: potential therapeutic targets. Science. 2012;338:68–72.

Duric V, Banasr M, Stockmeier CA, Simen AA, Newton SS, Overholser JC, Jurjus GJ, Dieter L, Duman RS Altered expression of synapse and glutamate related genes in post-mortem hippocampus of depressed subjects. Int J Neuropsychopharmacol. 2013;16:69–82.

Elizalde N, Garcia-Garcia AL, Totterdell S, Gendive N, Venzala E, Ramirez MJ, et al. Sustained stress-induced changes in mice as a model for chronic depression. Psychopharmacology (Berl). 2010b;201:393–406.

Elizalde N, Pastor PM, Garcia-Garcia AL, Serres F, Venzala E, Huarte J, et al. Regulation of markers of synaptic function in mouse models of depression: chronic mild stress and decreased expression of VGLUT1. J Neurochem. 2010a;114:1302–14.

Erlander MG, Tillakaratne NJ, Feldblum S, Patel N, Tobin AJ. Two genes encode distinct glutamate decarboxylases. Neuron. 1991;7:91–100.

Ernst C, Nagy C, Kim S, Yang JP, Deng X, Hellstrom IC, et al. Dysfunction of astrocyte connexins 30 and 43 in dorsal lateral prefrontal cortex of suicide completers. Biol Psychiatry. 2011;70:312–9.

Feyissa AM, Chandran A, Stockmeier CA, Karolewicz B. Reduced levels of NR2A and NR2B subunits of NMDA receptor and PSD-95 in the prefrontal cortex in major depression. Prog Neuropsychopharmacol Biol Psychiatry. 2009;33:70–5.

Furuta A, Takashima S, Yokoo H, Rothstein JD, Wada K, Iwaki T. Expression of glutamate transporter subtypes during normal human corticogenesis and type II lissencephaly. Brain Res Dev Brain Res. 2005;155:155–64.

Garcia-Garcia AL, Elizalde N, Matrov D, Harro J, Wojcik SM, Venzala E, et al. Increased vulnerability to depressive-like behavior of mice with decreased expression of VGLUT1. Biol Psychiatry. 2009;66:275–82.

Giaume C, Theis M. Pharmacological and genetic approaches to study connexin-mediated channels in glial cells of the central nervous system. Brain Res Rev. 2010;63:160–76.

Gibbons AS, Brooks L, Scarr E, Dean B. AMPA receptor expression is increased post-mortem samples of the anterior cingulate from subjects with major depressive disorder. J Affect Disord. 2012;136:1232–7.

Gittins RA, Harrison PJ. A morphometric study of glia and neurons in the anterior cingulate cortex in mood disorder. J Affect Disord. 2011;133:328–32.

Gosselin RD, Gibney S, O'Malley D, Dinan TG, Cryan JF. Region specific decrease in glial fibrillary acidic protein immunoreactivity in the brain of a rat model of depression. Neuroscience. 2009;159:915–25.

Goswami DB, Jernigan CS, Chandran A, Iyo AH, May WL, Austin MC, et al. Gene expression analysis of novel genes in the prefrontal cortex of major depressive disorder subjects. Prog Neuropsychopharmacol Biol Psychiatry. 2013;43:126–33.

Gronli J, Fiske E, Murison R, Bjorvatn B, Sorensen E, Ursin R, et al. Extracellular levels of serotonin and GABA in the hippocampus after chronic mild stress in rats. A microdialysis study in an animal model of depression. Behav Brain Res. 2007;181:42–51.

Hammen C. Stress and depression. Annu Rev Clin Psychol. 2005;1:293–319.

Hashimoto K, Sawa A. Iyo M. Increased levels of glutamate in brains from patients with mood disorders. Biol Psychiatry. 2007;62:1310–6.

Hasler G, van der Veen JW, Tumonis T, Meyers N, Shen J, Drevets WC. Reduced prefrontal glutamate/glutamine and gamma-aminobutyric acid levels in major depression determined using proton magnetic resonance spectroscopy. Arch Gen Psychiatry. 2007;64:193–200.

Herberg LJ, Rose IC. Excitatory amino acid pathways in brain-stimulation reward. Behav Brain Res. 1990;39:230–39.

Herman JP, Larson BR. Differential regulation of forebrain glutamic acid decarboxylase mrna expression by aging and stress. Brain Res. 2001;912:60–6.

Herman JP, Renda A, Bodie B. Norepinephrine-gamma-aminobutyric acid (GABA) interaction in limbic stress circuits: effects of reboxetine on gabaergic neurons. Biol Psychiatry. 2003;53:166–74.

Hill MN, Hellemans KG, Verma P, Gorzalka BB, Weinberg J. Neurobiology of chronic mild stress: parallels to major depression. Neurosci Biobehav Rev. 2012;36:2085–117.

Hughes EG, Maguire JL, McMinn MT, Scholz RE, Sutherland ML. Loss of glial fibrillary acidic protein results in decreased glutamate transport and inhibition of PKA-induced EAAT2 cell surface trafficking. Brain Res Mol Brain Res. 2004;124:114–23.

Iwata M, Shirayama Y, Ishida H, Hazama GI, Nakagome K. Hippocampal astrocytes are necessary for antidepressant treatment of learned helplessness rats. Hippocampus. 2011;21:877–84.

Jacque CM, Vinner C, Kujas M, Raoul M, Racadot J, Baumann NA. Determination of glial fibrillary acidic protein (GFAP) in human brain tumors. J Neurol Sci. 1978;35:147–55.

Jernigan CS, Goswami DB, Austin MC, Iyo AH, Chandran A, Stockmeier CA, Karolewicz B. The mTOR signaling pathway in the prefrontal cortex is compromised in major depressive disorder. Prog Neuropsychopharmacol Biol Psychiatry. 2011;35:1774–9.

John CS, Smith KL, Van't Veer A, Gompf HS, Carlezon WA Jr, Cohen BM, et al. Blockade of astrocytic glutamate uptake in the prefrontal cortex induces anhedonia. Neuropsychopharmacol. 2012;37:2467–75.

Kang HJ, Voleti B, Hajszan T, Rajkowska G, Stockmeier CA, Licznerski P, et al. Decreased expression of synapse-related genes and loss of synapses in major depressive disorder. Nat Med. 2012;18:1413–7.

Karolewicz B, Maciag D, O'Dwyer G, Stockmeier CA, Feyissa AM, Rajkowska G. Reduced level of glutamic acid decarboxylase-67 kDa in the prefrontal cortex in major depression. Int J Neuropsychopharmacol. 2010;13:411–20.

Karolewicz B, Stockmeier CA, Ordway GA. Elevated levels of the NR2C subunit of the NMDA receptor in the locus coeruleus in depression. Neuropsychopharmacology. 2005 Aug;30(8):1557–67.

Karolewicz B, Szebeni K, Gilmore T, Maciag D, Stockmeier CA, Ordway GA. Elevated levels of NR2A and PSD-95 in the lateral amygdala in depression. Int J Neuropsychopharmacol. 2009;12:143–53.

Kaufman DL, Houser CR, Tobin AJ. Two forms of the gamma-aminobutyric acid synthetic enzyme glutamate decarboxylase have distinct intraneuronal distributions and cofactor interactions. J Neurochem. 1991;56:720–23.

Kendler KS, Thornton LM, Gardner CO. Genetic risk, number of previous depressive episodes, and stressful life events in predicting onset of major depression. Am J Psychiatry. 2001;158:582–86.

Kessler RC. The effects of stressful life events on depression. Annu Rev Psychol. 1997;48:191–214.

Khundakar A, Morris C, Oakley A, Thomas AJ. A morphometric examination of neuronal and glial cell pathology in the orbitofrontal cortex in late-life depression. Int Psychogeriatr. 2011a;23:132–40.

Khundakar AA, Morris CM, Oakley AE, Thomas AJ. Cellular pathology within the anterior cingulate cortex of patients with late-life depression: a morphometric study. Psychiatry Res. 2011b;194:184–9.

Khundakar AA, Thomas AJ. Morphometric changes in early- and late-life major depressive disorder: evidence from postmortem studies. Int Psychogeriatr. 2009;21:844–54.

Kimelberg HK. Water homeostasis in the brain: basic concepts. Neuroscience. 2004;129:851–60.

Knable MB, Torrey EF, Webster MJ, Bartko JJ. Multivariate analysis of prefrontal cortical data from the Stanley Foundation Neuropathology Consortium. Brain Res Bull. 2001;55:651–9.

Koehler RC, Roman RJ, Harder DR. Astrocytes and the regulation of cerebral blood flow. Trends Neurosci. 2009;32:160–9.

Kristiansen LV, Meador-Woodruff JH. Abnormal striatal expression of transcripts encoding NMDA interacting PSD proteins in schizophrenia, bipolar disorder and major depression. Schizophr Res. 2005;78:87–93.

Krugers HJ, Hoogenraad CC, Groc L. Stress hormones and AMPA receptor trafficking in synaptic plasticity and memory. Nat Rev Neurosci. 2010;11:675–81.

Law AJ, Deakin JF. Asymmetrical reductions of hippocampal NMDAR1 glutamate receptor mrna in the psychoses. Neuroreport. 2001;12:2971–4.

Lee Y, Gaskins D, Anand A, Shekhar A. Glia mechanisms in mood regulation: a novel model of mood disorders. Psychopharmacology (Berl). 2007;191:55–65.

Lee Y, Son H, Kim G, Kim S, Lee DH, Roh GS, et al. Glutamine deficiency in the prefrontal cortex increases depressive-like behaviours in male mice. J Psychiatry Neurosci. 2013;38:183–91.

Lepack A, Chowdhury GMI, Duric V, Maldonado-Aviles JG, Behar KL, Banasr M, et al. Chronic stress alters rates of GABA synthesis, and reduces expression of GAD67, calbindin and other GABA-related proteins in the frontal cortex of rats. Biol Psychiatry. 2013;73:123S.

Leventopoulos M, Rüedi-Bettschen D, Knuesel I, Feldon J, Pryce CR, Opacka-Juffry J. Long-term effects of early life deprivation on brain glia in Fischer rats. Brain Res. 2007;1142:119–26.

Li N, Lee B, Liu RJ, Banasr M, Dwyer JM, Iwata M, Li XY, et al. MTOR-dependent synapse formation underlies the rapid antidepressant effects of NMDA antagonists. Science. 2010;329:959–64.

Li N, Liu RJ, Dwyer JM, Banasr M, Lee B, Son H, et al. Glutamate N-methyl-D-aspartate receptor antagonists rapidly reverse behavioral and synaptic deficits caused by chronic stress exposure. Biol Psychiatry. 2011;69:754–61.

Li YK, Wang F, Wang W, Wu PF, Xiao JL, Hu ZL, et al. Aquaporin-4 deficiency impairs synaptic plasticity and associative fear memory in the lateral amygdala: involvement of downregulation of glutamate transporter-1 expression. Neuropsychopharmacology. 2012;37:1867–78.

Lou JS, Li CY, Yang XC, Fang J, Yang YX, Guo JY. Protective effect of gan mai da zao decoction in unpredictable chronic mild stress-induced behavioral and biochemical alterations. Pharm Biol. 2010;48:1328–36.

Maes M, Yirmyia R, Noraberg J, Uytterhoeven M, Vrydags N, Bosmans E. The inflammatory & neurodegenerative (I & ND) hypothesis of depression: leads for future research and new drug developments in depression. Metab Brain Dis. 2009;24:27–53.

Maciag D, Hughes J, O'Dwyer G, Pride Y, Stockmeier CA, Sanacora G, et al. Reduced density of calbindin immunoreactive GABAergic neurons in the occipital cortex in major depression: relevance to neuroimaging studies. Biol Psychiatry. 2010;67:465–70.

Mazure CM, Bruce ML, Maciejewski PK, Jacobs SC. Adverse life events and cognitive-personality characteristics in the prediction of major depression and antidepressant response. Am J Psychiatry. 2000;157:896–903.

Meshorer E, Biton IE, Ben-Shaul Y, Ben-Ari S, Assaf Y, Soreq H, et al. Chronic cholinergic imbalances promote brain diffusion and transport abnormalities. FASEB J. 2005;19:910–22.

Middeldorp J, Hol EM. GFAP in health and disease. Prog Neurobiol. 2011;93:421–43.

Michael N, Erfurth A, Ohrmann P, Arolt V, Heindel W, Pfleiderer B. Metabolic changes within the left dorsolateral prefrontal cortex occurring with electroconvulsive therapy in patients with treatment resistant unipolar depression. Psych Medicine. 2003;33:1277–84.

Miguel-Hidalgo JJ, Baucom C, Dilley G, Overholser JC, Meltzer HY, Stockmeier CA, et al. Glial fibrillary acidic protein immunoreactivity in the prefrontal cortex distinguishes younger from older adults in major depressive disorder. Biol Psychiatry. 2000;48:861–73.

Miguel-Hidalgo JJ, Waltzer R, Whittom AA, Rajkowska G, Stockmeier CA. Glial and glutamatergic markers in depression, alcoholism, and their comorbidity. J Affect Disord. 2010;127:230–40.

Miguel-Hidalgo JJ, Wilson BA, Meshram A, Rajkowska G, Hussain S, Stockmeier CA. Biochemical and immunohistochemical evidence for reduced Gap junction-forming connexin 43 in the orbitofrontal cortex in alcohol dependence and depression. Biol Psychiatry. 2012;71:55S.

Mirza Y, Tang J, Russell A, Banerjee SP, Bhandari R, Ivey J, et al. Reduced anterior cingulate cortex glutamatergic concentrations in childhood major depression. J Am Acad Child Adolesc Psychiatry. 2004;43:341–8.

Monroe SM, Torres LD, Guillaumot J, Harkness KL, Roberts JE, Frank E, et al. Life stress and the long-term treatment course of recurrent depression: III. Nonsevere life events predict recurrence for medicated patients over 3 years. J Consult Clin Psychol. 2006;74:112–20.

Müller MB, Lucassen PJ, Yassouridis A, Hoogendijk WJ, Holsboer F, Swaab DF. Neither major depression nor glucocorticoid treatment affects the cellular integrity of the human hippocampus. Eur J Neurosci. 2001;14:1603–12.

Nedergaard M, Verkhratsky A. Artifact versus reality-how astrocytes contribute to synaptic events. Glia. 2012;60:1013–23.

Nico B, Frigeri A, Nicchia GP, Quondamatteo F, Herken R, Errede M, et al. Role of aquaporin-4 water channel in the development and integrity of the blood-brain barrier. J Cell Sci. 2001;114:1297–07.

Nowak B, Zadrożna M, Ossowska G, Sowa-Kućma M, Gruca P, Papp M, et al. Alterations in hippocampal calcium-binding neurons induced by stress models of depression: a preliminary assessment. Pharmacol Rep. 2010;62:1204–10.

Nudmamud-Thanoi S, Reynolds GP. The NR1 subunit of the glutamate/NMDA receptor in the superior temporal cortex in schizophrenia and affective disorders. Neurosci Lett. 2004;372:173–7.

Ongür D, Drevets WC, Price JL. Glial reduction in the subgenual prefrontal cortex in mood disorders. Proc Natl Acad Sci USA. 1998;95:13290–5.

Ordway GA, Szebeni A, Chandley MJ, Stockmeier CA, Xiang L, Newton SS, et al. Low gene expression of bone morphogenetic protein 7 in brainstem astrocytes in major depression. Int J Neuropsychopharmacol. 2012;15:855–68.

Paradise MB, Naismith SL, Norrie LM, Graeber MB, Hickie IB. The role of glia in late-life depression. Int Psychogeriatr. 2012;24:1878–90.

Paulson OB, Newman EA. Does the release of potassium from astrocyte endfeet regulate cerebral blood flow? Science. 1987;237:896–98.

Pfleiderer B, Michael N, Erfurth A, Ohrmann P, Hohmann U, Wolgast M, et al. Effective electroconvulsive therapy reverses glutamate/glutamine deficit in the left anterior cingulum of unipolar depressed patients. Psych Research. 2003;122:185–92.

Pilc A, Wierońska JM, Skolnick P. Glutamate-Based Antidepressants: Preclinical Psychopharmacology. Biol Psychiatry. 2013 Feb 28. doi:pii: S0006-3223(13)00092-9. 10.1016/j.biopsych.2013.01.021. (Epub ahead of print).

Pittenger C, Duman RS. Stress, depression, and neuroplasticity: a convergence of mechanisms. Neuropsychopharmacology. 2008;33:88–109.

Rajkowska G, Hughes J, Stockmeier C, Miguel-Hidalgo JJ, Maciag D. Coverage of blood vessels by astrocytic endfeet is reduced in major depressive disorder. Biol Psychiatry. 2013;73:613–21.

Rajkowska G, Miguel-Hidalgo JJ. Gliogenesis and glial pathology in depression. CNS Neurol Disord Drug Targets. 2007;6:219–33.

Rajkowska G, Miguel-Hidalgo JJ, Wei J, Dilley G, Pittman SD, Meltzer HY, et al. Morphometric evidence for neuronal and glial prefrontal cell pathology in major depression. Biol Psychiatry. 1999;45:1085–98.

Rajkowska G, Miguel-Hidalgo JJ, Dubey P, Stockmeier CA, Krishnan KR. Prominent reduction in pyramidal neurons density in the orbitofrontal cortex of elderly depressed patients. Biol Psychiatry. 2005;58:297–306.

Rajkowska G, O'Dwyer G, Teleki Z, Stockmeier CA, Miguel-Hidalgo JJ. GABAergic neurons immunoreactive for calcium binding proteins are reduced in the prefrontal cortex in major depression. Neuropsychopharmacology. 2007;32:471–82.

Rajkowska G, Stockmeier CA. Astrocyte pathology in major depressive disorder: insights from human postmortem brain tissue. Curr Drug Targets. 2013 Oct;14(11):1225–36.

Raudensky J, Yamamoto BK. Effects of chronic unpredictable stress and methamphetamine on hippocampal glutamate function. Brain Res. 2007;1135:129–35.

Sanacora G, Banasr M. From pathophysiology to novel antidepressant drugs: glial contributions to the pathology and treatment of mood disorders. Biol Psychiatry. 2013;73:1172–9.

Sanacora G, Gueorguieva R, Epperson CN, Wu YT, Appel M, Rothman DL. Subtype-specific alterations of gammaaminobutyric acid and glutamate in patients with major depression. Arch Gen Psychiatry. 2004;61:705–13.

Sanacora G, Zarate CA, Krystal JH, Manji HK. Targeting the glutamatergic system to develop novel, improved therapeutics for mood disorders. Nat Rev Drug Discov. 2008;7:426–37.

Schipke CG, Heuser I, Peters O. Antidepressants act on glial cells: SSRIs and serotonin elicit astrocyte calcium signaling in the mouse prefrontal cortex. J Psychiatr Res. 2011;45:242–8.

Sequeira A, Mamdani F, Ernst C. *et al.* Global brain gene expression analysis links glutamatergic and GABAergic alterations to suicide and major depression. PLoS One. 2009;4:e6585.

Si X, Miguel-Hidalgo JJ, O'Dwyer G, Stockmeier CA, Rajkowska G. Age-dependent reductions in the level of glial fibrillary acidic protein in the prefrontal cortex in major depression. Neuropsychopharmacology. 2004;29:2088–96.

Stockmeier C, Cobb JA, O'Neill K, Milner JN, Simpson J, Lawrence TJ, et al. Astrocytic Alterations in Postmortem Hippocampus in Major Depressive Disorder (MDD). Biol Psychiatry. 2010;67:56S.

Sullivan SM, Lee A, Björkman ST, Miller SM, Sullivan RK, Poronnik P, et al. Cytoskeletal anchoring of GLAST determines susceptibility to brain damage: an identified role for GFAP. J Biol Chem. 2007;282:29414–23.

Toro CT, Hallak JE, Dunham JS, Deakin JF. Glial fibrillary acidic protein and glutamine synthetase in subregions of prefrontal cortex in schizophrenia and mood disorder. Neurosci Lett. 2006;404:276–81.

Torres-Platas D, Mackay D, Chana G, Beasley C, Landau S, Everall IP. Reduced neuronal size and glial cell density in area 9 of the dorsolateral prefrontal cortex in subjects with major depressive disorder. Cereb Cortex. 2002;12:386–94.

Torres-Platas SG, Hercher C. Davoli MA, Maussion G, Labonté B, Turecki G, et al. Astrocytic hypertrophy in anterior cingulate white matter of depressed suicides. Neuropsychopharmacology. 2011;36:2650–8.

Venzala E, García-García AL, Elizalde N, Tordera RM. Social vs. Environmental stress models of depression from a behavioural and neurochemical approach. Eur Neuropsychopharmacol. 2012;23:697–708.

Wang Q, Van Heerikhuize J, Aronica E, Kawata M, Seress L, Joels M, et al. Glucocorticoid receptor protein expression in human hippocampus; stability with age. Neurobiol Aging. 2013;34:1662–73.

Webster MJ, O'Grady J, Kleinman JE, Weickert CS. Glial fibrillary acidic protein mRNA levels in the cingulate cortex of individuals with depression, bipolar disorder and schizophrenia. Neuroscience. 2005;133:453–61.

Wierońska JM, Brański P, Szewczyk B, Pałucha A, Papp M, Gruca P, et al. Changes in the expression of metabotropic glutamate receptor 5 (mGluR5) in the rat hippocampus in an animal model of depression. Pol J Pharmacol. 2001;53:659–62.

Will CC, Aird F, Redei EE. Selectively bred Wistar-Kyoto rats: an animal model of depression and hyper-responsiveness to antidepressants. Mol Psychiatry. 2003;8:925–32.

Yang C, Hu YM, Zhou ZQ, Zhang GF, Yang JJ. Acute administration of ketamine in rats increases hippocampal BDNF and mTOR levels during forced swimming test. Ups J Med Sci. 2013;118:3–8.
Yuen EY, Wei J, Liu W, Zhong P, Li X, Yan Z. Repeated stress causes cognitive impairment by suppressing glutamate receptor expression and function in prefrontal cortex. Neuron. 2012;73(5):962–77.
Zadrozna M, Nowak B, Lason-Tyburkiewicz M, Wolak M, Sowa-Kucma M, Papp M, et al. Different pattern of changes in calcium binding proteins immunoreactivity in the medial prefrontal cortex of rats exposed to stress models of depression. Pharmacol Rep. 2011;63:1539–46.
Zarate CA Jr, Singh JB, Carlson PJ, Brutsche NE, Ameli R, Luckenbaugh DA, et al. A randomized trial of an N-methyl-D-aspartate antagonist in treatment-resistant major depression. Arch Gen Psychiatry. 2006;63:856–64.
Zeng XN, Sun XL, Gao L, Fan Y, Ding JH, Hu G. Aquaporin-4 deficiency down-regulates glutamate uptake and GLT-1 expression in astrocytes. Mol Cell Neurosci. 2007;34:34–9.

Chapter 15
Evidence of Glutamatergic Dysfunction in the Pathophysiology of Schizophrenia

J.C. Hammond, D. Shan, J.H. Meador-Woodruff and R.E. McCullumsmith

Abstract Abnormalities of the glutamate system are widely recognized to be involved in the pathophysiology of schizophrenia, though the exact mechanism is still unclear. Accumulating evidence from postmortem studies has implicated alterations in several components of glutamatergic synapses, including abnormalities of glutamate receptors and transporters. These data support the hypothesis that expression, trafficking, and downstream signaling pathways of *N*-methyl-D-aspartate (NMDA) receptors are altered in this illness. Changes in glutamate transporter expression suggest there may be chronic glutamate spillover from the synaptic cleft, leading to increased activation of extrasynaptic glutamate receptors. We propose that changes in NMDA-subtype glutamate receptor function and glutamate transporter expression are components of a common pathophysiological pathway leading to the schizophrenia phenotype.

Abbreviations

NMDA	N-methyl-D-aspartate
CRH	Corticotropin-releasing hormone
AMPAR	α-amino-3-hydroxy-5-methyl-4-isoxazolepropionic acid receptor
GluA	Glutamate receptor AMPA-subtype
EAAT	Excitatory amino acid transporter
CSF	Cerebrospinal fluid
PCP	Phencyclidine
LTP	Long-term potentiation
LTD	Long-term depression
GluN	Glutamate receptor NMDA-subtype
EEA1	Early endosome Antigen
VGLUT	Vesicular glutamate transporter

R.E. McCullumsmith (✉)
Department of Psychiatry and Behavioral Neuroscience, University of Cincinnat, Room E688B, 231 Albert Sabin Way, ML0583, PO Box 670583, Cincinnati, OH 45267-0583, USA
e-mail: robert.mccullumsmith@uc.edu

J.C. Hammond · D. Shan · J.H. Meador-Woodruff
Department of Psychiatry and Behavioral Neurobiology, University of Alabama at Birmingham, Birmingham, USA

M. Popoli et al. (eds.), *Synaptic Stress and Pathogenesis of Neuropsychiatric Disorders*, 265
DOI 10.1007/978-1-4939-1056-4_15, © Springer Science+Business Media New York 2014

xCT	Cystine-glutamate antiporter
PSD95	Postsynaptic density 95
SynGAP	Synaptic GTPase activating protein

15.1 Introduction

While the pathophysiology of schizophrenia has focused on dopamine abnormalities for decades, accumulating evidence suggests abnormalities of the glutamate system in this illness. Glutamate neurotransmission is typically tightly regulated and alterations in glutamate release, receptor activation or glutamate reuptake may result in altered synaptic function. Interestingly, environmental stress causes secretion of corticotropin-releasing hormone (CRH) and elevation of cortisol, which in turn alters glutamate neurotransmission (Lowy et al. 1993; Bagley and Moghaddam 1997; Heim et al. 2002; Thompson et al. 2007). For example, increases in stress and cortisol levels lead to increased presynaptic release of glutamate in preclinical models (Bagley and Moghaddam 1997; Musazzi et al. 2010). On the postsynaptic neuron, glutamate receptor expression is also altered in response to stress. Administration of corticosterone to neuronal cultures altered the trafficking of the α-amino-3-hydroxy-5-methyl-4-isoxazolepropionic acid (AMPA) receptor subunit GluA2 to the membrane (Groc et al. 2008). Additionally, stress may cause a reduction in synaptic spines (Chen et al. 2008). Further, expression of one of the excitatory amino acid transporters (EAATs), EAAT2, and removal of glutamate from the synapse may be altered by stress (Zink et al. 2010). Thus, environmental stress and hormone release affects the entire synapse and may regulate glutamate neurotransmission. It may be argued that persons with severe mental illness have chronic, unpredictable, stressful life episodes, often attributable to manifestations of their illness, such as psychosis. In this chapter, we will focus on the pathophysiology of schizophrenia, a severe mental illness associated with profound loss of function and recurring stressful psychotic episodes.

15.2 Overview of Schizophrenia

Schizophrenia is a severe mental illness that directly afflicts about 1 % of the adult US population, and many more people indirectly (Bhugra 2005; Wu et al. 2005). Development of schizophrenia at a relatively young age, late teens to early twenties for men and twenties to early thirties for women, creates significant burdens for the sick, their families, and society (Buchanan and Carpenter 2000). Patients with schizophrenia typically endure multiple hospitalizations, medication side effects, and psychotic symptoms that hinder their ability to live independently and cost society billions of dollars annually (Knapp et al. 2004).

Schizophrenia is characterized by a myriad of clinical findings, including positive, negative, and cognitive symptoms (Association 2000; Buchanan and Carpenter 2000). Positive symptoms include delusions, hallucinations, or agitation. Oftentimes patients report auditory hallucinations in the form of a running commentary of the patients thoughts and actions (Kay 1990; Badcock 2010). Negative symptoms, including lack of drive, social withdrawal, decreased eye contact, apathy, and diminished spontaneous movement, may be more debilitating to patients and are often not addressed by pharmacotherapy (Laruelle et al. 1999; Fleischhacker 2000). In addition to positive and negative symptoms, patients may have cognitive deficits such as disorganized thinking and deficits in executive functioning, verbal fluency, and working memory (Rajji and Mulsant 2008; Szoke et al. 2008; Wobrock et al. 2008; Potkin et al. 2009; Zanello et al. 2009).

15.2.1 Neurodevelopmental Hypothesis of Schizophrenia

While there are well-established criteria in place for making the diagnosis of schizophrenia, the cause is still unknown. Most recent evidence supports a combination of genetic and environmental factors contributing to the development of the disorder (Marenco and Weinberger 2000). To date, no one gene, single nucleotide polymorphism, or mutation has been consistently linked to the illness, and it is likely that multiple susceptibility genes create a predisposition to developing schizophrenia (Gershon et al. 2011; Fanous et al. 2012; Levinson et al. 2012). Hypotheses regarding the underlying pathophysiology of schizophrenia are primarily centered on abnormalities of neurodevelopment, brain structure, and neurotransmission (Javitt and Zukin 1991; Coyle 1996; Roy et al. 1998; Marenco and Weinberger 2000; Kraguljac et al. 2012b). However, the course of schizophrenia suggests that it should be viewed as a longitudinal developmental model, rather than as a static neurochemical model (Marenco and Weinberger 2000; Lewis and Levitt 2002). The neurodevelopmental hypothesis of schizophrenia suggests prenatal, perinatal, and postnatal events are associated with the development of the schizophrenia phenotype. Several studies suggest perinatal or early life stress or trauma increases the risk of schizophrenia in susceptible individuals (Li et al. 2009b; Holloway et al. 2013; Novak et al. 2013). These susceptible individuals may then develop positive and negative symptoms during late adolescence or early adulthood typically following a lengthy period of normal development through puberty (Alda et al. 1996; Holtzman et al. 2013). Patients diagnosed with schizophrenia may experience waxing and waning of symptoms throughout their lifetime, accompanied by a decline in social, occupational, and cognitive functioning.

15.2.2 Stress and Schizophrenia

People with schizophrenia have altered responses to stress, and significant environmental stress may trigger a relapse or necessitate hospitalization (Braff et al. 2001a, b; Marwaha and Johnson 2004). Relative risk of developing schizophrenia is increased by stress, including childhood trauma and increased use of marijuana prior to age 18 (Kristensen and Cadenhead 2007; Bossong and Niesink 2010; Brown 2011). Stress responses can increase cytokine levels in cerebrospinal fluid (CSF). Inflammatory stress, such as maternal infection during pregnancy, may increase the risk of developing schizophrenia (for the fetus), possibly due to cytokines crossing the placental barrier (Babulas et al. 2006; Brown 2006). Other work has investigated the role of immune modulation in rodent models of schizophrenia. For example, challenges to the immune system of pregnant rats will cause altered social behaviors in the offspring that may be restored by administration of dopamine antagonists (Richtand and McNamara 2008; Bronson et al. 2011). For patients with schizophrenia, development of a routine to lessen stress may help to prevent relapse for these individuals (Torrey 2006).

While the interplay of genetic susceptibility and environmental stressors may contribute to the development of schizophrenia, it is likely that there are common cellular and neurochemical changes in the pathophysiology of the illness (Fig. 15.1). These common pathophysiological pathways likely include abnormalities of neurotransmitters such as glutamate, which are found throughout the central nervous system. The effects of phencyclidine (PCP), an *N*-methyl-D-aspartate (NMDA)-subtype glutamate receptor antagonist, strongly implicate abnormalities of glutamate in this illness. PCP induces psychotic symptoms in naive subjects and exacerbates symptoms in subjects with schizophrenia (Javitt and Zukin 1991; Lahti et al. 1995; Lahti and Tamminga 1995; Tamminga 1999). Chronic administration of NMDA receptor antagonists, like PCP, can induce a persistent psychotic symptomatology (Morris et al. 2005; Reynolds et al. 2005). These data suggest a central role for glutamate transmission in schizophrenia.

The persistence of changes in a system in response to stimuli is referred to as plasticity (Gordon 1969). Neuroplasticity refers, in part, to the ability of the brain to learn and form new memories. Molecular correlates of learning and memory, including long-term potentiation (LTP) and long-term depression (LTD), facilitate the strengthening or weakening of synapses, shaping the functional connectivity of neurocircuits (Malenka and Nicoll 1999; McCullumsmith et al. 2004; Talbot et al. 2009). Interestingly, glutamate neurotransmission is central to LTP, LTD, and plasticity (Lewis et al. 2004; Deep-Soboslay et al. 2011; McCullumsmith and Meador-Woodruff 2011). The effects of PCP, taken together with the central role of glutamate transmission in neuroplasticity, have led to investigation of glutamate neurotransmission in schizophrenia. Considering schizophrenia as a disorder of neuroplasticity is one way to integrate the neurochemical and developmental hypotheses of the illness. In the next section, we will describe the critical components of glutamate transmission found within synapses and without.

15.3 Biology of Excitatory Glutamate Synapses

There are three cells involved in the release, activity, and reuptake of glutamate: presynaptic neurons, postsynaptic neurons, and astrocytes. Glutamate, released from vesicles in the presynaptic neuron may bind to and activate ionotropic

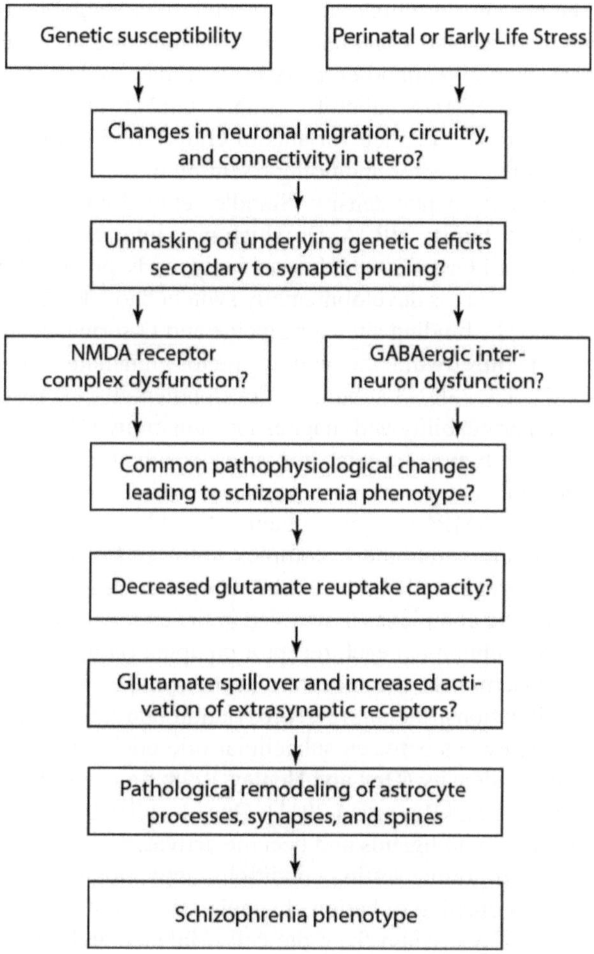

Fig. 15.1 Schematic of putative changes in glutamate neurotransmission in pathophysiology of schizophrenia. NMDA N-methyl-D-aspartate, GABA Gamma-aminobutyric acid

(NMDA, AMPA, Kainate) and metabotropic ($mGluR_1$–$mGluR_8$) glutamate receptors expressed on both neurons and astrocytes.

15.3.1 Glutamate Receptor Assembly and Function

There are two groups of glutamate receptors: metabotropic G protein-coupled receptors and ionotropic ligand-gated receptors. There are three subtypes of ionotropic receptors: kainate, AMPA, and NMDA receptors (Dingledine et al. 1999). These ionotropic receptors function as ion channels in response to the binding of a ligand. Each of the AMPA receptor subunits, GluA1–4, has a ligand-binding domain, located in the extracellular N-terminus and the extracellular loop between two transmembrane domains (Armstrong et al. 1998; Armstrong and Gouaux 2000).

The presence of GluA2 in the receptor confers gating of calcium through the pore, whereas receptors lacking GluA2 subunits are permeable to calcium, sodium, and potassium (Wenthold et al. 1996; Petralia et al. 1997; Swanson et al. 1997). There are three NMDA subunits: GluN1, GluN2, and GluN3 (Tuominen et al. 2005). There are eight splice variants of GluN1 which influence the subcellular localization of the receptor, including retention in the endoplasmic reticulum or expression at the postsynaptic density (Standley et al. 2000; Stephenson 2006). There are four GluN2 subunits: NR2A–D, which may interact with different signaling molecules (Ryan and Grant 2009). Interestingly, early postnatal brains have a predominance of NR2B that is developmentally switched to NR2A-containing receptors (Liu et al. 2004). The binding sites for glycine and D-serine, coagonists for NMDA receptors, are on GluN1 while the binding site for glutamate is on GluN2 (Johnson and Ascher 1987). The GluN3 subunit is developmentally expressed and is important for calcium permeability and magnesium sensitivity (Gallinat et al. 2007).

As with most proteins, the receptor subunits are synthesized in the endoplasmic reticulum before they are packaged and assembled into functional ion channels. The four AMPA receptor subunits, GluA1–4, are typically assembled as a dimer of dimers into a tetrameric complex in the endoplasmic reticulum (Rosenmund et al. 1998; Greger and Esteban 2007; Greger et al. 2007). The NMDA receptors are also tetrameric complexes assembled in the endoplasmic reticulum, with two obligatory GluN1 subunits in each receptor complex (Dingledine et al. 1999). Localization to and insertion of the receptors at the synapse is dependent upon posttranslational modifications, including glycosylation and phosphorylation, which facilitate trafficking events between subcellular microdomains, such as the Golgi and the postsynaptic density (Dev and Henley 1998; Song and Huganir 2002; Jiang et al. 2006; Gladding and Raymond 2011). Once localized to the plasma membrane, the receptors may bind ligands and become activated.

Under normal resting conditions, activation of NMDA-type glutamate receptors leads to opening of cation channels followed by influx of calcium and sodium and efflux of potassium from the cell (Malenka and Nicoll 1999; Nicoll and Malenka 1999). However, prior to activation of NMDA receptors, activation of nearby AMPA-type glutamate receptors provides the depolarization necessary to remove the magnesium blockade of the NMDA receptors (Malinow 2003; Boehm et al. 2006).

The close proximity of the AMPA-type glutamate receptors with NMDA receptors is a highly regulated process involving multiple pools of AMPA receptors. Recent studies have described insertion of the AMPA receptors either directly at the synapse or to extrasynaptic areas (Passafaro et al. 2001; Shi et al. 2001; Hirling 2008). The lateral movement of GluA1/GluA2-containing receptors from extrasynaptic sites to the synapse following induction of LTP is referred to as the regulated receptor pool (Contractor and Heinemann 2002; Triller and Choquet 2005; Hirling 2008; Kropf et al. 2008). The constitutive cycling pool of receptors includes GluA2/GluA3-containing receptors that are cycled between the synapse and an intracellular domain (Ashby et al. 2004, 2006; Hanley 2010). This cycling of the receptors occurs in specialized vesicles called endosomes—membrane-bound organelles comprised of lipid bilayers that usually form directly from the plasma membrane

(Kobayashi et al. 1998; Carroll et al. 1999; Beattie et al. 2000). In a clathrin-dependent process, a small pocket forms in the membrane, followed by invagination of the membrane, and a closing off of the newly formed endosome via the protein dynamin (Carroll et al. 1999). These endosomes formed from the membrane are called early endosomes and express the protein early endosome antigen (EEA1) (Rubino et al. 2000). From the early endosome, AMPA receptors can be sorted to late endosomes (Rab7 positive) for degradation or to recycling endosomes (Rab11 positive) for reinsertion into the plasma membrane (Ehlers 2000). The sorting of the AMPA receptors depends, in part, on the activation of the NMDA receptors (Sossa et al. 2006). Interestingly, changes in spine morphology with LTP induction are likely due to cycling and reinsertion of AMPA receptors into the membrane from recycling endosomes (Ehlers 2003; Park et al. 2004, 2006). This turnover of AMPA receptors is essential for receptor localization and glutamate neurotransmission.

15.3.2 Glutamate Release and Reuptake

The neurotransmitter glutamate is cycled in a well-regulated process between presynaptic and postsynaptic neurons and astrocytes. In the presynaptic neuron, glutamine is converted to glutamate by the enzyme glutaminase (Bellocchio et al. 2000). Glutamate is then packaged into vesicles via vesicular glutamate transporters (VGLUTs) for release into the synapse (Takamori et al. 2000). There are three isoforms of VGLUTs: VGLUT1–3, that are differentially located in the CNS. While VGLUT1 and VGLUT2 protein are found throughout the neocortex, VGLUT1 transcripts are expressed in layers I–III while VGLUT2 transcripts are localized mainly to layer IV (Fremeau et al. 2001; Fujiyama et al. 2001; Kaneko and Fujiyama 2002). VGLUT1 and VGLUT2 may be expressed in vesicles separately in some synapses or localized together in vesicles (Fremeau et al. 2004a; Herzog et al. 2006; Liguz-Lecznar and Skangiel-Kramska 2007). The glutamate binding affinity varies among the three VGLUTs. The binding affinities of VGLUT1 (K_m = 2–3.4 mM) and VGLUT2 (K_m = 1.27–4.7 mM) are higher than the binding affinity of VGLUT3 (K_m = 0.56–1.5 mM) (Bellocchio et al. 2000; Fremeau et al. 2001; Herzog et al. 2001; Fremeau et al. 2002; Gras et al. 2002). Like VGLUT1 and VGLUT2, VGLUT3 is also distributed throughout the CNS, including expression in the cortex (Fremeau et al. 2002). Unlike VGLUT1 and VGLUT2, VGLUT3 may be expressed postsynaptically in dendrites and cell bodies suggesting VGLUT3 may have roles other than packaging glutamate into vesicles (Fremeau et al. 2002, 2004b; Herzog et al. 2004). Vesicles loaded with glutamate bind with the presynaptic membrane, and release glutamate into the synaptic cleft, where it may bind with and activate the ionotropic receptors (Hollmann and Heinemann 1994; Hollmann et al. 1994).

Excitatory amino acid transporters (EAATs) rapidly remove the released glutamate from the synapse (Masson et al. 1999). In most brain regions, 90 % of glutamate reuptake is attributable to astroglial-localized EAAT2 (Danbolt 2001). EAAT2 is expressed at high concentrations in perisynaptic regions, where it acts to "trap" glutamate in the synapse by acting like a sponge. This effect is due to the high on/off rate of glutamate binding to the transporter as well as a transport efficiency of about 0.5 (Tzingounis and Wadiche 2007). Once transported, glutamate may be recycled back to the presynaptic terminal following conversion to glutamine, or it can enter the TCA cycle and act as an energy intermediate. Finally, there is accumulating evidence that astrocytes may release glutamate through a vesicular mechanism and/or the cystine/glutamate antiporter (xCT) (Patel et al. 2001; McKenna 2011).

15.3.3 Glial Glutamate Transporters

There are multiple types of EAATs with specific cellular localization. EAAT1 and EAAT2 are primarily localized to astroglia while EAAT3 and EAAT4 are primarily localized to neurons (reviewed, O'Shea 2002). EAAT3 is located postsynaptically and is present early in development, suggesting it is involved in the development of the neuronal response to glutamate (Nieoullon et al. 2006). EAAT5 is primarily localized to the retina and will not be further discussed. EAATs function as homomers to transport 3Na^+, 1H^+, and 1 glutamate into the cell and 1K^+out of the cell (Zerangue and Kavanaugh 1996; Levy et al. 1998; Danbolt 2001). Importantly, Na^+/K^+ ATPase, which is necessary to maintain this gradient, is tightly coupled to glutamate transporters (Rose et al. 2009). Glutamate may also be exchanged with cystine via the xCT, which transports cystine into astrocytes for glutathione synthesis (Bridges et al. 2012). In rodent brain tissue, clusters of proteins have been identified in astrocytes which function as complexes to facilitate glutamate transport. One complex contained GLT1 (rodent EAAT2), Na^+/K^+ ATPase, hexokinase, and intact mitochondria (Genda et al. 2011). A similar complex that is predicted to facilitate glutamate transport contained the Na^+/K^+ ATPase, the water channel aquaporin 4, and mGluR5 (Illarionova et al. 2010). There is also evidence that adenosine signaling may regulate EAAT2 and aquaporin 4 expression in astrocytes (Lee et al. 2013). Together, these data indicate that glutamate reuptake is a complex and tightly regulated process.

15.3.4 Alterations of Glutamate Transmission in Schizophrenia

Evidence for dysfunction of the neurotransmitter glutamate and the expression of glutamate receptors in schizophrenia has long been supported by the observation that PCP and related compounds like ketamine, which are NMDA receptor antagonists, can induce both positive and negative symptoms of schizophrenia as well as cognitive deficits (Javitt and Zukin 1991; Tamminga 1999). These compounds also exacerbate positive and negative symptoms in persons with schizophrenia (Lahti et al. 1995). These findings led to the NMDA receptor hypofunction hypothesis that postulated that there was a deficit of NMDA receptor protein in schizophrenia. However, abnormalities of receptor function may not simply be a problem of too many or too few receptors, but a problem with localization and the interaction of the receptor complex with signaling pathways. We discuss these and other ideas in the sections below, where we have divided our discussion of these data based on synaptic versus extrasynaptic localization of the dependent measure.

15.4 Synaptic Alterations in Schizophrenia

15.4.1 Alterations in Glutamate Release

As previously discussed, glutamate is packaged by VGLUTs and released from the presynaptic terminal. Several laboratories have examined expression of protein and mRNA of the VGLUTs. In one report, there was a decrease in VGLUT1 protein expression and an increase in mRNA expression in the anterior cingulate cortex (ACC) in schizophrenia (Oni-Orisan et al. 2008). As these studies were done using homogenates, the disparate results may be explained by changes in intrinsic excitatory neurons or extrinsic presynaptic neurons (Oni-Orisan et al. 2008). Alternatively, it is hypothesized that the discrepancy may be explained by the presence of riboswitch RNAs, which regulate mRNA expression by sensing the need for their protein product (Blencowe and Khanna 2007; Cheah et al. 2007). Another group reported a decrease in VGLUT1 mRNA expression in the hippocampus (Eastwood and Harrison 2005). VGLUT2 mRNA expression in the inferior temporal gyrus was increased in schizophrenia (Uezato et al. 2009). In this same study, there were not any changes in VGLUT1–3 mRNA expression in the hippocampus (Uezato et al. 2009).

Binding of the vesicles to the presynaptic membrane is controlled by the SNARE (soluble NSF attachment protein/SNAP; receptor) complex. There are several reports of decreased expression of proteins in the SNARE complex including SNAP25 and syntaxin (Fatemi et al. 2001; Honer et al. 2002; Halim et al. 2003). However, there are also reports of unchanged expression of synaptophysin, syntaxin, and SNAP25 in the frontal cortex in schizophrenia (Gabriel et al. 1997; Scarr et al. 2006). Taken together, the VGLUT and SNARE complex data suggest an abnormality of presynaptic glutamate release in schizophrenia.

15.4.2 Alterations in Glutamate Reuptake

EAAT3 is an EAAT localized to neurons (Shashidharan et al. 1997; Holmseth et al. 2012). Importantly, EAAT3 regulates the amplitude of NMDA receptor currents and may limit the activation of nearby AMPA receptors at the synapse (Diamond 2001; Levenson et al. 2002; Zuo and Fang 2005). Interestingly, there are alterations in EAAT3 mRNA and protein expression in schizophrenia. Our laboratory has reported increased expression of EAAT3 mRNA and protein in the ACC (Bauer et al. 2008). However, another study found no change in protein expression of the neuronal glutamate transporter EAAT3 or presynaptic VGLUT1–2 in the superior temporal gyrus (Shan et al. 2013). While there was no alteration in EAAT3 mRNA expression in the thalamus, there was a decrease in EAAT3 mRNA expression in the striatum (Smith et al. 2001; McCullumsmith and Meador-Woodruff 2002). Other studies have reported increased EAAT3 mRNA expression in the frontal cortex, decreased expression in the striatum, and no change in the dorsolateral prefrontal cortex (DLPFC; McCullumsmith and Meador-Woodruff 2002; Lauriat et al. 2006; Nudmamud-Thanoi et al. 2007; Horiuchi et al. 2012; Rao et al. 2012). The inconsistencies in these findings are similar to those of the glutamate receptor expression and may reflect cell-specific alterations. These data suggest neuronal reuptake of glutamate may be altered in glutamate synapses in schizophrenia.

15.4.3 Alterations in Glutamate Receptor Expression

It was postulated that a loss or hypofunction of NMDA receptor activity would be present in patients with schizophrenia. However, the multiple studies (>20) of NMDA receptor expression in postmortem brains from patients with schizophrenia have yielded inconsistent findings (reviewed in McCullumsmith et al. 2012). With the notable exception of AMPA receptor subunits in hippocampus, studies of AMPA and kainate are also generally inconsistent with divergent findings across multiple brain regions and levels of gene expression (Meador-Woodruff et al. 2001).

15.4.4 Alterations in Glutamate Receptor Trafficking Proteins

Glutamate receptors interact with several proteins with myriad functions including trafficking of the receptors, stabilization of the receptors within the synapse, and downstream signaling pathways. Trafficking of AMPA receptors is regulated, in part, by cornichons and transmembrane AMPAR-regulatory proteins (TARPs) (Schwenk et al. 2009; Kato et al. 2010). Our laboratory has found profound abnormalities in mRNA transcript expression of several cornichon proteins in schizophrenia (Drummond et al. 2012).

AMPA receptor trafficking is complex and includes pools of receptors that may translocate to and from the synapse or be turned over in endosomes. Clathrin-mediated endocytosis of AMPA receptors is also regulated by multiple proteins. There are reports of alterations in dynamin-1, amphiphysin, and AP-2, proteins involved in receptor endocytosis, in schizophrenia (Pennington et al. 2008; English et al. 2009; Focking et al. 2011). Work in our laboratory has examined subcellular localization of the AMPA receptors in endosome compartments. While the expression of AMPA receptor subunits in late endosomes, typically destined for degradation, was not changed in this illness, we found increased expression of one AMPA receptor subunit in early endosomes (Hammond et al. 2010, 2011). There are also several reports of decreased expression of AMPA receptor trafficking proteins in schizophrenia (Dev et al. 1999, Mirnics et al. 2000, Toyooka et al. 2002a, b; Whiteheart and Matveeva 2004; Lu and Ziff 2005; Beneyto and Meador-Woodruff 2006). Taken together, these data are consistent with the hypothesis that it is receptor trafficking and signaling, not global receptor expression levels, that are abnormal in this illness.

There is also evidence of abnormal NMDA receptor trafficking in schizophrenia. In the thalamus of patients with schizophrenia, there was reduced transcript expression of NR1 exon 22-containing isoforms, which regulate intracellular distribution and cell surface expression of NMDA receptors (Ehlers et al. 1995; Okabe et al. 1999; Clinton et al. 2003). Using postmortem human brain homogenate, our laboratory isolated the endoplasmic reticulum and found decreased protein expression of postsynaptic density 95 (PSD95) and the NR2B subunit in this fraction, suggesting an increased rate of transit through the endoplasmic reticulum (Kristiansen et al. 2010a). Further, trafficking of NR2B-containing NMDA receptors is controlled in part by association of the receptor with a microtubule-associated complex consisting of several proteins (including CASK, ABPA1, and mLin7) bound to the microtubule-associated ATPase, KIF17 (Setou et al. 2000). There was increased expression of transcripts for CASK, ABPA1, and mLin7 and decreased expression of protein for CASK and mLin7 in schizophrenia, suggesting NR2B-containing NMDA receptor transport may be altered in schizophrenia (Kristiansen et al. 2010b). Together, these data implicate altered trafficking of NMDA receptors in schizophrenia.

Unlike AMPA receptors, NMDA receptors are typically not turned over as rapidly at the synapse. However, NMDA receptor clustering and synaptic localization is associated with a complex of proteins (Sheng and Lee 2000). Profound abnormalities of proteins in this complex have been described in schizophrenia (Toyooka et al. 2002b; Clinton et al. 2003; Kristiansen et al. 2006; Beneyto and Meador-Woodruff 2008; Kristiansen et al. 2010b; Sodhi et al. 2011). In particular, PSD95 and synaptic GTPase activating protein (SynGAP) are decreased (Funk et al. 2009). Posttranslational modifications of NMDA receptors are essential for localization to and functioning of the receptors at the synapse. For example, mice with decreased expression of the phosphorylation site serine 897 on NR1 exhibit impaired incorporation of the NMDA receptor at the synapse and impaired LTP (Li et al. 2009a). In postmortem studies of brains from patients with schizophrenia, there is decreased phosphorylation of the NR1 subunit at this serine residue (Emamian et al. 2004). These alterations in glutamate receptor trafficking and localization suggest dysfunction of glutamate receptor signaling in schizophrenia.

15.4.5 Alterations in Downstream Glutamate Signaling

Signaling pathways that are downstream of glutamate receptor activation have also been implicated in schizophrenia. The protein SynGAP, which is decreased in this illness, couples with PSD95 and NMDA receptors to regulate downstream signaling of the MAP/ERK pathway, which is important for NMDA receptor localization, cell growth and apoptosis (Komiyama et al. 2002). Phosphorylation and expression of signaling molecules in the MAP signaling pathway and the cyclic adenosine monophosphate (cAMP) signaling pathway are also altered in schizophrenia (Funk et al. 2012). For example, expression of the signaling proteins Rack1, Fyn, and Cdk5 as well as phosphorylation of PSD95 at serine 295 and NR2B at Y1336 were increased in the DLPFC in schizophrenia (Funk et al. 2012). Interestingly, in this same study, expression of the proteins Rap2, JNK1, JNK2, and PSD-95, and phosphorylation of JNK1/2 at threonine 183/tyrosine185 and PSD-95 at serine 295 were decreased in the ACC (Funk et al. 2012).

Alterations were also found in the Duo/Ras-related C3 botulinum toxin substrate 1/p21-activated kinase 1 (PAK1) pathway. The proteins Duo and Cdc42 phosphorylate PAK1, which modifies the activity of regulatory myosin light chain (MLC) and cofilin (Rex, Chen et al. 2009). Alterations of MLC and cofilin phosphorylation may alter dendritic spine maintenance via alterations of actin cytoskeleton dynamics (Hotulainen and Hoogenraad 2010). In schizophrenia, expression of Duo and phosphorylation of PAK1 were decreased in the ACC and DLPFC (Rubio et al. 2012). Cdc42 protein was decreased and phosphorylation of MLC was increased in the ACC, but not the DLPFC in schizophrenia (Rubio et al. 2012). These results suggest that there are region-specific differences in signal transduction pathways in this illness.

Other signaling pathways have been implicated as well. Several studies have detailed alterations in proteins and phosphoproteins in the neuregulin1-ErbB4 pathway, which modulates LTP, neuronal migration and synaptic activity (Anton et al. 2004; Li et al. 2007). One group reported decreased sarcoma (Src) kinase activity following ErbB4 activation associated with postsynaptic densities isolated from brain tissues from subjects with schizophrenia (Hahn et al. 2006; Hahn 2011). Signaling through the dopamine- and cAMP-regulated neuronal phosphoprotein (DARPP-32) protein is also implicated. Interestingly, stimulation of the dopamine receptor phosphorylates DARPP32 through the cAMP pathway, while stimulation of the NMDA receptor dephosphorylates phospho-DARPP32 through activation of calcineurin (Walaas and Greengard 1984; Wang et al. 1988; Halpain et al. 1990). DARPP32, in its phosphorylated or dephosphorylated state, then activates or deactivates the next protein in the signaling pathway. Ultimately, these signaling pathways regulate cellular functions including transcription and translation, DNA methylation, protein trafficking, and cellular metabolism (Lalli and Sassone-Corsi 1994; Markiv et al. 2012).

15.5 Extrasynaptic Alterations in Schizophrenia

15.5.1 Alterations in Glial Glutamate Reuptake

In addition to alterations in glutamate receptors and downstream signaling, alterations in the EAATs have been found in schizophrenia. In general, there is a decrease in the regional expression of the glial transporters EAAT1 and EAAT2 (Ohnuma et al. 1998; Smith et al. 2001; McCullumsmith 2002; Lauriat et al. 2006; Shan et al.

2013). Knockout of glutamate aspartate transporter (GLAST) (EAAT1) in mice causes schizophrenia-like behavioral endophenotypes, locomotor hyperactivity, and abnormal social behaviors, which are reversed with administration of antipsychotic medication (Karlsson et al. 2008, 2009). Interestingly, the GLAST (EAAT1) knockout animals, which likely have a subtle increase of synaptic glutamate, have cognitive and behavioral impairment, but no obvious neurotoxic abnormalities (Watase et al. 1998; Karlsson et al. 2008, 2009). One region where total EAAT2 protein was not changed is the DLPFC. However, there is a large increase in a putative negative regulator of EAAT2 protein expression in this region, and changes in EAAT glycosylation suggest decreased surface expression of these transporters in the frontal cortex (Bauer et al. 2008, 2010).

Enzymes involved in the cycling of glutamate and glutamine are also altered in schizophrenia. For example, glutaminase is increased while glutamine synthetase and carboxypeptidase II are decreased in the thalamus and frontal cortex, respectively (Burbaeva et al. 1999; Goff and Coyle 2001; Gluck et al. 2002; Laruelle et al. 2003; Ghose et al. 2004; Bruneau et al. 2005). These studies support the hypothesis of impaired glutamate synthesis and cycling in schizophrenia.

15.5.2 Extrasynaptic Receptors

Extrasynaptic receptors are located near the synapse and may modulate glutamate neurotransmission. The metabotropic glutamate receptors (mGluRs), which are located extrasynaptically, may increase or decrease NMDA receptor activity (Ambrosini et al. 1995; Skeberdis et al. 2001; Lea et al. 2002). One study examined expression of these receptors and found increased expression of mGluR1 and mGluR2/3 in the prefrontal cortex, but no changes in the striatum in schizophrenia (Gupta et al. 2005). In another study, mRNA expression of the mGluRs was unchanged in the thalamus in schizophrenia (Richardson-Burns et al. 2000). Expression of mGluR5 in pyramidal cells of Brodmann area 11 of the frontal cortex is also increased in this illness (Ohnuma et al. 1998). These data suggest that alterations in extrasynaptic mGluR receptors contributing to the pathophysiology of schizophrenia may be region specific.

The metabotropic glutamate receptors likely work in concert with the glutamate transporters. There is an increase in the ratio of mGluR5 mRNA expression to EAAT2 mRNA expression in the parahippocampal gyrus (Ohnuma et al. 2000a). An increased ratio of mGluRs to EAAT2 is also described in the prefrontal cortex suggesting dysfunction of glutamate reuptake (Ohnuma et al. 1998). Further work has been done with the group II mGluR receptors and the xCT. Protein expression of xCT is elevated in postmortem human DLPFC (Baker et al. 2008). Interestingly, N-acetylcysteine, a cysteine prodrug, blunts psychotomimetic effects in rodents treated with PCP (Baker et al. 2008). These data suggest that the xCT may be an extrasynaptic target for treatment.

NMDA-type glutamate receptors are also located extrasynaptically. Typically, these extrasynaptic NMDA receptors are distributed along the sides of the spine and the surface of the dendrites where they are associated and have contacts with axons and glia (Aoki et al. 1994; Kharazia and Weinberg 1999; Takumi et al. 1999; Petralia et al. 2010). These extrasynaptic NMDA receptors usually contain NR2B, and activation of these receptors in rodent hippocampal neurons induces LTD (Scimemi et al. 2004; Alamilla and Gillespie 2011; Liu et al. 2012). There are several reports of alterations in the NR2B subunit in particular in schizophrenia. One group reported a possible shift in relative subunit mRNA expression in the prefrontal and parietotemporal cortices in schizophrenia, without alterations in total subunit expression (Akbarian et al. 1996). Another group used radio-ligand binding and found increased NR2B-containing receptors in the superior temporal cortex (Grimwood et al. 1999). There is also an increase in NR2B subunit expression in the hippocampus in schizophrenia (Gao et al. 2000). Together, these data suggest alterations in extrasynaptic receptors may be present in schizophrenia.

15.5.3 Kynurenic Acid

Emerging evidence also describes the role of decreased activity of the enzyme for tryptophan/kynurenine metabolism, indoleamine 2,3-dioxygenase (IDO), in the development of schizophrenia (Muller et al. 2012; Steiner et al. 2012; Anderson and Maes 2013; Carlborg et al. 2013). It has been hypothesized that patients with schizophrenia may have a dysfunction in their immune response and may have decreased levels of IDO, which results in increased levels of kynurenic acid (Muller et al. 2012). In preclinical models, rodents with increased levels of kynurenic acid exhibited neurocognitive defects, including impairment in learning and memory and altered prepulse inhibition (Wonodi and Schwarcz 2010). Interestingly, kynurenic acid acts as an antagonist of NMDA-subtype glutamate receptors (Muller 2008).

15.5.4 Glutamate Receptor Modulators

The NMDA receptor has a coagonist binding site, where compounds such as D-serine and glycine may bind to positively modulate the receptor. While there is ongoing debate about which of these amino acids is the endogenous ligand for the NMDA receptor, one recent study argues that synaptic NMDA receptors preferentially bind D-serine, while extrasynaptic receptors bind glycine (Henneberger et al. 2010; Papouin et al. 2012). If schizophrenia is a problem of NMDA receptor hypofunction, it follows that treatment or adjunct treatment of patients with NMDA receptor coagonists may improve the symptoms of this illness. Supporting this notion, patients with schizophrenia have decreased serum levels of D-serine (Hashimoto et al. 2003). In one small clinical trial, glycine administration reduced negative symptoms of patients with schizophrenia (Heresco-Levy et al. 2004). Other studies using D-serine or cycloserine had encouraging results, but a recent large clinical trial did not show an effect of D-serine on positive, negative, or cognitive symptoms or global assessment of functioning scores (Tsai et al. 1998; Heresco-Levy et al. 2005; Kantrowitz et al. 2010; Lane et al. 2010; Tsai and Lin 2010). In another recent study, patients who were treated with clozapine had worsening of negative symptoms or exacerbation of positive symptoms with D-cycloserine or glycine adjunct therapy (Goff et al. 1996, 1999; Potkin et al. 1999). One explanation for worsening of symptoms with adjunct coagonist treatment is that D-cycloserine or glycine may be selectively activating extrasynaptic NMDA receptors (Watson et al. 1990; Lane et al. 2006). Regardless of the mechanism, the idea that activation of synaptic versus extrasynaptic NMDA receptors differentially affects synapses may be an important development towards understanding the role of NMDA receptors in this illness.

15.6 Summary and Conclusions

While there is mounting evidence for alterations in glutamate neurotransmission in schizophrenia, there is no clear or consistent pattern of alterations in glutamate receptor subunit expression (McCullumsmith et al. 2012). In contrast, several studies have consistently found changes in glutamate receptor trafficking molecules (Hammond et al. 2010; Funk et al. 2012). For example, at least six different studies found decreased PSD95 mRNA or protein (Ohnuma et al. 2000b; Clinton and Meador-Woodruff 2004; Toro and Deakin 2005; Kristiansen et al. 2006; Funk et al. 2009, 2012). These data suggest the hypothesis that there is not a problem, for example, of too much or too little NMDA receptor, but a problem of how NMDA receptors are localized. It follows that linkage of receptors to their intracellular signaling partners may be impaired as well, if the receptors themselves are not properly localized. Converging evidence supports this prediction, including one study that found decreased Src kinase activity associated with PSDs isolated from subjects with schizophrenia (Moghaddam and Adams 1998; Funk et al. 2009; Kantrowitz and Javitt 2010; Pitcher et al. 2011). Taken together, these data support the hypothesis that the NMDA receptor signaling complex is "sick" in this illness.

Abnormalities of the NMDA receptor signaling complex could be secondary to impaired gamma-aminobutyric acid (GABA) interneuron function. Several studies have demonstrated deficiencies of parvalbumin positive interneurons in the frontal cortex in schizophrenia, and together these data suggest a shift in the excitatory/inhibitory balance towards excitation (Lewis et al. 2004, 2008). Such aberrant modulation of NMDA receptors could yield the pathophysiological receptor changes found in postmortem brain described above.

Accumulating evidence also suggests that there is diminished glutamate reuptake capacity in schizophrenia. EAAT2 protein expression is decreased in several brain areas, and a negative regulator of glutamate reuptake is elevated in the frontal cortex. Due to its dual role as a buffer and a transporter of glutamate, decreased density or altered localization of a glial glutamate transporter could lead to spillover of glutamate out of synapses, activating extrasynaptic receptors and altering input specificity of cortical circuits. Other data suggest increased levels of the xCT as a change that could also increase extrasynaptic glutamate levels, as this molecule transports glutamate out of the astrocyte. Glutamate spillover may lead to activation of cell death pathways and loss of input specificity and ultimately the schizophrenia phenotype (Kullmann and Asztely 1998; Hardingham et al. 2002; Tsvetkov et al. 2004; Marcaggi and Attwell 2007). Taken together, these data suggest that there may be increased levels of glutamate and glutamine cycling between astrocytes and neurons.

Several studies have tried to address the question of glutamate levels in living patients. One study found decreased glutamate levels in the CSF in schizophrenia, while other studies found no change in glutamate levels in CSF or serum (Kim et al. 1980; Perry 1982; Gattaz et al. 1985; Korpi et al. 1987; Alfredsson and Wiesel 1989). Antipsychotic treatment may account for these inconsistencies (Goff et al. 1996). Perhaps a better approach to address this question is magnetic resonance spectroscopy (MRS). MRS data have been inconsistent as well, but this approach is limited as there is no way to know how (synapse, presynaptic terminal, extrasynaptic space, astrocyte) the glutamate or glutamine is partitioned in the brain. However, using ratios of glutamate:n-acetyl-aspartate or glutamate:glutamine gives a better picture of possible shifts in these metabolic pathways (Clark et al. 2006; Kraguljac et al. 2012b). Interestingly, recent work suggests that there is a high correlation between n-acetyl-aspartate:creatine and glutamate + glutamine (Glx):creatine ratios in normal and treated schizophrenics, while in untreated subjects this correlation is lost (Kraguljac et al. 2012b). These data argue for a nuanced view of abnormal glutamate/glutamine cycling that is compatible with chronic, low level glutamate spillover in brain circuits where the normally tight regulation of extrasynaptic glutamate is disrupted.

While large increases in glutamate levels at the synapse may cause neurotoxicity, subtle changes over time may cause synaptic stress leading to plastic changes and remodeling, consistent with the concept that schizophrenia is a chronic illness (Olney 1982; McCullumsmith et al. 2004; Lewis and Gonzalez-Burgos 2008; Lau and Tymianski 2010). We propose that chronic glutamate spillover may contribute to remodeling of synapses, astrocytic processes, as well as the nature and structure of interactions between neurons and glia. Supporting this hypothesis, several studies have reported decreased numbers of synaptic spines and increased packing density in schizophrenia (Selemon et al. 1995; Rajkowska et al. 1998, 2002).

There are some notable limitations to the postmortem data described in this chapter. Most of the studies relied on tissue samples from brain regions, where all of the cell types and extracellular matrix are blended together into the same sample. Data derived from these samples represent the net effects of changes in the dependent measures from different cell types and/or subcellular partitions. While many studies include unmedicated subjects or antipsychotic-treated rodents, it is difficult to model a lifetime of severe psychiatric illness and antipsychotic treatment, as both these factors may significantly impact neurochemistry. Finally, postmortem studies do not capture the illness at its so-called first break. Instead, brains from afflicted subjects are typically collected years or even decades after onset of the illness, and represent a "matured" phase of the disease. Despite these limitations, the postmortem approach has a direct translational nature that is difficult to simulate in an animal model. How does one ask a rodent if it hears voices?

15.6.1 Summary

In this chapter, we have conceptualized schizophrenia as a disorder of neuroplasticity that is likely caused by a combination of genetic susceptibility and perinatal or early life stress. These putative etiologies lead to changes in neurochemistry, including abnormalities of synaptic and extrasynaptic elements of glutamatergic neurocircuits. Interestingly, the pathological changes found in schizophrenia are very similar to those induced by stress in animal models. Finally, regardless of the initial insult(s) or lesion, we propose that there are common pathophysiological pathways that lead to the schizophrenia phenotype, and these pathways include profound abnormalities of NMDA receptor and glutamate transporter expression and function.

References

Akbarian S, Sucher NJ, Bradley D, Tafazzoli A, Trinh D, Hetrick WP, Potkin SG, Sandman CA, Bunney WE Jr, Jones EG. Selective alterations in gene expression for NMDA receptor subunits in prefrontal cortex of schizophrenics. J Neurosci. 1996;16(1):19–30.

Alamilla J, Gillespie DC. Glutamatergic inputs and glutamate-releasing immature inhibitory inputs activate a shared postsynaptic receptor population in lateral superior olive. Neuroscience. 2011;196:285–96.

Alda M, Ahrens B, Lit W, Dvorakova M, Labelle A, Zvolsky P, Jones B. Age of onset in familial and sporadic schizophrenia. Acta Psychiatr Scand. 1996;93(6):447–50.

Alfredsson G, Wiesel FA. Monoamine metabolites and amino acids in serum from schizophrenic patients before and during sulpiride treatment. Psychopharmacology (Berl). 1989;99(3):322–7.

Ambrosini A, Bresciani L, Fracchia S, Brunello N, Racagni G. Metabotropic glutamate receptors negatively coupled to adenylate cyclase inhibit N-methyl-D-aspartate receptor activity and prevent neurotoxicity in mesencephalic neurons in vitro. Mol Pharmacol. 1995;47(5):1057–64.

Anderson G, Maes M. Schizophrenia: Linking prenatal infection to cytokines, the tryptophan catabolite (TRYCAT) pathway, NMDA receptor hypofunction, neurodevelopment and neuroprogression. Prog Neuropsychopharmacol Biol Psychiatry. 2013;42:5–19.

Anton ES, Ghashghaei HT, Weber JL, McCann C, Fischer TM, Cheung ID, Gassmann M, Messing A, Klein R, Schwab MH, Lloyd KC, Lai C. Receptor tyrosine kinase ErbB4 modulates neuroblast migration and placement in the adult forebrain. Nat Neurosci. 2004;7(12):1319–28.

Aoki C, Venkatesan C, Go CG, Mong JA, Dawson TM. Cellular and subcellular localization of NMDA-R1 subunit immunoreactivity in the visual cortex of adult and neonatal rats. J Neurosci. 1994;14(9):5202–22.

Armstrong N, Gouaux E. Mechanisms for activation and antagonism of an AMPA-sensitive glutamate receptor: crystal structures of the GluR2 ligand binding core. Neuron. 2000;28(1):165–81.

Armstrong N, Sun Y, Chen G, Gouaux E. Structure of a glutamate-receptor ligand-binding core in complex with kainate. Nature. 1998;395(6705):913–7.

Ashby M, De La Rue S, Ralph G, Uney J, Collingridge G, Henley J. Removal of AMPA receptors (AMPARs) from synapses is preceded by transient endocytosis of extrasynaptic AMPARs. J Neurosci. 2004;24(22):5172–6.

Ashby MC, Maier SR, Nishimune A, Henley JM. Lateral diffusion drives constitutive exchange of AMPA receptors at dendritic spines and is regulated by spine morphology. J Neurosci. 2006;26(26):7046–55.
Association PA. Diagnostic and statistical manual of mental disorders. Washington, D.C.: American Psychiatric Association; 2000.
Babulas V, Factor-Litvak P, Goetz R, Schaefer CA, Brown AS. Prenatal exposure to maternal genital and reproductive infections and adult schizophrenia. Am J Psychiatry. 2006;163(5):927–9.
Badcock JC. The cognitive neuropsychology of auditory hallucinations: a parallel auditory pathways framework. Schizophr Bull. 2010;36(3):576–84.
Bagley J, Moghaddam B. Temporal dynamics of glutamate efflux in the prefrontal cortex and in the hippocampus following repeated stress: effects of pretreatment with saline or diazepam. Neuroscience. 1997;77(1):65–73.
Baker DA, Madayag A, Kristiansen LV, Meador-Woodruff JH, Haroutunian V, Raju I. Contribution of cystine-glutamate antiporters to the psychotomimetic effects of phencyclidine. Neuropsychopharmacology. 2008;33(7):1760–72.
Bauer D, Gupta D, Harotunian V, Meador-Woodruff JH, McCullumsmith RE. Abnormal expression of glutamate transporter and transporter interacting molecules in prefrontal cortex in elderly patients with schizophrenia. Schizophr Res. 2008;104(1–3):108–20.
Bauer D, Haroutunian V, Meador-Woodruff JH, McCullumsmith RE. Abnormal glycosylation of EAAT1 and EAAT2 in prefrontal cortex of elderly patients with schizophrenia. Schizophr Res. 2010;117(1):92–8.
Beattie E, Carroll R, Yu X, Morishita W, Yasuda H, von Zastrow M, Malenka R. Regulation of AMPA receptor endocytosis by a signaling mechanism shared with LTD. Nat Neurosci. 2000;3(12):1291–300.
Bellocchio EE, Reimer RJ, Fremeau RT Jr, Edwards RH. Uptake of glutamate into synaptic vesicles by an inorganic phosphate transporter. Science. 2000;289(5481):957–60.
Beneyto M, Meador-Woodruff JH. Lamina-specific abnormalities of AMPA receptor trafficking and signaling molecule transcripts in the prefrontal cortex in schizophrenia. Synapse. 2006;60(8): 585–98.
Beneyto M, Meador-Woodruff JH. Lamina-specific abnormalities of NMDA receptor-associated postsynaptic protein transcripts in the prefrontal cortex in schizophrenia and bipolar disorder. Neuropsychopharmacology. 2008;33(9):2175–86.
Bhugra D. The global prevalence of schizophrenia. PLoS Med. 2005;2(5):e151 (quiz e175).
Blencowe BJ, Khanna M. Molecular biology: RNA in control. Nature. 2007;447(7143):391–3.
Boehm J, Kang M, Johnson R, Esteban J, Huganir R, Malinow R. Synaptic incorporation of AMPA receptors during LTP is controlled by a PKC phosphorylation site on GluR1. Neuron. 2006;51(2):213–25.
Bossong MG, Niesink RJ. Adolescent brain maturation, the endogenous cannabinoid system and the neurobiology of cannabis-induced schizophrenia. Prog Neurobiol. 2010;92(3):370–85.
Braff DL, Geyer MA, Light GA, Sprock J, Perry W, Cadenhead KS, Swerdlow NR. Impact of prepulse characteristics on the detection of sensorimotor gating deficits in schizophrenia. Schizophr Res. 2001a;49(1–2):171–8.
Braff DL, Geyer MA, Swerdlow NR. Human studies of prepulse inhibition of startle: normal subjects, patient groups, and pharmacological studies. Psychopharmacology (Berl). 2001b;156(2–3):234–58.
Bridges R, Lutgen V, Lobner D, Baker AM. Thinking outside the cleft to understand synaptic activity: Contribution of the cystine-glutamate antiporter (system xc-) to normal and pathological glutamatergic signaling. Pharmacol Rev. 2012;64(3).
Bronson SL, Ahlbrand R, Horn PS, Kern JR, Richtand NM. Individual differences in maternal response to immune challenge predict offspring behavior: contribution of environmental factors. Behav Brain Res. 2011;220(1):55–64.
Brown AS. Prenatal infection as a risk factor for schizophrenia. Schizophr Bull. 2006;32(2):200–2.
Brown AS. The environment and susceptibility to schizophrenia. Prog Neurobiol. 2011;93(1):23–58.

Bruneau EG, McCullumsmith RE, Haroutunian V, Davis KL, Meador-Woodruff JH. Increased expression of glutaminase and glutamine synthetase mRNA in the thalamus in schizophrenia. Schizophr Res. 2005;75(1):27–34.

Buchanan RW, Carpenter WT. Schizophrenia: introduction and overview. In: Sadock BJ, Sadock VA, Editors. Comprehensive textbook of psychiatry. Vol. 1. Philadelphia: Lippincott, Williams, and Wilkins; 2000. pp. 1096–110.

Burbaeva G, Savushkina OK, Dmitriev AD. Brain isoforms of creatine kinase in health and mental diseases: Alzheimer's disease and schizophrenia. Vestn Ross Akad Med Nauk. 1999;1:20–24.

Carlborg A, Jokinen J, Jonsson EG, Erhardt S, Nordstrom P. CSF kynurenic acid and suicide risk in schizophrenia spectrum psychosis. Psychiatry Res. 2013;205(1–2):165–7.

Carroll R, Beattie E, Xia H, L Cüscher, Altschuler Y, Nicoll R, Malenka R, von Zastrow M. Dynamin-dependent endocytosis of ionotropic glutamate receptors. Proc Natl Acad Sci U S A. 1999;96(24):14112–17.

Cheah MT, Wachter A, Sudarsan N, Breaker RR. Control of alternative RNA splicing and gene expression by eukaryotic riboswitches. Nature. 2007;447(7143):497–500.

Chen Y, Dube CM, Rice CJ, Baram TZ. Rapid loss of dendritic spines after stress involves derangement of spine dynamics by corticotropin-releasing hormone. J Neurosci. 2008;28(11):2903–11.

Clark JF, Doepke A, Filosa JA, Wardle RL, Lu A, Meeker TJ, Pyne-Geithman GJ. N-acetylaspartate as a reservoir for glutamate. Med Hypotheses. 2006;67(3):506–12.

Clinton SM, Haroutunian V, Davis KL, Meador-Woodruff JH. Altered transcript expression of NMDA receptor-associated postsynaptic proteins in the thalamus of subjects with schizophrenia. Am J Psychiatry. 2003;160(6):1100–9.

Clinton SM, Meador-Woodruff JH. Abnormalities of the NMDA receptor and associated intracellular molecules in the thalamus in schizophrenia and bipolar disorder. Neuropsychopharmacology. 2004;29(7):1353–62.

Contractor A, Heinemann SF. Glutamate receptor trafficking in synaptic plasticity. Sci STKE. 2002;156:re14.

Coyle J. The glutamatergic dysfunction hypothesis for schizophrenia. Harv Rev Psychiatry. 1996;3(5):241–53.

Danbolt NC. Glutamate uptake. Prog Neurobiol. 2001;65(1):1–105.

Deep-Soboslay A, Benes FM, Haroutunian V, Ellis JK, Kleinman JE, Hyde TM. Psychiatric brain banking: three perspectives on current trends and future directions. Biol Psychiatry. 2011;69(2):104–12.

Dev KK, Henley JM. The regulation of AMPA receptor-binding sites. Mol Neurobiol. 1998;17(1–3):33–58.

Dev KK, Nishimune A, Henley JM, Nakanishi S. The protein kinase C alpha binding protein PICK1 interacts with short but not long form alternative splice variants of AMPA receptor subunits. Neuropharmacology. 1999;38(5):635–44.

Diamond JS. Neuronal glutamate transporters limit activation of NMDA receptors by neurotransmitter spillover on CA1 pyramidal cells. J Neurosci. 2001;21(21):8328–38.

Dingledine R, Borges K, Bowie D, Traynelis S. The glutamate receptor ion channels. Pharmacol Rev. 1999;51(1):7–61.

Drummond JB, Simmons M, Haroutunian V, Meador-Woodruff JH. Upregulation of cornichon transcripts in the dorsolateral prefrontal cortex in schizophrenia. Neuroreport. 2012;23(17):1031–4.

Eastwood SL, Harrison PJ. Decreased expression of vesicular glutamate transporter 1 and complexin II mRNAs in schizophrenia: further evidence for a synaptic pathology affecting glutamate neurons. Schizophr Res. 2005;73(2–3):159–72.

Ehlers M. Reinsertion or degradation of AMPA receptors determined by activity-dependent endocytic sorting. Neuron. 2000;28(2):511–25.

Ehlers M. Activity level controls postsynaptic composition and signaling via the ubiquitin-proteasome system. Nat Neurosci. 2003;6(3):231–42.

Ehlers MD, Tingley WG, Huganir RL. Regulated subcellular distribution of the NR1 subunit of the NMDA receptor. Science. 1995;269(5231):1734–7.

Emamian ES, Karayiorgou M, Gogos JA. Decreased phosphorylation of NMDA receptor type 1 at serine 897 in brains of patients with schizophrenia. J Neurosci. 2004;24(7):1561–4.

English JA, Dicker P, Focking M, Dunn MJ, Cotter DR. 2-D DIGE analysis implicates cytoskeletal abnormalities in psychiatric disease. Proteomics. 2009;9(12):3368–82.

Fanous AH, Zhou B, Aggen SH, Bergen SE, Amdur RL, Duan J, Sanders AR, Shi J, Mowry BJ, Olincy A, Amin F, Cloninger CR, Silverman JM, Buccola NG, Byerley WF, Black DW, Freedman R, Dudbridge F, Holmans PA, Ripke S, Gejman PV, Kendler KS, Levinson DF, Schizophrenia C, Psychiatric Genome-Wide Association Study. Genome-wide association study of clinical dimensions of schizophrenia: polygenic effect on disorganized symptoms. Am J Psychiatry. 2012;169(12):1309–17.

Fatemi SH, Earle JA, Stary JM, Lee S, Sedgewick J. Altered levels of the synaptosomal associated protein SNAP-25 in hippocampus of subjects with mood disorders and schizophrenia. Neuroreport. 2001;12(15):3257–62.

Fleischhacker W. Negative symptoms in patients with schizophrenia with special reference to the primary versus secondary distinction. Encephale 2000;26(Spec No 1):12–4.

Focking M, Dicker P, English JA, Schubert KO, Dunn MJ, Cotter DR. Common proteomic changes in the hippocampus in schizophrenia and bipolar disorder and particular evidence for involvement of cornu ammonis regions 2 and 3. Arch Gen Psychiatry. 2011;68(5):477–88.

Fremeau RT Jr, Troyer MD, Pahner I, Nygaard GO, Tran CH, Reimer RJ, Bellocchio EE, Fortin D, Storm J-Mathisen, Edwards RH. The expression of vesicular glutamate transporters defines two classes of excitatory synapse. Neuron. 2001;31(2):247–60.

Fremeau RT Jr, Burman J, Qureshi T, Tran CH, Proctor J, Johnson J, Zhang H, Sulzer D, Copenhagen DR, Storm J-Mathisen, Reimer RJ, Chaudhry FA, Edwards RH. The identification of vesicular glutamate transporter 3 suggests novel modes of signaling by glutamate. Proc Natl Acad Sci U S A. 2002;99(22):14488–93.

Fremeau RT Jr, Kam K, Qureshi T, Johnson J, Copenhagen DR, Storm J-Mathisen, Chaudhry FA, Nicoll RA, Edwards RH. Vesicular glutamate transporters 1 and 2 target to functionally distinct synaptic release sites. Science. 2004a;304(5678):1815–9.

Fremeau RY Jr, Voglmaier S, Seal RP, Edwards RH. VGLUTs define subsets of excitatory neurons and suggest novel roles for glutamate. Trends Neurosci. 2004b;27(2):98–103.

Fujiyama F, Furuta T, Kaneko T. Immunocytochemical localization of candidates for vesicular glutamate transporters in the rat cerebral cortex. J Comp Neurol. 2001;435(3):379–87.

Funk AJ, McCullumsmith RE, Haroutunian V, Meador-Woodruff JH. Abnormal activity of the MAPK- and cAMP-associated signaling pathways in frontal cortical areas in postmortem brain in schizophrenia. Neuropsychopharmacology. 2012;37(4):896–905.

Funk AJ, Rumbaugh G, Harotunian V, McCullumsmith RE, Meador-Woodruff JH. Decreased expression of NMDA receptor-associated proteins in frontal cortex of elderly patients with schizophrenia. Neuroreport. 2009;20(11):1019–22.

Gabriel SM, Haroutunian V, Powchik P, Honer WG, Davidson M, Davies P, Davis KL. Increased concentrations of presynaptic proteins in the cingulate cortex of subjects with schizophrenia. Arch Gen Psychiatry. 1997;54(6):559–66.

Gallinat J, Gotz T, Kalus P, Bajbouj M, Sander T, Winterer G. Genetic variations of the NR3A subunit of the NMDA receptor modulate prefrontal cerebral activity in humans. J Cogn Neurosci. 2007;19(1):59–68.

Gao XM, Sakai K, Roberts RC, Conley RR, Dean B, Tamminga CA. Ionotropic glutamate receptors and expression of N-methyl-D-aspartate receptor subunits in subregions of human hippocampus: effects of schizophrenia. Am J Psychiatry. 2000;157(7):1141–9.

Gattaz WF, Gasser T, Beckmann H. Multidimensional analysis of the concentrations of 17 substances in the CSF of schizophrenics and controls. Biol Psychiatry. 1985;20(4):360–6.

Genda EN, Jackson JG, Sheldon AL, Locke SF, Greco TM, O'Donnell JC, Spruce LA, Xiao R, Guo W, Putt M, Seeholzer S, Ischiropoulos H, Robinson MB. Co-compartmentalization of the astroglial glutamate transporter, GLT-1, with glycolytic enzymes and mitochondria. J Neurosci. 2011;31(50):18275–88.

Gershon ES, Alliey-Rodriguez N, Liu C. After GWAS: searching for genetic risk for schizophrenia and bipolar disorder. Am J Psychiatry. 2011;168(3):253–6.
Ghose S, Weickert CS, Colvin SM, Coyle JT, Herman MM, Hyde TM, Kleinman JE. Glutamate carboxypeptidase II gene expression in the human frontal and temporal lobe in schizophrenia. Neuropsychopharmacology. 2004;29(1):117–25.
Gladding CM, Raymond LA. Mechanisms underlying NMDA receptor synaptic/extrasynaptic distribution and function. Mol Cell Neurosci. 2011;48(4):308–20.
Gluck MR, Thomas RG, Davis KL, Haroutunian V. Implications for altered glutamate and GABA metabolism in the dorsolateral prefrontal cortex of aged schizophrenic patients. Am J Psychiatry. 2002;159(7):1165–73.
Goff DC, Coyle JT. The emerging role of glutamate in the pathophysiology and treatment of schizophrenia. Am J Psychiatry. 2001;158(9):1367–77.
Goff DC, Tsai G, Manoach DS, Flood J, Darby DG, Coyle JT. D-cycloserine added to clozapine for patients with schizophrenia. Am J Psychiatry. 1996;153(12):1628–30.
Goff DC, Henderson DC, Evins AE, Amico E. A placebo-controlled crossover trial of D-cycloserine added to clozapine in patients with schizophrenia. Biol Psychiatry. 1999;45(4):512–4.
Gordon MW. Neuronal plasticity and memory. Am J Orthopsychiatry. 1969;39(4):578–94.
Gras C, Herzog E, Bellenchi GC, Bernard V, Ravassard P, Pohl M, Gasnier B, Giros B, Mestikawy SEl. A third vesicular glutamate transporter expressed by cholinergic and serotoninergic neurons. J Neurosci. 2002;22(13):5442–51.
Greger I, Esteban J. AMPA receptor biogenesis and trafficking. Curr Opin Neurobiol. 2007;17(3):289–97.
Greger I, Ziff E, Penn A. Molecular determinants of AMPA receptor subunit assembly. Trends Neurosci. 2007;30(8):407–16.
Grimwood S, Slater P, Deakin JF, Hutson PH. NR2B-containing NMDA receptors are up-regulated in temporal cortex in schizophrenia. Neuroreport. 1999;10(3):461–5.
Groc L, Choquet D, Chaouloff F. The stress hormone corticosterone conditions AMPAR surface trafficking and synaptic potentiation. Nat Neurosci. 2008;11(8):868–70.
Gupta DS, McCullumsmith RE, Beneyto M, Haroutunian V, Davis KL, Meador-Woodruff JH. Metabotropic glutamate receptor protein expression in the prefrontal cortex and striatum in schizophrenia. Synapse. 2005;57(3):123–31.
Hahn CG. A Src link in schizophrenia. Nat Med. 2011;17(4):425–7.
Hahn CG, Wang HY, Cho DS, Talbot K, Gur RE, Berrettini WH, Bakshi K, Kamins J, Borgmann-Winter KE, Siegel SJ, Gallop RJ, Arnold SE. Altered neuregulin 1-erbB4 signaling contributes to NMDA receptor hypofunction in schizophrenia. Nat Med. 2006;12(7):824–8.
Halim ND, Weickert CS, McClintock BW, Hyde TM, Weinberger DR, Kleinman JE, Lipska BK. Presynaptic proteins in the prefrontal cortex of patients with schizophrenia and rats with abnormal prefrontal development. Mol Psychiatry. 2003;8(9):797–810.
Halpain S, Girault JA, Greengard P. Activation of NMDA receptors induces dephosphorylation of DARPP-32 in rat striatal slices. Nature. 1990;343(6256):369–72.
Hammond JC, McCullumsmith RE, Funk AJ, Haroutunian V, Meador JH-Woodruff. Evidence for abnormal forward trafficking of AMPA receptors in frontal cortex of elderly patients with schizophrenia. Neuropsychopharmacology. 2010;35(10):2110–9.
Hammond JC, McCullumsmith RE, Haroutunian V, Meador JH-Woodruff. Endosomal trafficking of AMPA receptors in frontal cortex of elderly patients with schizophrenia. Schizophr Res. 2011;130(1–3):260–5.
Hanley JG. Endosomal sorting of AMPA receptors in hippocampal neurons. Biochem Soc Trans. 2010;38(2):460–5.
Hardingham GE, Fukunaga Y, Bading H. Extrasynaptic NMDARs oppose synaptic NMDARs by triggering CREB shut-off and cell death pathways. Nat Neurosci. 2002;5(5):405–14.
Hashimoto K, Fukushima T, Shimizu E, Komatsu N, Watanabe H, Shinoda N, Nakazato M, Kumakiri C, Okada S, Hasegawa H, Imai K, Iyo M. Decreased serum levels of D-serine in patients with schizophrenia: evidence in support of the N-methyl-D-aspartate receptor hypofunction hypothesis of schizophrenia. Arch Gen Psychiatry. 2003;60(6):572–6.

Heim C, Newport DJ, Wagner D, Wilcox MM, Miller AH, Nemeroff CB. The role of early adverse experience and adulthood stress in the prediction of neuroendocrine stress reactivity in women: a multiple regression analysis. Depress Anxiety. 2002;15(3):117–25.

Henneberger C, Papouin T, Oliet SH, Rusakov DA. Long-term potentiation depends on release of D-serine from astrocytes. Nature. 2010;463(7278):232–6.

Heresco-Levy U, Ermilov M, Lichtenberg P, Bar G, Javitt DC. High-dose glycine added to olanzapine and risperidone for the treatment of schizophrenia. Biol Psychiatry. 2004;55(2):165–71.

Heresco-Levy U, Javitt DC, Ebstein R, Vass A, Lichtenberg P, Bar G, Catinari S, Ermilov M. D-serine efficacy as add-on pharmacotherapy to risperidone and olanzapine for treatment-refractory schizophrenia. Biol Psychiatry. 2005;57(6):577–85.

Herzog E, Bellenchi GC, Gras C, Bernard V, Ravassard P, Bedet C, Gasnier B, Giros B, El SM. The existence of a second vesicular glutamate transporter specifies subpopulations of glutamatergic neurons. J Neurosci. 2001;21(22):RC181.

Herzog E, Gilchrist J, Gras C, Muzerelle A, Ravassard P, Giros B, Gaspar P, Mestikawy SEl. Localization of VGLUT3, the vesicular glutamate transporter type 3, in the rat brain. Neuroscience. 2004;123(4):983–1002.

Herzog E, Takamori S, Jahn R, Brose N, Wojcik SM. Synaptic and vesicular co-localization of the glutamate transporters VGLUT1 and VGLUT2 in the mouse hippocampus. J Neurochem. 2006;99(3):1011–8.

Hirling H. Endosomal trafficking of AMPA-type glutamate receptors. Neuroscience. 2008;158:36–44.

Hollmann M, Heinemann S. Cloned glutamate receptors. Annu Rev Neurosci. 1994;17:31–108.

Hollmann M, Maron C, Heinemann S. N-glycosylation site tagging suggests a three transmembrane domain topology for the glutamate receptor GluR1. Neuron. 1994;13(6):1331–43.

Holloway T, Moreno JL, Umali A, Rayannavar V, Hodes GE, Russo SJ, Gonzalez-Maeso J. Prenatal stress induces schizophrenia-like alterations of serotonin 2A and metabotropic Glutamate 2 receptors in the adult offspring: role of maternal immune system. J Neurosci. 2013;33(3):1088–98.

Holmseth S, Dehnes Y, Huang YH, Follin VV-Arbelet, Grutle NJ, Mylonakou MN, Plachez C, Zhou Y, Furness DN, Bergles DE, Lehre KP, Danbolt NC. The density of EAAC1 (EAAT3) glutamate transporters expressed by neurons in the mammalian CNS. J Neurosci. 2012;32(17):6000–13.

Holtzman CW, Trotman HD, Goulding SM, Ryan AT, McDonald AN, Shapiro DI, Brasfield JL, Walker EF. Stress and neurodevelopmental processes in the emergence of psychosis. Neuroscience. 2013;249:172–91.

Honer WG, Falkai P, Bayer TA, Xie J, Hu L, Li HY, Arango V, Mann JJ, Dwork AJ, Trimble WS. Abnormalities of SNARE mechanism proteins in anterior frontal cortex in severe mental illness. Cereb Cortex. 2002;12(4):349–56.

Horiuchi Y, Iida S, Koga M, Ishiguro H, Iijima Y, Inada T, Watanabe Y, Someya T, Ujike H, Iwata N, Ozaki N, Kunugi H, Tochigi M, Itokawa M, Arai M, Niizato K, Iritani S, Kakita A, Takahashi H, Nawa H, Arinami T. Association of SNPs linked to increased expression of SLC1A1 with schizophrenia. Am J Med Genet B Neuropsychiatr Genet. 2012;159B(1):30–7.

Hotulainen P, Hoogenraad CC. Actin in dendritic spines: connecting dynamics to function. J Cell Biol. 2010;189(4):619–29.

Illarionova BN, Gunnarson E, Li Y, Brismar H, Bondar A, Zelenin S, Aperia A. Functional and molecular interactions between aquaporins and Na, K-ATPase. Neuroscience. 2010;168(4):915–25.

Javitt DC, Zukin SR. Recent advances in the phencyclidine model of schizophrenia. Am J Psychiatry. 1991;148(10):1301–8.

Jiang J, Suppiramaniam V, Wooten M. Posttranslational modifications and receptor-associated proteins in AMPA receptor trafficking and synaptic plasticity. Neurosignals. 2006;15(5):266–82.

Johnson JW, Ascher P. Glycine potentiates the NMDA response in cultured mouse brain neurons. Nature. 1987;325(6104):529–31.

Kaneko T, Fujiyama F. Complementary distribution of vesicular glutamate transporters in the central nervous system. Neurosci Res. 2002;42(4):243–50.

Kantrowitz JT, Javitt DC. N-methyl-d-aspartate (NMDA) receptor dysfunction or dysregulation: the final common pathway on the road to schizophrenia? Brain Res Bull. 2010;83(3–4):108–21.

Kantrowitz JT, Malhotra AK, Cornblatt B, Silipo G, Balla A, Suckow RF, C'Souza D, Saksa J, Woods SW, Javitt DC. High dose D-serine in the treatment of schizophrenia. Schizophr Res. 2010;121(1–3):125–30.

Karlsson RM, Tanaka K, Heilig M, Holmes A. Loss of glial glutamate and aspartate transporter (excitatory amino acid transporter 1) causes locomotor hyperactivity and exaggerated responses to psychotomimetics: rescue by haloperidol and metabotropic glutamate 2/3 agonist. Biol Psychiatry. 2008;64(9):810–4.

Karlsson RM, Tanaka K, Saksida LM, Bussey TJ, Heilig M, Holmes A. Assessment of glutamate transporter GLAST (EAAT1)-deficient mice for phenotypes relevant to the negative and executive/cognitive symptoms of schizophrenia. Neuropsychopharmacology. 2009;34(6):1578–89.

Kato AS, Gill MB, Yu H, Nisenbaum ES, Bredt DS. TARPs differentially decorate AMPA receptors to specify neuropharmacology. Trends Neurosci. 2010;33(5):241–8.

Kay SR. Significance of the positive-negative distinction in schizophrenia. Schizophr Bull. 1990;16(4):635–52.

Kharazia VN, Weinberg RJ. Immunogold localization of AMPA and NMDA receptors in somatic sensory cortex of albino rat. J Comp Neurol. 1999;412(2):292–302.

Kim JS, Kornhuber HH, Schmid-Burgk W, Holzmuller B. Low cerebrospinal fluid glutamate in schizophrenic patients and a new hypothesis on schizophrenia. Neurosci Lett. 1980;20(3):379–82.

Knapp M, Mangalore R, Simon J. The global costs of schizophrenia. Schizophr Bull. 2004;30(2):279–93.

Kobayashi T, Stang E, Fang KS, de Moerloose P, Parton RG, Gruenberg J. A lipid associated with the antiphospholipid syndrome regulates endosome structure and function. Nature. 1998;392(6672):193–7.

Komiyama NH, Watabe AM, Carlisle HJ, Porter K, Charlesworth P, Monti J, Strathdee DJ, O'Carroll CM, Martin SJ, Morris RG, O'Dell TJ, Grant SG. SynGAP regulates ERK/MAPK signaling, synaptic plasticity, and learning in the complex with postsynaptic density 95 and NMDA receptor. J Neurosci. 2002;22(22):9721–32.

Korpi ER, Kaufmann CA, Marnela KM, Weinberger DR. Cerebrospinal fluid amino acid concentrations in chronic schizophrenia. Psychiatry Res. 1987;20(4):337–45.

Kraguljac NV, Reid MA, White DM, den Hollander J, Lahti AC. Regional decoupling of N-acetyl-aspartate and glutamate in schizophrenia. Neuropsychopharmacology. 2012b;37(12):2635–42.

Kristensen K, Cadenhead KS. Cannabis abuse and risk for psychosis in a prodromal sample. Psychiatry Res. 2007;151(1–2):151–4.

Kristiansen L, Beneyto M, Haroutunian V, Meador-Woodruff J. Changes in NMDA receptor subunits and interacting PSD proteins in dorsolateral prefrontal and anterior cingulate cortex indicate abnormal regional expression in schizophrenia. Mol Psychiatry. 2006;11(8):737–47, 705.

Kristiansen L, Bakir B, Haroutunian V, Meador-Woodruff J. Expression of the NR2B-NMDA receptor trafficking complex in prefrontal cortex from a group of elderly patients with schizophrenia. Schizophr Res. 2010a;119:198–209.

Kristiansen L, Patel S, Haroutunian V, Meador-Woodruff J. Expression of the NR2B-NMDA receptor subunit and its Tbr-1/CINAP regulatory proteins in postmortem brain suggest altered receptor processing in schizophrenia. Synapse. 2010b;64(7):495–502.

Kropf M, Rey G, Glauser L, Kulangara K, Johnsson K, Hirling H. Subunit-specific surface mobility of differentially labeled AMPA receptor subunits. Eur J Cell Biol. 2008;87(10):763–78.

Kullmann DM, Asztely F. Extrasynaptic glutamate spillover in the hippocampus: evidence and implications. Trends Neurosci. 1998;21(1):8–14.

Lahti AC, Tamminga CA. Recent developments in the neuropharmacology of schizophrenia. Am J Health Syst Pharm. 1995;52(3 Suppl 1):S5–8.

Lahti AC, Holcomb HH, Medoff DR, Tamminga CA. Ketamine activates psychosis and alters limbic blood flow in schizophrenia. Neuroreport. 1995;6(6):869–72.
Lalli E, Sassone-Corsi P. Signal transduction and gene regulation: the nuclear response to cAMP. J Biol Chem. 1994;269(26):17359–62.
Lane HY, Huang CL, Wu PL, Liu YC, Chang YC, Lin PY, Chen PW, Tsai G. Glycine transporter I inhibitor, N-methylglycine (sarcosine), added to clozapine for the treatment of schizophrenia. Biol Psychiatry. 2006;60(6):645–9.
Lane HY, Lin CH, Huang YJ, Liao CH, Chang YC, Tsai GE. A randomized, double-blind, placebo-controlled comparison study of sarcosine (N-methylglycine) and D-serine add-on treatment for schizophrenia. Int J Neuropsychopharmacol. 2010;13(4):451–60.
Laruelle M, Abi-Dargham A, Gil R, Kegeles L, Innis R. Increased dopamine transmission in schizophrenia: relationship to illness phases. Biol Psychiatry. 1999;46(1):56–72.
Laruelle M, Slifstein M, Huang Y. Relationships between radiotracer properties and image quality in molecular imaging of the brain with positron emission tomography. Mol Imaging Biol. 2003;5(6):363–75.
Lau A, Tymianski M. Glutamate receptors, neurotoxicity and neurodegeneration. Pflugers Arch. 2010;460(2):525–42.
Lauriat TL, Dracheva S, Chin B, Schmeidler J, McInnes LA, Haroutunian V. Quantitative analysis of glutamate transporter mRNA expression in prefrontal and primary visual cortex in normal and schizophrenic brain. Neuroscience. 2006;137(3):843–51.
Lea P, Custer S, Vicini S, Faden A. Neuronal and glial mGluR5 modulation prevents stretch-induced enhancement of NMDA receptor current. Pharmacol Biochem Behav. 2002;73(2):287–98.
Lee MR, Ruby CL, Hinton DJ, Choi S, Adams CA, Young KN, Choi DS. Striatal adenosine signaling regulates EAAT2 and astrocytic AQP4 expression and alcohol drinking in mice. Neuropsychopharmacology. 2013;38(3):437–45.
Levenson J, Weeber E, Selcher JC, Kategaya LS, Sweatt JD, Eskin A. Long-term potentiation and contextual fear conditioning increase neuronal glutamate uptake. Nat Neurosci. 2002;5(2):155–61.
Levinson DF, Shi J, Wang K, Oh S, Riley B, Pulver AE, Wildenauer DB, Laurent C, Mowry BJ, Gejman PV, Owen MJ, Kendler KS, Nestadt G, Schwab SG, Mallet J, Nertney D, Sanders AR, Williams NM, Wormley B, Lasseter VK, Albus M, Godard S-Bauche, Alexander M, Duan J, O'Donovan MC, Walsh D, O'Neill A, Papadimitriou GN, Dikeos D, Maier W, Lerer B, Campion D, Cohen D, Jay M, Fanous A, Eichhammer P, Silverman JM, Norton N, Zhang N, Hakonarson H, Gao C, Citri A, Hansen M, Ripke S, Schizophrenia Psychiatric GC, Dudbridge F, Holmans PA. Genome-wide association study of multiplex schizophrenia pedigrees. Am J Psychiatry. 2012;169(9):963–73.
Levy LM, Warr O, Attwell D. Stoichiometry of the glial glutamate transporter GLT-1 expressed inducibly in a Chinese hamster ovary cell line selected for low endogenous Na + -dependent glutamate uptake. J Neurosci. 1998;18(23):9620–8.
Lewis DA, Gonzalez-Burgos G. Neuroplasticity of neocortical circuits in schizophrenia. Neuropsychopharmacology. 2008;33(1):141–65.
Lewis DA, Levitt P. Schizophrenia as a disorder of neurodevelopment. Annu Rev Neurosci. 2002;25:409–32.
Lewis DA, Volk DW, Hashimoto T. Selective alterations in prefrontal cortical GABA neurotransmission in schizophrenia: a novel target for the treatment of working memory dysfunction. Psychopharmacology (Berl). 2004;174(1):143–50.
Lewis DA, Hashimoto T, Morris HM. Cell and receptor type-specific alterations in markers of GABA neurotransmission in the prefrontal cortex of subjects with schizophrenia. Neurotox Res. 2008;14(2–3):237–48.
Li B, Woo RS, Mei L, Malinow R. The neuregulin-1 receptor erbB4 controls glutamatergic synapse maturation and plasticity. Neuron. 2007;54(4):583–97.
Li B, Devidze N, Barengolts D, Prostak N, Sphicas E, Apicella AJ, Malinow R, Emamian ES. NMDA receptor phosphorylation at a site affected in schizophrenia controls synaptic and behavioral plasticity. J Neurosci. 2009a;29(38):11965–72.

Li Q, Cheung C, Wei R, Hui ES, Feldon J, Meyer U, Chung S, Chua SE, Sham PC, Wu EX, McAlonan GM. Prenatal immune challenge is an environmental risk factor for brain and behavior change relevant to schizophrenia: evidence from MRI in a mouse model. PLoS One. 2009b;4(7):e6354.

Liguz-Lecznar M, Skangiel-Kramska J. Vesicular glutamate transporters (VGLUTs): the three musketeers of glutamatergic system. Acta Neurobiol Exp (Wars). 2007;67(3):207–18.

Liu DD, Yang Q, Li ST. Activation of extrasynaptic NMDA receptors induces LTD in rat hippocampal CA1 neurons. Brain Res Bull. 2012;93:10–6.

Liu XB, Murray KD, Jones EG. Switching of NMDA receptor 2 A and 2B subunits at thalamic and cortical synapses during early postnatal development. J Neurosci. 2004;24(40):8885–95.

Lowy MT, Gault L, Yamamoto BK. Adrenalectomy attenuates stress-induced elevations in extracellular glutamate concentrations in the hippocampus. J Neurochem. 1993;61(5):1957–60.

Lu W, Ziff EB. PICK1 interacts with ABP/GRIP to regulate AMPA receptor trafficking. Neuron. 2005;47(3):407–21.

Malenka RC, Nicoll RA. Long-term potentiation-a decade of progress? Science. 1999;285(5435):1870–4.

Malinow R. AMPA receptor trafficking and long-term potentiation. Philos Trans R Soc Lond B Biol Sci. 2003;358(1432):707–14.

Marcaggi P, Attwell D. Short- and long-term depression of rat cerebellar parallel fibre synaptic transmission mediated by synaptic crosstalk. J Physiol. 2007;578(Pt 2):545–50.

Marenco S, Weinberger DR. The neurodevelopmental hypothesis of schizophrenia: following a trail of evidence from cradle to grave. Dev Psychopathol. 2000;12(3):501–27.

Markiv A, Rambaruth ND, Dwek MV. Beyond the genome and proteome: targeting protein modifications in cancer. Curr Opin Pharmacol. 2012;12(4):408–13.

Marwaha S, Johnson S. Schizophrenia and employment—a review. Soc Psychiatry Psychiatr Epidemiol. 2004;39(5):337–49.

Masson J, Sagne C, Hamon M, Mestikawy SEl. Neurotransmitter transporters in the central nervous system. Pharmacol Rev. 1999;51(3):439–64.

McCullumsmith RE, Meador-Woodruff JH. Striatal excitatory amino acid transporter transcript expression in schizophrenia, bipolar disorder, and major depressive disorder. Neuropsychopharmacology. 2002;26(3):368–75.

McCullumsmith RE, Meador-Woodruff JH. Novel approaches to the study of postmortem brain in psychiatric illness: old limitations and new challenges. Biol Psychiatry. 2011;69(2):127–33.

McCullumsmith RE, Clinton SM, Meador-Woodruff JH. Schizophrenia as a disorder of neuroplasticity. Int Rev Neurobiol. 2004;59:19–45.

McCullumsmith RE, Hammond J, Funk A, Meador-Woodruff JH. Recent advances in targeting the ionotropic glutamate receptors in treating schizophrenia. Curr Pharm Biotechnol. 2012;13(8):1535–42.

McKenna MC. Glutamate dehydrogenase in brain mitochondria: Do lipid modifications and transient metabolon formation influence enzyme activity?. Neurochem Int. 2011;59(4):525–33.

Meador-Woodruff JH, Hogg AJ Jr, Smith RE. Striatal ionotropic glutamate receptor expression in schizophrenia, bipolar disorder, and major depressive disorder. Brain Res Bull. 2001;55(5):631–40.

Mirnics K, Middleton F, Marquez A, Lewis D, Levitt P. Molecular characterization of schizophrenia viewed by microarray analysis of gene expression in prefrontal cortex. Neuron. 2000;28(1):53–67.

Moghaddam B, Adams BW. Reversal of phencyclidine effects by a group II metabotropic glutamate receptor agonist in rats. Science. 1998;281(5381):1349–52.

Morris BJ, Cochran SM, Pratt JA. PCP: from pharmacology to modelling schizophrenia. Curr Opin Pharmacol. 2005;5(1):101–6.

Muller N. Inflammation and the glutamate system in schizophrenia: implications for therapeutic targets and drug development. Expert Opin Ther Targets. 2008;12(12):1497–507.

Muller N, Myint AM, Schwarz MJ. Inflammation in schizophrenia. Adv Protein Chem Struct Biol. 2012;88:49–68.

Musazzi L, Milanese M, Farisello P, Zappettini S, Tardito D, Barbiero VS, Bonifacino T, Mallei A, Baldelli P, Racagni G, Raiteri M, Benfenati F, Bonanno G, Popoli M. Acute stress increases depolarization-evoked glutamate release in the rat prefrontal/frontal cortex: the dampening action of antidepressants. PLoS ONE. 2010;5(1):e8566.

Nicoll RA, Malenka RC. Expression mechanisms underlying NMDA receptor-dependent long-term potentiation. Ann N Y Acad Sci. 1999;868:515–25.

Nieoullon A, Canolle B, Masmejean F, Guillet B, Pisano P, Lortet S. The neuronal excitatory amino acid transporter EAAC1/EAAT3: does it represent a major actor at the brain excitatory synapse? J Neurochem. 2006;98(4):1007–18.

Novak G, Fan T, O'Dowd BF, George SR. Postnatal maternal deprivation and pubertal stress have additive effects on dopamine D2 receptor and CaMKII beta expression in the striatum. Int J Dev Neurosci. 2013;31:189–95.

Nudmamud-Thanoi S, Piyabhan P, Harte MK, Cahir M, Reynolds GP. Deficits of neuronal glutamatergic markers in the caudate nucleus in schizophrenia. J Neural Transm. 2007;Suppl(72):281–5.

O'Shea RD. Roles and regulation of glutamate transporters in the central nervous system. Clin Exp Pharmacol Physiol. 2002;29(11):1018–23.

Ohnuma T, Augood SJ, Arai H, McKenna PJ, Emson PC. Expression of the human excitatory amino acid transporter 2 and metabotropic glutamate receptors 3 and 5 in the prefrontal cortex from normal individuals and patients with schizophrenia. Brain Res Mol Brain Res. 1998;56(1–2):207–17.

Ohnuma T, Kato H, Arai H, Faull RL, McKenna PJ, Emson PC. Gene expression of PSD95 in prefrontal cortex and hippocampus in schizophrenia. Neuroreport. 2000a;11(14):3133–7.

Ohnuma T, Tessler S, Arai H, Faull RL, McKenna PJ, Emson PC. Gene expression of metabotropic glutamate receptor 5 and excitatory amino acid transporter 2 in the schizophrenic hippocampus. Brain Res Mol Brain Res. 2000b;85(1–2):24–31.

Okabe S, Miwa A, Okado H. Alternative splicing of the C-terminal domain regulates cell surface expression of the NMDA receptor NR1 subunit. J Neurosci. 1999;19(18):7781–92.

Olney JW. The toxic effects of glutamate and related compounds in the retina and the brain. Retina. 1982;2(4):341–59.

Oni-Orisan A, Kristiansen L, Haroutunian V, Meador-Woodruff J, McCullumsmith R. Altered vesicular glutamate transporter expression in the anterior cingulate cortex in schizophrenia. Biol Psychiatry. 2008;63(8):766–75.

Papouin T, Ladepeche L, Ruel J, Sacchi S, Labasque M, Hanini M, Groc L, Pollegioni L, Mothet JP, Oliet SH. Synaptic and extrasynaptic NMDA receptors are gated by different endogenous coagonists. Cell. 2012;150(3):633–46.

Park M, Penick E, Edwards J, Kauer J, Ehlers M. Recycling endosomes supply AMPA receptors for LTP. Science. 2004;305(5692):1972–5.

Park M, Salgado JM, Ostroff L, Helton TD, Robinson CG, Harris KM, Ehlers MD. Plasticity-induced growth of dendritic spines by exocytic trafficking from recycling endosomes. Neuron. 2006;52(5):817–30.

Passafaro M, Piëch V, Sheng M. Subunit-specific temporal and spatial patterns of AMPA receptor exocytosis in hippocampal neurons. Nat Neurosci. 2001;4(9):917–26.

Patel DR, Young AM, Croucher MJ. Presynaptic alpha-amino-3-hydroxy-5-methyl-4-isoxazole propionate receptor-mediated stimulation of glutamate and GABA release in the rat striatum in vivo: a dual-label microdialysis study. Neuroscience. 2001;102(1):101–11.

Pennington K, Beasley CL, Dicker P, Fagan A, English J, Pariante CM, Wait R, Dunn MJ, Cotter DR. Prominent synaptic and metabolic abnormalities revealed by proteomic analysis of the dorsolateral prefrontal cortex in schizophrenia and bipolar disorder. Mol Psychiatry. 2008;13(12):1102–17.

Perry TL. Normal cerebrospinal fluid and brain glutamate levels in schizophrenia do not support the hypothesis of glutamatergic neuronal dysfunction. Neurosci Lett. 1982;28(1):81–5.

Petralia RS, Wang YX, Mayat E, Wenthold RJ. Glutamate receptor subunit 2-selective antibody shows a differential distribution of calcium-impermeable AMPA receptors among populations of neurons. J Comp Neurol. 1997;385(3):456–76.
Petralia RS, Wang YX, Hua F, Yi Z, Zhou A, Ge L, Stephenson FA, Wenthold RJ. Organization of NMDA receptors at extrasynaptic locations. Neuroscience. 2010;167(1):68–87.
Pitcher GM, Kalia LV, Ng D, Goodfellow NM, Yee KT, Lambe EK, Salter MW. Schizophrenia susceptibility pathway neuregulin 1-ErbB4 suppresses Src upregulation of NMDA receptors. Nat Med. 2011;17(4):470–8.
Potkin SG, Jin Y, Bunney BG, Costa J, Gulasekaram B. Effect of clozapine and adjunctive high-dose glycine in treatment-resistant schizophrenia. Am J Psychiatry. 1999;156(1):145–7.
Potkin SG, Turner JA, Brown GG, McCarthy G, Greve DN, Glover GH, Manoach DS, Belger A, Diaz M, Wible CG, Ford JM, Mathalon DH, Gollub R, Lauriello J, O'Leary D, van Erp TG, Toga AW, Preda A, Lim KO, FBIRN. Working memory and DLPFC inefficiency in schizophrenia: the FBIRN study. Schizophr Bull. 2009;35(1):19–31.
Rajji TK, Mulsant BH. Nature and course of cognitive function in late-life schizophrenia: a systematic review. Schizophr Res. 2008;102(1–3):122–40.
Rajkowska G, Selemon LD, Goldman-Rakic PS. Neuronal and glial somal size in the prefrontal cortex: a postmortem morphometric study of schizophrenia and Huntington disease. Arch Gen Psychiatry. 1998;55(3):215–24.
Rajkowska G, Miguel JJ-Hidalgo, Makkos Z, Meltzer H, Overholser J, Stockmeier C. Layer-specific reductions in GFAP-reactive astroglia in the dorsolateral prefrontal cortex in schizophrenia. Schizophr Res. 2002;57(2–3):127–38.
Rao JS, Kellom M, Reese EA, Rapoport SI, Kim HW. Dysregulated glutamate and dopamine transporters in postmortem frontal cortex from bipolar and schizophrenic patients. J Affect Disord. 2012;136(1–2):63–71.
Rex CS, Chen LY, Sharma A, Liu J, Babayan AH, Gall CM, Lynch G. Different Rho GTPase-dependent signaling pathways initiate sequential steps in the consolidation of long-term potentiation. J Cell Biol. 2009;186(1):85–97.
Reynolds LM, Cochran SM, Morris BJ, Pratt JA, Reynolds GP. Chronic phencyclidine administration induces schizophrenia-like changes in N-acetylaspartate and N-acetylaspartylglutamate in rat brain. Schizophr Res. 2005;73(2–3):147–52.
Richardson-Burns SM, Haroutunian V, Davis KL, Watson SJ, Meador-Woodruff JH. Metabotropic glutamate receptor mRNA expression in the schizophrenic thalamus. Biol Psychiatry. 2000;47(1):22–8.
Richtand NM, McNamara RK. Serotonin and dopamine interactions in psychosis prevention. Prog Brain Res. 2008;172:141–53.
Rose ME, Koo JC, Antflick JE, Ahmed SM, Angers S, Hampson DR. Glutamate transporter coupling to Na, K-ATPase. J Neurosci. 2009;29(25):8143–55.
Rosenmund C, Stern-Bach Y, Stevens C. The tetrameric structure of a glutamate receptor channel. Science. 1998;280(5369):1596–9.
Roy PD, Zipursky RB, Saint-Cyr JA, Bury A, Langevin R, Seeman MV. Temporal horn enlargement is present in schizophrenia and bipolar disorder. Biol Psychiatry. 1998;44(6):418–22.
Rubino M, Miaczynska M, Lippé R, Zerial M. Selective membrane recruitment of EEA1 suggests a role in directional transport of clathrin-coated vesicles to early endosomes. J Biol Chem. 2000;275(6):3745–8.
Rubio MD, Haroutunian V, Meador-Woodruff JH. Abnormalities of the Duo/Ras-related C3 botulinum toxin substrate 1/p21-activated kinase 1 pathway drive myosin light chain phosphorylation in frontal cortex in schizophrenia. Biol Psychiatry. 2012;71(10):906–14.
Ryan TJ, Grant SG. The origin and evolution of synapses. Nat Rev Neurosci. 2009;10(10):701–12.
Scarr E, Gray L, Keriakous D, Robinson PJ, Dean B. Increased levels of SNAP-25 and synaptophysin in the dorsolateral prefrontal cortex in bipolar I disorder. Bipolar Disord. 2006;8(2):133–43.
Schwenk J, Harmel N, Zolles G, Bildl W, Kulik A, Heimrich B, Chisaka O, Jonas P, Schulte U, Fakler B, Klocker N. Functional proteomics identify cornichon proteins as auxiliary subunits of AMPA receptors. Science. 2009;323(5919):1313–9.

Scimemi A, Fine A, Kullmann DM, Rusakov DA. NR2B-containing receptors mediate cross talk among hippocampal synapses. J Neurosci. 2004;24(20):4767–77.
Selemon LD, Rajkowska G, Goldman-Rakic PS. Abnormally high neuronal density in the schizophrenic cortex. A morphometric analysis of prefrontal area 9 and occipital area 17. Arch Gen Psychiatry. 1995;52(10):805–18 (discussion 819–20).
Setou M, Nakagawa T, Seog DH, Hirokawa N. Kinesin superfamily motor protein KIF17 and mLin-10 in NMDA receptor-containing vesicle transport. Science. 2000;288(5472):1796–802.
Shan D, Lucas EK, Drummond JB, Haroutunian V, Meador-Woodruff JH, McCullumsmith RE. Abnormal expression of glutamate transporters in temporal lobe areas in elderly patients with schizophrenia. Schizophr Res. 2013;144:1–8.
Shashidharan P, Huntley GW, Murray JM, Buku A, Moran T, Walsh MJ, Morrison JH, Plaitakis A. Immunohistochemical localization of the neuron-specific glutamate transporter EAAC1 (EAAT3) in rat brain and spinal cord revealed by a novel monoclonal antibody. Brain Res. 1997;773(1–2):139–48.
Sheng M, Lee SH. Growth of the NMDA receptor industrial complex. Nat Neurosci. 2000;3(7):633–5.
Shi S, Hayashi Y, Esteban JA, Malinow R. Subunit-specific rules governing AMPA receptor trafficking to synapses in hippocampal pyramidal neurons. Cell. 2001;105(3):331–43.
Skeberdis V, Lan J, Opitz T, Zheng X, Bennett M, Zukin R. mGluR1-mediated potentiation of NMDA receptors involves a rise in intracellular calcium and activation of protein kinase C. Neuropharmacology. 2001;40(7):856–65.
Smith RE, Haroutunian V, Davis KL, Meador-Woodruff JH. Expression of excitatory amino acid transporter transcripts in the thalamus of subjects with schizophrenia. Am J Psychiatry. 2001;158(9):1393–9.
Sodhi MS, Simmons M, McCullumsmith R, Haroutunian V, Meador-Woodruff JH. Glutamatergic gene expression is specifically reduced in thalamocortical projecting relay neurons in schizophrenia. Biol Psychiatry. 2011;70(7):646–54.
Song I, Huganir R. Regulation of AMPA receptors during synaptic plasticity. Trends Neurosci. 2002;25(11):578–88.
Sossa K, Court B, Carroll R. NMDA receptors mediate calcium-dependent, bidirectional changes in dendritic PICK1 clustering. Mol Cell Neurosci. 2006;31(3):574–85.
Standley S, Roche KW, McCallum J, Sans N, Wenthold RJ. PDZ domain suppression of an ER retention signal in NMDA receptor NR1 splice variants. Neuron. 2000;28(3):887–98.
Steiner J, Bogerts B, Sarnyai Z, Walter M, Gos T, Bernstein HG, Myint AM. Bridging the gap between the immune and glutamate hypotheses of schizophrenia and major depression: Potential role of glial NMDA receptor modulators and impaired blood-brain barrier integrity. World J Biol Psychiatry. 2012;13(7):482–92.
Stephenson FA. Structure and trafficking of NMDA and GABAA receptors. Biochem Soc Trans. 2006;34(Pt 5):877–81.
Swanson GT, Kamboj SK, Cull-Candy SG. Single-channel properties of recombinant AMPA receptors depend on RNA editing, splice variation, and subunit composition. J Neurosci. 1997;17(1):58–69.
Szoke A, Meary A, Trandafir A, Bellivier F, Roy I, Schurhoff F, Leboyer M. Executive deficits in psychotic and bipolar disorders - implications for our understanding of schizoaffective disorder. Eur Psychiatry. 2008;23(1):20–5.
Takamori S, Riedel D, Jahn R. Immunoisolation of GABA-specific synaptic vesicles defines a functionally distinct subset of synaptic vesicles. J Neurosci. 2000;20(13):4904–11.
Takumi Y, Ramirez-Leon V, Laake P, Rinvik E, Ottersen OP. Different modes of expression of AMPA and NMDA receptors in hippocampal synapses. Nat Neurosci. 1999;2(7):618–24.
Talbot K, Ong WY, Blake DJ, Tang J, Louneva N, Carlson GC, Arnold SE. Dysbindin-1 and its protein family. In: Lajtha A, Javitt D, Kantrowitz J, Editors. Handbook of neurochemistry and molecular neurobiology. USA: Springer US; 2009. pp. 107–241.
Tamminga C. Glutamatergic aspects of schizophrenia. Br J Psychiatry. 1999; Suppl(37):2–15.

Thompson KN, Phillips LJ, Komesaroff P, Yuen HP, Wood SJ, Pantelis C, Velakoulis D, Yung AR, McGorry PD. Stress and HPA-axis functioning in young people at ultra high risk for psychosis. J Psychiatr Res. 2007;41(7):561–9.

Toro C, Deakin JF. NMDA receptor subunit NRI and postsynaptic protein PSD-95 in hippocampus and orbitofrontal cortex in schizophrenia and mood disorder. Schizophr Res. 2005;80(2–3):323–30.

Torrey EF. Surviving schizophrenia: a manual for families, patients and providers. New York: Harper Collins; 2006.

Toyooka K, Asama K, Watanabe Y, Muratake T, Takahashi M, Someya T, Nawa H. Decreased levels of brain-derived neurotrophic factor in serum of chronic schizophrenic patients. Psychiatry Res. 2002a;110(3):249–57.

Toyooka K, Iritani S, Makifuchi T, Shirakawa O, Kitamura N, Maeda K, Nakamura R, Niizato K, Watanabe M, Kakita A, Takahashi H, Someya T, Nawa H. Selective reduction of a PDZ protein, SAP-97, in the prefrontal cortex of patients with chronic schizophrenia. J Neurochem. 2002b;83(4):797–806.

Triller A, Choquet D. Surface trafficking of receptors between synaptic and extrasynaptic membranes: and yet they do move! Trends Neurosci. 2005;28(3):133–9.

Tsai G, Yang P, Chung LC, Lange N, Coyle JT. D-serine added to antipsychotics for the treatment of schizophrenia. Biol Psychiatry. 1998;44(11):1081–9.

Tsai EG, Lin PY. Strategies to enhance N-methyl-D-aspartate receptor-mediated neurotransmission in schizophrenia, a critical review and meta-analysis. Curr Pharm Des. 2010;16(5):522–37.

Tsvetkov E, Shin RM, Bolshakov VY. Glutamate uptake determines pathway specificity of long-term potentiation in the neural circuitry of fear conditioning. Neuron. 2004;41(1):139–51.

Tuominen H, Tiihonen J, Wahlbeck K. Glutamatergic drugs for schizophrenia: a systematic review and meta-analysis. Schizophr Res. 2005;72(2–3):225–34.

Tzingounis AV, Wadiche JI. Glutamate transporters: confining runaway excitation by shaping synaptic transmission. Nat Rev Neurosci. 2007;8(12):935–47.

Uezato A, Meador-Woodruff JH, McCullumsmith RE. Vesicular glutamate transporter mRNA expression in the medial temporal lobe in major depressive disorder, bipolar disorder, and schizophrenia. Bipolar Disord. 2009;11(7):711–25.

Walaas IS, Greengard P. DARPP-32, a dopamine- and adenosine 3′:5′-monophosphate-regulated phosphoprotein enriched in dopamine-innervated brain regions. Regional I and cellular distribution in the rat brain. J Neurosci. 1984;4(1):84–98.

Wang JK, Walaas SI, Greengard P. Protein phosphorylation in nerve terminals: comparison of calcium/calmodulin-dependent and calcium/diacylglycerol-dependent systems. J Neurosci. 1988;8(1):281–8.

Watase K, Hashimoto K, Kano M, Yamada K, Watanabe M, Inoue Y, Okuyama S, Sakagawa T, Ogawa S, Kawashima N, Hori S, Takimoto M, Wada K, Tanaka K. Motor discoordination and increased susceptibility to cerebellar injury in GLAST mutant mice. Eur J Neurosci. 1998;10(3):976–88.

Watson GB, Bolanowski MA, Baganoff MP, Deppeler CL, Lanthorn TH. D-cycloserine acts as a partial agonist at the glycine modulatory site of the NMDA receptor expressed in Xenopus oocytes. Brain Res. 1990;510(1):158–60.

Wenthold RJ, Petralia RS, Blahos J II, Niedzielski AS. Evidence for multiple AMPA receptor complexes in hippocampal CA1/CA2 neurons. J Neurosci. 1996;16(6):1982–9.

Whiteheart S, Matveeva E. Multiple binding proteins suggest diverse functions for the N-ethylmaleimide sensitive factor. J Struct Biol. 2004;146(1–2):32–43.

Wobrock T, Schneider M, Kadovic D, Schneider-Axmann T, Ecker UK, Retz W, Rosler M, Falkai P. Reduced cortical inhibition in first-episode schizophrenia. Schizophr Res. 2008;105(1–3):52–261.

Wonodi I, Schwarcz R. Cortical kynurenine pathway metabolism: a novel target for cognitive enhancement in Schizophrenia. Schizophr Bull. 2010;36(2):211–8.

Wu EQ, Birnbaum HG, Shi L, Ball DE, Kessler RC, Moulis M, Aggarwal J. The economic burden of schizophrenia in the United States in 2002. J Clin Psychiatry. 2005;66(9):1122–9.

Zanello A, Curtis L, Badan Ba M, Merlo MC. Working memory impairments in first-episode psychosis and chronic schizophrenia. Psychiatry Res. 2009;165(1–2):10–8.
Zerangue N, Kavanaugh MP. Flux coupling in a neuronal glutamate transporter. Nature. 1996;383(6601):634–7.
Zink M, Vollmayr B, Gebicke-Haerter PJ, Henn FA. Reduced expression of glutamate transporters vGluT1, EAAT2 and EAAT4 in learned helpless rats, an animal model of depression. Neuropharmacology. 2010;58(2):465–73.
Zuo Z, Fang H. Glutamate transporter type 3 attenuates the activation of N-methyl-D-aspartate receptors co-expressed in Xenopus oocytes. J Exp Biol. 2005;208(Pt 11):2063–70.

Chapter 16
Metabolic Stress and Neuropsychiatric Disorders

Claudia A. Grillo and Lawrence P. Reagan

Abstract The complications of metabolic disorders like diabetes, obesity, and the metabolic syndrome (MetS) are well characterized in peripheral tissues, but there is a growing appreciation that the complications of metabolic disorders extend to the central nervous system (CNS). Interestingly, the structural, electrophysiological, neurochemical, and anatomical underpinnings responsible for neuroplasticity deficits associated with metabolic disorders are strikingly similar to those observed in animals subjected to chronic stress, as well as in patients with stress-related psychiatric illnesses such as major depressive disorder. This has led to the hypothesis that diabetes, obesity, and MetS may be considered chronic metabolic stressors and led to the suggestion that common mechanistic mediators are responsible for the neurological complications associated with both metabolic disorders and neuropsychiatric disorders. The goal of this chapter is to provide an overview of stress neurobiology, with a particular emphasis on the causes and consequences of the metabolic stress in the CNS. This will include a discussion of the development and progression of mood disorders in patients with metabolic disorders, as well as a discussion of a novel model of obesity/MetS developed in our laboratory that is helping to elucidate the underlying mechanistic mediators of comorbid depression and obesity.

Abbreviations

AD	Alzheimer's disease
BBB	Blood–brain barrier
BDNF	Brain-derived neurotrophic factor
CNS	Central nervous system
EPM	Elevated plus maze
FST	Forced swim test
HFS	High frequency stimulation
HPA axis	Hypothalamic-pituitary-adrenal axis
IR	Insulin receptor
LTP	Long-term potentiation
MetS	Metabolic syndrome

L. P. Reagan (✉) · C. A. Grillo
Department of Pharmacology, Physiology and Neuroscience, University of South Carolina School of Medicine, 6439 Garners Ferry Rd, Columbia, SC 29208, USA
e-mail: lawrence.reagan@uscmed.sc.edu

M. Popoli et al. (eds.), *Synaptic Stress and Pathogenesis of Neuropsychiatric Disorders*, DOI 10.1007/978-1-4939-1056-4_16,

PTSD Posttraumatic stress disorder
pSTAT3 Phosphorylated of signal transducer and activator of transcription 3

16.1 Introduction

Acute exposure to stress activates the hypothalamic-pituitary-adrenal (HPA) axis, leading to the release of epinephrine and glucocorticoids from the adrenal gland. Once released, these hormones activate a variety of responses in the periphery and central nervous system (CNS) that are proposed to be adaptive in nature. These responses are initiated by activation of the HPA axis. In the CNS, stress hormones play a critical role in the facilitation and consolidation of strong emotional memories in limbic regions such as the hippocampus and amygdala (Conrad 2005; Roozendaal et al. 2009). Unlike these adaptive responses to acute stressful stimuli, exposure to chronic stress often results in maladaptive responses that are proposed to contribute to the pathology of cardiovascular disease, hypertension, cancer metastasis, gastrointestinal disorders, and immune dysfunction, among others. In the CNS, exposure to stressful life events has been proposed to play an important role in the etiology and progression of neuropsychiatric disorders such as depressive illness, anxiety disorders, and posttraumatic stress disorder (PTSD) (Diamond et al. 2004; McEwen 2008). Beyond stressful life experiences, HPA axis dysfunction is also observed in metabolic disorders such as diabetes mellitus and obesity and the metabolic syndrome (MetS) (De Nicola et al. 1976; Leedom et al. 1987; Meehan et al. 1986; Oster et al. 1988; Plotsky et al. 1992; Scribner et al. 1991; Winocur et al. 2005). This has led to the concept that diabetes and obesity act as chronic metabolic stressors in the CNS (Dallman et al. 2006), a concept that is supported by studies indicating that the neurological consequences of metabolic disorders is strikingly similar to the effects of chronic stress (Reagan 2012). Indeed, clinical studies indicate that there is an association between metabolic disorders and mood disorders (Andersen 2010; Anderson 2001 and Anderson 2010; Fabricatore and Wadden 2006; Luppino et al. 2010; McElroy et al. 2004; Simon et al. 2006; Stunkard et al. 2003) and ongoing preclinical studies are investigating the underlying mechanisms that link neuropsychiatric disorders, obesity, and diabetes. This chapter will review these relationships between metabolic and mood disorders, but we will begin with a discussion of more general issues related to experimental models of stress.

16.2 Experimental Models of Stress: Controversies Versus Consensus

A review of the literature will quickly determine that there is controversy related to the causes and consequences of chronic stress. However, closer examination of these studies provides several explanations for these disparate findings. An obvious source of these sometimes equivocal findings is the variety of stress paradigms employed by investigators, which have particular advantages as well as disadvantages as it relates to their translational potential. For example, many investigators examine the effects of early life stress, including the effects of prenatal stress, postnatal handling or maternal separation. Such paradigms may be particularly useful for the examination of the potential role of epigenetic mechanisms in the development of stress-related mood disorders. Stress paradigms performed in adult animals may include restraint stress, exposure to variable or unpredictable stress, psychosocial stress such as resident intruder stress, and social hierarchy stress such as the visible burrow system. An advantage of these various stress paradigms is their ability to elicit neuroplasticity deficits that are similar to those observed in patients with neuropsychiatric disorders, such as neuroanatomical alterations and deficits in cognitive performance. A major limitation of these studies is reproducibility from laboratory to laboratory. However, the inability of stress paradigms to result in universally consistent findings should not be unexpected given the fundamental differences in how the paradigms are performed, the duration of the various paradigms, the choice of animal in the studies, and the endpoint measures that are used to evaluate the effects of stress. For example, the duration of a "chronic" stress paradigm can vary from several days to several weeks to several months depending on the laboratory performing the studies. There are also variable findings from endpoint measures ranging from molecular assays to behavioral analyses, which may be related to the experimental approaches utilized by a given laboratory. For example, the effects of repeated stress on neurochemical parameters such as measurement of extracellular levels of the excitatory amino acid neurotransmitter glutamate may be assessed through superfusion assays of synaptosomal preparations or via in vivo microdialysis. Since extracellular glutamate may originate from the vesicular pool or the metabolic pool (Timmerman and Westerink 1997), an advantage of the superfusion approach is that it can directly examine the readily releasable pool of glutamate in response to stress and antidepressant treatment, as shown by Popoli and colleagues (Barbiero et al. 2007; Bonanno et al. 2005; Musazzi et al. 2010). Conversely, microdialysis allows for the analysis of the effects of stress in vivo through the real-time assessment of glutamate efflux in relatively discrete brain regions (Bagley and Moghaddam 1997; Lowy et al. 1993). In this regard, our prior in vivo microdialysis studies indicate that the effects of acute versus chronic stress differentially impact extracellular glutamate efflux and that some but not all antidepressants may inhibit the effects of stress (Piroli et al. 2013; Reagan et al. 2012; Reznikov et al. 2007). Although these studies employed slightly different stress paradigms of dissimilar durations, used different experimental approaches (superfusion of synaptosomes vs. in vivo microdialysis), and also examined the effects of different antidepressants, the important take-home message is the same: stress adversely affects glutamate neurochemistry in stress responsive regions like the hippocampus, amygdala, and prefrontal cortex, findings that may be directly relevant to the clinical setting (McEwen et al. 2010; Popoli et al. 2012). As such, these results are representative of the sometimes equivocal findings from experimental models of stress. More importantly, these findings are consistent with the heterogeneity in the clinical features and differential responses to antidepressant treatments observed in patients with mood disorders. Beyond the effects of stressful life events, it is also clear that metabolic stress associated with diabetes and obesity is associated with increased risk of developing mood disorders, thereby providing another level of complexity in determining the mechanistic mediators of neuropsychiatric disorders.

16.3 Neuroplasticity Deficits in Metabolic Disorders

The hypothesis that activation of insulin receptor (IR) signaling improves cognitive performance has been supported by both clinical and preclinical studies. For example, it has been established that insulin enhances cognitive performance in healthy subjects (Benedict et al. 2004, 2007), in aged subjects (Manning et al. 1998), and in Alzheimer's Disease (AD) patients (Craft et al. 1999, 2012; Reger et al. 2008). Animal studies also support the hypothesis that insulin enhances behavioral performance (Park et al. 2000). For example, icv injection of insulin enhances spatial memory in male rats in a dose-dependent fashion (Haj-ali et al. 2009), whereas intra-CA1 insulin microinjections have been shown to improve behavioral performance in the water maze (Moosavi et al. 2007). Studies that have examined the behavioral consequences of decreasing CNS IRs also support the hypothesis that insulin promotes cognitive function (Nistico et al. 2012). Interestingly, decreases in insulin activity observed in diabetes, obesity, and the MetS elicit neuroplasticity deficits that are similar to those observed in experimental models of stress. These observations provide possible causes for the increased risk of comorbid depressive illness in patients with metabolic disorders (see Reagan 2012). For example, changes in the metabolic and endocrine milieu, including impairments in HPA axis activity, hyperglycemia, insulin and leptin resistance, and increased productions of pro-inflammatory cytokines represent potential causes for the neurological consequences of metabolic disorders, including neurochemical, electrophysiological, and neuroanatomical deficits that ultimately lead to cognitive impairments. Indeed, there is a large body of work supporting the consensus that metabolic disorders adversely affect neuroplasticity. For instance, defective insulin signaling is a characteristic feature of the AD brain (Talbot et al. 2012) and as noted above intranasal insulin administration promotes cognitive function in adults with early-stage AD (Craft et al. 2012). Undoubtedly, defective insulin signaling contributes to AD pathogenesis, as Hoyer proposed nearly 25 years ago (Hoyer and Nitsch 1989). In experimental models of diabetes, the morphological deficits in the hippocampus include neuronal atrophy (Magariños and McEwen 2000; Martinez-Tellez et al. 2005), decreases in neuronal density (Beauquis et al. 2006), synaptic reorganization (Grillo et al. 2005), as well as decreases in neurogenesis/cell proliferation (Beauquis et al. 2008; Kim et al. 2003; Stranahan et al. 2008). Additional neuroplasticity deficits include decreases in synaptic transmission (Alzoubi et al. 2005; Artola et al. 2005; Biessels et al. 1996; Gerges et al. 2003; Izumi et al. 2003; Kamal et al. 1999; Oomura et al. 2006; Stranahan et al. 2008; Valastro et al. 2002), which may result from changes in glutamate receptor expression and trafficking (Chabot et al. 1997; Di Luca et al. 1999; Gagne et al. 1997; Gardoni et al. 2002), as well as increases in oxidative stress mediators (Grillo et al. 2003; Reagan et al. 2008; Tuzcu and Baydas 2006). Ultimately, the long-term consequence of diabetes-induced neuroplasticity deficits is cognitive impairments (see Biessels et al. 2008; Reagan 2012).

Beyond deficits in spatial learning, changes in anxiety-like behaviors are among the earliest behavioral changes observed in experimental models of metabolic disorders. For example, decreases in social interactions and fear-related behaviors are observed in type 1 diabetic rats, including increases in passive avoidance, defensive postures, and submissive-like behaviors (Leedom et al. 1987; Meehan et al. 1986). Anxiety-like behaviors, such as decreases in open arm time or open arm entries in the elevated plus maze (EPM) or reduced behaviors in the open field test, are also observed in diabetic rodents (Asakawa et al. 2003; Miyata et al. 2007; Ramanathan et al. 1998; Sharma et al. 2010; Thorre et al. 1997). Deficits in the forced swim test (FST) have also been reported in leptin-deficient *ob/ob* mice (Collin et al. 2000; Yamada et al. 2011), leptin receptor deficient *db/db* mice (Sharma et al. 2010), and in rodents fed a high fat diet (Yamada et al. 2011). In summary, the clinical and preclinical literature indicate that metabolic disorders impair neuroplasticity, which includes deficits in behavioral performance and the development of depressive-like and anxiety-like behaviors. While there may be consensus regarding the neurological consequences of metabolic disorders, the underlying mechanisms responsible for these deficits remain a subject of debate.

16.4 Disentangling the Causes and Consequences of Metabolic Stress

The wide variety of endocrine and metabolic changes associated with obesity and diabetes is an obvious obstacle in accurately identifying the mechanistic links between metabolic stress and neuropsychiatric disorders. Due to the absence of good pharmacological tools such as an IR antagonist, we have developed an alternative molecular strategy to more selectively examine the role of IRs in neuroplasticity deficits observed in diabetes and obesity phenotypes. In this regard, we have developed and characterized a lentivirus vector that produces an antisense RNA selective for the insulin receptor (IRAS) and performed site-specific injections of this virus to differentiate between the functional activities of different IR populations in the rat brain. Our initial studies focused on the hypothalamus due to the well-described role of hypothalamic IRs in the regulation of body weight, body composition, food intake, and metabolism (see Schwartz et al. 2000). When injected into the hypothalamus to target IRs expressed in the arcuate nucleus (Hypo-IRAS), the LV-IRAS construct decreases the expression and activity of hypothalamic IRs, while not affecting IR expression or activity in the hippocampus. In agreement with previous studies using different molecular approaches (Bruning et al. 2000; Obici et al. 2002), downregulation of hypothalamic IRs produced significant increases in body weight gain and body adiposity, as well as increases in plasma leptin levels and plasma triglyceride levels (Grillo et al. 2007, 2011a). Subsequent studies determined that downregulation of hypothalamic IRs elicited leptin resistance (Grillo et al. 2011b) and hepatic insulin resistance (Paranjape et al. 2011) while not affecting HPA axis function or plasma adiponectin, estrogen or testosterone levels (Grillo et al. 2007, 2011b). Collectively, these endocrine and metabolic changes are consistent with features of the MetS and as such the Hypo-IRAS rat provides a unique model system to examine the deleterious consequences of obesity on the CNS.

Since diabetes/obesity phenotypes are associated with decreases in hippocampal synaptic plasticity, we compared several endpoint measures of neuroplasticity in the hippocampus of Hypo-IRAS rats to rats that received the LV-Control construct in the hypothalamus (Hypo-Con). In this regard, while high frequency stimulation (HFS) of the Schaffer collaterals elicited long-term potentiation (LTP) in CA1 pyramidal neurons in the hippocampus of Hypo-Con rats, HFS failed to produce LTP in CA1 pyramidal neurons of Hypo-IRAS rats (Grillo et al. 2011a). Paired pulse facilitation was similar in both Hypo-IRAS and Hypo-Con rats, suggesting that the deficits in synaptic transmission were specific for the postsynaptic side. Subsequent immunoblot analysis determined that the phosphorylation of Ser845 on the GluA1 receptor subunit was significantly reduced in the hippocampus of Hypo-IRAS rats compared to Hypo-Con rats (Grillo et al. 2011a), thereby providing a potential mechanistic basis for these electrophysiological deficits. We also measured dendritic morphology via confocal immunofluorescence using the presynaptic protein synaptophysin and the postsynaptic protein PSD-95 in the hippocampus of Hypo-IRAS. Similar to our previous observations in type 1 diabetes rats (Grillo et al. 2005), Hypo-IRAS rats exhibited significant redistribution and clustering of synaptophysin and PSD-95 immunoreactivity, suggesting the obesity/MetS phenotype elicits changes in hippocampal synaptic organization and dendritic morphology. Lastly, we examined contextual fear conditioned responses in Hypo-IRAS rats and Hypo-Con rats as a measure of hippocampal-dependent performance. While unconditioned freezing and freezing during the acquisition period were the same in both groups, Hypo-IRAS rats exhibited a significant reduction in retention freezing behaviors compared to Hypo-Con rats (Grillo et al. 2011b). These behavioral deficits were associated with decreases in behaviorally induced fos-like immunoreactivity in the CA1 region of Hypo-IRAS rats, thereby providing another indicator of decreased functional activity in the CA1 region of Hypo-IRAS rats. Importantly, these changes in retention freezing behaviors occurred in the absence of changes in locomotor activity, illustrating that the obesity/MetS phenotype does not elicit generalized behavioral deficits. Collectively, these data demonstrate that the obesity/MetS phenotype elicited by the downregulation of hypothalamic IRs impairs hippocampal synaptic plasticity in a similar manner as has been observed in experimental models of diabetes and obesity. However, it is important to note that unlike our previous studies in type 1 diabetes rats (McEwen and Reagan 2004; Piroli et al. 2004; 2007) or obese Zucker rats (Winocur et al. 2005), hippocampal IR expression and/or activity is unaffected in Hypo-IRAS rats, suggesting that the neuroplasticity deficits in Hypo-IRAS rats result from changes in the endocrine and metabolic milieu and not from deficits in hippocampal IR activity. Moreover, several endocrine measures, including HPA axis activity, are unaffected in Hypo-IRAS rats compared to Hypo-Con rats (Grillo et al. 2007, 2011c), demonstrating that Hypo-IRAS rats exhibit more selective metabolic and endocrine changes compared to experimental models of diabetes or obesity. As a result, the Hypo-IRAS model allows for a more discrete examination of the potential metabolic and endocrine causes of hippocampal neuroplasticity deficits in metabolic disorders.

16.5 Mechanistic Links Between Metabolic Disorders and Neuropsychiatric Disorders

In view of the increased risk of neuropsychiatric disorders in patients with obesity and diabetes (Andersen et al. 2010; Anderson et al. 2001, 2010; Fabricatore and Wadden 2006; Luppino et al. 2010; McElroy et al. 2004; Simon et al. 2006; Stunkard et al. 2003), we examined whether Hypo-IRAS rats exhibit depressive-like and anxiety-like behaviors. Specifically, we examined behavioral performance of Hypo-IRAS rats and Hypo-Con rats in the FST, the sucrose preference test and the EPM. In the FST (Porsolt et al. 1977, 1978), behaviors are considered to be either "active" (i.e., swimming or climbing) or "immobility" (little or no movement). In the pretest of the FST, both Hypo-IRAS and Hypo-Con rats exhibited similar levels of immobility and active behaviors. However, in the test phase of the FST performed 24 h later, Hypo-IRAS rats exhibited a significant increase in immobility behaviors with a corresponding decrease in active behaviors when compared to Hypo-Con rats. This included a significant decrease in the latency to exhibit immobility behavior in Hypo-IRAS rats (Fig. 16.1). Collectively, these behavioral changes indicate that rats with the obesity/MetS phenotype are exhibiting "behavioral despair" (Grillo et al. 2011c). As another measure of "depressive-like behaviors," we examined sucrose preference in Hypo-IRAS and Hypo-Con rats. While total fluid intake did not change, Hypo-IRAS rats exhibited a significant decrease in sucrose consumption, indicating that these rats are exhibiting anhedonia. Lastly, Hypo-IRAS rats exhibited significant decreases in open arm time in the EPM in the absence of differences in locomotor activity or total distance traveled in the maze. Such results suggest that Hypo-IRAS rats are exhibiting "anxiety-like behaviors" (Grillo et al. 2011c).

While these studies indicate that Hypo-IRAS rats develop a depressive-like and anxiety-like phenotype, the question that remains to be answered is what are the potential mechanistic links between obesity and mood disorders? Our ongoing studies are beginning to address these questions. An obvious candidate is the adipocyte derived hormone leptin. While leptin is known to facilitate hippocampal synaptic plasticity under physiological settings (for review, see Harvey 2007), leptin resistance (i.e., decreases in leptin signalin and/or leptin transport across the blood–brain barrier) is a hallmark feature of metabolic disorders (Banks et al. 1999; Banks 2004;

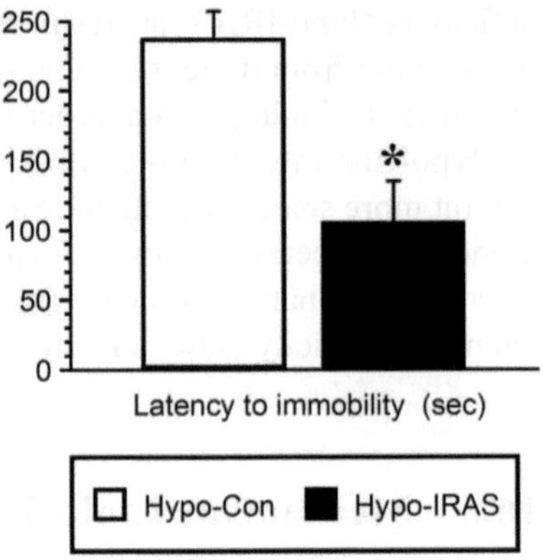

Fig. 16.1 Hypo-*IRAS* rats exhibit behavioral despair in the FST. In addition to increases in immobility behaviors (Grillo et al. 2011c), Hypo-*IRAS* rats also exhibit significant decreases in latency to exhibit immobility behaviors when compared to Hypo-Con rats. See text for details. *$p < 0.05$ compared to Hypo-Con rats; data based upon at least 10 rats/group.

Burguera et al. 2000; Levin et al. 2004; Levin and Dunn-Meynell 2002). These observations have led to the suggestion that reduced CNS leptin activity may be a mechanistic link between obesity and major depressive illness (Lu 2007); our studies provide support for this hypothesis. For example, Hypo-IRAS rats exhibit decreases in leptin-stimulated phosphorylation of STAT3 (pSTAT3), which may result from a combination of decreased leptin transport and/or leptin signaling (Grillo et al. 2011b). It is also interesting to note that studies by Banks and coworkers have shown that increases in plasma triglycerides, a characteristic feature of obesity, directly inhibits BBB leptin transport (Banks et al. 2004; Farr et al. 2008). As such, impairments in hippocampal plasticity and development of behavioral deficits in obesity phenotypes may result from a combination of increases in plasma leptin and triglyceride levels. One way to begin to address this question would be to return plasma leptin and plasma triglyceride levels to those observed in Hypo-Con rats. To achieve this objective, we subjected Hypo-IRAS rats to two different food restriction paradigms to more selectively examine whether normalization of plasma leptin and triglycerides levels would restore hippocampal synaptic plasticity. In one group of rats, a mild food restriction paradigm was initiated prior to the development of the obesity/MetS phenotype (Prevention group); in the second group of Hypo-IRAS rats, we allowed the obesity/MetS phenotype develop before initiation of food restriction (Reversal group). As expected, these food restriction paradigms effectively inhibited (Prevention) or reversed (Reversal) the Hypo-IRAS-induced increases in plasma leptin and triglyceride levels. These food restriction paradigms also restored synaptic transmission and phosphorylation state of GluA1 receptors in the hippocampus of Hypo-IRAS rats (Grillo et al. 2011a). Collectively, these data suggest that central leptin resistance, perhaps facilitated by increases in plasma triglyceride levels, is a key mechanistic mediator of comorbid obesity and depressive illness. In addition, data from the literature suggest that triglycerides may directly impair hippocampal plasticity (Farr et al. 2008) and thereby also serve as a link between obesity and mood disorders.

Beyond leptin and triglycerides, there is also a potential role for pro-inflammatory cytokines. For example, clinical studies indicate that plasma levels of IL-6 and

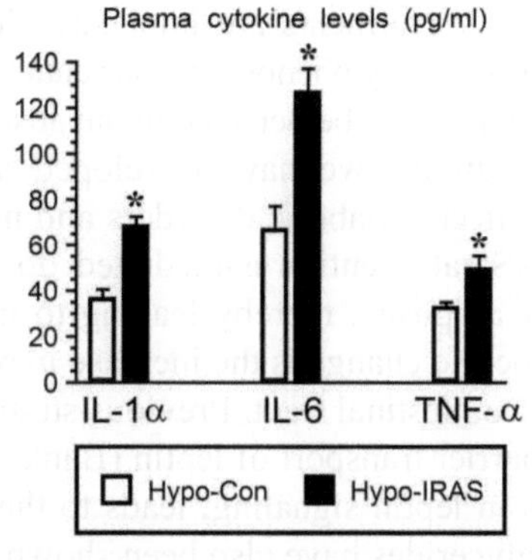

Fig. 16.2 Hypo-*IRAS* rats exhibit significant increases in plasma interleukin (*IL*)-1α, *IL*-6, and tumor necrosis factor (*TNF*)-α levels. Plasma levels of the pro-inflammatory cytokines *IL*-1α, *IL*-6, and *TNF*-α are increased in Hypo-*IRAS* rats that develop the MetS/obesity phenotype compared to Hypo-Con rats, thereby providing a potential cause of the neurological consequences of metabolic disorders, including the increased risk for the development and progression of neuropsychiatric disorders. See text for details. *$p < 0.05$ compared to Hypo-Con rats; data based upon at least 10 rats/group.

tumor necrosis factor (TNF)-α are elevated in patients with depression and pro-inflammatory cytokines are linked to treatment-resistant depression (Raison et al. 2006). Moreover, preclinical studies demonstrate that pro-inflammatory cytokines elicit depressive-like symptoms in animals (Capuron and Miller 2011). In obesity phenotypes, macrophage accumulation in adipose tissue leads to increased secretion of pro-inflammatory cytokines and as a result chronic mild inflammation is a characteristic feature of obesity (Lumeng and Saltiel 2011). Interestingly, we have found that plasma levels of IL-1α, IL-6, and TNF-α are increased in Hypo-IRAS rats (Fig. 16.2), suggesting that adipocyte derived pro-inflammatory cytokines may also be mechanistic links between obesity and mood disorders. Mechanistically, pro-inflammatory cytokines are proposed to impair the activity of neural networks implicated in the pathology of depressive illness, in part by decreasing brain-derived neurotrophic factor (BDNF) levels (Capuron and Miller 2011). In support of this hypothesis, BDNF protein expression is reduced in the plasma, hippocampus and amygdala of Hypo-IRAS rats (Grillo et al. 2011c).

While these results identify leptin resistance, increases in triglycerides and pro-inflammatory cytokines as potential mechanistic links between metabolic disorders and neuropsychiatric disorders, obviously there are other endocrine and/or metabolic changes that may contribute these comorbidities. As noted above, impairments in HPA axis activity are often observed in metabolic disorders and HPA axis activity may be correlated with the degree of glycemic control in diabetes patients (Oltmanns et al. 2006). In this context, our findings that HPA axis dysfunction is not observed in Hypo-IRAS rats that develop a depressive-like phenotype is somewhat puzzling. However, a recent clinical study identified associations between inflammation, dyslipidemia, and obesity in patients with depressive illness, but did not identify an association with HPA axis activity (Reedt Dortland et al. 2013).

Therefore, while HPA axis impairments are implicated in the pathophysiology of mood disorders and diabetes/obesity phenotypes, our data in Hypo-IRAS rats suggest that obesity-induced anhedonia may be detected in the absence of HPA axis deficits.

Based on these observations, we have developed a working model of the mechanistic links that connect metabolic disorders and neuropsychiatric disorders (Fig. 16.3). In hypo-IRAS rats, lentivirus-mediated downregulation of hypothalamic IRs increases body adiposity, thereby leading to increases in plasma leptin levels. An additional endocrine change is the increase in plasma triglyceride levels, presumably from the gastrointestinal tract. Previous studies indicate that triglycerides impair blood-brain barrier transport of leptin (Banks et al. 2004), which when combined with decreases in leptin signaling, leads to the development of a CNS-deficient leptin state. Triglycerides have also been shown to directly impact hippocampal synaptic transmission and the performance of hippocampal-dependent behaviors (Farr et al. 2008). Increases in adiposity will also facilitate macrophage recruitment, which will lead to increased synthesis and secretion of adipocyte-derived pro-inflammatory cytokines implicated in the pathogenesis of depressive illness, like IL-1α, IL-6, and TNF-α. Collectively, the changes in endocrine, metabolic, and inflammatory milieu are at least in part responsible for neuroplasticity deficits in the neural circuits implicated in the pathophysiology of neuropsychiatric disorders. For example, decreases in CNS leptin activity (Harvey et al. 2006), as well as increases in triglyceride levels (Farr et al. 2008), may directly impair glutamatergic function and hippocampal synaptic transmission. Deficient CNS leptin activity is also associated with hippocampal morphological changes, including decreases in spine density (Stranahan et al. 2009) and synaptic reorganization (Grillo et al. 2011b). While the exact mechanisms remain to be determined, pro-inflammatory cytokines are proposed to downregulate neurotrophic factor levels and also negatively affect neurotransmitter synthesis and activity (Raison et al. 2006), a hypothesis that has been extended to comorbid depression and obesity (Capuron et al. 2008). Beyond these identified causes in Hypo-IRAS rats, the neurological consequences of metabolic disorders may also result from changes in HPA axis activity (Reagan et al. 2008), as well as from cerebrovascular complications (Biessels et al. 2008).

16.6 Conclusions and Future Directions

The clinical and epidemiological data clearly indicate that the development and progression of neuropsychiatric disorders is a long-term complication of metabolic disorders like diabetes, obesity, and MetS. Indeed, these patients populations are two- to threefold more likely to develop comorbid depression when compared to nondiabetic individuals, have a more severe course of illness, and exhibit a tenfold increased risk of suicide (Ali et al. 2006; Anderson et al. 2001). The positive view from the evaluation of these studies is that there appear to be common mechanistic mediators in the development of comorbid depression and obesity/diabetes phenotypes. However, the pessimistic perspective is that given the wide variety of

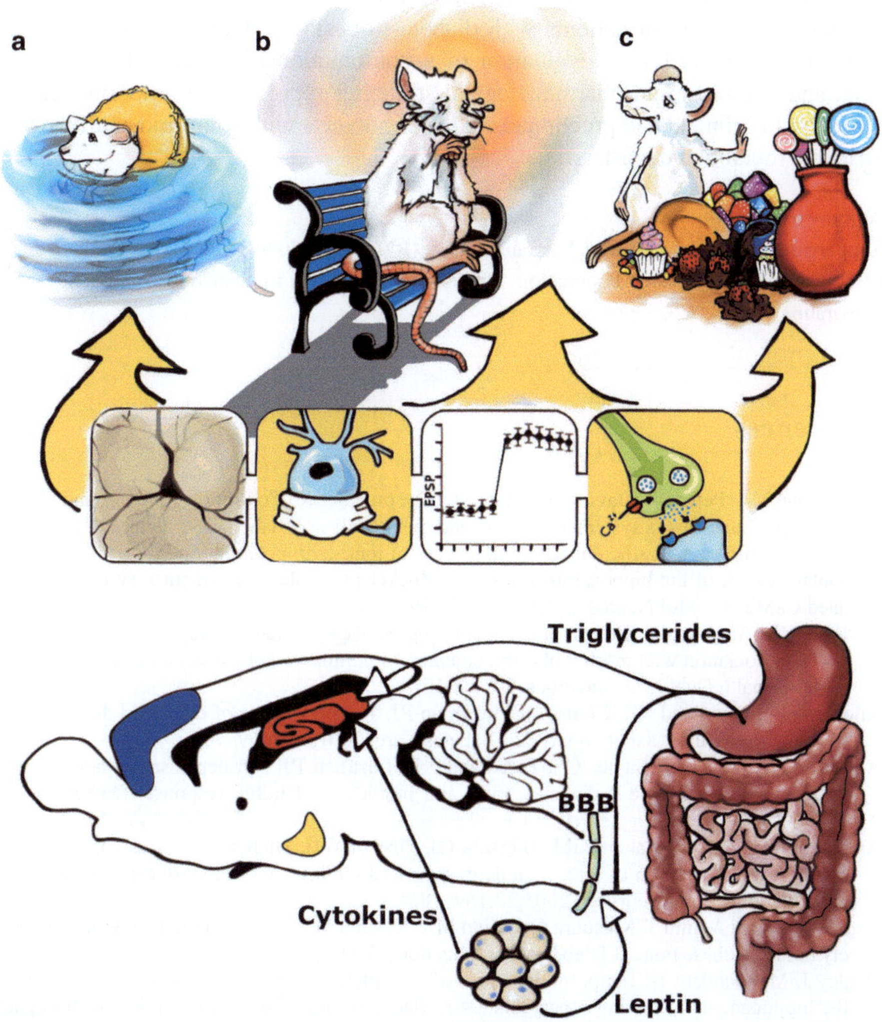

Fig. 16.3 Changes in the metabolic, endocrine, and inflammatory milieu are mechanistic links in comorbid neuropsychiatric disorders and metabolic disorders. Leptin resistance, involving decreases in leptin signaling and triglyceride-mediated decreases in blood–brain barrier (*BBB*) leptin transport, is a hallmark feature of metabolic disorders and impairs hippocampal synaptic plasticity. Beyond effects at the BBB, triglycerides may act directly in the hippocampus to adversely affect synaptic transmission and behavior. Increases in adiposity associated with metabolic disorders will lead to macrophage recruitment, which will lead to the increased synthesis and secretion of pro-inflammatory cytokines. When combined with additional alterations, such as deficits in HPA axis function (not shown), these changes will reduce morphological, electrophysiological, and neurochemical plasticity in the brain regions such as the hippocampus (shown in *red*), prefrontal cortex (*blue*), and the amygdala (*yellow*), and thereby increase the risk of comorbid mood disorders in patients with diabetes, obesity, and MetS. See text for details. (Figure adapted from Fadel et al. 2013 and Grillo et al. 2011b)

potential mechanistic links, development of a specific strategy to successfully manage mood disorders in patients with diabetes, obesity or MetS will be extremely challenging. Accordingly, evaluation of a combination of lifestyle interventions (diet and exercise) and pharmacological strategies represents an important future direction for clinical and preclinical studies in subjects with comorbid neuropsychiatric and metabolic disorders.

Acknowledgments
Supported by the Department of Veterans Affairs (IO1 BX001804-01; LPR) and the University of South Carolina Research Foundation. The authors would like to thank Victoria Macht for the preparation of Fig. 16.3.

References

Ali S, Stone MA, Peters JL, Davies MJ, Khunti K. The prevalence of co-morbid depression in adults with type 2 diabetes: a systematic review and meta-analysis. Diabet Med. 2006;23:1165–73.
Alzoubi KH, Aleisa AM, Alkadhi KA. Impairment of long-term potentiation in the CA1, but not dentate gyrus, of the hippocampus in Obese Zucker rats: role of calcineurin and phosphorylated CaMKII. J Mol Neurosci. 2005;27:337–46.
Andersen JR, Aasprang A, Bergsholm P, Sletteskog N, Vage V, Natvig GK. Anxiety and depression in association with morbid obesity: changes with improved physical health after duodenal switch. Health Qual Life Outcomes. 2010;8:52.
Anderson RJ, Freedland KE, Clouse RE, Lustman PJ. The prevalence of comorbid depression in adults with diabetes: a meta-analysis. Diabetes Care. 2001;24:1069–78.
Anderson RJ, Gott BM, Sayuk GS, Freedland KE, Lustman PJ. Antidepressant pharmacotherapy in adults with type 2 diabetes: rates and predictors of initial response. Diabetes Care. 2010;33:485–9.
Artola A, Kamal A, Ramakers GM, Biessels GJ, Gispen WH. Diabetes mellitus concomitantly facilitates the induction of long-term depression and inhibits that of long-term potentiation in hippocampus. Eur J Neurosci. 2005;22:169–78.
Asakawa A, Inui A, Inui T, Katsuura G, Fujino MA, Kasuga M. Leptin treatment ameliorates anxiety in ob/ob obese mice. J Diabetes Complications. 2003;17:105–7.
Bagley J, Moghaddam B. Temporal dynamics of glutamate efflux in the prefrontal cortex and in the hippocampus following repeated stress: effects of pretreatment with saline or diazepam. Neuroscience. 1997;77:65–73.
Banks WA. The many lives of leptin. Peptides. 2004;25:331–8.
Banks WA, Coon AB, Robinson SM, Moinuddin A, Shultz JM, Nakaoke R, Morley JE. Triglycerides induce leptin resistance at the blood-brain barrier. Diabetes. 2004;53:1253–60.
Banks WA, DiPalma CR, Farrell CL. Impaired transport of leptin across the blood-brain barrier in obesity. Peptides. 1999;20:1341–5.
Barbiero VS, Giambelli R, Musazzi L, Tiraboschi E, Tardito D, Perez J, Drago F, Racagni G, Popoli M. Chronic antidepressants induce redistribution and differential activation of alphaCaM kinase II between presynaptic compartments. Neuropsychopharmacology. 2007;32:2511–9.
Beauquis J, Roig P, Homo-Delarche F, De Nicola A, Saravia F. Reduced hippocampal neurogenesis and number of hilar neurones in streptozotocin-induced diabetic mice: reversion by antidepressant treatment. Eur J Neurosci. 2006;23:1539–46.
Beauquis J, Saravia F, Coulaud J, Roig P, Dardenne M, Homo-Delarche F, De Nicola A. Prominently decreased hippocampal neurogenesis in a spontaneous model of type 1 diabetes, the nonobese diabetic mouse. Exp Neurol. 2008;210:359–67.

Benedict C, Hallschmid M, Hatke A, Schultes B, Fehm HL, Born J, Kern W. Intranasal insulin improves memory in humans. Psychoneuroendocrinology. 2004;29:1326–34.

Benedict C, Hallschmid M, Schmitz K, Schultes B, Ratter F, Fehm HL, Born J, Kern W. Intranasal insulin improves memory in humans: superiority of insulin aspart. Neuropsychopharmacology. 2007;32:239–43.

Biessels GJ, Deary IJ, Ryan CM. Cognition and diabetes: a lifespan perspective. Lancet Neurol. 2008;7:184–90.

Biessels G-J, Kamal A, Ramakers GM, Urban IJ, Spruijt BM, Erkelens DW, Gispen WH. Place learning and hippocampal synaptic plasticity in streptozotocin-induced diabetic rats. Diabetes. 1996;45:1259–66.

Bonanno G, Giambelli R, Raiteri L, Tiraboschi E, Zappettini S, Musazzi L, Raiteri M, Racagni G, Popoli M. Chronic antidepressants reduce depolarization-evoked glutamate release and protein interactions favoring formation of SNARE complex in hippocampus. J Neurosci. 2005;25:3270–9.

Bruning JC, Gautam D, Burks DJ, Gillette J, Schubert M, Orban PC, Klein R, Krone W, Muller-Wieland D, Kahn CR. Role of brain insulin receptor in control of body weight and reproduction. Science. 2000;289:2122–5.

Burguera B, Couce ME, Curran GL, Jensen MD, Lloyd RV, Cleary MP, Poduslo JF. Obesity is associated with a decreased leptin transport across the blood-brain barrier in rats. Diabetes. 2000;49:1219–23.

Capuron L, Miller AH. Immune system to brain signaling: neuropsychopharmacological implications. Pharmacol Ther. 2011;130:226–38.

Capuron L, Su S, Miller AH, Bremner JD, Goldberg J, Vogt GJ, Maisano C, Jones L, Murrah NV, Vaccarino V. Depressive symptoms and metabolic syndrome: is inflammation the underlying link? Biol Psychiatry. 2008;64:896–900.

Chabot C, Massicotte G, Milot M, Trudeau F, Gagne J. Impaired modulation of AMPA receptors by calcium-dependent processes in streptozotocin-induced diabetic rats. Brain Res. 1997;768:249–56.

Collin M, Hakansson-Ovesjo ML, Misane I, Ogren SO, Meister B. Decreased 5-HT transporter mRNA in neurons of the dorsal raphe nucleus and behavioral depression in the obese leptin-deficient ob/ob mouse. Brain Res Mol Brain Res. 2000;81:51–61.

Conrad CD. The relationship between acute glucocorticoid levels and hippocampal function depends upon task aversivenes and memory processing stage. Nonlinearity Biol Toxicol Med. 2005;3:57–78.

Craft S, Asthana S, Newcomer JW, Wilkinson CW, Matos IT, Baker LD, Cherrier M, Lofgreen C, Latendresse S, Petrova A, Plymate S, Raskind M, Grimwood K, Veith RC. Enhancement of memory in Alzheimer Disease with insulin and somatostatin, but not glucose. Arch Gen Psychiatry. 1999;56:1135–40.

Craft S, Baker LD, Montine TJ, Minoshima S, Watson GS, Claxton A, Arbuckle M, Callaghan M, Tsai E, Plymate SR, Green PS, Leverenz J, Cross D, Gerton B. Intranasal insulin therapy for Alzheimer disease and amnestic mild cognitive impairment: a pilot clinical trial. Arch Neurol. 2012;69:29–38.

Dallman MF, Pecoraro NC, la Fleur SE, Warne JP, Ginsberg AB, Akana SF, Laugero KC, Houshyar H, Strack AM, Bhatnagar S, Bell ME. Glucocorticoids, chronic stress, and obesity. Prog Brain Res. 2006;153:75–105.

De Nicola AF, Fridman O, Del Castillo EJ, Foglia VG. The influence of streptozotocin diabetes on adrenal function in male rats. Horm Metab Res. 1976;8:388–92.

Di Luca M, Ruts L, Gardoni F, Cattabeni F, Biessels G-J, Gispen WH. NMDA receptor subunits are modified transcriptionally and post-translationally in the brain of streptozotocin-diabetic rats. Diabetologia. 1999;42:693–701.

Diamond DM, Campbell A, Park CR, Vouimba RM. Preclinical research on stress, memory, and the brain in the development of pharmacotherapy for depression. Eur Neuropsychopharmacol. 2004;14(Suppl 5):S491–5.

Fabricatore AN, Wadden TA. Obesity. Annu Rev Clin Psychol. 2006;2:357–377.

Fadel JR, Jolivalt CG, Reagan LP. Food for thought: the role of appetitive peptides in age-related cognitive decline. Ageing Res Rev. 2013;12(3):764–76

Farr SA, Yamada KA, Butterfield DA, Abdul HM, Xu L, Miller NE, Banks WA, Morley JE. Obesity and hypertriglyceridemia produce cognitive impairment. Endocrinology. 2008;149:2628–36.

Gagne J, Milot M, Gelinas S, Lahsaini A, Trudeau F, Martinoli MG, Massicotte G. Binding properties of glutamate receptors in streptozotocin-induced diabetes in rats. Diabetes. 1997;46:841–6.

Gardoni F, Kamal A, Bellone C, Biessels GJ, Ramakers GM, Cattabeni F, Gispent WH, Di Luca M. Effects of streptozotocin-diabetes on the hippocampal NMDA receptor complex in rats. J Neurochem. 2002;80:438–47.

Gerges NZ, Aleisa AM, Alkadhi KA. Impaired long-term potentiation in obese zucker rats: possible involvement of presynaptic mechanism. Neuroscience. 2003;120:535–9.

Grillo CA, Piroli GG, Evans AN, Macht VA, Wilson SP, Scott KA, Sakai RR, Mott DD, Reagan LP. Obesity/hyperleptinemic phenotype adversely affects hippocampal plasticity: effects of dietary restriction. Physiol Behav. 2011a;104:235–41.

Grillo CA, Piroli GG, Junor L, Wilson SP, Mott DD, Wilson MA, Reagan LP. Obesity/hyperleptinemic phenotype impairs structural and functional plasticity in the rat hippocampus. Physiol Behav. 2011b;105:138–44.

Grillo CA, Piroli GG, Kaigler KF, Wilson SP, Wilson MA, Reagan LP. Downregulation of hypothalamic insulin receptor expression elicits depressive-like behaviors in rats. Behav Brain Res. 2011c;222:230–5.

Grillo CA, Piroli GG, Rosell DR, Hoskin EK, McEwen BS, Reagan LP. Region specific increases in oxidative stress and superoxide dismutase in the hippocampus of diabetic rats subjected to stress. Neuroscience. 2003;121:133–40.

Grillo CA, Piroli GG, Wood GE, Reznikov LR, McEwen BS, Reagan LP. Immunocytochemical analysis of synaptic proteins provides new insights into diabetes-mediated plasticity in the rat hippocampus. Neuroscience. 2005;136:477–86.

Grillo CA, Tamashiro KL, Piroli GG, Melhorn S, Gass JT, Newsom RJ, Reznikov LR, Smith A, Wilson SP, Sakai RR, Reagan LP. Lentivirus-mediated downregulation of hypothalamic insulin receptor expression. Physiol Behav. 2007;92:691–701.

Haj-ali V, Mohaddes G, Babri SH. Intracerebroventricular insulin improves spatial learning and memory in male Wistar rats. Behav Neurosci. 2009;123:1309–14.

Harvey J. Leptin regulation of neuronal excitability and cognitive function. Curr Opin Pharmacol. 2007;7:643–7.

Harvey J, Solovyova N, Irving A. Leptin and its role in hippocampal synaptic plasticity. Prog Lipid Res. 2006;45:369–78.

Hoyer S, Nitsch R. Cerebral excess release of neurotransmitter amino acids subsequent to reduced cerebral glucose metabolism in early-onset dementia of Alzheimer type. J Neural Transm. 1989;75:227–32.

Izumi Y, Yamada KA, Matsukawa M, Zorumski CF. Effects of insulin on long-term potentiation in hippocampal slices from diabetic rats. Diabetologia. 2003;46:1007–12.

Kamal A, Biessels G-J, Urban IJA, Gispen WH. Hippocampal synaptic plasticity in streptozotocin-diabetic rats: impairment of long-term potentiation and facilitation of long-term depression. Neuroscience. 1999;90:737–45.

Kim HB, Jang MH, Shin MC, Lim BV, Kim YP, Kim KJ, Kim EH, Kim CJ. Treadmill exercise increases cell proliferation in dentate gyrus of rats with streptozotocin-induced diabetes. J Diabetes Complications. 2003;17:29–33.

Leedom LJ, Meehan WP, Zeidler A. Avoidance responding in mice with diabetes mellitus. Physiol Behav. 1987;40:447–51.

Levin BE, Dunn-Meynell AA. Reduced central leptin sensitivity in rats with diet-induced obesity. Am J Physiol Regul Integr Comp Physiol. 2002;283:R941–R8.

Levin BE, Dunn-Meynell AA, Banks WA. Obesity-prone rats have normal blood-brain barrier transport but defective central leptin signaling before obesity onset. Am J Physiol Regul Integr Comp Physiol. 2004;286:R143–R50.

Lowy MT, Gault L, Yamamoto BK. Adrenalectomy attenuates stress-induced elevations in extracellular glutamate concentrations in the hippocampus. J Neurochem. 1993;61:1957–60.
Lu XY. The leptin hypothesis of depression: a potential link between mood disorders and obesity? Curr Opin Pharmacol. 2007;7:648–52.
Lumeng CN, Saltiel AR. Inflammatory links between obesity and metabolic disease. J Clin Invest. 2011;121:2111–7.
Luppino FS, de Wit LM, Bouvy PF, Stijnen T, Cuijpers P, Penninx BW, Zitman FG. Overweight, obesity, and depression: a systematic review and meta-analysis of longitudinal studies. Arch Gen Psychiatry. 2010;67:220–9.
Magariños AM, McEwen BS. Experimental diabetes in rats causes hippocampal dendritic and synaptic reorganization and increased glucocorticoid reactivity to stress. Proc Natl Acad Sci U S A. 2000;97:11056–61.
Manning CA, Stone WS, Korol DL, Gold PE. Glucose enhancement of 24-h memory retrival in healthy elderly humans. Behav Brain Res. 1998;93:71–6.
Martinez-Tellez R, Gomez-Villalobos MJ, Flores G. Alteration in dendritic morphology of cortical neurons in rats with diabetes mellitus induced by streptozotocin. Brain Res. 2005;1048:108–115.
McElroy SL, Kotwal R, Malhotra S, Nelson EB, Keck PE, Nemeroff CB. Are mood disorders and obesity related? A review for the mental health professional. J Clin Psychiatry. 2004;65:634–51 (quiz).
McEwen BS. Central effects of stress hormones in health and disease: understanding the protective and damaging effects of stress and stress mediators. Eur J Pharmacol. 2008;583:174–85.
McEwen BS, Chattarji S, Diamond DM, Jay TM, Reagan LP, Svenningsson P, Fuchs E. The neurobiological properties of tianeptine (Stablon): from monoamine hypothesis to glutamatergic modulation. Mol Psychiatry. 2010;15:237–49.
McEwen BS, Reagan LP. Glucose transporter expression in the central nervous system: relationship to synaptic function. Eur J Pharmacol. 2004;490:13–24.
Meehan WP, Leedom LJ, Nagayama T, Zeidler A. Agonistic behavior patterns in mice with streptozotocin-induced diabetes mellitus. Physiol Behav. 1986;38:301–6.
Miyata S, Yamada N, Hirano S, Tanaka S, Kamei J. Diabetes attenuates psychological stress-elicited 5-HT secretion in the prefrontal cortex but not in the amygdala of mice. Brain Res. 2007;1147:233–9.
Moosavi M, Naghdi N, Choopani S. Intra CA1 insulin microinjection improves memory consolidation and retrieval. Peptides. 2007;28:1029–34.
Musazzi L, Milanese M, Farisello P, Zappettini S, Tardito D, Barbiero VS, Bonifacino T, Mallei A, Baldelli P, Racagni G, Raiteri M, Benfenati F, Bonanno G, Popoli M. Acute stress increases depolarization-evoked glutamate release in the rat prefrontal/frontal cortex: the dampening action of antidepressants. PLoS ONE. 2010;5:e8566.
Nistico R, Cavallucci V, Piccinin S, Macri S, Pignatelli M, Mehdawy B, Blandini F, Laviola G, Lauro D, Mercuri NB, D'Amelio M. Insulin receptor beta-subunit haploinsufficiency impairs hippocampal late-phase LTP and recognition memory. Neuromolecular Med. 2012;14(4):262–9.
Obici S, Feng Z, Karkanias G, Baskin DG, Rossetti L. Decreasing hypothalamic insulin receptors causes hyperphagia and insulin resistance in rats. Nat Neurosci. 2002;5:566–72.
Oltmanns KM, Dodt B, Schultes B, Raspe HH, Schweiger U, Born J, Fehm HL, Peters A. Cortisol correlates with metabolic disturbances in a population study of type 2 diabetic patients. Eur J Endocrinol. 2006;154:325–31.
Oomura Y, Hori N, Shiraishi T, Fukunaga K, Takeda H, Tsuji M, Matsumiya T, Ishibashi M, Aou S, Li XL, Kohno D, Uramura K, Sougawa H, Yada T, Wayner MJ, Sasaki K. Leptin facilitates learning and memory performance and enhances hippocampal CA1 long-term potentiation and CaMK II phosphorylation in rats. Peptides. 2006;27:2738–49.
Oster MH, Castonguay TM, Keen CL, Stern JS. Circadian rhythm of corticosterone in diabetic rats. Life Sci. 1988;43:1643–5.
Paranjape SA, Chan O, Zhu W, Horblitt AM, Grillo C, Wilson S, Reagan L, Sherwin RS. Chronic reductions of insulin receptors in the ventromedial hypothalamus produces glucose intolerance and islet dysfunction in the absence of weight gain. Am J Physiol. 2011;301:E978–E83.

Park CR, Seely RJ, Craft S, Woods SC. Intracerebroventricular insulin enhances memory in a passive-avoidance task. Physiol Behav. 2000;68:509–14.
Piroli GG, Grillo CA, Reznikov LR, Reagan LP (2007) Expression and functional activities of glucose transporters in the central nervous system. In: Lajtha A, Editor. Handbook of neurochemistry and molecular neurobiology. New York: Springer; 2007. p. 387–404.
Piroli GG, Grillo CA, Charron MJ, McEwen BS, Reagan LP. Biphasic effects of stress upon GLUT8 glucose transporter expression and trafficking in the diabetic rat hippocampus. Brain Res. 2004;1006:28–35.
Piroli GG, Reznikov LR, Grillo CA, Hagar JM, Fadel JR, Reagan LP. Tianeptine modulates amygdalar glutamate neurochemistry and synaptic proteins in rats subjected to repeated stress. Exp Neurol. 2013;241:184–93.
Plotsky PM, Thrivikraman KV, Watts AG, Hauger RL. Hypothalamic-pituitary-adrenal axis function in the Zucker obese rat. Endocrinology. 1992;130:1931–41.
Popoli M, Yan Z, McEwen BS, Sanacora G. The stressed synapse: the impact of stress and glucocorticoids on glutamate transmission. Nat Rev Neurosci. 2012;13:22–37.
Porsolt RD, Anton G, Blavet N, Jalfre M. Behavioural despair in rats: a new model sensitive to antidepressant treatments. Eur J Pharmacol. 1978;47:379–91.
Porsolt RD, Le Pichon M, Jalfre M. Depression: a new animal model sensitive to antidepressant treatments. Nature. 1977;266:730–2.
Raison CL, Capuron L, Miller AH. Cytokines sing the blues: inflammation and the pathogenesis of depression. Trends Immunol. 2006;27:24–31.
Ramanathan M, Jaiswal AK, Bhattacharya SK. Differential effects of diazepam on anxiety in streptozotocin induced diabetic and non-diabetic rats. Psychopharmacology (Berl). 1998;135:361–7.
Reagan LP. Diabetes as a chronic metabolic stressor: causes, consequences and clinical complications. Exp Neurol. 2012;233:68–78.
Reagan LP, Grillo CA, Piroli GG. The As and Ds of stress: metabolic, morphological and behavioral consequences. Eur J Pharmacol. 2008;585:64–75.
Reagan LP, Reznikov LR, Evans AN, Gabriel C, Mocaer E, Fadel JR. The antidepressant agomelatine inhibits stress-mediated changes in amino acid efflux in the rat hippocampus and amygdala. Brain Res. 2012;1466:91–8.
Reedt Dortland AK, Vreeburg SA, Giltay EJ, Licht CM, Vogelzangs N, van Veen T, de Geus EJ, Penninx BW, Zitman FG. The impact of stress systems and lifestyle on dyslipidemia and obesity in anxiety and depression. Psychoneuroendocrinology. 2013;38:209–18.
Reger MA, Watson GS, Green PS, Wilkinson CW, Baker LD, Cholerton B, Fishel MA, Plymate SR, Breitner JC, DeGroodt W, Mehta P, Craft S. Intranasal insulin improves cognition and modulates beta-amyloid in early AD. Neurology. 2008;70:440–8.
Reznikov LR, Grillo CA, Piroli GG, Pasumarthi RK, Reagan LP, Fadel J. Acute stress-mediated increases in extracellular glutamate levels in the rat amygdala: differential effects of antidepressant treatment. Eur J Neurosci. 2007;25:3109–14.
Roozendaal B, McEwen BS, Chattarji S. Stress, memory and the amygdala. Nat Rev Neurosci. 2009;10:423–33.
Schwartz MW, Woods SC, Porte D Jr, Seeley RJ, Baskin DG. Central nervous system control of food intake. Nature. 2000;404:661–71.
Scribner KA, Walker CD, Cascio CS, Dallman MF. Chronic streptozotocin diabetes in rats facilitates the acute stress response without altering pituitary or adrenal responsiveness to secretagogues. Endocrinology. 1991;129:99–108.
Sharma AN, Elased KM, Garrett TL, Lucot JB. Neurobehavioral deficits in db/db diabetic mice. Physiol Behav. 2010;101:381–8.
Simon GE, Von Korff M, Saunders K, Miglioretti DL, Crane PK, van Belle G, Kessler RC. Association between obesity and psychiatric disorders in the US adult population. Arch Gen Psychiatry. 2006;63:824–30.
Stranahan AM, Arumugam TV, Cutler RG, Lee K, Egan JM, Mattson MP. Diabetes impairs hippocampal function through glucocorticoid-mediated effects on new and mature neurons. Nat Neurosci. 2008;11:309–17.

Stranahan AM, Lee K, Martin B, Maudsley S, Golden E, Cutler RG, Mattson MP. Voluntary exercise and caloric restriction enhance hippocampal dendritic spine density and BDNF levels in diabetic mice. Hippocampus. 2009;19:951–61.

Stunkard AJ, Faith MS, Allison KC. Depression and obesity. Biol Psychiatry. 2003;54:330–7.

Talbot K, Wang HY, Kazi H, Han LY, Bakshi KP, Stucky A, Fuino RL, Kawaguchi KR, Samoyedny AJ, Wilson RS, Arvanitakis Z, Schneider JA, Wolf BA, Bennett DA, Trojanowski JQ, Arnold SE. Demonstrated brain insulin resistance in Alzheimer's disease patients is associated with IGF-1 resistance, IRS-1 dysregulation, and cognitive decline. J Clin Invest. 2012;122:1316–38.

Thorre K, Chaouloff F, Sarre S, Meeusen R, Ebinger G, Michotte Y. Differential effects of restraint stress on hippocampal 5-HT metabolism and extracellular levels of 5-HT in streptozotocin-diabetic rats. Brain Res. 1997;772:209–16.

Timmerman W, Westerink BH. Brain microdialysis of GABA and glutamate: what does it signify? Synapse. 1997;27:242–61.

Tuzcu M, Baydas G. Effect of melatonin and vitamin E on diabetes-induced learning and memory impairment in rats. Eur J Pharmacol. 2006;537:106–10.

Valastro B, Cossette J, Lavoie N, Gagnon S, Trudeau F, Massicotte G. Up-regulation of glutamate receptors is associated with LTP defects in the early stages of diabetes mellitus. Diabetologia. 2002;45:642–50.

Winocur G, Greenwood CE, Piroli GG, Grillo CA, Reznikov LR, Reagan LP, McEwen BS. Memory impairment in obese Zucker rats: an investigation of cognitive function in an animal model of insulin resistance and obesity. Behav Neurosci. 2005;119:1389–95.

Yamada N, Katsuura G, Ochi Y, Ebihara K, Kusakabe T, Hosoda K, Nakao K. Impaired CNS leptin action is implicated in depression associated with obesity. Endocrinology. 2011;152:2634–43.

Chapter 17
Using Our Understanding of Stress-Related Effects on Glutamate Neurotransmission to Guide the Development of Novel Treatment Strategies

Carly Kiselycznyk and Gerard Sanacora

Abstract The majority of treatments for neuropsychiatric disorders have been based on serendipitous discoveries, with little understanding of the pathogenic and pathophysiological mechanisms underlying these disorders. As many of these disorders are sensitive to stress, an understanding of the physiology of stress is important in avoiding and reversing stress-sensitive disorders. Increased understanding of the glutamatergic synapse has revealed a system that is affected by both stress and multiple neuropsychiatric treatments, suggesting a possible convergent target in these disorders. This chapter reviews how traditional neuropsychiatric treatments affect the glutamatergic synapse, and how future therapies may be developed to more directly target this system.

17.1 Introduction

The biological and behavioral responses to stress can be beneficial or, as is the case with most neuropsychiatric illnesses, maladaptive and pathogenic. These dual effects of stress can in part be explained by the similarly dual effects of the neurotransmitter glutamate on the strength of synaptic connections between neurons. This chapter details how knowledge of stress-induced glutamatergic dysregulation can be used to develop novel therapeutics to effectively treat neuropsychiatric disorders.

The brain is both the control center for the response to stress, as well as a target for its effects. Along with refocusing energy to organs and muscles needed for escape, stress can help increase cognitive performance in the face of a challenge (Barha et al. 2007; Yuen et al. 2011, 2009). The cognitive effects of stress can be explained by the response of the glutamatergic neurotransmitter system as glucocorticoid stress hormones are known to cause rapid increases in extracellular gluta-

G. Sanacora (✉) · C. Kiselycznyk
Department of Psychiatry, Yale University, 100 York St. Suite 2J, New Haven, CT 06511
e-mail: gerard.sanacora@yale.edu

C. Kiselycznyk
Department of Psychiatry, Yale University, 300 George Street, Suite 901, New Haven CT 06511, USA

M. Popoli et al. (eds.), *Synaptic Stress and Pathogenesis of Neuropsychiatric Disorders*,
DOI 10.1007/978-1-4939-1056-4_17,

mate release (Groeneweg et al. 2011; Stein-Behrens et al. 1994; Venero and Borrell 1999). Diverse types of behavioral stress also increase extracellular glutamate levels in the prefrontal cortex (PFC), hippocampus, and amygdala, as well as the striatum (Moghaddam 1993, 2002; Reznikov et al. 2007; Rutherford et al. 2007; Tardito et al. 2010), and this is dependent on glucocorticoid activation (Lowy et al. 1993).

The timing and amount of glutamate transmission is thought to influence cognitive function through the strengthening or weakening of the synapse. In certain conditions, stress-induced glutamate release is followed by increases in synaptic strength and long-term potentiation (LTP) (Luine et al. 1996), as well as corresponding increases in glutamate receptors at the synapse in the hippocampus and PFC (Groc et al. 2008; Karst and Joëls 2005; Krugers et al. 2010; Yuen et al. 2011, 2009). These alterations in synaptic plasticity are similarly tied to morphological changes as LTP stimulation leads to new and larger dendritic spines (Engert and Bonhoeffer 1999; Matsuzaki et al. 2004).

While stress is an everyday part of life that can boost cognitive and physical performance, it is also a known risk factor for multiple psychiatric conditions (Anisman and Zacharko 1990; Kessler et al. 2012). Exposure to an extreme stress can lead to symptoms of posttraumatic stress disorder (PTSD), characterized by heightened fear memory of a stressful event and parallels increased synaptic strengthening after stress. More common though, it is exposure to chronic unpredictable stress (CUS) that is a risk factor for multiple illnesses such as depression, anxiety, bipolar, schizophrenia, addiction, among others (Caspi et al. 2003; Hammen 2005; Kendler et al. 1999a, 1999b; Lupien et al. 2009; Schneiderman et al. 2005; Sinha 2008). While these illnesses have historically given rise to distinct treatments, their common sensitivity to chronic stress suggests underlying similarities in etiology that can be useful guides in the development of future therapies.

While acute stress increases glutamate release, the effects of chronic exposure to stress on glutamatergic transmission and synaptic strength are still poorly understood. There are complicated adaptations to additional exposures to stress that vary between, and even within, brain regions. Extracellular glutamate levels remained elevated in the hippocampus, but not PFC or striatum, after repeated tail pinch in the same day (Bagley and Moghaddam 1997; Rutherford et al. 2007) and within the PFC there are diverse responses between populations of neurons (Jackson and Moghaddam 2006). Previous exposure to a 21-day chronic restraint stress (CRS), led to longer lasting elevations of glutamate in the face of a novel acute stress challenge. Additionally, CUS leads to reduced glutamate cycling in the PFC as measured by ^{13}C-acetate metabolism (Banasr et al. 2010). While acute increases in glutamatergic transmission can lead to synaptic potentiation, excessive glutamate release can lead to excitotoxicity or cell damage (Sapolsky 2000, 2003). The potentially damaging effects of glutamate lead to a U-shaped curve of glutamate release on synaptic health, with acute instances of stress leading to synaptic potentiation and increased performance on some tasks, and chronic or excessive stress leads to reduced LTP, cell damage, morphological changes and behavioral deficits (Kim and Diamond 2002; Luine et al. 1996).

Many of these changes are dependent on glutamatergic receptors, supporting the role of excessive glutamate in mediating these effects. Once released to the extracellular space, glutamate can be bound by ionotropic and metabotropic glutamate receptors. Ionotropic receptors include *N*-methyl-D-aspartate receptors (NMDARs), alpha-amino-3-hydroxy-5-methyl-4-isoxazolepropionic acid receptors (AMPARs), and kainate receptors, while metabotropic receptors are composed of subunits mGluR1–8. Subunit composition, phosphorylation, kinetics and the location of these receptors play important roles in modulating the receptors' effects on postsynaptic cells and synaptic plasticity.

In rodents, both CRS and CUS, as well as treatment with chronic glucocorticoids, leads to dendritic atrophy and spine loss in pyramidal cells of the CA3 region of the hippocampus (Magariños and McEwen 1995a, 1995b; Sapolsky 2000). These functional and morphological effects of stress are blocked by drugs reducing glutamate release (Watanabe et al. 1992) and by NMDA, but not AMPAR antagonists (Kim et al. 1996; Magariños and McEwen 1995b; Martin and Wellman 2011). Similar changes are observed in select regions of the PFC, where even relatively mild repeated stressors can lead to dendritic retraction and spine loss and this is blocked by the presence of NMDAR antagonists (Izquierdo et al. 2006; Li et al. 2010; Martin and Wellman 2011). CUS also leads to a loss of synaptic proteins, such as the AMPAR subunit GluA1 and synaptic proteins PSD-95 and synapsin, as would be expected with a loss of spines (Li et al. 2010). These morphological changes potentially parallel the reduced neuronal size observed in patient populations (Rajkowska et al. 1999; Stockmeier et al. 2004), but this has not been directly tested.

Together, this evidence suggests that dysregulation of glutamate transmission at the synapse can link chronic stress exposure to psychiatric illness and can guide future therapies. In the past, accidental discoveries with poor understanding of the true mechanisms of action have characterized the development of novel treatments for neuropsychiatric disorders. However, reexamination of traditional therapies such as monoaminergic antidepressants has revealed convergent effects on glutamatergic targets at the synapse that may reverse changes observed after stress. Similarly, drugs developed to directly target the glutamatergic system for nonpsychiatric disorders have shown off-label efficacy in many of these illnesses. This chapter summarizes how knowledge of the stressed synapse relates to established traditional neuropsychiatric therapies and what future therapies might be developed to target the glutamatergic synapse more directly (see Fig. 17.1 for overview of therapies targeting the glutamatergic synapse).

17.2 Therapies Regulating Presynaptic Release of Glutamate

The risk of excitotoxicity after stress suggests that therapies reducing glutamate release could ameliorate the development of stress-sensitive disorders. In fact multiple established antidepressants have now been found to reduce stimulated glutamate release, and drugs directly targeting glutamate have efficacy in neuropsychiatric disorders. However, glutamate release and stress can both play positive and

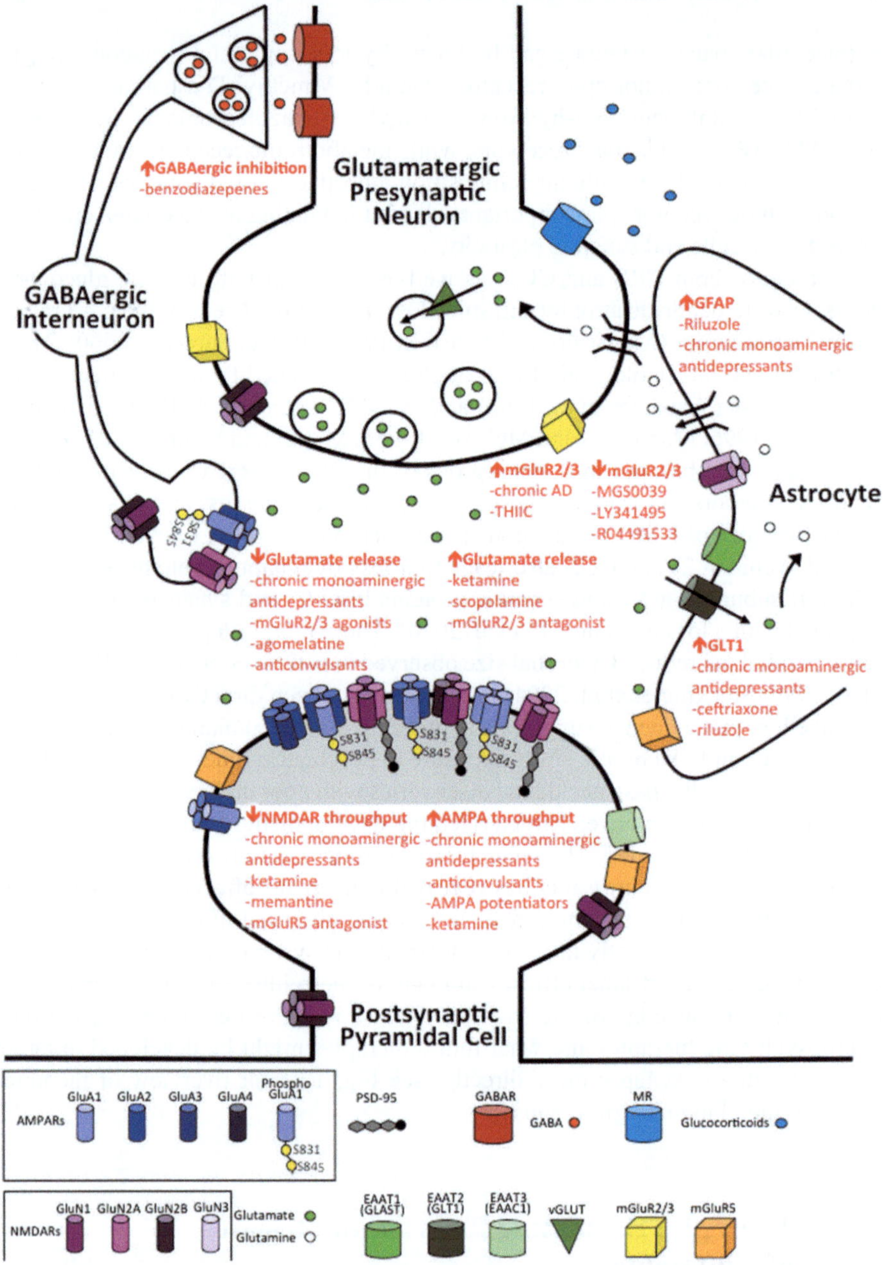

Fig. 17.1 Glutmatergic targets for antidepressant and antistress drug development. *AMPA* α-amino-3-hydroxy-5-methyl-4-isoxazolepropionic acid, *EAAT* excitatory amino acid transporter (1, 2, and 3), *EAAC1* excitatory amino-acid carrier 1, *GABA* γ-aminobutyric acid, *GABAR* γ-aminobutyric acid receptor, *GFAP* glial fibrillary acidic protein, *GLAST* glutamate aspartate transporter, *GLT1* glutamate transporter 1, *mGluR2/3* metabotropic glutamate receptors 2 and 3, *mGluR5* metabotropic glutamate receptor 5, *MR* mineralocorticoid receptor, *PSD-95* postsynaptic density protein 95, *THIIC* N-(4-{[3-hydroxy-4-(2-methylpropanoyl)-2-(trifluoromethyl)phenoxy]methyl}benzyl)-1-methyl-1 H-imidazole-4-carboxamide, *vGLUT* vesicular glutamate transporters, *NMDAR* N-methyl-D-aspartate receptor, *AMPAR* amino-3-hydroxy-5-methyl-4-isoxazolepropionic acid receptor

negative roles in synaptic strength. Recent work on novel antidepressant therapies show the therapeutic effects of pharmacologically modulated glutamate release are complex, suggesting the timing, amplitude, and duration of glutamatergic excitation may all be critical factors in determining the relative benefits and harmful effects in relation to neuropsychiatric disorders.

17.2.1 Traditional Neuropsychiatric Therapies Modulate Presynaptic Glutamate Release

Many traditional antidepressants, such as chronic fluoxetine and desipramine, have been reassessed for effects on glutamate release and found to reduce stimulated glutamate release after chronic treatment (Bonanno et al. 2005; Musazzi et al. 2010; (for a review, see Musazzi et al. 2012). Treatment with the atypical antidepressant tianeptine can block stress-induced glutamate release and, correspondingly, morphological changes in the hippocampus and amygdala (Czéh et al. 2001; Magariños et al. 1999; McEwen et al. 2010; Reznikov et al. 2007), and corresponding increases in anxiety-like behavior (McEwen et al. 2010). Similar reductions in glutamate release are seen in antidepressants of nonmonoaminergic mechanisms. The antidepressant agomelatine, which targets the melatonergic MT(1) and MT(2) receptors, as well as a 5-HT(2 C), reduces stress-induced glutamate release in the PFC (Milanese et al. 2013; Popoli 2009; Tardito et al. 2010, 2012) and can reverse the effects of prenatal stress in rats (Morley-Fletcher et al. 2011). Treatment with chronic antidepressants also increased expression of the metabotropic glutamatergic receptor, mGluR2/3, activation of which suppresses presynaptic glutamate release (Matrisciano et al. 2002) and chronic treatment with amitriptyline, a tricyclic antidepressant, reversed decreases in mGluR2/3 observed in the hippocampus after olfactory bulbectomy (Wieroñska et al. 2001).

Similarly, anxiolytics can reduce stress-induced increases in glutamate in the hippocampus and PFC (Bagley and Moghaddam 1997) and reduce hippocampal atrophy (Magariños et al. 1999). Anxiolytics such as diazepam and other benzodiazepines increase GABAergic cell transmission, increasing inhibition on glutamatergic cells that effectively leads to reductions in glutamate release (see Fig. 17.1).

17.2.2 Treatments Targeting Glutamate Release Have Efficacy in Psychiatric Illnesses

Drugs originally developed to reduce stimulated glutamate release, such as anticonvulsants or treatments for amyotrophic lateral sclerosis (ALS), have demonstrated effects in preclinical rodent models and efficacy in mood disorders. In the preclinical literature, the antiepileptic drug phenytoin is known to reduce glutamate release and, when administered during chronic stress, blocks the dendritic atrophy observed in the hippocampus (Watanabe et al. 1992), but it has not been fully investigated in clinical mood disorder trials. Other anticonvulsants, such as valproate and lamotrigine are FDA approved for use in the treatment of bipolar disorder, and are used as off-label treatments for other mood disorders (Calabrese et al. 1999; Du et al. 2007; McElroy et al. 2004; van der Loos et al. 2009) (for a review of anticonvulsants in psychiatry, see Ettinger and Argoff 2007; Mula et al. 2007). The drug riluzole, which has anticonvulsant properties in addition to providing clinical benefit in the treatment of ALS, also appears to have clinical benefits in relation to anxiety, mood disorders, OCD in several small nonplacebo controlled clinical trials (Coric et al. 2005; Pittenger et al. 2008; Sanacora et al. 2004). Riluzole has also been shown to have antidepressant-like properties in several rodent models (Banasr et al. 2010), and the details of these studies are discussed below. Riluzole is known to reduce glutamatergic transmission, though it is not clear if it works on presynaptic glutamate release or through other mechanisms affecting extracellular glutamate levels.

17.2.3 Novel Neuropsychiatric Treatments and Glutamate Release

In sum, the evidence suggests a number of therapies with diverse structures, but seemingly convergent effects on presynaptic glutamate release in regions implicated in neuropsychiatric disorders, posses antidepressant-like properties in rodent models and in the clinic. While this may suggest presynaptic glutamate release as an ideal target for many of these stress-sensitive disorders, a new class of effective antidepressants suggests that the story is more complicated. Efforts to create fast-acting therapies that directly target the glutamatergic system have led to the discovery of the antidepressant properties of drugs that appear to acutely increase glutamate release such as the NMDAR antagonist ketamine.

Evidence suggesting that antidepressants downregulate NMDAR expression led to the testing of NMDAR antagonists in preclinical models of depression (Trullas and Skolnick 1990). NMDAR antagonists have been found to have fast-acting antidepressant activity in preclinical and clinical studies (Diazgranados et al. 2010; Ibrahim et al. 2011; Skolnick et al. 2001, 2009) and the NMDAR antagonist ketamine has in particular demonstrated efficacy in clinical trials (Berman et al. 2000; Zarate et al. 2006a) (for a review of ketamine in depression, see Mathews and Zarate 2013). The subanesthetic doses at which ketamine has been shown to have antidepressant-like effects are also known to induce a sharp increase of glutamate efflux in the PFC and hippocampus as measured in microdialysis (Moghaddam et al. 1997). More recently these same doses were found to increase glutamate cycling in the PFC (Chowdhury et al. 2012) and to stimulate a series of cellular processes that are associated with changes in synaptic plasticity (Autry et al. 2011; Li et al. 2010). Preclinical work has suggested that these antidepressant-like effects are dependent on AMPA/kainate receptor activity, indicating a requirement for increased synaptic transmission (Autry et al. 2011; Koike et al. 2011; Maeng et al. 2008). Interestingly, a similar mechanism involving a rapid increase in glutamate release and activation of AMPA receptors has also been shown to be related to the rapid antidepressant-like effects of scopolamine (Voleti et al. 2013). However, it is critical to note that the increased glutamate efflux produced by these treatments appears to be of short duration, and to have completely dissipated by the time the antidepressant-like behavioral effects are observed. Recent work demonstrating that ketamine treatment reduces expression of presynaptic release machinery over a period of hours (Müller et al. 2013), suggests the overall effect of the treatments on glutamate release is complex and may vary with time.

The complex role of glutamate release in antidepressant therapies is further demonstrated in the case of mGluR2/3-related treatments. The metabotropic mGluR2/3-containing glutamate receptor is predominantly located presynaptically (Tamaru et al. 2001) and its activation exerts negative feedback on additional glutamate release (Anwyl 1999; Cartmell and Schoepp 2000; Tamaru et al. 2001). mGluR2/3 expression is altered in depressed patients and preclinical models of depression (Feyissa et al. 2010; Matrisciano et al. 2008; Wierońska et al. 2008) and had been proposed as a novel target for depression (Sanacora et al. 2008; Witkin et al. 2007). As noted earlier, treatment with chronic monoaminergic-based antidepressants increases mGlur2/3 expression (Matrisciano et al. 2002). However, pharmacological strategies both increasing or decreasing mGluR2/3 activation have demonstrated preclinical efficacy as anxiolytics and antidepressants (Palucha and Pilc 2007; Pilc et al. 2008). For a review of metabotropic receptors in psychiatry, see Chaki et al. (2013).

Potentiating these receptors can dampen excessive glutamate release and therefore may be beneficial in mediating stress-induced pathophysiology. Administration of a low dose mGluR2/3 agonist shortens the latency to therapeutic effects of chronic antidepressant treatments in preclinical models (Matrisciano et al. 2005, 2007) and agonists of the mGluR2/3 receptor have antidepressant-like efficacy in preclinical tests (DD and Marek 2002; Swanson et al. 2005). A new positive allosteric modulator of mGluR2/3 called N-(4-{[3-hydroxy-4-(2-methylpropanoyl)-2-(trifluoromethyl)phenoxy]methyl}benzyl)-1-methyl-1 H-imidazole-4-carboxamide (THIIC) has robust preclinical antidepressant-like effects (Fell et al. 2011; Johnson et al. 2005). THIIC and other allosteric modulators only activate metabotropic glutamate receptors under conditions of excessive glutamate release to reduce glutamate release (Johnson et al. 2005).

However, treatments with opposing effects on the mGluR2/3 receptor have similar antidepressant-like efficacy. Various mGluR2/3 antagonists including MGS0039, LY341495, and RO4491533 have demonstrated efficacy in the rodent forced swim test (FST) (Chaki et al. 2004; Pałucha-Poniewiera et al. 2010; Yoshimizu et al. 2006). Similar to ketamine, these preclinical effects are dependent on AMPAR throughput suggesting that an increase in glutamatergic transmission is necessary for its effects (Dwyer et al. 2012; Koike et al. 2011).

While the efficacy of NMDAR and mGluR2/3-based treatments supports a role of glutamatergic transmission in antidepressant therapy, it casts doubt on the hypothesis that a simple stress-induced hyperglutamatergic state is the sole contributor to the pathophysiology of mood disorders, and that simply reducing presynaptic glutamate release is necessary and sufficient to mitigate and reverse stress-sensitive disorders. Instead, it now appears that different pathophysiological processes may predominate at different stages in the evolution of the illness. It is possible that the early phases in the evolution of stress sensitive disorders are associated with excessive glutamate efflux and sustained elevation of extracellular glutamate concentrations. This is consistent with reports showing elevated glutamate to be associated with hippocampal toxicity (Sapolsky 2000). However, once the disorder has developed, compensatory changes resulting in diminished synaptic glutamatergic neurotransmission may dominate in relation to the cognitive, behavioral, and emotional symptoms associated with the disorder. This may explain why in the case of some fast-acting antidepressants, decreased glutamate release may actually block any therapeutic effect. As the number of ketamine clinical trials grows, some data suggest that drugs reducing glutamate release, such as with anesthetics, may reduce the efficacy of ketamine (Abdallah et al. 2012). However, this remains to be confirmed in more definite studies. As will be further discussed below, while blocking NMDARs has antidepressant action, therapies that instead boost AMPAR throughput may be associated with similar types of antidepressant-like therapeutic effects (Chappell et al. 2007; Knapp et al. 2002; Li et al. 2001; Lindholm et al. 2012; Nations et al. 2012). Correspondingly stress-induced pyramidal cell atrophy can be reduced by blocking NMDAR activation, but not by blocking AMPA receptors (Magariños and McEwen 1995b). Together these studies suggest that the stress-induced increase in glutamate release is not in itself harmful, but its subsequent postsynaptic effects resulting from the relative activation of the various glutamatergic receptors may be more important for determining the physiological or pathophysiological consequences of the enhanced release.

17.3 Therapies Regulating Extracellular Glutamate Uptake

The ubiquitous nature of glutamate, along with its ability to cause excitotoxicity, necessitates a tightly regulated system controlling its release and extracellular levels. Once released to the extracellular space, glutamate is not broken down but is instead taken up by neighboring glia or neurons via excitatory amino acid transporters (EAAT1–5 in humans) (O'Shea 2002). EAAT 1 and 2 (GLAST and GLT1 in rodents) mainly transport glutamate to astrocytes where it can be converted to glutamine, while EAAT3 (EAAC1 in rodents), transports glutamate to neurons (Anderson and Swanson 2000; Arriza et al. 1994). EAATs and astrocytes placed near the synapse play a critical role in regulating extracellular glutamate levels and risk of excitotoxicity (Arriza et al. 1994; Shigeri et al. 2004; Zarate et al. 2002; Zheng et al. 2008).

While the role of glia cells in neuropsychiatry has been understudied in the past, there is now an increased understanding of their complexity and roles in glutamatergic transmission and dysfunction. Astrocytes are ideally placed to help restrict extracellular glutamate to the synapse and limit glutamate to spillover to peri- or extrasynaptic sites, where transmission is thought to weaken synaptic strength and damage the cell (Hardingham and Bading 2010). While normally associated with glutamate uptake, astrocytes may also release glutamate to the extracellular space in certain conditions (Malarkey and Parpura 2008), leading to increased glutamate at these potentially damaging extrasynaptic locations (Talantova et al. 2013). As a single astrocyte can cover multiple synapses, the disruption of individual astrocytes can have wide-reaching effects (Bushong et al. 2002).

Glial cells also express glutamatergic receptors though their subunit composition, expression patterns, and function are not well studied (Hansson and Rönnbäck 2004; Verkhratsky and Kirchhoff 2007). For example, unlike neuronal NMDARS, astrocytic NMDARs are unblocked by magnesium at baseline, suggesting NMDARS containing NR3 subunits (Palygin et al. 2011), while NR2B containing receptors may be expressed after injury or stress such as ischemia (Krebs et al. 2003). A weak magnesium blockade at baseline suggests that these receptors are more sensitive than neurons to increases in extracellular glutamate (Lalo et al. 2006; Palygin et al. 2011). The expression of metabotropic glutamatergic receptors on glia is also debated, with recent evidence that mGluR5 receptors are only expressed in younger animals (Sun et al. 2013). Functionally, the unique properties of glial NMDARs likely explains the differential effects of various NMDAR antagonists on glia and neurons, with the NR2B-selective antagonist ifenprodil selectively blocking an inward current and Ca^{2+} influx in neurons, but not astrocytes, and memantine and MK-801 blocking both cell types (Palygin et al. 2011).

Interestingly, a reduction in glial density and number is observed in the PFC of major depressive disorder (MDD) patients (Cotter 2001; Cotter et al. 2002; Ongur et al. 1998; Rajkowska and Miguel-Hidalgo 2007; Rajkowska et al. 1999; Uranova et al. 2004) (for recent reviews, see Rajkowska and Stockmeier 2013; Sanacora and Banasr 2013). These reductions have been observed in depression, bipolar disorder, and in some cases, schizophrenia, in regions implicated in these disorders such as Brodmanns area 24, the orbitofrontal cortex, and the dorsolateral PFC (Gittins and Harrison 2011; Ongur et al. 1998; Rajkowska et al. 1999). Glial fibrillary acidic protein (GFAP), a marker of astrocytes, reveals reductions in the hippocampus, PFC and amygdala, with the effect in the PFC is the most consistent in patient populations (Altshuler et al. 2010; Johnston-Wilson et al. 2000; Müller et al. 2013; Webster et al. 2001). Changes in glutamate transporters such as EAAT 2 are similarly observed in depressed patients (Bernard et al. 2011; Choudary et al. 2005; McCullumsmith and Meador-Woodruff 2002; Sequeira et al. 2009).

Changes in glutamatergic uptake and cycling are observed after stress, indicative of adaptations to increased exposure to glutamate. After CUS, rodents show decreases in glutamine cycle rate (Banasr et al. 2010), which could relate to observed CUS or corticosterone-induced loss of glia in the PFC (Alonso 2000; Banasr et al. 2010; Banasr and Duman 2007). Some forms of stress or corticosterone exposure have been demonstrated to increase expression of GLT-1 (but not GLAST) in the PFC and hippocampus of rodents therefore increasing glutamate uptake (Autry et al.

2006; Zink et al. 2010; Zschocke et al. 2005), possibly as a neuroprotective countermeasure to stress-induced increases in glutamate efflux. Illustrating the potential pathological behavioral effects of impaired glutamate clearance from the extracellular space, application of a glial toxin to the PFC in rodents, leads to depressive-like behaviors after one exposure to stress as opposed to weeks of chronic stress (Banasr and Duman 2008). Providing additional support to the hypothesis that impaired glial-mediated glutamate uptake can be associated with depressive-like behaviors, rats bred for higher levels of learned helplessness showed a significantly suppressed expression of GLT1 in hippocampus and cerebral cortex compared to nonhelpless littermates (Zink et al. 2010).

17.3.1 Traditional Neuropsychiatric Therapies on Glutamate Uptake and Glia

There is evidence that traditional antidepressants have glio-protective effects and thus can influence uptake of extracellular glutamate. In preclinical studies, treatment with fluoxetine reduced the stress-induced loss of hippocampal GFAP in tree shrews, but had no effect in nonstressed animals (Czéh et al. 2006). Similarly, chronic administration of the tricyclic antidepressant clomipramine increased GFAP expression in stressed animals (Liu et al. 2009). Another study found that chronic fluoxetine increased GLT1 expression in the hippocampus and cortex in rats, while desipramine and an monoamine oxidase inhibitor (MAOI) showed more modest effects (Zink et al. 2011), though this study did not test these antidepressants in relation to stress. Similarly, the antidepressant paroxetine increased hippocampal GFAP expression (Sillaber et al. 2008). In contrast, it should be noted that some studies have failed to find an ability of antidepressants to reverse stress-induced GFAP loss in the hippocampus (Araya-Callís et al. 2012) or cortex (Fatemi et al. 2008).

17.3.2 Treatments Targeting Glial and Glutamate Uptake Have Efficacy in Psychiatric Illnesses

While traditional antidepressants have some glio-protective properties, drugs with well-established effects on glial cell function and glutamate uptake might serve as more attractive therapies in looking to reverse stress-induced glial loss. Treatment with B-lactam antibiotics such as ceftriaxone has been shown to increase GLT1 function (Rothstein et al. 2005). Considering this property of ceftriaxone, several studies have since demonstrated its ability to modify several forms of behavior believed to be modulated by glutamatergic activation (Trantham-Davidson et al. 2012), including reducing depressive and anxiety-like behaviors in mice (Mineur et al. 2007). However, side effects and the difficulties related to the delivery of ceftriaxone have prohibited larger clinical trials in psychiatric patients to date.

As described earlier, the neuroprotective drug riluzole reduces glutamatergic transmission; however, the exact mechanism of action of riluzole remains unclear. More recent studies suggest much of riluzoles neuroprotective effects could be mediated through the effects on GLT1 expression (Fumagalli et al. 2008; Yoshizumi et al. 2012). As also described earlier, multiple open label (nonplacebo controlled) clinical studies have found riluzole to be effective in the treatment of MDD, bipolar disorder (BPD), anxiety and obsessive–compulsive disorder (OCD) (Coric et al. 2005; Pittenger et al. 2008; Sanacora et al. 2004). Additionally, riluzole was shown to alter glutamine/glutamate cycling in BPD patients (Brennan et al. 2010). Preclinical studies show an antidepressant-like action of chronic riluzole, and an ability to reverse chronic-stress induced depression-like behaviors and loss of GLT-1 and GFAP (Banasr et al. 2010; Gourley et al. 2012). Ongoing studies will test if GLT-1 or glial activity is necessary for the antidepressant-like activity of riluzole.

17.3.3 Novel Neuropsychiatric Treatments on Glutamate Uptake and Glia

As mentioned previously, some NMDAR antagonists such as ketamine have antidepressant efficacy, however the exact mechanism behind these treatments still remains to be elucidated. Most hypotheses have focused on a blockade of NMDARs on neuronal cells; however, Mitterauer recently suggested a key role of astrocytic NMDARs in ketamines therapeutic effect (Mitterauer 2012). However, it should be noted that subunit-specific antagonists such as NR2B-selective antagonists, also have clinical and preclinical antidepressant-like effects (Li et al. 2010; Maeng et al. 2008; Preskorn et al. 2008), and previous work found that the NR2B selective antagonist ifenprodil selectively affects neurons and not astrocytes (Palygin et al. 2011). NR2B-containing receptors may only be expressed in glia after adverse events such as ischemia, possibly leading to differential activity of NMDAR antagonists in healthy and diseased states (Krebs et al. 2003).

17.4 Therapies Regulating Postsynaptic Effects of Glutamate Release

As discussed above, effective antidepressant therapies are known to both increase, and decrease extracellular glutamate levels. In the case of NMDAR antagonists and mGluR2/3 antagonists, the antidepressant-like effect appears to require a transient increase in glutamatergic transmission through AMPARs to initiate the cascade of cellular changes that have been associated with the antidepressant-like effects (Dwyer et al. 2012; Li et al. 2010; Maeng et al. 2008). This suggests that the type of postsynaptic glutamatergic transmission is critical in generating a rapid treatment response and possibly also in determining the responses to stress. The administration of NMDAR, but not AMPAR, antagonists blocks stress-induced morphological and plasticity-related changes in pyramidal cells in the hippocampus (Magariños and McEwen 1995b), with similar results in the mPFC (Martin and Wellman 2011). Administration of NMDAR-antagonists also blocks stress-induced alterations of hippocampal LTP (Kim et al. 1996). Hippocampal CA3 and CA1 pyramidal cell atrophy caused by restraint stress was blocked by CA3 pyramidal cell-specific conditional knockout of GluN1, suggesting that these effects are mediated specifically by pyramidal cell NMDARs (Christian et al. 2011), similarly despair behavior during chronic swim stress was reduced by a pyramidal cell knockout of NR2B in the cortex and CA1 (Kiselycznyk et al. 2011). These findings suggest that increased postsynaptic NMDAR-mediated glutamate transmission has a critical role in mediating the effects of stress, and AMPAR throughput is necessary for an antidepressant response.

The importance of postsynaptic NMDAR throughput in mediating excitotoxic insults is consistent with the aforementioned work showing that excessive glutamate transmission through NMDARs, particularly, extrasynaptic sites, can lead to cell damage (for a review, see Hardingham and Bading 2010). While activation of synaptic NMDARs leads to activation of **CREB (cAMP response element-binding** protein) and increases in synaptic strength, spillover of glutamate to extrasynaptic sites enables activation of extrasynaptic NMDARs that decreases CREB signaling and lead to long-term depression and activation of cell death pathways (Hardingham et al. 2002). In adulthood, extrasynaptic NMDARs are thought to be mainly NR2B-containing receptors, while synaptic NMDARs are mainly composed of NR2A receptors. Whether the location or subunit composition of NMDARs is important to cell growth pathways is unclear, however recent work suggests it is the C-terminal tail on the NR2B subunit that is responsible for activating downstream pathways mediating cell death pathways (Martel et al. 2012). Interestingly, this C-terminal tail of NR2B can be cleaved in the presence of calpain (Guttmann et al. 2001, 2002), a signaling molecule known to be increased by extrasynaptic NMDAR throughput (Xu et al. 2009) leaving a functional NR2B-containing NMDAR with possibly altered trafficking to extrasynaptic sites and altered interaction with downstream signaling pathways (Gladding and Raymond 2011). Activation of synaptic versus extrasynaptic NMDARs is linked to nuclear CREB signaling through the messenger protein Jacob. Extrasynaptic receptors increase Jacob trafficking to the nucleus where Jacob triggers CREB shut-off pathways and cell death, while synaptic NMDARs phosphorylate Jacob to block its effects on CREB (Dieterich et al. 2008; Karpova et al. 2013).

17.4.1 Traditional Antidepressants' Effect on Postsynaptic Sites

Chronic, but not acute, treatment with some antidepressants reduces NMDAR transmission (Paul and Skolnick 2003; Reynolds and Miller 1988; Skolnick et al. 1996). These same treatments appear to augment AMPAR transmission, as multiple chronic antidepressant treatments increased phosphorylation of the GluA1 subunit (McEwen et al. 2010; Svenningsson et al. 2007), and synaptic GluA1 and GluA2 levels (Du et al. 2007). Similarly, anticonvulsants with primarily antidepressant activity such as lamotrigine and riluzole increase GluA1 and GluA2 subunits in the hippocampus, while therapies with primarily antimanic properties, such as lithium and valproate, however, reduce GluA1 and GluA2 levels (Du et al. 2007). However, the efficacy of acute imipramine in the rodent forced swim test is not blocked in mice lacking the phosphorylation sites on GluA1 that are increased after antidepressant treatment (Kiselycznyk et al. 2013).

17.4.2 Therapies Targeting Postsynaptic Glutamatergic Receptors

As mentioned earlier, NMDAR antagonists such as ketamine act as fast-acting antidepressants in treatment-resistant patients (Berman et al. 2000; Diazgranados et al. 2010; Mathew et al. 2010; Murrough et al. 2013; Valentine et al. 2011; Zarate et al. 2006, 2012) and preclinical assays, such as the FST (Autry et al. 2011; Li et al. 2010; Maeng et al. 2008). Compared with traditional antidepressants that can take weeks or months to reduce symptoms, ketamine is effective in a matter of hours and one infusion can reduce depressive symptoms for days to weeks in some patients. NMDAR antagonists also have been reported to have anxiety-reducing effects (Cryan and Dev 2007; Barkus et al. 2011). In preclinical models, ketamine leads to long-term increases in synaptic strength in the PFC, increasing synaptic proteins like GluA1 and increasing spine density. These same NMDAR antagonists can reverse CUS-induced spine loss and behavioral changes (Li et al. 2010), and are known to increase expression of BDNF, and neurogenesis (Gould and Cameron 1997; Metsis et al. 1993).

Acute systemic administration of pharmacological antagonists specific to the GluN2B-subunit is sufficient to produce the antidepressant-like effects seen with nonsubunit-selective NMDAR antagonists such as ketamine, both clinically (Preskorn et al. 2008) and preclinically (Li et al. 2010; Maeng et al. 2008). Administration of selective GluN2B antagonists such as Ro 25–6981 produces no effect on anxiety-like behavior in the mouse elevated plus maze (EPM) (Mathur et al. 2009), but anxiolytic-like in the novelty-suppressed feeding (NSF) task (Li et al. 2010). Administration of another GluN2B antagonist, ifenprodil, was also anxiolytic-like effects in the rat EPM (Fraser et al. 1996).

However, attempts to selectively delete NMDAR subunits with genetic techniques have not produced depression-related behaviors. Constitutive genetic deletion of the obligatory GluN1 subunit are lethal, however viable conditional knockouts of this subunit have been generated with postnatal deletion in specific regions and cell types. Mice with a restricted deletion of GluN1 to pyramidal cells of the CA3 region of the hippocampus displayed no differences from control mice in HPA-axis activation or anxiety-like behavior in the EPM (Christian et al. 2011; Cravens et al. 2006). Similar deletion of the subunit GluN2B in corticohippocampal pyramidal cells displayed no alterations in the FST and anxiety (Kiselycznyk et al.

2011). The lack of depression-related effect in these selective deletions could be due to alterations in cell-type, region, or age; however, it suggests that deletion of NMDA receptors is not enough to induce an antidepressant-like response.

Metabotropic glutamate receptors containing mGluR5 are located postsynaptically and typically located near NMDARs (Brakeman et al. 1997; Lujan et al. 1996; Tu et al. 1999). mGluR5 activity is tied to NMDARs and help regulate NMDAR throughput and activation of mGluR5 receptors increases NMDAR transmission, while blocking mGluR5 reduces NMDAR throughput (Attucci et al. 2001; Awad et al. 2000; Doherty et al. 2000; Pisani et al. 2001). Similarly, repeated treatment with the mGluR5 antagonist 3-((2-Methyl-4-thiazolyl)ethynyl)pyridine (MTEP) also reduces NR1 expression (Cowen et al. 2005). mGlur5 is decreased in depressed patients, as well as rats bred for depression-related phenotypes (Kovačević et al. 2012). It is similarly decreased in the hippocampus after chronic treatment with corticosterone in rodents (Iyo et al. 2010).

However, mGlu5 knockout mice appear to have reduced depression-related behavior in the FST (Li et al. 2006). Additionally, mGluR5 antagonists such as **2-Methyl-6-(phenylethynyl)pyridine (MPEP)** or MTEP have antidepressant- and anxiolytic-like efficacy in preclinical models (Belozertseva et al. 2007; Brodkin et al. 2002; Busse et al. 2004; Li et al. 2006; Molina-Hernández et al. 2006; Pałucha et al. 2005; Pilc et al. 2002; Tatarczyńska et al. 2001; Wieronska et al. 2002). Similar antidepressant-like effects are observed with the negative allosteric modulator (GRN-529), in preclinical tests (Hughes et al. 2012). While the majority of support for mGluR5-related therapies has been preclinical, some clinical trials have shown efficacy for the mGluR5 treatment Fenobam in anxiety (Pecknold et al. 1982; Porter et al. 2005). As blockade of mGluR5 decreases NMDAR transmission, the antidepressant-like mGluR5 antagonists effects correspond to the efficacy of NMDAR antagonists.

Administration of drugs blocking non-NMDARs (i.e., AMPAR and kainate receptors), do not affect depression-related activity in the FST (Maeng et al. 2008), unlike NMDAR antagonists. Administration of drugs selectively targeting AMPARs, such as GYKI 52466 or LY32635, have been found to cause anxiolytic-like (Alt et al. 2006; Kapus et al. 2008; Kotlinska and Liljequist 1998; Matheus and Guimarães 1997), anxiogenic-like (Vekovischeva et al. n.d.), or no changes (Fitzgerald et al. 2010; Kapus et al. 2008) in anxiety-related behaviors, depending on the rodent species tested or behavioral paradigm used. Mice lacking key phosphorylation sites on the AMPA subunit GluA1 also demonstrate decreases in anxiety (Kiselycznyk et al. 2013).

However, as mentioned earlier, treatments like ketamine transiently increase extracellular glutamate release and are dependent on AMPAR transmission to generate the antidepressant-like response, suggesting that increased AMPAR throughput is necessary for the effect. Additionally, AMPAR potentiators or AMPAkines appear to have efficacy as antidepressants in a variety of rodent models (Knapp et al. 2002; Li et al. 2001; Lindholm et al. 2012). In clinical studies, the AMPA potentiator LY451395 helped relieve depressive symptoms in Alzheimers patients (Chappell et al. 2007). More recently, a small phase Ib study was completed with the AMPA potentiating drug Org 26576. While the study demonstrated good safety and tolerability of the drug, along with some numerical advantages in terms of depressive severity and cognitive functioning, the differences did not reach the level of statistical significance in this exploratory study with limited power (Nations et al. 2012).

Together with the studies in ketamine, the findings suggest that increased AMPAR throughput while blocking NMDARs is necessary for an antidepressant-like response. Deletion of NMDARs without concurrent increased AMPAR throughput would therefore not be predicted to have antidepressant-like activity. Ketamine achieves this by increasing extracellular glutamate while blocking NMDARs and leaving AMPARs free. Traditional antidepressants, while they may reduce presynaptic glutamate release, also increase levels of AMPAR subunits and reduce NMDAR transmission. This proposed relationship between synaptic AMPARs and NMDARs (especially extrasynaptic NMDARs) in regulating synaptic strength suggests multiple new directions for the development of future therapeutics.

17.4.3 Future Directions Targeting Postsynaptic Sites

Increasing AMPAR throughput has antidepressant efficacy and could reverse stress-induced deficits, and AMPARs have a mainly synaptic localization in adulthood. Together with the knowledge that extrasynaptic, but not synaptic, NMDARs can mediate cell-death after excessive glutamate release, this suggests that the balance between synaptic and extrasynaptic glutamate transmission is a convergent target for stress-sensitive disorders. Ketamine has the benefit of both increasing glutamate release and boosting synaptic throughput while simultaneously blocking possibly extrasynaptic NMDARSs. However, knowledge of the effects of synaptic versus extrasynaptic throughput may enable the development of better strategies.

As extrasynaptic NMDARs are thought to be NR2B-rich, a potential strategy to target extrasynaptic receptors would be to use NR2B-selective compounds. NR2B-selective antagonists have preclinical and clinical efficacy (Li et al. 2010; Maeng et al. 2008; Preskorn et al. 2008). However, while GluN2B-containing receptors may be preferentially expressed at extrasynaptic sites, it is not a clear division. The NMDAR memantine selectively targets extrasynaptic NMDARs at selective dose range (Xia et al. 2010). Memantine has shown efficacy in preclinical (Moryl et al. 1993; Rogóz et al. 2002) and clinical studies (Muhonen et al. 2008), though memantine had no effect in some clinical studies of depression (Zarate et al. 2006). Recently a more selective version of memantine has been developed called nitromemantine that is reported to more selectively target extrasynaptic receptors (Lipton 2006).

Negative results in memantines antidepressant efficacy may be due to memantines lack of effect on extracellular glutamate release. As memantine is not known to increase glutamate release, its blockade of extrasynaptic receptors could reduce the negative effects of stress over time. However without the burst of synaptic throughput it would not be expected to have antidepressant-like effects at baseline. Instead, memantine could be combined with another source of stimulated glutamate release to generate a more rapidly acting treatment. Combining the selective blockade of extrasynaptic NMDARs by memantine with the glutamate release of acute stress exposure could lead to novel treatment strategies. This strategy would block cell death pathways in only regions and instances of stress, allowing for selectivity not available with current pharmacological treatments, also known as a pathologically activated therapeutic (PAT, Lipton 2006). The combination of NMDAR antagonists such as memantine and stress has previously been shown to have effects on morphology not seen with either treatment alone. Administration of the NMDAR antagonists CPP during CRS lead to a significant increase in PFC spine density not observed in NMDAR antagonist treatment alone (Martin and Wellman 2011). Similarly administration of tianeptine during CUS caused hippocampal hypertrophy compared to nonstressed animals (Czéh et al. 2001).

17.5 Conclusions

Converging evidence suggests that alterations in the glutamatergic neurotransmitter system play a key role in the pathogenesis and pathophysiology of stress-induced psychiatric disorders. The stress-induced release of glutamate, its uptake by glial cells, and its postsynaptic effects are all potential therapeutic targets (see Fig. 17.1 for overview of convergent targets). However, knowledge of both the positive and negative effects of stress and glutamate release may guide the design of therapies that better allow for prophylaxis and recovery from glutamate dysregulation after stress. We have seen that while multiple antidepressants help block stress-induced glutamate release, many new classes of antidepressants acutely have the exact opposite result and in fact appear to require increases in extracellular glutamate release to generate the antidepressant-like effect. Knowing that glutamate transmission focused to synaptic sites can activate cell growth versus cell death pathways, attempting to reduce stress-induced glutamate release would block these potentially beneficial effects and restrict recovery. However, regulation of the negative feedback on glutamate release through mGluR2/3 receptors may allow for a therapy selectively activated in cases of stress and extreme glutamate release and present another PAT, as with memantine.

Therapies targeting glutamate uptake may present better targets with fewer side effects. Supporting the health of glial cells to regulate extracellular glutamate levels could help reduce extrasynaptic activation while maintaining synaptic throughput. As loss of glial is one of the most consistent findings in depression, and a single astrocyte can affect many neurons, therapies targeting glial could have broad effects. However, it is unclear if increasing glutamate uptake itself will only block subsequent exposures to high glutamate levels or will be able to reverse effects of stress to have a fast-acting antidepressant effect. The ability of glia to release glutamate into the extracellular space could be used to induce glutamate release similar to ketamine; however, it would most likely cause an increase of glutamate release to extrasynaptic, versus synaptic areas, leading to activation of cell death versus cell death pathways.

Finally, therapies targeting postsynaptic sites present opportunities to activate both cell death and cell growth pathways depending on the activation of synaptic or extrasynaptic receptors. Therapies designed to target these postsynaptic glutamatergic receptors pose a substantial risk for side effects if selectivity cannot be achieved. Memantine, and now nitromemantine have been reported to selectively block extrasynaptic sites but have a narrow dose range to target extrasynaptic receptors. Future studies examining the effects of selective blockade of extrasynaptic NMDARs, if possible, would be extremely interesting. However, blocking extrasynaptic receptors would require concomitant increases in glutamate release to have a burst of synaptic throughput and a fast-acting antidepressant response. Alternatively, AMPARs could be targeted directly with AMPA potentiators to activate synaptic throughput, however this strategy would not allow for selectivity to regions activated in pathological scenarios.

The increased understanding of the relationship between the glutamatergic system and stress has illuminated potential pathways of regulating synaptic glutamate transmission to develop novel treatment strategies for stress-sensitive neuropsychiatric disorders. While current antidepressant therapies, such as monoaminergic-based treatments, have been tied to mediators of synaptic activity their exact mechanism of action remains unclear. Traditional antidepressant treatments leading to increased AMPAR levels may increase synaptic transmission, but not selectively in regions or cell-types activated in depression. Additionally, mechanisms reducing glutamate release could protect against the negative consequences of stress, but also block the increased synaptic transmission possibly needed for recovery. In total, a lack of understanding of the mechanisms behind current therapies could explain their lack of consistent effects that ultimately leads to the large gap between the number of patients prescribed antidepressants and those successfully treated. Instead, the converging evidence on novel glutamatergic and plasticity-related therapeutic targets supports a new generation of mechanistically based treatments that can more directly and consistently address the numerous challenges of treating stress-related neuropsychiatric illnesses.

References

Abdallah CG, Fasula M, Kelmendi B, Sanacora G, Ostroff R. Rapid antidepressant effect of ketamine in the electroconvulsive therapy setting. J ECT. 2012;28(3):157–61. doi:10.1097/YCT.0b013e31824f8296.

Alonso G. Prolonged corticosterone treatment of adult rats inhibits the proliferation of oligodendrocyte progenitors present throughout white and gray matter regions of the brain. Glia. 2000;31(3):219–31.

Alt A, Weiss B, Ogden AM, Li X, Gleason SD, Calligaro DO, Bleakman D, Witkin JM. In vitro and in vivo studies in rats with LY293558 suggest AMPA/kainate receptor blockade as a novel potential mechanism for the therapeutic treatment of anxiety disorders. Psychopharmacology (Berl). 2006;185(2):240–7. doi:10.1007/s00213-005-0292-0.

Altshuler LL, Abulseoud OA, Foland-Ross L, Bartzokis G, Chang S, Mintz J, Hellemann G, Vinters HV. Amygdala astrocyte reduction in subjects with major depressive disorder but not bipolar disorder. Bipolar Disord. 2010;12(5):541–9. doi:10.1111/j.1399-5618.2010.00838.x.

Anderson CM, Swanson RA. Astrocyte glutamate transport: review of properties, regulation, and physiological functions. Glia. 2000;32(1):1–14.

Anisman H, Zacharko RM. Multiple neurochemical and behavioral consequences of stressors: implications for depression. Pharmacol Ther. 1990;46(1):119–36.

Anwyl R. Metabotropic glutamate receptors: electrophysiological properties and role in plasticity. Brain Res Brain Res Rev. 1999;29(1):83–120.

Araya-Callís C, Hiemke C, Abumaria N, Flugge G. Chronic psychosocial stress and citalopram modulate the expression of the glial proteins GFAP and NDRG2 in the hippocampus. Psychopharmacology (Berl). 2012;224(1):209–22. doi:10.1007/s00213-012-2741-x.

Arriza JL, Fairman WA, Wadiche JI, Murdoch GH, Kavanaugh MP, Amara SG. Functional comparisons of three glutamate transporter subtypes cloned from human motor cortex. J Neurosci. 1994;14(9):5559–69.

Attucci S, Carlà V, Mannaioni G, Moroni F. Activation of type 5 metabotropic glutamate receptors enhances NMDA responses in mice cortical wedges. Br J Pharmacol. 2001;132(4):799–806. doi:10.1038/sj.bjp.0703904.

Autry AE, Grillo CA, Piroli GG, Rothstein JD, McEwen BS, Reagan LP. Glucocorticoid regulation of GLT-1 glutamate transporter isoform expression in the rat hippocampus. Neuroendocrinology. 2006;83(5-6):371–9. doi:10.1159/000096092.

Autry AE, Adachi M, Nosyreva E, Na ES, Los MF, Cheng P, Kavalali ET, Monteggia LM. NMDA receptor blockade at rest triggers rapid behavioural antidepressant responses. Nature. 2011;475(7354):91–5. doi:10.1038/nature10130.

Awad H, Hubert GW, Smith Y, Levey AI, Conn PJ. Activation of metabotropic glutamate receptor 5 has direct excitatory effects and potentiates NMDA receptor currents in neurons of the subthalamic nucleus. J Neurosci. 2000;20(21):7871–9.

Bagley J, Moghaddam B. Temporal dynamics of glutamate efflux in the prefrontal cortex and in the hippocampus following repeated stress: effects of pretreatment with saline or diazepam. Neuroscience. 1997;77(1):65–73.

Banasr M, Duman RS. Regulation of neurogenesis and gliogenesis by stress and antidepressant treatment. CNS Neurol Disord Drug Targets. 2007;6(5):311–20.

Banasr M, Duman RS. Glial loss in the prefrontal cortex is sufficient to induce depressive-like behaviors. Biol Psychiatry. 2008;64(10):863–70. doi:10.1016/j.biopsych.2008.06.008.

Banasr M, Chowdhury GMI, Terwilliger R, Newton SS, Duman RS, Behar KL, Sanacora G. Glial pathology in an animal model of depression: reversal of stress-induced cellular, metabolic and behavioral deficits by the glutamate-modulating drug riluzole. Mol Psychiatry. 2010a;15(5):501–11. doi:10.1038/mp.2008.106.

Barha CK, Pawluski JL, Galea LAM. Maternal care affects male and female offspring working memory and stress reactivity. Physiol Behav. 2007;92(5):939–50. doi:10.1016/j.physbeh.2007.06.022.

Barkus C, Feyder M, Graybeal C, Wright T, Wiedholz L, Izquierdo A, et al. Do GluA1 knockout mice exhibit behavioral abnormalities relevant to the negative or cognitive symptoms of schizophrenia and schizoaffective disorder? Neuropharmacology. 2011;62(3):1263–72. doi:10.1016/j.neuropharm.2011.06.005.

Belozertseva IV, Kos T, Popik P, Danysz W, Bespalov AY. Antidepressant-like effects of mGluR1 and mGluR5 antagonists in the rat forced swim and the mouse tail suspension tests. European Neuropsychopharmacol. 2007;17(3):172–9. doi:10.1016/j.euroneuro.2006.03.002.

Berman RM, Cappiello A, Anand A, Oren DA, Heninger GR, Charney DS, Krystal JH. Antidepressant effects of ketamine in depressed patients. Biol Psychiatry. 2000;47(4):351–4.

Bernard R, Kerman IA, Thompson RC, Jones EG, Bunney WE, Barchas JD, et al. Altered expression of glutamate signaling, growth factor, and glia genes in the locus coeruleus of patients with major depression. Mol Psychiatry. 2011;16(6):634–46. doi:10.1038/mp.2010.44.

Bonanno G, Giambelli R, Raiteri L, Tiraboschi E, Zappettini S, Musazzi L, et al. Chronic antidepressants reduce depolarization-evoked glutamate release and protein interactions favoring formation of SNARE complex in hippocampus. J Neurosci. 2005;25(13):3270–9. doi:10.1523/JNEUROSCI.5033-04.2005.

Brakeman PR, Lanahan AA, O'Brien R, Roche K, Barnes CA, Huganir RL, Worley PF. Homer: a protein that selectively binds metabotropic glutamate receptors. Nature. 1997;386(6622):284–8. doi:10.1038/386284a0.

Brennan BP, Hudson JI, Jensen JE, McCarthy J, Roberts JL, Prescot AP, et al. Rapid enhancement of glutamatergic neurotransmission in bipolar depression following treatment with riluzole. Neuropsychopharmacology. 2010;35(3):834–46. doi:10.1038/npp.2009.191.

Brodkin J, Busse C, Sukoff SJ, Varney MA. Anxiolytic-like activity of the mGluR5 antagonist MPEP: a comparison with diazepam and buspirone. Pharmacol Biochem Behav. 2002;73(2):359–66.

Bushong EA, Martone ME, Jones YZ, Ellisman MH. Protoplasmic astrocytes in CA1 stratum radiatum occupy separate anatomical domains. J Neurosci. 2002;22(1):183–92.

Busse CS, Brodkin J, Tattersall D, Anderson JJ, Warren N, Tehrani L, et al. The behavioral profile of the potent and selective mGlu5 receptor antagonist 3-[(2-methyl-1,3-thiazol-4-yl)ethynyl] pyridine (MTEP) in rodent models of anxiety. Neuropsychopharmacology. 2004;29(11):1971–9. doi:10.1038/sj.npp.1300540.

Calabrese JR, Bowden CL, Sachs GS, Ascher JA, Monaghan E, Rudd GD. A double-blind placebo-controlled study of lamotrigine monotherapy in outpatients with bipolar I depression. Lamictal 602 Study Group. J Clin Psychiatry. 1999;60(2):79–88.

Cartmell J, Schoepp DD. Regulation of neurotransmitter release by metabotropic glutamate receptors. J Neurochem. 2000;75(3):889–907.

Caspi A, Sugden K, Moffitt TE, Taylor A, Craig IW, Harrington H, et al. Influence of life stress on depression: moderation by a polymorphism in the *5-HTT* gene. Science(New York NY). 2003;301(5631):386–9. doi:10.1126/science.1083968.

Chaki S, Yoshikawa R, Hirota S, Shimazaki T, Maeda M, Kawashima N, et al. MGS0039: a potent and selective group II metabotropic glutamate receptor antagonist with antidepressant-like activity. Neuropharmacology. 2004;46(4):457–67. doi:10.1016/j.neuropharm.2003.10.009.

Chaki S, Ago Y, Palucha-Paniewiera A, Matrisciano F, Pilc A. mGlu2/3 and mGlu5 receptors: potential targets for novel antidepressants. Neuropharmacology. 2013;66:40–52. doi:10.1016/j.neuropharm.2012.05.022.

Chappell AS, Gonzales C, Williams J, Witte MM, Mohs RC, Sperling R. AMPA potentiator treatment of cognitive deficits in Alzheimer disease. Neurology. 2007;68(13):1008–12. doi:10.1212/01.wnl.0000260240.46070.7c.

Choudary PV, Molnar M, Evans SJ, Tomita H, Li JZ, Vawter MP, et al. Altered cortical glutamatergic and GABAergic signal transmission with glial involvement in depression. Proc Natl Acad Sci U S A. 2005;102(43):15653–8. doi:10.1073/pnas.0507901102.

Chowdhury GMI, Behar KL, Cho W, Thomas Ma, Rothman DL, Sanacora G. ^{1}H-[^{13}C]-nuclear magnetic resonance spectroscopy measures of ketamine's effect on amino acid neurotransmitter metabolism. Biol Psychiatry. 2012;71(11):1022–5. doi:10.1016/j.biopsych.2011.11.006.

Christian KM, Miracle AD, Wellman CL, Nakazawa K. Chronic stress-induced hippocampal dendritic retraction requires CA3 NMDA receptors. Neuroscience. 2011;174:26–36. doi:10.1016/j.neuroscience.2010.11.033.

Coric V, Taskiran S, Pittenger C, Wasylink S, Mathalon DH, Valentine G, et al. Riluzole augmentation in treatment-resistant obsessive-compulsive disorder: an open-label trial. Biol Psychiatry. 2005;58(5):424–8. doi:10.1016/j.biopsych.2005.04.043.

Cotter D. Reduced glial cell density and neuronal size in the anterior cingulate cortex in major depressive disorder. Arch Gen Psychiatry. 2001;58(6):545–53. doi:10.1001/archpsyc.58.6.545.

Cotter D, Mackay D, Chana G, Beasley C, Landau S, Everall IP. Reduced neuronal size and glial cell density in area 9 of the dorsolateral prefrontal cortex in subjects with major depressive disorder. Cerebral Cortex. 2002;12(4):386–94. (New York, N.Y. : 1991).

Cowen MS, Djouma E, Lawrence AJ. The metabotropic glutamate 5 receptor antagonist 3-[(2-methyl-1,3-thiazol-4-yl)ethynyl]-pyridine reduces ethanol self-administration in multiple strains of alcohol-preferring rats and regulates olfactory glutamatergic systems. J Pharmacol Exp Ther. 2005;315(2):590–600. doi:10.1124/jpet.105.090449.

Cravens CJ, Vargas-Pinto N, Christian KM, Nakazawa K. CA3 NMDA receptors are crucial for rapid and automatic representation of context memory. Eur J Neurosci. 2006;24(6):1771–80. doi:10.1111/j.1460-9568.2006.05044.x.

Cryan JF, Dev KK. Role of glutamate in anxiety In: *Handbook of fear and anxiety.* Blanchard DC, Blanchard RM (eds). 2007.

Czéh B, Michaelis T, Watanabe T, Frahm J, de Biurrun G, van Kampen M, et al. Stress-induced changes in cerebral metabolites, hippocampal volume, and cell proliferation are prevented by antidepressant treatment with tianeptine. Proc Natl Acad Sci U S A. 2001;98(22):12796–801. doi:10.1073/pnas.211427898.

Czéh B, Simon M, Schmelting B, Hiemke C, Fuchs E. Astroglial plasticity in the hippocampus is affected by chronic psychosocial stress and concomitant fluoxetine treatment. Neuropsychopharmacology. 2006;31(8):1616–26. doi:10.1038/sj.npp.1300982.

DD DDS, Marek GJ. Preclinical pharmacology of mGlu2/3 receptor agonists: novel agents for schizophrenia? Curr Drug Targets CNS Neurol Disord. 2002;1(2):215–25.

Diazgranados N, Ibrahim L, Brutsche NE, Newberg A, Kronstein P, Khalife S, et al. A randomized add-on trial of an N-methyl-D-aspartate antagonist in treatment-resistant bipolar depression. Arch Gen Psychiatry. 2010;67(8):793–802. doi:10.1001/archgenpsychiatry.2010.90.

Dieterich DC, Karpova A, Mikhaylova M, Zdobnova I, König I, Landwehr M, et al. Caldendrin-Jacob: a protein liaison that couples NMDA receptor signalling to the nucleus. PLoS Biol. 2008;6(2):e34. doi:10.1371/journal.pbio.0060034.

Doherty AJ, Palmer MJ, Bortolotto ZA, Hargreaves A, Kingston AE, Ornstein PL, et al. A novel, competitive mGlu(5) receptor antagonist (LY344545) blocks DHPG-induced potentiation of NMDA responses but not the induction of LTP in rat hippocampal slices. Br J Pharmacol. 2000;131(2):239–44. doi:10.1038/sj.bjp.0703574.

Du J, Suzuki K, Wei Y, Wang Y, Blumenthal R, Chen Z, et al. The anticonvulsants lamotrigine, riluzole, and valproate differentially regulate AMPA receptor membrane localization: relationship to clinical effects in mood disorders. Neuropsychopharmacology. 2007;32(4):793–802. doi:10.1038/sj.npp.1301178.

Dwyer JM, Lepack AE, Duman RS. mTOR activation is required for the antidepressant effects of $mGluR_2/_3$ blockade. Int J Neuropsychopharmacol. 2012;15(4):429–34. doi:10.1017/S1461145711001702.

Engert F, Bonhoeffer T. Dendritic spine changes associated with hippocampal long-term synaptic plasticity. Nature. 1999;399(6731):66–70. doi:10.1038/19978.

Ettinger AB, Argoff CE. Use of antiepileptic drugs for nonepileptic conditions: psychiatric disorders and chronic pain. Neurotherapeutics. 2007;4(1):75–83. doi:10.1016/j.nurt.2006.10.003.

Fatemi SH, Folsom TD, Reutiman TJ, Pandian T, Braun NN, Haug K. Chronic psychotropic drug treatment causes differential expression of connexin 43 and GFAP in frontal cortex of rats. Schizophr Res. 2008;104(1–3):127–34. doi:10.1016/j.schres.2008.05.016.

Fell MJ, Witkin JM, Falcone JF, Katner JS, Perry KW, Hart J, et al. N-(4-((2-(trifluoromethyl)-3-hydroxy-4-(isobutyryl)phenoxy)methyl)benzyl)-1-methyl-1H-imidazole-4-carboxamide (THIIC), a novel metabotropic glutamate 2 potentiator with potential anxiolytic/antidepressant properties: in vivo profiling suggests a link between behavioral and central nervous system neurochemical changes. J Pharmacol Exp Ther. 2011;336(1):165–77. doi:10.1124/jpet.110.172957.

Feyissa AM, Woolverton WL, Miguel-Hidalgo JJ, Wang Z, Kyle PB, Hasler G, et al. Elevated level of metabotropic glutamate receptor 2/3 in the prefrontal cortex in major depression. Prog Neuropsychopharmacol Biol Psychiatry. 2010;34(2):279–83. doi:10.1016/j.pnpbp.2009.11.018.

Fitzgerald PJ, Barkus C, Feyder M, Wiedholz LM, Chen Y-C, Karlsson R-M, … Holmes A. Does gene deletion of AMPA GluA1 phenocopy features of schizoaffective disorder? Neurobiol Dis. 2010;40(3):608–21. doi:10.1016/j.nbd.2010.08.005.

Fraser CM, Cooke MJ, Fisher A, Thompson ID, Stone TW. Interactions between ifenprodil and dizocilpine on mouse behaviour in models of anxiety and working memory. Eur Neuropsychopharmacol. 1996;6(4):311–6.

Fumagalli E, Funicello M, Rauen T, Gobbi M, Mennini T. Riluzole enhances the activity of glutamate transporters GLAST, GLT1 and EAAC1. Eur J Pharmacol. 2008;578(2–3):171–6. doi:10.1016/j.ejphar.2007.10.023.

Gittins RA, Harrison PJ. A morphometric study of glia and neurons in the anterior cingulate cortex in mood disorder. J Affect Disord. 2011;133(1–2):328–32. doi:10.1016/j.jad.2011.03.042.

Gladding CM, Raymond LA. Mechanisms underlying NMDA receptor synaptic/extrasynaptic distribution and function. Mol Cell Neurosci. 2011;48(4):308–20. doi:10.1016/j.mcn.2011.05.001.

Gould E, Cameron HA. Early NMDA receptor blockade impairs defensive behavior and increases cell proliferation in the dentate gyrus of developing rats. Behav Neurosci. 1997;111(1):49–56.

Gourley SL, Espitia JW, Sanacora G, Taylor JR. Antidepressant-like properties of oral riluzole and utility of incentive disengagement models of depression in mice. Psychopharmacology (Berl). 2012;219(3):805–14. doi:10.1007/s00213-011-2403-4.

Groc L, Choquet D, Chaouloff F. The stress hormone corticosterone conditions AMPAR surface trafficking and synaptic potentiation. Nat Neurosci. 2008;11(8):868–70. doi:10.1038/nn.2150.

Groeneweg FL, Karst H, de Kloet ER, Joëls M. Rapid non-genomic effects of corticosteroids and their role in the central stress response. J Endocrinol. 2011;209(2):153–67. doi:10.1530/JOE-10-0472.

Guttmann RP, Baker D, Seifert KM, Cohen AS, Coulter DA, Lynch DR. Specific proteolysis of the NR2 B subunit—at multiple sites by calpain. J Neurochem. 2001;78(5): 1083–93.

Guttmann RP, Sokol S, Baker D, Simpkins KL, Dong Y, Lynch DR. Proteolysis of the N-methyld-aspartate receptor by calpain in situ. J. Pharmacol. Exp. Ther. 2002;302(3):1023–30.

Hammen C. Stress and depression. Annu Rev Clin Psychol. 2005;1:293–319. doi:10.1146/annurev.clinpsy.1.102803.143938.

Hansson E, Rönnbäck L. Altered neuronal-glial signaling in glutamatergic transmission as a unifying mechanism in chronic pain and mental fatigue. Neurochem Res. 2004;29(5):989–96.

Hardingham GE, Bading H. Synaptic versus extrasynaptic NMDA receptor signalling: implications for neurodegenerative disorders. Nat Rev Neurosci. 2010;11(10):682–96. doi:10.1038/nrn2911.

Hardingham GE, Fukunaga Y, Bading H. Extrasynaptic NMDARs oppose synaptic NMDARs by triggering CREB shut-off and cell death pathways. Nat Neurosci. 2002;5(5):405–14. doi:10.1038/nn835.

Hughes ZA, Neal SJ, Smith DL, Sukoff Rizzo SJ, Pulicicchio CM, Lotarski S, et al. Negative allosteric modulation of metabolic glutamate receptor 5 results in broad spectrum activity relevant to treatment resistant depression. Neuropharmacology. 2012;66:202–14. doi:10.1016/j.neuropharm.2012.04.007.

Ibrahim L, Diazgranados N, Luckenbaugh DA, Machado-Vieira R, Baumann J, Mallinger AG, Zarate CA. Rapid decrease in depressive symptoms with an N-methyl-d-aspartate antagonist in ECT-resistant major depression. Prog Neuropsychopharmacol Biol Psychiatry. 2011;35(4):1155–9. doi:10.1016/j.pnpbp.2011.03.019.

Iyo AH, Feyissa AM, Chandran A, Austin MC, Regunathan S, Karolewicz B. Chronic corticosterone administration down-regulates metabotropic glutamate receptor 5 protein expression in the rat hippocampus. Neuroscience. 2010;169(4):1567–74. doi:10.1016/j.neuroscience.2010.06.023

Izquierdo A, Wellman CL, Holmes A. Brief uncontrollable stress causes dendritic retraction in infralimbic cortex and resistance to fear extinction in mice. J Neurosci. 2006;26(21):5733–8. doi:10.1523/JNEUROSCI.0474-06.2006

Jackson ME, Moghaddam B. Distinct patterns of plasticity in prefrontal cortex neurons that encode slow and fast responses to stress. Eur J Neurosci. 2006;24(6):1702–10. doi:10.1111/j.1460-9568.2006.05054.x.

Johnson MP, Barda D, Britton TC, Emkey R, Hornback WJ, Jagdmann GE, et al. Metabotropic glutamate 2 receptor potentiators: receptor modulation, frequency-dependent synaptic activity, and efficacy in preclinical anxiety and psychosis model(s). Psychopharmacology (Berl). 2005;179(1):271–83. doi:10.1007/s00213-004-2099-9.

Johnston-Wilson NL, Sims CD, Hofmann J-P, Anderson L, Shore AD, Torrey EF, Yolken RH. Disease-specific alterations in frontal cortex brain proteins in schizophrenia, bipolar disorder, and major depressive disorder. Mol Psychiatry. 2000;5(2):142–9. doi:10.1038/sj.mp.4000696.

Kapus GL, Gacsályi I, Vegh M, Kompagne H, Hegedus E, Leveleki C, et al. Antagonism of AMPA receptors produces anxiolytic-like behavior in rodents: effects of GYKI 52466 and its novel analogues. Psychopharmacology (Berl). 2008;198(2):231–41. doi:10.1007/s00213-008-1121-z.

Karpova A, Mikhaylova M, Bera S, Bär J, Reddy PP, Behnisch T, et al. Encoding and transducing the synaptic or extrasynaptic origin of NMDA receptor signals to the nucleus. Cell. 2013;152(5):1119–33. doi:10.1016/j.cell.2013.02.002.

Karst H, Joëls M. Corticosterone slowly enhances miniature excitatory postsynaptic current amplitude in mice CA1 hippocampal cells. J Neurophysiol. 2005;94(5):3479–86. doi:10.1152/jn.00143.2005.

Kendler KS, Karkowski LM, Prescott CA. Causal relationship between stressful life events and the onset of major depression. Am J Psychiatry. 1999a;156(6):837–41.

Kendler KS, Karkowski LM, Prescott CA. The assessment of dependence in the study of stressful life events: validation using a twin design. Psychol Med. 1999b;29(6):1455–60.

Kessler RC, Avenevoli S, McLaughlin KA, Green JG, Lakoma MD, Petukhova M, et al. Lifetime co-morbidity of DSM-IV disorders in the US national comorbidity survey replication adolescent supplement (NCS-A). Psychol Med. 2012;42(9):1997–2010. doi:10.1017/S0033291712000025.

Kim JJ, Diamond DM. The stressed hippocampus, synaptic plasticity and lost memories. Nat Rev Neurosci. 2002;3(6):453–62. doi:10.1038/nrn849.

Kim JJ, Foy MR, Thompson RF. Behavioral stress modifies hippocampal plasticity through N-methyl-D-aspartate receptor activation. Proc Natl Acad Sci U S A. 1996;93(10):4750–3.

Kiselycznyk C, Svenningsson P, Delpire E, Holmes A. Genetic, pharmacological and lesion analyses reveal a selective role for corticohippocampal GLUN2B in a novel repeated swim stress paradigm. Neuroscience. 2011;193:259–268. doi:10.1016/j.neuroscience.2011.06. 015.

Kiselycznyk C, Zhang X, Huganir RL, Holmes A, Svenningsson P. Reduced phosphorylation of GluA1 subunits relates to anxiety-like behaviours in mice. Int J Neuropsychopharmacol. 2013;16(4):919–24. doi:10.1017/S1461145712001174.

Knapp RJ, Goldenberg R, Shuck C, Cecil A, Watkins J, Miller C, et al. Antidepressant activity of memory-enhancing drugs in the reduction of submissive behavior model. Eur J Pharmacol. 2002;440(1):27–35.

Koike H, Iijima M, Chaki S. Involvement of AMPA receptor in both the rapid and sustained antidepressant-like effects of ketamine in animal models of depression. Behav Brain Res. 2011;224(1):107–11. doi:10.1016/j.bbr.2011.05.035.

Kotlinska J, Liljequist S. The putative AMPA receptor antagonist, LY326325, produces anxiolytic-like effects without altering locomotor activity in rats. Pharmacol Biochem Behav. 1998;60(1):119–24.

Kovačević T, Skelin I, Minuzzi L, Rosa-Neto P, Diksic M. Reduced metabotropic glutamate receptor 5 in the Flinders sensitive line of rats, an animal model of depression: an autoradiographic study. Brain Res Bull. 2012;87(4–5):406–12. doi:10.1016/j.brainresbull.2012.01.010.

Krebs C, Fernandes HB, Sheldon C, Raymond LA, Baimbridge KG. Functional NMDA receptor subtype 2B is expressed in astrocytes after ischemia in vivo and anoxia in vitro. J Neurosci. 2003;23(8):3364–72.

Krugers HJ, Hoogenraad CC, Groc L. Stress hormones and AMPA plasticity and memory. Nat Rev Neurosci. 2010;10:675–81.

Lalo U, Pankratov Y, Kirchhoff F, North RA, Verkhratsky A. NMDA receptors mediate neuron-to-glia signaling in mouse cortical astrocytes. J Neurosci. 2006;26(10):2673–83. doi:10.1523/JNEUROSCI.4689-05.2006.
Li X, Tizzano JP, Griffey K, Clay M, Lindstrom T, Skolnick P. Antidepressant-like actions of an AMPA receptor potentiator (LY392098). Neuropharmacology. 2001;40(8):1028–33.
Li X, Need AB, Baez M, Witkin JM. Metabotropic glutamate 5 receptor antagonism is associated with antidepressant-like effects in mice. J Pharmacol Exp Ther. 2006;319(1):254–9. doi:10.1124/jpet.106.103143.
Li N, Lee B, Liu R-J, Banasr M, Dwyer JM, Iwata M, et al. mTOR-dependent synapse formation underlies the rapid antidepressant effects of NMDA antagonists. Science. 2010;329(5994):959–64. doi:10.1126/science.1190287. (New York NY).
Lindholm JSO, Autio H, Vesa L, Antila H, Lindemann L, Hoener MC, et al. The antidepressant-like effects of glutamatergic drugs ketamine and AMPA receptor potentiator LY 451646 are preserved in bdnf^{+}/$^{-}$ heterozygous null mice. Neuropharmacology. 2012;62(1):391–7. doi:10.1016/j.neuropharm.2011.08.015.
Lipton SA. NMDA receptors, glial cells, and clinical medicine. Neuron. 2006;50(1):9–11.
Liu Q, Li B, Zhu H-Y, Wang Y-Q, Yu J, Wu G-C. Clomipramine treatment reversed the glial pathology in a chronic unpredictable stress-induced rat model of depression. Eur Neuropsychopharmacol. 2009;19(11):796–805. doi:10.1016/j.euroneuro.2009.06.010.
Lowy MT, Gault L, Yamamoto BK. Adrenalectomy attenuates stress-induced elevations in extracellular glutamate concentrations in the hippocampus. J Neurochem. 1993;61(5):1957–60.
Luine V, Martinez C, Villegas M, Magariños AM, McEwen BS. Restraint stress reversibly enhances spatial memory performance. Physiol Behav. 1996;59(1):27–32.
Lujan R, Nusser Z, Roberts JD, Shigemoto R, Somogyi P. Perisynaptic location of metabotropic glutamate receptors mGluR1 and mGluR5 on dendrites and dendritic spines in the rat hippocampus. Eur J Neurosci. 1996;8(7):1488–500.
Lupien SJ, McEwen BS, Gunnar MR, Heim C. Effects of stress throughout the lifespan on the brain, behaviour and cognition. Nat Rev Neurosci. 2009;10(6):434–45. doi:10.1038/nrn2639.
Maeng S, Zarate CA, Du J, Schloesser RJ, McCammon J, Chen G, Manji HK. Cellular mechanisms underlying the antidepressant effects of ketamine: role of alpha-amino-3-hydroxy-5-methylisoxazole-4-propionic acid receptors. Biol Psychiatry. 2008;63(4):349–52. doi:10.1016/j.biopsych.2007.05.028.
Magariños AM, McEwen BS. Stress-induced atrophy of apical dendrites of hippocampal CA3c neurons: comparison of stressors. Neuroscience. 1995a;69(1):83–8.
Magariños AM, McEwen BS. Stress-induced atrophy of apical dendrites of hippocampal CA3c neurons: involvement of glucocorticoid secretion and excitatory amino acid receptors. Neuroscience. 1995b;69(1):89–98.
Magariños AM, Deslandes A, McEwen BS. Effects of antidepressants and benzodiazepine treatments on the dendritic structure of CA3 pyramidal neurons after chronic stress. Eur J Pharmacol. 1999;371(2-3):113–22.
Malarkey EB, Parpura V. Mechanisms of glutamate release from astrocytes. Neurochem Int. 2008;52(1–2):142–54. doi:10.1016/j.neuint.2007.06.005.
Martel M-A, Ryan TJ, Bell KFS, Fowler JH, McMahon A, Al-Mubarak B, et al. The subtype of GluN2 C-terminal domain determines the response to excitotoxic insults. Neuron. 2012;74(3):543–56. doi:10.1016/j.neuron.2012.03.021.
Martin KP, Wellman CL. NMDA receptor blockade alters stress-induced dendritic remodeling in medial prefrontal cortex. Cerebral Cortex. 2011;21(10):2366–73. doi:10.1093/cercor/bhr021. (New York, N.Y. : 1991)
Matheus MG, Guimarães FS. Antagonism of non-NMDA receptors in the dorsal periaqueductal grey induces anxiolytic effect in the elevated plus maze. Psychopharmacology (Berl). 1997;*132*(1):14–8.
Mathew SJ, Murrough JW, aan het Rot M, Collins KA, Reich DL, Charney DS. Riluzole for relapse prevention following intravenous ketamine in treatment-resistant depression: a pilot ran-

domized, placebo-controlled continuation trial. Int J Neuropsychopharmacol. 2010;13(1):71–82. doi:10.1017/S1461145709000169.

Mathews DC, Zarate J. Current status of ketamine and related compounds for depression. J Clin Psychiatry. 2013;74(05):516–7. doi:10.4088/JCP.13ac08382.

Mathur P, Graybeal C, Feyder M, Davis MI, Holmes A. Fear memory impairing effects of systemic treatment with the NMDA NR2B subunit antagonist, Ro 25-6981, in mice: attenuation with ageing. Pharmacol Biochem Behav. 2009;91(3):453–60. doi:10.1016/j.pbb.2008.08.028.

Matrisciano F, Storto M, Ngomba RT, Cappuccio I, Caricasole A, Scaccianoce S, et al. Imipramine treatment up-regulates the expression and function of mGlu2/3 metabotropic glutamate receptors in the rat hippocampus. Neuropharmacology. 2002;42(8):1008–15.

Matrisciano F, Scaccianoce S, Del Bianco P, Panaccione I, Canudas AM, Battaglia G, et al. Metabotropic glutamate receptors and neuroadaptation to antidepressants: imipramine-induced down-regulation of beta-adrenergic receptors in mice treated with metabotropic glutamate 2/3 receptor ligands. J Neurochem. 2005;93(5):1345–52. doi:10.1111/j.1471-4159.2005.03141.x.

Matrisciano F, Panaccione I, Zusso M, Giusti P, Tatarelli R, Iacovelli L, et al. Group-II metabotropic glutamate receptor ligands as adjunctive drugs in the treatment of depression: a new strategy to shorten the latency of antidepressant medication? Mol Psychiatry. 2007;12(8):704–6. doi:10.1038/sj.mp.4002005.

Matrisciano F, Caruso A, Orlando R, Marchiafava M, Bruno V, Battaglia G, et al. Defective group-II metaboropic glutamate receptors in the hippocampus of spontaneously depressed rats. Neuropharmacology. 2008;55(4):525–31. doi:10.1016/j.neuropharm.2008.05.014.

Matsuzaki M, Honkura N, Ellis-Davies GCR, Kasai H. Structural basis of long-term potentiation in single dendritic spines. Nature. 2004;429(6993):761–6. doi:10.1038/nature02617.

McCullumsmith RE, Meador-Woodruff JH. Striatal excitatory amino acid transporter transcript expression in schizophrenia, bipolar disorder, and major depressive disorder. Neuropsychopharmacology. 2002;26(3):368–75. doi:10.1016/S0893-133X(01)00370-0.

McElroy SL, Zarate CA, Cookson J, Suppes T, Huffman RF, Greene P, Ascher J. A 52-week, open-label continuation study of lamotrigine in the treatment of bipolar depression. J Clin Psychiatry. 2004;*65*(2):204–10.

McEwen BS, Chattarji S, Diamond DM, Jay TM, Reagan LP, Svenningsson P, Fuchs E. The neurobiological properties of tianeptine (Stablon): from monoamine hypothesis to glutamatergic modulation. Mol Psychiatry. 2010;15(3):237–49. doi:10.1038/mp.2009.80.

Metsis M, Timmusk T, Arenas E, Persson H. Differential usage of multiple brain-derived neurotrophic factor promoters in the rat brain following neuronal activation. Proc Natl Acad Sci U S A. 1993;90(19):8802–6.

Milanese M, Tardito D, Musazzi L, Treccani G, Mallei A, Bonifacino T, et al. Chronic treatment with agomelatine or venlafaxine reduces depolarization-evoked glutamate release from hippocampal synaptosomes. BMC Neurosci. 2013;14:75. doi:10.1186/1471-2202-14-75.

Mineur YS, Picciotto MR, Sanacora G. Antidepressant-like effects of ceftriaxone in male C57BL/6J mice. Biol Psychiatry. 2007;61(2):250–2. doi:10.1016/j.biopsych.2006.04.037.

Mitterauer BJ. Ketamine may block NMDA receptors in astrocytes causing a rapid antidepressant effect. Front Synaptic Neurosci. 2012;4:8. doi:10.3389/fnsyn.2012.00008.

Moghaddam B. Stress preferentially increases extraneuronal levels of excitatory amino acids in the prefrontal cortex: comparison to hippocampus and basal ganglia. J Neurochem. 1993;60(5):1650–7.

Moghaddam B. Stress activation of glutamate neurotransmission in the prefrontal cortex: implications for dopamine-associated psychiatric disorders. Biol Psychiatry. 2002;51(10):775–87.

Moghaddam B, Adams B, Verma A, Daly D. Activation of glutamatergic neurotransmission by ketamine: a novel step in the pathway from NMDA receptor blockade to dopaminergic and cognitive disruptions associated with the prefrontal cortex. J Neurosci. 1997;17(8):2921–7.

Molina-Hernández M, Tellez-Alcántara NP, Pérez-García J, Olivera-Lopez JI, Jaramillo MT. Antidepressant-like and anxiolytic-like actions of the mGlu5 receptor antagonist MTEP, microinjected into lateral septal nuclei of male Wistar rats. Prog Neuropsychopharmacol Biol Psychiatry. 2006;30(6):1129–35. doi:10.1016/j.pnpbp.2006.04.022.

Morley-Fletcher S, Mairesse J, Soumier A, Banasr M, Fagioli F, Gabriel C,et al. Chronic agomelatine treatment corrects behavioral, cellular, and biochemical abnormalities induced by prenatal stress in rats. Psychopharmacology (Berl). 2011;217(3):301–13. doi:10.1007/s00213-011-2280-x.

Moryl E, Danysz W, Quack G. Potential antidepressive properties of amantadine, memantine and bifemelane. Pharmacol Toxicol. 1993;72(4–5):394–7. doi:10.1111/j.1600-0773.1993.tb01351.x.

Muhonen LH, Lönnqvist J, Juva K, Alho H. Double-blind, randomized comparison of memantine and escitalopram for the treatment of major depressive disorder comorbid with alcohol dependence. J Clin Psychiatry. 2008;69(3):392–9.

Mula M, Pini S, Cassano GB. The role of anticonvulsant drugs in anxiety disorders: a critical review of the evidence. J Clin Psychopharmacol. 2007;27(3):263–72. doi:10.1097/jcp.0b013e318059361a.

Müller HK, Wegener G, Liebenberg N, Zarate CA, Popoli M, Elfving B. Ketamine regulates the presynaptic release machinery in the hippocampus. J Psychiatr Res. 2013;47(7):892–9. doi:10.1016/j.jpsychires.2013.03.008.

Murrough JW, Iosifescu DV, Chang LC, Al Jurdi RK, Green CM, Perez AM, et al. Antidepressant efficacy of ketamine in treatment-resistant major depression: a two-site randomized controlled trial. Am J Psychiatry. 2013;170(10):1134–42. doi:10.1176/appi.ajp.2013.13030392.

Musazzi L, Milanese M, Farisello P, Zappettini S, Tardito D, Barbiero VS, et al. Acute stress increases depolarization-evoked glutamate release in the rat prefrontal/frontal cortex: the dampening action of antidepressants. PloS ONE. 2010;5(1):e8566. doi:10.1371/journal.pone.0008566.

Musazzi L, Treccani G, Mallei A, Popoli M. The action of antidepressants on the glutamate system: regulation of glutamate release and glutamate receptors. Biol Psychiatry. 2012;73(12):1180–8. doi:10.1016/j.biopsych.2012.11.009.

Nations KR, Dogterom P, Bursi R, Schipper J, Greenwald S, Zraket D, et al. Examination of Org 26576, an AMPA receptor positive allosteric modulator, in patients diagnosed with major depressive disorder: an exploratory, randomized, double-blind, placebo-controlled trial. J Psychopharmacol. 2012;26(12):1525–39. doi:10.1177/0269881112458728.

O'Shea RD. Roles and regulation of glutamate transporters in the central nervous system. Clin Exp Pharmacol Physiol. 2002;29(11):1018–23.

Ongur D, Drevets WC, Price JL. Glial reduction in the subgenual prefrontal cortex in mood disorders. Proc Natl Acad Sci U S A. 1998;95(22):13290–5. doi:10.1073/pnas.95.22.13290.

Palucha A, Pilc A. Metabotropic glutamate receptor ligands as possible anxiolytic and antidepressant drugs. Pharmacol Ther. 2007;115(1):116–47. doi:10.1016/j.pharmthera.2007.04.007.

Pałucha A, Brański P, Szewczyk B, Wierońska JM, Kłak K, Pilc A. Potential antidepressant-like effect of MTEP, a potent and highly selective mGluR5 antagonist. Pharmacol Biochem Behav. 2005;81(4):901–6. doi:10.1016/j.pbb.2005.06.015.

Pałucha-Poniewiera A, Wierońska JM, Brański P, Stachowicz K, Chaki S, Pilc A. On the mechanism of the antidepressant-like action of group II mGlu receptor antagonist, MGS0039. Psychopharmacology (Berl). 2010;212(4):523–35. doi:10.1007/s00213-010-1978-5.

Palygin O, Lalo U, Pankratov Y. Distinct pharmacological and functional properties of NMDA receptors in mouse cortical astrocytes. Br J Pharmacol. 2011;163(8):1755–66. doi:10.1111/j.1476-5381.2011.01374.x.

Paul IA, Skolnick P. Glutamate and depression: clinical and preclinical studies. Ann N Y Acad Sci. 2003;1003:250–72.

Pecknold JC, McClure DJ, Appeltauer L, Wrzesinski L, Allan T. Treatment of anxiety using fenobam (a nonbenzodiazepine) in a double-blind standard (diazepam) placebo-controlled study. J Clin Psychopharmacol. 1982;2(2):129–33.

Pilc A, Kłodzińska A, Brański P, Nowak G, Pałucha A, Szewczyk B, et al. Multiple MPEP administrations evoke anxiolytic- and antidepressant-like effects in rats. Neuropharmacology. 2002;43(2):181–7.

Pilc A, Chaki S, Nowak G, Witkin JM. Mood disorders: regulation by metabotropic glutamate receptors. Biochem Pharmacol. 2008;75(5):997–1006. doi:10.1016/j.bcp.2007.09.021.

Pisani A, Gubellini P, Bonsi P, Conquet F, Picconi B, Centonze D, et al. Metabotropic glutamate receptor 5 mediates the potentiation of N-methyl-D-aspartate responses in medium spiny striatal neurons. Neuroscience. 2001;106(3):579–87.

Pittenger C, Coric V, Banasr M, Bloch M, Krystal JH, Sanacora G. Riluzole in the treatment of mood and anxiety disorders. CNS Drugs. 2008;22(9):761–86.

Popoli M. Agomelatine: innovative pharmacological approach in depression. CNS Drugs. 2009;23(Suppl 2):27–34. doi:10.2165/11318640-000000000-00000.

Porter RHP, Jaeschke G, Spooren W, Ballard TM, Büttelmann B, Kolczewski S, et al. Fenobam: a clinically validated nonbenzodiazepine anxiolytic is a potent, selective, and noncompetitive mGlu5 receptor antagonist with inverse agonist activity. J Pharmacol Exp Ther. 2005;315(2):711–21. doi:10.1124/jpet.105.089839.

Preskorn SH, Baker B, Kolluri S, Menniti FS, Krams M, Landen JW. An innovative design to establish proof of concept of the antidepressant effects of the NR2B subunit selective N-methyl-D-aspartate antagonist, CP-101,606, in patients with treatment-refractory major depressive disorder. J Clin Psychopharmacol. 2008;28(6):631–7. doi:10.1097/JCP.0b013e31818a6cea.

Rajkowska G, Miguel-Hidalgo JJ. Gliogenesis and glial pathology in depression. CNS Neurol Disord Drug Targets. 2007;6(3):219–33.

Rajkowska G, Stockmeier CA. Astrocyte pathology in major depressive disorder: insights from human postmortem brain tissue. Curr Drug Targets. 2013;14(11):1225–36.

Rajkowska G, Miguel-Hidalgo JJ, Wei J, Dilley G, Pittman SD, Meltzer HY, … Stockmeier CA (1999). Morphometric evidence for neuronal and glial prefrontal cell pathology in major depression. Biol Psychiatry. 45(9):1085–98. doi:10.1016/S0006-3223(99)00041-4.

Reynolds IJ, Miller RJ. Tricyclic antidepressants block N-methyl-D-aspartate receptors: similarities to the action of zinc. Br J Pharmacol. 1988;95(1):95–102.

Reznikov LR, Grillo CA, Piroli GG, Pasumarthi RK, Reagan LP, Fadel J. Acute stress-mediated increases in extracellular glutamate levels in the rat amygdala: differential effects of antidepressant treatment. Eur J Neurosci. 2007;25(10):3109–14. doi:10.1111/j.1460-9568.2007.05560.x.

Rogóz Z, Skuza G, Maj J, Danysz W. Synergistic effect of uncompetitive NMDA receptor antagonists and antidepressant drugs in the forced swimming test in rats. Neuropharmacology. 2002;42(8):1024–30.

Rothstein JD, Patel S, Regan MR, Haenggeli C, Huang YH, Bergles DE, et al. Beta-lactam antibiotics offer neuroprotection by increasing glutamate transporter expression. Nature. 2005;433(7021):73–7. doi:10.1038/nature03180.

Rutherford EC, Pomerleau F, Huettl P, Strömberg I, Gerhardt GA. Chronic second-by-second measures of L-glutamate in the central nervous system of freely moving rats. J Neurochem. 2007;102(3):712–22. doi:10.1111/j.1471-4159.2007.04596.x.

Sanacora G, Banasr M. From pathophysiology to novel antidepressant drugs: glial contributions to the pathology and treatment of mood disorders. Biol Psychiatry. 2013;73(12):1172–9. doi:10.1016/j.biopsych.2013.03.032.

Sanacora G, Kendell SF, Fenton L, Coric V, Krystal JH. Riluzole augmentation for treatment-resistant depression. Am J Psychiatry. 2004;161(11):2132. doi:10.1176/appi.ajp.161.11.2132.

Sanacora G, Zarate CA, Krystal JH, Manji HK. Targeting the glutamatergic system to develop novel, improved therapeutics for mood disorders. Nat Rev Drug Discov. 2008;7(5):426–37. doi:10.1038/nrd2462.

Sapolsky RM. The possibility of neurotoxicity in the hippocampus in major depression: a primer on neuron death. Biol Psychiatry. 2000;48(8):755–65.

Sapolsky RM. Neuroprotective gene therapy against acute neurological insults. Nat Rev Neurosci. 2003;4(1):61–9. doi:10.1038/nrn1006.

Schneiderman N, Ironson G, Siegel SD. Stress and health: psychological, behavioral, and biological determinants. Annu Rev Clin Psychol. 2005;1:607–28. doi:10.1146/annurev.clinpsy.1.102803.144141.

Sequeira A, Mamdani F, Ernst C, Vawter MP, Bunney WE, Lebel V, et al. Global brain gene expression analysis links glutamatergic and GABAergic alterations to suicide and major depression. PloS ONE. 2009;4(8):e6585. doi:10.1371/journal.pone.0006585.

Shigeri Y, Seal RP, Shimamoto K. Molecular pharmacology of glutamate transporters, EAATs and VGLUTs. Brain Res Brain Res Rev. 2004;45(3):250–65. doi:10.1016/j.brainresrev.2004.04.004.

Sillaber I, Panhuysen M, Henniger MSH, Ohl F, Kühne C, Pütz B, et al. Profiling of behavioral changes and hippocampal gene expression in mice chronically treated with the SSRI paroxetine. Psychopharmacology (Berl). 2008;200(4):557–72. doi:10.1007/s00213-008-1232-6.

Sinha R. Chronic stress, drug use, and vulnerability to addiction. Ann N Y Acad Sci. 2008;1141:105–30. doi:10.1196/annals.1441.030.

Skolnick P, Layer RT, Popik P, Nowak G, Paul IA, Trullas R. Adaptation of N-methyl-D-aspartate (NMDA) receptors following antidepressant treatment: implications for the pharmacotherapy of depression. Pharmacopsychiatry. 1996;29(1):23–6. doi:10.1055/s-2007-979537.

Skolnick P, Legutko B, Li X, Bymaster FP. Current perspectives on the development of non-biogenic amine-based antidepressants. Pharmacol Res. 2001;43(5):411–23. doi:10.1006/phrs.2000.0806.

Skolnick P, Popik P, Trullas R. Glutamate-based antidepressants: 20 years on. Trends Pharmacol Sci. 2009;30(11):563–9. doi:10.1016/j.tips.2009.09.002.

Stein-Behrens B, Mattson MP, Chang I, Yeh M, Sapolsky R. Stress exacerbates neuron loss and cytoskeletal pathology in the hippocampus. J Neurosci. 1994;14(9):5373–80.

Stockmeier CA, Mahajan GJ, Konick LC, Overholser JC, Jurjus GJ, Meltzer HY, et al. Cellular changes in the postmortem hippocampus in major depression. Biol Psychiatry. 2004;56(9):640–50. doi:10.1016/j.biopsych.2004.08.022.

Sun W, McConnell E, Pare J-F, Xu Q, Chen M, Peng W, et al. Glutamate-dependent neuroglial calcium signaling differs between young and adult brain. Science. 2013;339(6116):197–200. doi:10.1126/science.1226740. (New York NY).

Svenningsson P, Bateup H, Qi H, Takamiya K, Huganir RL, Spedding M, et al. Involvement of AMPA receptor phosphorylation in antidepressant actions with special reference to tianeptine. Eur J Neurosci. 2007;26(12):3509–17. doi:10.1111/j.1460-9568.2007.05952.x.

Swanson CJ, Bures M, Johnson MP, Linden A-M, Monn JA, Schoepp DD. Metabotropic glutamate receptors as novel targets for anxiety and stress disorders. Nat Rev Drug Discov. 2005;4(2):131–44. doi:10.1038/nrd1630.

Talantova M, Sanz-Blasco S, Zhang X, Xia P, Akhtar MW, Okamoto S, et al. A induces astrocytic glutamate release, extrasynaptic NMDA receptor activation, and synaptic loss. Proc Natl Acad Sci U S A. 2013;110(27):E2518–27. doi:10.1073/pnas.1306832110.

Tamaru Y, Nomura S, Mizuno N, Shigemoto R. Distribution of metabotropic glutamate receptor mGluR3 in the mouse CNS: differential location relative to pre- and postsynaptic sites. Neuroscience. 2001;106(3):481–503.

Tardito D, Milanese M, Bonifacino T, Musazzi L, Grilli M, Mallei A, et al. Blockade of stress-induced increase of glutamate release in the rat prefrontal/frontal cortex by agomelatine involves synergy between melatonergic and 5-HT2C receptor-dependent pathways. BMC Neurosci. 2010;11:68. doi:10.1186/1471-2202-11-68.

Tardito D, Molteni R, Popoli M, Racagni G. Synergistic mechanisms involved in the antidepressant effects of agomelatine. Eur Neuropsychopharmacol. 2012;22(Suppl 3):S482–6. doi:10.1016/j.euroneuro.2012.06.016.

Tatarczyńska E, Klodzińska A, Chojnacka-Wójcik E, Palucha A, Gasparini F, Kuhn R, Pilc A. Potential anxiolytic- and antidepressant-like effects of MPEP, a potent, selective and systemically active mGlu5 receptor antagonist. Br J Pharmacol. 2001;132(7):1423–30. doi:10.1038/sj.bjp.0703923.

Trantham-Davidson H, LaLumiere RT, Reissner KJ, Kalivas PW, Knackstedt LA. Ceftriaxone normalizes nucleus accumbens synaptic transmission, glutamate transport, and export following cocaine self-administration and extinction training. J Neurosci. 2012;32(36):12406–10. doi:10.1523/JNEUROSCI.1976-12.2012.

Trullas R, Skolnick P. Functional antagonists at the NMDA receptor complex exhibit antidepressant actions. Eur J Pharmacol. 1990;185(1):1–10.

Tu JC, Xiao B, Naisbitt S, Yuan JP, Petralia RS, Brakeman P, et al. Coupling of mGluR/Homer and PSD-95 complexes by the Shank family of postsynaptic density proteins. Neuron. 1999;23(3):583–92.

Uranova NA, Vostrikov VM, Orlovskaya DD, Rachmanova VI. Oligodendroglial density in the prefrontal cortex in schizophrenia and mood disorders: a study from the Stanley neuropathology consortium. Schizophr Res. 2004;67(2–3), 269–75. doi:10.1016/S0920-9964(03)00181-6.

Valentine GW, Mason GF, Gomez R, Fasula M, Watzl J, Pittman B, et al. The antidepressant effect of ketamine is not associated with changes in occipital amino acid neurotransmitter content as measured by [(1)H]-MRS. Psychiatry Res. 2011;191(2):122–7. doi:10.1016/j.pscychresns.2010.10.009.

Van der Loos MLM, Mulder PGH, Hartong EGTM, Blom MBJ, Vergouwen AC, de Keyzer HJUEM, et al. Efficacy and safety of lamotrigine as add-on treatment to lithium in bipolar depression: a multicenter, double-blind, placebo-controlled trial. J Clin Psychiatry. 2009;70(2):223–31.

Vekovischeva OY, Aitta-aho T, Verbitskaya E, Sandnabba K, Korpi ER. Acute effects of AMPA-type glutamate receptor antagonists on intermale social behavior in two mouse lines bidirectionally selected for offensive aggression. Pharmacol Biochem Behav. n.d.;87(2):241–9. doi:10.1016/j.pbb.2007.04.020.

Venero C, Borrell J. Rapid glucocorticoid effects on excitatory amino acid levels in the hippocampus: a microdialysis study in freely moving rats. Eur J Neurosci. 1999;11(7):2465–73.

Verkhratsky A, Kirchhoff F. NMDA receptors in glia. Neuroscientist. 2007;13(1):28–37. doi:10.1177/1073858406294270.

Voleti B, Navarria A, Liu R-J, Banasr M, Li N, Terwilliger R, et al. Scopolamine rapidly increases mammalian target of rapamycin complex 1 signaling, synaptogenesis, and antidepressant behavioral responses. Biol Psychiatry. 2013;74(10):742–9. doi:10.1016/j.biopsych.2013.04.025.

Watanabe Y, Gould E, Cameron HA, Daniels DC, McEwen BS. Phenytoin prevents stress- and corticosterone-induced atrophy of CA3 pyramidal neurons. Hippocampus. 1992;2(4):431–5. doi:10.1002/hipo.450020410.

Webster MJ, Knable MB, Johnston-Wilson N, Nagata K, Inagaki M, Yolken RH. Immunohistochemical localization of phosphorylated glial fibrillary acidic protein in the prefrontal cortex and hippocampus from patients with schizophrenia, bipolar disorder, and depression. Brain Behav Immun. 2001;15(4):388–400. doi:10.1006/brbi.2001.0646.

Wieroñska JM, Brañski P, Szewczyk B, Papp M, Gruca P, Moryl E, Pilc A. Preliminary communication changes in the expression of metabotropic glutamate receptor 5 (mglur5) in the rat hippocampus in an animal model of depression. Pol J Pharmacol. 2001;5:5–8.

Wieronska JM, Szewczyk B, Branski P, Palucha A, Pilc A. Antidepressant-like effect of MPEP, a potent, selective and systemically active mGlu5 receptor antagonist in the olfactory bulbectomized rats. Amino Acids. 2002;23(1–3):213–6. doi:10.1007/s00726-001-0131-5.

Wierońska JM, Legutko B, Dudys D, Pilc A. Olfactory bulbectomy and amitriptyline treatment influences mGlu receptors expression in the mouse brain hippocampus. Pharmacol Rep. 2008;60(6):844–55.

Witkin JM, Marek GJ, Johnson BG, Schoepp DD. Metabotropic glutamate receptors in the control of mood disorders. CNS Neurol Disord Drug Targets. 2007;6(2):87–100.

Xia P, Chen HV, Zhang D, Lipton SA. Memantine preferentially blocks extrasynaptic over synaptic NMDA receptor currents in hippocampal autapses. J Neurosci. 2010;30(33):11246–50. doi:10.1523/JNEUROSCI.2488-10.2010.

Xu J, Kurup P, Zhang Y, Goebel-Goody SM, Wu PH, Hawasli AH, et al. Extrasynaptic NMDA receptors couple preferentially to excitotoxicity via calpain-mediated cleavage of STEP. J Neurosci. 2009;29(29):9330–43. doi:10.1523/JNEUROSCI.2212-09.2009.

Yoshimizu T, Shimazaki T, Ito A, Chaki S. An mGluR2/3 antagonist, MGS0039, exerts antidepressant and anxiolytic effects in behavioral models in rats. Psychopharmacology (Berl). 2006;186(4):587–93. doi:10.1007/s00213-006-0390-7.

Yoshizumi M, Eisenach JC, Hayashida K. Riluzole and gabapentinoids activate glutamate transporters to facilitate glutamate-induced glutamate release from cultured astrocytes. Eur J Pharmacol. 2012;677(1–3):87–92. doi:10.1016/j.ejphar.2011.12.015.

Yuen EY, Liu W, Karatsoreos IN, Feng J, McEwen BS, Yan Z. Acute stress enhances glutamatergic transmission in prefrontal cortex and facilitates working memory. Proc Natl Acad Sci U S A. 2009;106(33):14075–9. doi:10.1073/pnas.0906791106.

Yuen EY, Liu W, Karatsoreos IN, Ren Y, Feng J, McEwen BS, Yan Z. Mechanisms for acute stress-induced enhancement of glutamatergic transmission and working memory. Mol Psychiatry. 2011;16(2):156–70. doi:10.1038/mp.2010.50.

Zarate CA, Quiroz J, Payne J, Manji HK. Modulators of the glutamatergic system: implications for the development of improved therapeutics in mood disorders. Psychopharmacol Bull. 2002;36(4):35–83.

Zarate CA, Singh JB, Carlson PJ, Brutsche NE, Ameli R, Luckenbaugh DA, et al. A randomized trial of an N-methyl-D-aspartate antagonist in treatment-resistant major depression. Arch Gen Psychiatry. 2006a;63(8):856–64. doi:10.1001/archpsyc.63.8.856.

Zarate CA, Singh JB, Quiroz JA, De Jesus G, Denicoff KK, Luckenbaugh DA, et al. A double-blind, placebo-controlled study of memantine in the treatment of major depression. Am J Psychiatry. 2006b;163(1):153–5. doi:10.1176/appi.ajp.163.1.153.

Zarate CA, Brutsche NE, Ibrahim L, Franco-Chaves J, Diazgranados N, Cravchik A, et al. Replication of ketamine's antidepressant efficacy in bipolar depression: a randomized controlled add-on trial. Biol Psychiatry. 2012;71(11):939–46. doi:10.1016/j.biopsych.2011.12.010.

Zheng K, Scimemi A, Rusakov DA. Receptor actions of synaptically released glutamate: the role of transporters on the scale from nanometers to microns. Biophys J. 2008;95(10):4584–96. doi:10.1529/biophysj.108.129874.

Zink M, Vollmayr B, Gebicke-Haerter PJ, Henn FA. Reduced expression of glutamate transporters vGluT1, EAAT2 and EAAT4 in learned helpless rats, an animal model of depression. Neuropharmacology. 2010;58(2):465–73. doi:10.1016/j.neuropharm.2009.09.005.

Zink M, Rapp S, Donev R, Gebicke-Haerter PJ, Thome J. Fluoxetine treatment induces EAAT2 expression in rat brain. J Neural Trans. 2011;118(6):849–55. doi:10.1007/s00702-010-0536-y. (Vienna, Austria: 1996).

Zschocke J, Bayatti N, Clement AM, Witan H, Figiel M, Engele J, Behl C. Differential promotion of glutamate transporter expression and function by glucocorticoids in astrocytes from various brain regions. J Biol Chem. 2005;280(41):34924–32. doi:10.1074/jbc.M502581200.

Index

M. Popoli et al. (eds.), *Synaptic Stress and Pathogenesis of Neuropsychiatric Disorders*, DOI 10.1007/978-1-4939-1056-4, © Springer Science+Business Media New York 2014

H

Q

R

MIX
Papier aus verantwortungsvollen Quellen
Paper from responsible sources
FSC® C105338
FSC
www.fsc.org

If you have any concerns about our products,
you can contact us on
ProductSafety@springernature.com

In case Publisher is established outside the EU,
the EU authorized representative is:
Springer Nature Customer Service Center GmbH
Europaplatz 3, 69115 Heidelberg, Germany

Printed by Libri Plureos GmbH
in Hamburg, Germany